Instructor's Solution

to accompany

Advanced Str
and
Applied Stress Analysis

Second Edition

Richard G. Budynas
Rochester Institute of Technology

WCB McGraw-Hill

**Boston Burr Ridge, IL Dubuque, IA Madison, WI New York San Francisco St. Louis
Bangkok Bogotá Caracas Lisbon London Madrid
Mexico City Milan New Delhi Seoul Singapore Sydney Taipei Toronto**

WCB/McGraw-Hill

A Division of The McGraw-Hill Companies

Instructor's Solutions Manual to accompany
ADVANCED STRENGTH AND APPLIED STRESS ANALYSIS

1 2 3 4 5 6 7 8 9 0 QSR/QSR 9 3 2 1 0 9 8

ISBN 0-07-365942-8

http://www.mhhe.com

TO THE INSTRUCTOR:

This solutions manual was assembled by the author to aid instructors in using the textbook, ADVANCED STRENGTH AND APPLIED STRESS ANALYSIS, 2nd ed. There are a total of 505 exercise problems in the text for which complete solutions are presented in this manual. This does not include the additional 70 *Computer Problems* given at the end of each chapter. However, the solutions to some of the non-open-ended computer problems are provided.

The instructor may wish to present different approaches to students other than that provided herein. However, an examination of the solutions presented may help to support whatever approach used.

At the time of this writing computational tools such as personal computers and programmable calculators help to make the manual solutions of problems much easier for the student. Computer aided engineering programs such as CAD packages and finite element programs make completely automatic solutions possible as well. I believe that a healthy learning approach is to use these programs to check the manual solutions when attempting to learn the underlying theories involved. It is a challenge to blend these tools appropriately to achieve understanding and competence within this engineering discipline.

The solutions presented were worked entirely by the author, and although every effort was made to insure correctness, there is no absolute guarantee that this is the case for any given problem. If any glaring errors are uncovered, the author would certainly appreciate being notified. It is hoped that these solutions will give further insight and provide additional support to the material presented in the text.

Richard G. Budynas, June 24, 1998

Solutions Manual - Advanced Strength and Applied Stress Analysis, 2nd Ed.

by Richard G. Budynas

TABLE OF CONTENTS

1.1 Separating trapezoid distribution to uniform and triangular

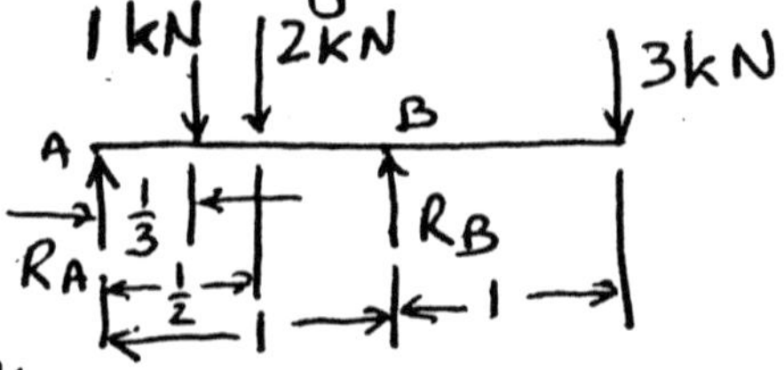

$\Sigma M_A = 0$ yields $R_B = 7.33$ kN

$\Sigma M_B = 0$ yields $R_A = -1.33$ kN

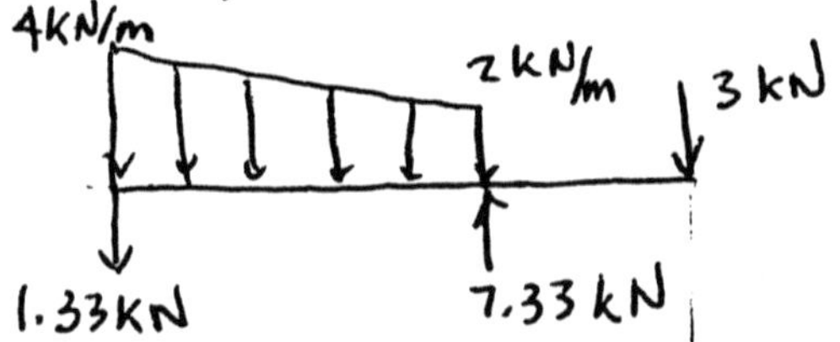

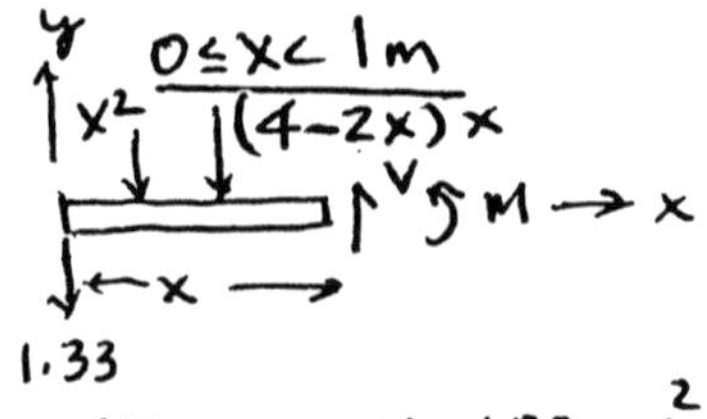

$\Sigma F_y = 0 \quad V - 1.33 - x^2 - (4-2x)x = 0$

$V = -x^2 + 4x + 1.33$ kN

$(\Sigma M_z)_x = 0$

$M + 1.33x + (\frac{2}{3}x)x^2 + (\frac{x}{2})(4-2x)x = 0$

$M = \frac{x^3}{3} - 2x^2 - 1.33x$ kN·m

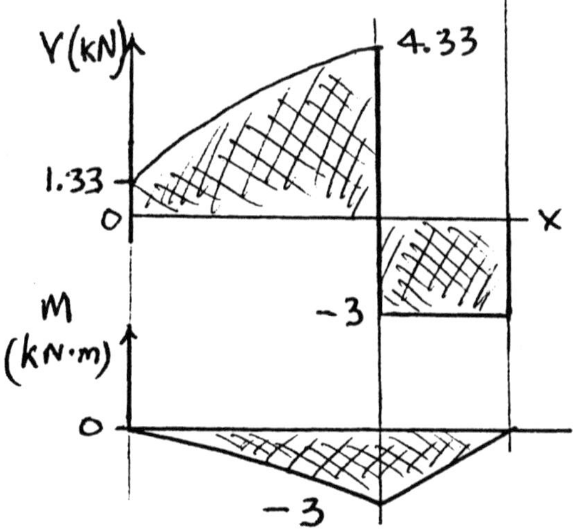

1.2

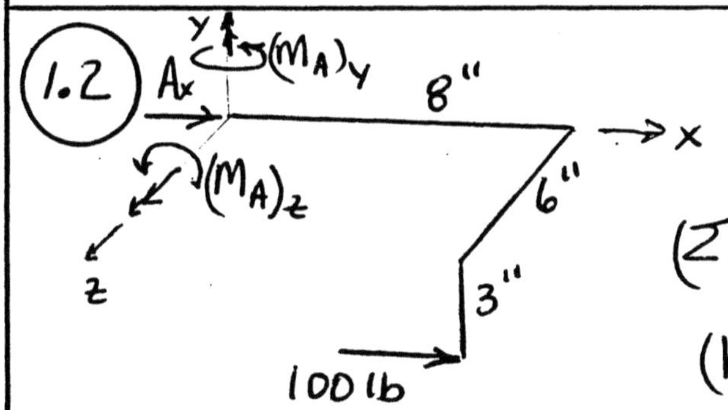

$\Sigma F_x = 0$

$100 + A_x = 0 \quad \therefore A_x = 100$ lb

$(\Sigma M_A)_z = 0$

$(M_A)_y + 6(100) = 0$

$\therefore (M_A)_y = -600$ lb·in

$(\Sigma M_A)_z = 0$

$(M_A)_z + 3(100) = 0$

$\therefore (M_A)_z = -300$ lb·in

1.3

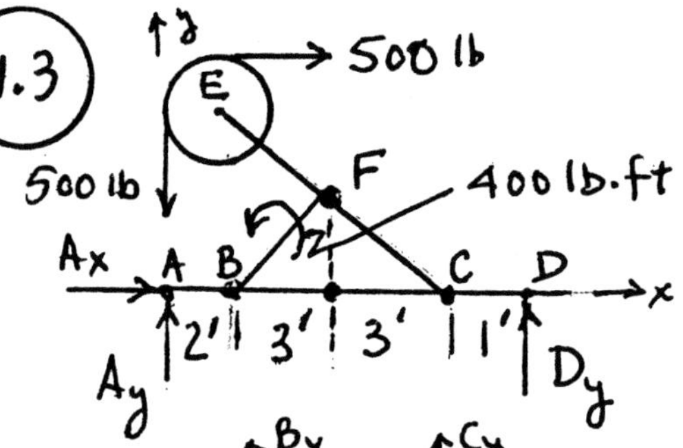

$\Sigma F_x = 0 \quad \underline{A_x = -500 \text{ lb}}$

$\Sigma M_A = 0$ yields $\underline{D_y = 511 \text{ lb}}$

$\Sigma F_y = 0$ yields $\underline{A_y = -11 \text{ lb}}$

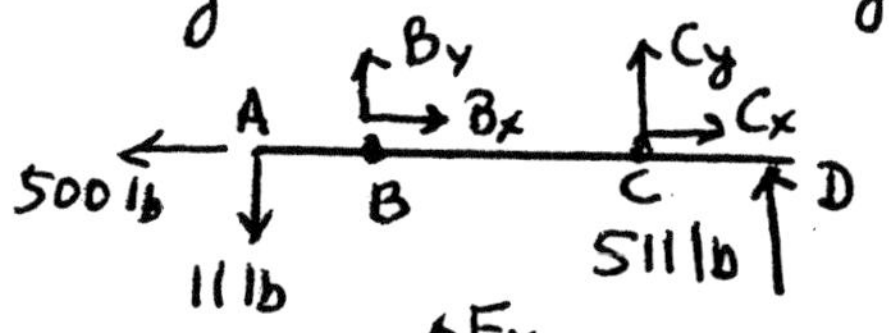

$\Sigma M_C = 0$ yields $\underline{B_y = 100 \text{ lb}}$

$\Sigma M_B = 0$ yields $\underline{C_y = -600 \text{ lb}}$

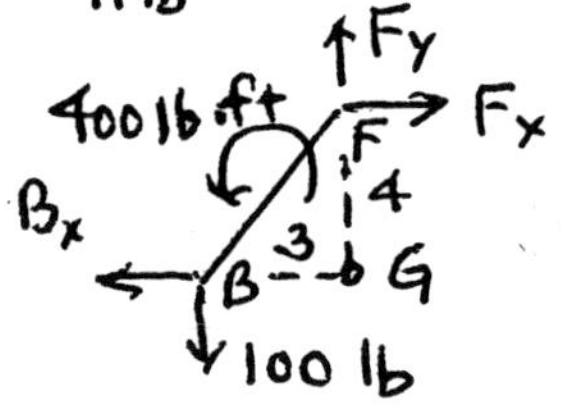

$\Sigma M_G = 0 \quad (100)(3) + 400 - 4F_x = 0$

$\underline{F_x = 175 \text{ lb}}$

$\Sigma F_x = 0 \quad \therefore \quad \underline{B_x = 175 \text{ lb}}$

Return to ABCD $\quad \Sigma F_x = 0$ yields $\underline{C_x = 325 \text{ lb}}$

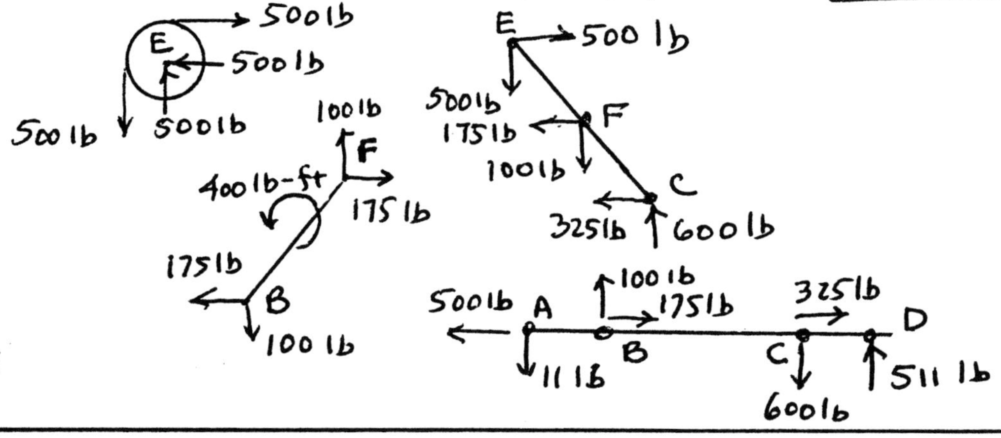

1.4

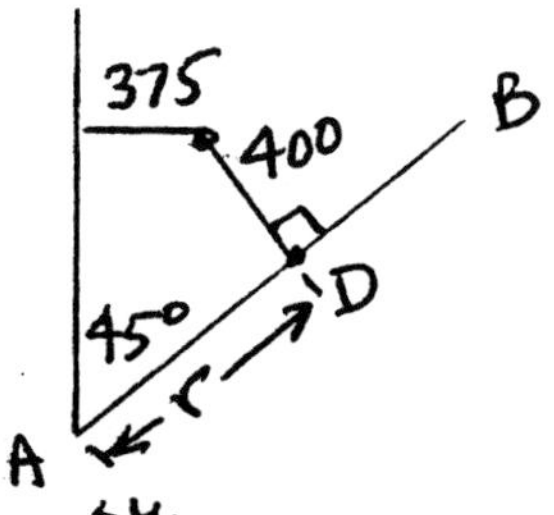

$375 + 400 \sin 45 = r \sin 45$

$375 + 200\sqrt{2} = (\sqrt{2}/2)\, r$

$r = 930.33 \text{ mm}$

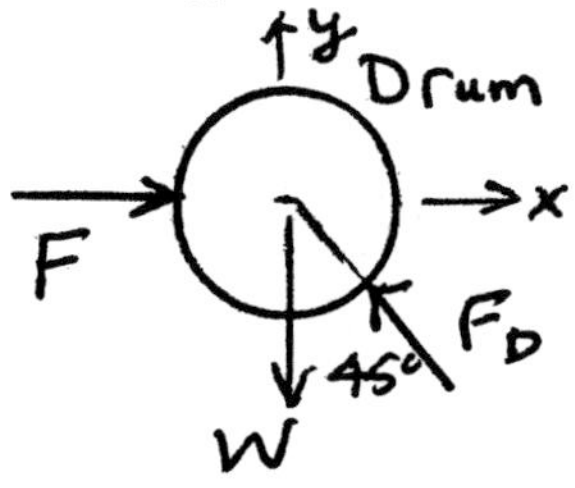

$W = 250(9.81) = 2453 \text{ N}$

$\Sigma F_y = 0 \quad F_D = \dfrac{2453}{\cos 45} = \underline{3468 \text{ N}}$

$\Sigma F_x = 0 \quad F = F_D \sin 45 = \underline{2453 \text{ N}}$

(continued)

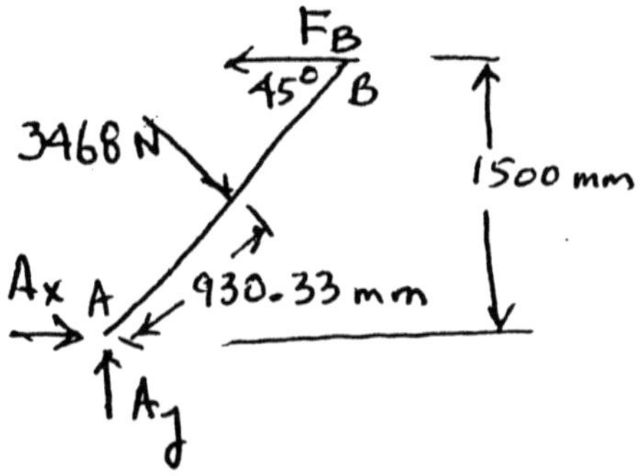

$$\Sigma M_A = 0 \quad 1500 F_B - 3468(930.33) = 0$$
$$\underline{F_B = 2151 \text{ N}}$$
$$\Sigma F_x = 0 \quad A_x + 3468 \cos 45 - 2151 = 0$$
$$\underline{A_x = -301.2 \text{ N}}$$
$$\Sigma F_y = 0 \quad A_y - 3468 \sin 45 = 0$$
$$\underline{A_y = 2452 \text{ N}}$$

1.5 Unit vector from B to C

$$\bar{e}_{BC} = \frac{-0.8\bar{i} + 0.5\bar{j} - 0.1\bar{k}}{[0.8^2 + 0.5^2 + 0.1^2]^{1/2}} = -0.8433\bar{i} + 0.5270\bar{j} - 0.1054\bar{k}$$

Likewise,

$$\bar{e}_{BD} = \frac{-0.8\bar{i} + 0.5\bar{j} - 1.1\bar{k}}{[0.8^2 + 0.5^2 + 1.1^2]^{1/2}} = -0.5521\bar{i} + 0.3450\bar{j} - 0.7591\bar{k}$$

$$\Sigma \bar{M}_A = \bar{0}$$

$$(0.4\bar{i} + 0.3\bar{k}) \times (-3\bar{j}) + (0.8\bar{i} + 0.6\bar{k}) \times \Big[(-0.8433\bar{i} + 0.5270\bar{j} - 0.1054\bar{k})F_{BC} + (-0.5521\bar{i} + 0.3450\bar{j} - 0.7591\bar{k})F_{BD}\Big] = \bar{0}$$

$\bar{i}$ comp.: $\quad 0.3162 F_{BC} + 0.2070 F_{BD} = 0.9 \quad$ ①

$\bar{j}$ comp.: $\quad -0.4217 F_{BC} + 0.2760 F_{BD} = 0 \quad$ ②

$\bar{k}$ comp.: $\quad 0.4216 F_{BC} + 0.2760 F_{BD} = 1.2 \quad$ ③

Eqs. ① & ③ are identical.
From Eqs ① & ② $\quad F_{BC} = 1.423$ kN & $F_{BD} = 2.174$ kN

<u>Reaction at C</u>

$$F_C = 1.423(-0.8433\bar{i} + 0.5270\bar{j} - 0.1054\bar{k}) = \underline{-1.20\bar{i} + 0.750\bar{j} - 0.150\bar{k} \text{ kN}}$$

<u>Reaction at D</u>

$$F_D = 2.174(-0.5521\bar{i} + 0.3450\bar{j} - 0.7591\bar{k}) = \underline{-1.20\bar{i} + 0.750\bar{j} - 1.650\bar{k} \text{ kN}}$$

<u>Reaction at A</u> $\quad \Sigma \bar{F} = \bar{0}$

$$A_x\bar{i} + A_y\bar{j} + A_z\bar{k} - 3\bar{j} - 1.2\bar{i} + 0.75\bar{j} - 0.15\bar{k} - 1.2\bar{i} + 0.75\bar{j} - 1.65\bar{k} = \bar{0}$$

$$\therefore \underline{\bar{A} = 2.4\bar{i} + 1.5\bar{j} + 1.8\bar{k} \text{ kN}}$$

1.6 Force on Pressure plate

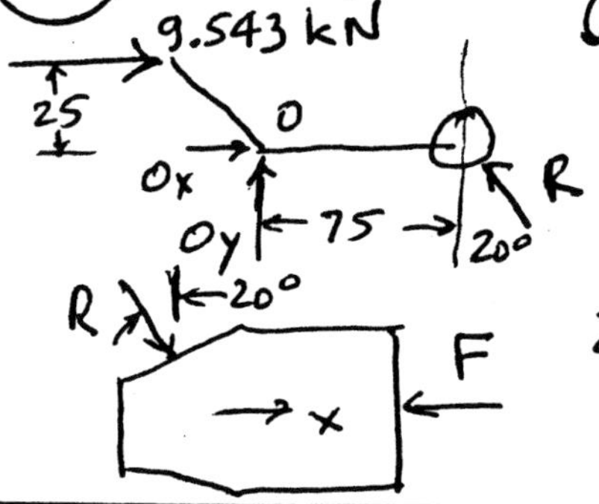

$$P = 180(10^3)\pi(150^2 - 75^2)(10^{-6}) = 9.543(10^3)\,N$$

$$\circlearrowleft + \Sigma M_O = 0$$

$$75R\cos 20 - 9.543(25) = 0$$

$$R = 3.385\ kN$$

$$\Sigma F_x = 0$$

$$F = 3.385\sin 20 = \underline{1.158\ kN}$$

1.7 Gear F $\phi = 20°$

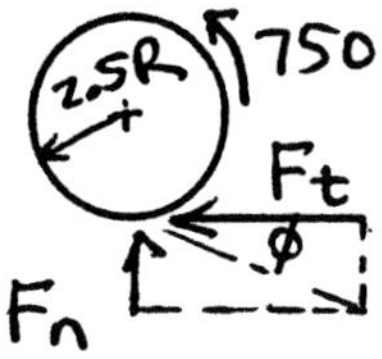

$$2.5F_t = 750 \quad \therefore F_t = 300\ lb$$

$$F_n = 300\tan 20 = 109.2\ lb$$

Shaft ABCD

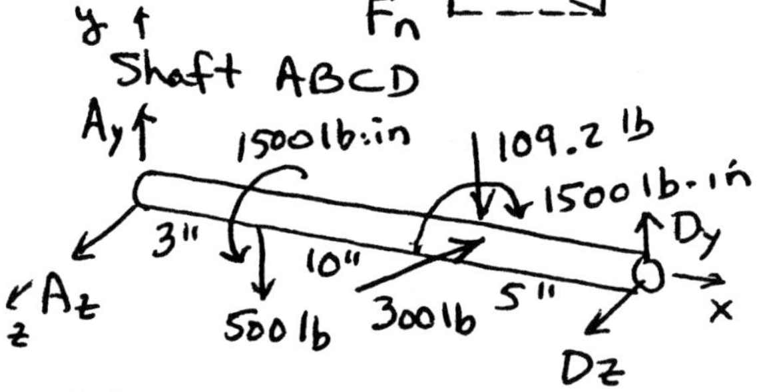

$$(\Sigma M_A)_z = 0$$

$$18D_y - 109.2(13) - 500(3) = 0$$

$$\underline{D_y = 162.1\ lb}$$

$$(\Sigma M_A)_y = 0 \quad -18D_z + 13(300) = 0 \Rightarrow \underline{D_z = 216.7\ lb}$$

$$(\Sigma M_D)_z = 0 \quad -18A_y + 500(15) + 109.2(5) = 0 \Rightarrow \underline{A_y = 446.9\ lb}$$

$$(\Sigma M_D)_y = 0 \quad 18A_z - 5(300) = 0 \Rightarrow \underline{A_z = 83.3\ lb}$$

1.8

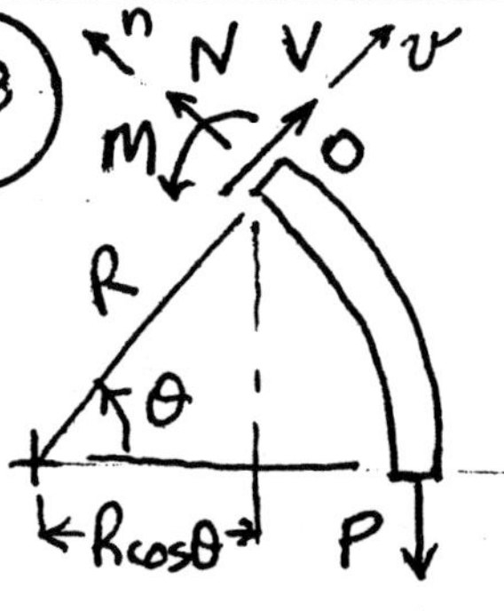

$$\Sigma F_n = 0$$

$$N - P\cos\theta = 0 \Rightarrow \underline{N = P\cos\theta}$$

$$\Sigma F_v = 0$$

$$V - P\sin\theta = 0 \Rightarrow \underline{V = P\sin\theta}$$

$$\Sigma M_O = 0$$

$$M - P(R - R\cos\theta) = 0$$

$$\underline{M = PR(1 - \cos\theta)}$$

1.9 $F_x = \int_{-h/2}^{h/2} \sigma_x b\,dy = \int_{-1.5}^{1.5} (2000y+500)(1)dy = \underline{1500 \text{ lb}}$

$M_z = -\int_{-h/2}^{h/2} y\sigma_x b\,dy = -\int_{-1.5}^{1.5} y(2000y+500)(1)dy = \underline{-4500 \text{ lb}\cdot\text{in}}$

1.10 $F_x = \int_{-h/2}^{h/2} \sigma_x b\,dy = \int_{-1.5}^{1.5} (500+2000y-30y^3)(1)dy = \underline{1500 \text{ lb}}$

$M_z = -\int_{-h/2}^{h/2} y\sigma_x b\,dy = -\int_{-1.5}^{1.5} y(500+2000y-3y^3)(1)dy = \underline{-4409 \text{ lb}\cdot\text{in}}$

1.11 $\Sigma F = 0$ is obvious as load distribution is antisymmetric. For the moment, since load dist. is antisym., then if M due to loading from $0<y<c$ is zero then total is zero. Thus, does $M = \int_0^c y\sigma_{res}\,dA = 0$?

$$M = \int_0^c S_y \frac{y^2}{c}\left[1+\frac{h}{c}-\frac{1}{2}\left(\frac{h}{c}\right)^2\right]b\,dy - \int_0^{c-h} S_y \frac{y^2}{c-h}b\,dy - \int_{c-h}^{c} S_y y\,b\,dy$$

$$= S_y b\left\{\frac{c^2}{3}\left[1+\frac{h}{c}-\frac{1}{2}\left(\frac{h}{c}\right)^2 - \frac{1}{3}(c-h)^2 - \frac{1}{2}\left[c^2-(c-h)^2\right]\right]\right\}$$

$$= \frac{S_y b}{6}\left(2c^2+2ch-h^2-2c^2+4ch-2h^2-3c^2+3c^2-6ch+3h^2\right) = \underline{\underline{0}}$$

1.12 Eqs. (1.4-5)

$\sigma_x = \dfrac{209(10^9)}{1-0.29}\left[-90+0.29(-30)\right](10^{-6}) = -22.5(10^6)\,\text{N/m}^2 = \underline{-22.5 \text{ MPa}}$

$\sigma_y = \dfrac{209(10^9)}{1-0.29}\left[-30+0.29(-90)\right](10^{-6}) = -12.8(10^6)\,\text{N/m}^2 = \underline{-12.8 \text{ MPa}}$

$\tau_{xy} = \dfrac{209(10^9)}{2(1+0.29)}\,120(10^{-6}) = 9.72(10^6)\,\text{N/m}^2 = \underline{9.72 \text{ MPa}}$

Eq(1.4-4c) $\epsilon_z = \dfrac{-0.29}{209(10^9)}\left[-22.5-12.8\right](10^6) = 49.0(10^{-6}) = \underline{49.0\,\mu}$

Thus,

$$[\sigma] = \begin{bmatrix} -22.5 & 9.72 & 0 \\ 9.72 & -12.8 & 0 \\ 0 & 0 & 0 \end{bmatrix} \text{MPa} \qquad [\epsilon] = \begin{bmatrix} -90 & 120 & 0 \\ 120 & -30 & 0 \\ 0 & 0 & 49 \end{bmatrix} \mu$$

<u>ANS</u>

1.13 Eqs(1.4-5)

$$\sigma_x = \frac{15(10^6)}{1-0.3^2}\left[300+0.3(-100)\right](10^{-6}) = \underline{4451\text{ psi}}$$

$$\sigma_y = \frac{15(10^6)}{1-0.3^2}\left[-100+0.3(300)\right](10^{-6}) = \underline{-165\text{ psi}}$$

$$\tau_{xy} = \frac{15(10^6)}{2(1+0.3)}\,200(10^{-6}) = \underline{1154\text{ psi}}$$

Eq.(1.4-4c) $\epsilon_z = \frac{-0.3}{15(10^6)}(4451-165) = -85.7(10^{-6}) = \underline{-85.7\mu}$

Thus,

$$[\sigma] = \begin{bmatrix} 4451 & 1154 & 0 \\ 1154 & -165 & 0 \\ 0 & 0 & 0 \end{bmatrix}\text{psi} \quad [\epsilon] = \begin{bmatrix} 300 & 200 & 0 \\ 200 & -100 & 0 \\ 0 & 0 & -85.7 \end{bmatrix}\mu \quad \underline{\text{ANS}}$$

1.14 $\epsilon_x = \frac{1}{E}(\sigma_x - \nu\sigma_y) \Rightarrow \sigma_x - \nu\sigma_y = E\epsilon_x$

(a) $\epsilon_y = \frac{1}{E}(\sigma_y - \nu\sigma_x) \Rightarrow \nu\sigma_x - \sigma_y = -E\epsilon_y$

$$\sigma_x = \frac{\begin{vmatrix} E\epsilon_x & -\nu \\ -E\epsilon_y & -1 \end{vmatrix}}{\begin{vmatrix} 1 & -\nu \\ \nu & 1 \end{vmatrix}} = \frac{-E\epsilon_x - \nu E\epsilon_y}{-1+\nu^2} = \frac{E}{1-\nu^2}(\epsilon_x + \nu\epsilon_y) \quad \underline{\text{ANS}}$$

$$\sigma_y = \frac{\begin{vmatrix} 1 & E\epsilon_x \\ \nu & -E\epsilon_y \end{vmatrix}}{-1+\nu^2} = \frac{-E\epsilon_y - E\nu\epsilon_x}{-1+\nu^2} = \frac{E}{1-\nu^2}(\epsilon_y + \nu\epsilon_x)$$

(b) Sub ϵ_z into Eqs.(1.4-3)

$$\sigma_x = \frac{E}{(1+\nu)(1-2\nu)}\left[(1-\nu)\epsilon_x + \nu\epsilon_y\right]$$

$$\sigma_y = \frac{E}{(1+\nu)(1-2\nu)}\left[(1-\nu)\epsilon_y + \nu\epsilon_x\right] \quad \underline{\text{ANS}}$$

The multiplier $\frac{E}{(1+\nu)(1-2\nu)}$ is slightly higher than $\frac{E}{1-\nu^2} = \frac{E}{(1+\nu)(1-\nu)}$ and the effect of the primary strain is reduced by ν (30% for metals)

1.15 Given: $\sigma_x = -6\text{ MPa}$, $\epsilon_y = \epsilon_z \approx 0$

Eq. (1.4-3a)

$$-6(10^6) = \frac{35(10^9)}{(1+0.4)(1-0.8)}\left[(1-0.4)\epsilon_x + 0.4(0)\right] \quad \text{solve for } \epsilon_x$$

$$\epsilon_x = -80(10^{-6}) = \underline{-80\mu}$$

Remaining Eqs (1.4-3)

$$\sigma_y = \frac{35(10^9)}{(1+0.4)(1-0.8)}\left[0.4(-80)(10^{-6})\right] = -4(10^6)\text{ N/m}^2 = \underline{-4\text{ MPa}} = \sigma_z$$

1.16 Eqs (1.4-3) and (1.4-8)

$$\sigma_x = \frac{3(10^7)}{(1.3)(1-0.6)}\left[(1-0.3)5 + (0.3)(-3+2)\right](10^{-4}) = \underline{18{,}460\text{ psi}}$$

$$\sigma_y = \frac{3(10^7)}{(1.3)(0.4)}\left[(1-0.3)(-3) + (0.3)(2+5)\right](10^{-4}) = \underline{0}$$

$$\sigma_z = \frac{3(10^7)}{(1.3)(0.4)}\left[(1-0.3)(2) + (0.3)(5-3)\right](10^{-4}) = \underline{11{,}540\text{ psi}}$$

$$\tau_{xy} = \frac{3(10^7)}{2(1.3)}(-2)(10^{-4}) = \underline{-2310\text{ psi}}$$

$$\tau_{yz} = \frac{3(10^7)}{2(1.3)}(1)(10^{-4}) = \underline{1154\text{ psi}}$$

$$\tau_{zx} = \frac{3(10^7)}{2(1.3)}(3)(10^{-4}) = \underline{3460\text{ psi}}$$

$$[\sigma] = \begin{bmatrix} 18.5 & -2.31 & 3.46 \\ -2.31 & 0 & 1.15 \\ 3.46 & 1.15 & 11.5 \end{bmatrix} \text{ kpsi} \quad \underline{\text{ANS}}$$

1.17 Eqs. (1.4-3) and (1.4-8)

$$\sigma_x = \frac{70(10^9)}{(1.33)(0.34)}\left[(1-0.33)(3) + (0.33)(0+5)\right](10^{-4}) = 56.7(10^6)\text{ N/m}^2 = \underline{56.7\text{ MPa}}$$

$$\sigma_y = \frac{70(10^9)}{(1.33)(0.34)}\left[(1-0.33)(0) + (0.33)(3+5)\right](10^{-4}) = 40.9(10^6)\text{ N/m}^2 = \underline{40.9\text{ MPa}}$$

$$\sigma_z = \frac{(70)(10^9)}{(1.33)(0.34)}\left[(1-0.33)(5) + (0.33)(3+0)\right](10^{-4}) = 67.2(10^6)\text{ N/m}^2 = \underline{67.2\text{ MPa}}$$

$$\tau_{xy} = \frac{70(10^9)}{2(1.33)}(-1)(10^{-4}) = -2.63(10^6)\text{ N/m}^2 = \underline{-2.63\text{ MPa}}$$

(continued)

$$\tau_{yz} = \frac{70(10^9)}{2(1.33)}(-4)(10^{-4}) = -10.5(10^6)\,N/m^2 = \underline{-10.5\ MPa}$$

$$\tau_{zx} = \frac{70(10^9)}{2(1.33)}(2)(10^{-4}) = 5.26\,(10^6)\,N/m^2 = \underline{5.26\,MPa}$$

$$[\sigma] = \begin{bmatrix} 56.7 & -2.63 & 5.26 \\ -2.63 & 40.9 & -10.5 \\ 5.26 & -10.5 & 67.2 \end{bmatrix} MPa \qquad \underline{ANS}$$

1.18 (a) x face: $(\sigma_x)_{net} = \sqrt{(12)^2+3^2+(-4)^2} = \underline{13\ kpsi}$

y face: $(\sigma_y)_{net} = \sqrt{3^2+5^2+1^2} = \underline{5.92\ kpsi}$

z face: $(\sigma_z)_{net} = \sqrt{(-4)^2+1^2+(-8)^2} = \underline{9\ kpsi}$

(b) $(\tau_x)_{net} = \sqrt{3^2+(-4)^2} = \underline{5\ kpsi}$

$(\tau_y)_{net} = \sqrt{3^2+1^2} = \underline{3.16\ kpsi}$

$(\tau_z)_{net} = \sqrt{(-4)^2+1^2} = \underline{4.12\ kpsi}$

1.19 (a) $$\epsilon_r = \frac{1}{29(10^6)}\left[625 - \frac{5625}{r^2} - 0.29\left(625+\frac{5625}{r^2}\right)\right] = 15.30 - \frac{250.2}{r^2}\ \mu \quad \underline{ANS}$$

$$\epsilon_\theta = \frac{1}{29(10^6)}\left[625 + \frac{5625}{r^2} - 0.29\left(625-\frac{5625}{r^2}\right)\right] = 15.30 + \frac{250.2}{r^2}\ \mu \quad \underline{ANS}$$

(b) $u_r = r\epsilon_\theta$ at $r=3$ $\epsilon_\theta = 43.1(10^{-6})$, $u_r = 1.293(10^{-4})$ in

$\Delta C = 2\pi\Delta r = 2\pi u_r = \underline{8.12(10^{-4})\ in}$

$C_i = 2\pi(3) = \underline{18.849556\ in}$

$C_o = 2\pi(3+1.293(10^{-4})) = \underline{18.850368\ in}$

1.20 (a) $$\epsilon_x = \frac{\partial u}{\partial x} = \frac{P}{6EI}[6y(L-x)] = \frac{Py}{EI}(L-x)$$

$$\epsilon_y = \frac{\partial v}{\partial y} = -\frac{P}{6EI}[6\nu y(L-x)] = -\frac{P\nu y}{EI}(L-x)$$

Eq. (1.4-5)

$$\sigma_x = \frac{E}{1-\nu^2}\left[\frac{Py}{EI}(L-x) + \nu\left(-\frac{P\nu y}{EI}\right)(L-x)\right] = \frac{Py}{I}(L-x) \qquad \underline{ANS}$$

$$\sigma_y = \frac{E}{1-\nu^2}\left[-\frac{P\nu y}{EI}(L-x) + \nu\frac{Py}{EI}(L-x)\right] = 0 \qquad \underline{ANS}$$

(continued)

$$\gamma_{xy} = \frac{\partial v}{\partial x} + \frac{\partial u}{\partial y} = -\frac{P}{6EI}\left[-3\nu y^2 + 3x(2L-x) + 3(1+\nu)c^2\right]$$
$$+ \frac{P}{6EI}\left[3x(2L-x) - 3(1+\nu)c^2 + 3(2+\nu)y^2\right]$$
$$= -\frac{P}{EI}(1+\nu)(c^2-y^2)$$

Eq (1.4-8) $\tau_{xy} = \frac{E}{2(1+\nu)}\left[-\frac{P}{EI}(1+\nu)(c^2-y^2)\right] = -\frac{P}{2I}(c^2-y^2)$ ANS

σ_x & σ_y agree with basic bending theory. For τ_{xy} see Example 3.4-3 Eq (a) where $V_y = -P$. The above agrees.

(b) At $y=0$ $u=0$, $v = -\frac{P}{6EI}\left[\underbrace{x^2(3L-x)}_{\text{bending}} + \underbrace{3(1+\nu)c^2 x}_{\text{shear}}\right]$ ①

Rotation

$$\theta = \frac{1}{2}\left(\frac{\partial v}{\partial x} - \frac{\partial u}{\partial y}\right) = \frac{1}{2}\left\{-\frac{P}{6EI}\left[-3\nu y^2 + 3x(2L-x) + 3(1+\nu)c^2\right]\right.$$
$$\left. - \frac{P}{6EI}\left[3x(2L-x) - 3(1+\nu)c^2 + 3(2+\nu)y^2\right]\right\}$$

$$\theta = -\frac{P}{2EI}\left[x(2L-x) + y^2\right]$$

$$\theta(x,0) = -\frac{P}{2EI}(2L-x) \quad ②$$

Eqs ① & ② agree with Example 1.5-1 except for shear term in Eq ①

(1.21) (a) $\epsilon_x = \frac{\partial u}{\partial x} = \frac{\rho g}{2E}(2b-2x) = \frac{\rho g}{E}(b-x)$

$\epsilon_y = \frac{\partial v}{\partial y} = -\frac{\nu \rho g}{E}(b-x)$ sub ϵ_x & ϵ_y into Eq. (1.4-5)

$$\sigma_x = \frac{E}{1-\nu^2}\left[\frac{\rho g}{E}(b-x) - \nu^2\frac{\rho g}{E}(b-x)\right] = \rho g(b-x) \quad ①$$

$$\sigma_y = \frac{E}{1-\nu^2}\left[-\frac{\nu \rho g}{E}(b-x) + \frac{\nu \rho g}{E}(b-x)\right] = 0$$

$$\gamma_{xy} = \frac{\partial v}{\partial x} + \frac{\partial u}{\partial y} = \frac{\nu \rho g}{E}y + \left(-\frac{\nu \rho g}{E}y\right) = 0 \quad \therefore \tau_{xy} = 0$$

(continued)

rod hanging under its own weight

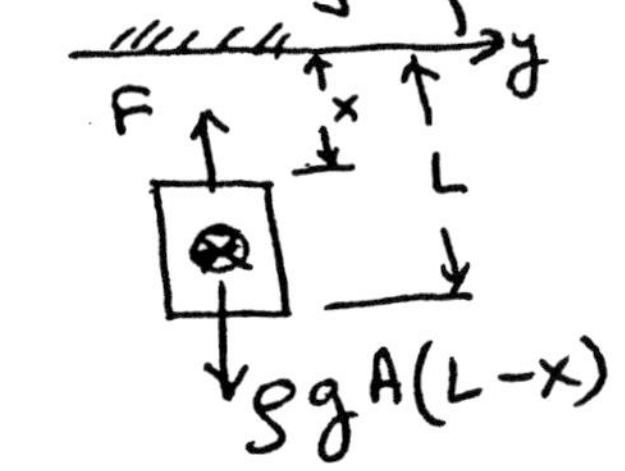

$\sigma_x = \frac{F}{A} = \rho g(L-x)$ agrees with ①

$\sigma_y = \tau_{xy} \approx 0$

(b)

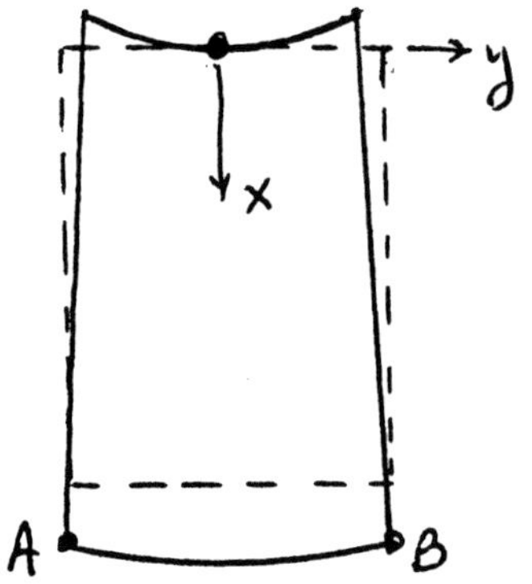

(c) $\theta = \frac{1}{2}\left(\frac{\partial v}{\partial x} - \frac{\partial u}{\partial y}\right) = \frac{1}{2}\left[\nu \frac{\rho g}{E} y - \left(-\nu \frac{\rho g}{E} y\right)\right] = \frac{\nu \rho g}{E} y$

Point A: $y = -a$, $\theta = -\frac{\nu \rho g a}{E}$ ANS

Note: rotation in figure is negative

Point B: $y = a$ $\theta = \frac{\nu \rho g a}{E}$ ANS

Note: rotation in figure is positive.

(1.22)

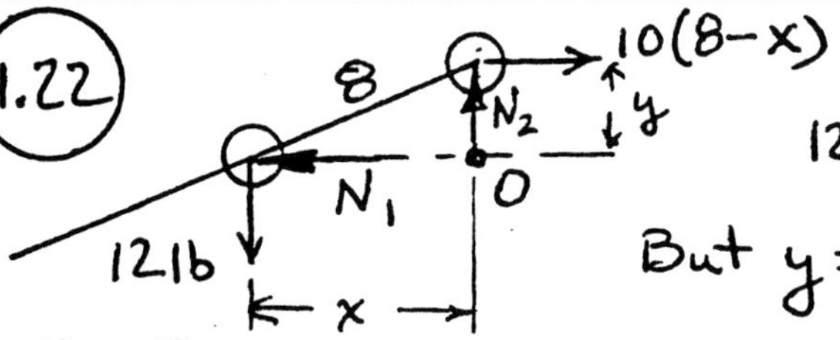

$\Sigma M_0 = 0$

$12x - 10(8-x)y = 0$ ①

But $y = \sqrt{8^2 - x^2} = \sqrt{64 - x^2}$ ②

Sub ② into ①

$12x = 10(8-x)\sqrt{64-x^2}$ square both sides

$144x^2 = 100(64 - 16x + x^2)(64 - x^2)$

Simplifying results in

$x^4 - 16x^3 + 1.44x^2 + 1024x - 4096 = 0$

Two methods of solving this are:

1. Plot function vs x in Excel. The first zero of the function is found when $x = 6.4$ in.
2. Solve polynomial using an HP calculator, MathCad, etc. Yields two complex roots, a negative root of −7.978 and 6.4 in.

Solution is $x = 6.4$ in and

$y = \sqrt{64 - 6.4^2} = 4.8$ in

(2.1)

$$\begin{Bmatrix} x_1 \\ y_1 \\ z_1 \end{Bmatrix} = \begin{bmatrix} 1 & 0 & 0 \\ 0 & \cos\theta & \sin\theta \\ 0 & -\sin\theta & \cos\theta \end{bmatrix} \begin{Bmatrix} x \\ y \\ z \end{Bmatrix} \qquad \theta = 45^\circ$$

$$\begin{Bmatrix} x' \\ y' \\ z' \end{Bmatrix} = \begin{bmatrix} \cos\phi & \sin\phi & 0 \\ -\sin\phi & \cos\phi & 0 \\ 0 & 0 & 1 \end{bmatrix} \begin{Bmatrix} x_1 \\ y_1 \\ z_1 \end{Bmatrix} \qquad \phi = -45^\circ$$

$$\begin{Bmatrix} x' \\ y' \\ z' \end{Bmatrix} = \begin{bmatrix} \cos\phi & \sin\phi & 0 \\ -\sin\phi & \cos\phi & 0 \\ 0 & 0 & 1 \end{bmatrix} \begin{bmatrix} 1 & 0 & 0 \\ 0 & \cos\theta & \sin\theta \\ 0 & -\sin\theta & \cos\theta \end{bmatrix} \begin{Bmatrix} x \\ y \\ z \end{Bmatrix} = \begin{bmatrix} \cos\phi & (\sin\phi\cos\theta) & (\sin\phi\sin\theta) \\ -\sin\phi & (\cos\phi\cos\theta) & (\cos\phi\sin\theta) \\ 0 & -\sin\theta & \cos\theta \end{bmatrix} \begin{Bmatrix} x \\ y \\ z \end{Bmatrix}$$

$$= \frac{1}{2} \begin{bmatrix} \sqrt{2} & -1 & -1 \\ \sqrt{2} & 1 & 1 \\ 0 & -\sqrt{2} & \sqrt{2} \end{bmatrix} \begin{Bmatrix} x \\ y \\ z \end{Bmatrix}$$

(a) Using Eqs. (2.1-4)

$$\sigma_{x'} = (0)(\tfrac{1}{2}) + (-40)(\tfrac{1}{4}) + 10(\tfrac{1}{4}) + 2(-30)(\tfrac{\sqrt{2}}{2})(-\tfrac{1}{2}) + 2(-15)(-\tfrac{1}{2})(-\tfrac{1}{2}) + 2(25)(-\tfrac{1}{2})(\tfrac{\sqrt{2}}{2})$$

$$= \underline{-11.46 \text{ MPa}}$$

$$\tau_{x'y'} = 0(\tfrac{1}{2})(\tfrac{\sqrt{2}}{2}) + (-40)(-\tfrac{1}{2})(\tfrac{1}{2}) + 10(-\tfrac{1}{2})(\tfrac{1}{2}) + (-30)\left[(\tfrac{\sqrt{2}}{2})(\tfrac{1}{2}) + (-\tfrac{1}{2})(\tfrac{\sqrt{2}}{2})\right]$$
$$+ (-15)\left[(-\tfrac{1}{2})(\tfrac{1}{2}) + (-\tfrac{1}{2})(\tfrac{1}{2})\right] + (25)\left[(\tfrac{\sqrt{2}}{2})(\tfrac{1}{2}) + (-\tfrac{1}{2})(\tfrac{\sqrt{2}}{2})\right] = \underline{15 \text{ MPa}}$$

$$\tau_{z'x'} = (0)(\tfrac{\sqrt{2}}{2})(0) + (-40)(-\tfrac{1}{2})(-\tfrac{\sqrt{2}}{2}) + 10(-\tfrac{1}{2})(\tfrac{\sqrt{2}}{2}) + (-30)\left[(\tfrac{\sqrt{2}}{2})(-\tfrac{\sqrt{2}}{2}) + (-\tfrac{1}{2})(0)\right]$$
$$+ (-15)\left[(-\tfrac{1}{2})(\tfrac{\sqrt{2}}{2}) + (-\tfrac{1}{2})(-\tfrac{\sqrt{2}}{2})\right] + (25)\left[\tfrac{\sqrt{2}}{2}(\tfrac{\sqrt{2}}{2}) + (-\tfrac{1}{2})(0)\right] = \underline{9.82 \text{ MPa}}$$

$$\sigma_{y'} = (0)(\tfrac{1}{2}) + (-40)(\tfrac{1}{4}) + (10)(\tfrac{1}{4}) + 2(-30)(\tfrac{\sqrt{2}}{2})(\tfrac{1}{2}) + 2(-15)(\tfrac{1}{2})(\tfrac{1}{2}) + 2(25)(\tfrac{1}{2})(\tfrac{\sqrt{2}}{2})$$

$$= \underline{-18.54 \text{ MPa}}$$

$$\tau_{y'z'} = (0)(\tfrac{\sqrt{2}}{2})(0) + (-40)(\tfrac{1}{2})(-\tfrac{\sqrt{2}}{2}) + 10(\tfrac{1}{2})(\tfrac{\sqrt{2}}{2}) + (-30)\left[\tfrac{\sqrt{2}}{2}(-\tfrac{\sqrt{2}}{2}) + \tfrac{1}{2}(0)\right]$$
$$+ (-15)\left[\tfrac{1}{2}(\tfrac{\sqrt{2}}{2}) + \tfrac{1}{2}(-\tfrac{\sqrt{2}}{2})\right] + (25)\left[(\tfrac{\sqrt{2}}{2})(\tfrac{\sqrt{2}}{2}) + \tfrac{1}{2}(0)\right] = \underline{45.18 \text{ MPa}}$$

$$\sigma_{z'} = 0(0) + (-40)(\tfrac{1}{2}) + 10(\tfrac{1}{2}) + 2(-30)(0)(-\tfrac{\sqrt{2}}{2}) + 2(-15)(-\tfrac{\sqrt{2}}{2})(\tfrac{\sqrt{2}}{2}) + 2(25)(\tfrac{\sqrt{2}}{2})(0)$$

$$= \underline{0}$$

$$[\sigma'] = \begin{bmatrix} -11.46 & 15 & 9.82 \\ 15 & -18.54 & 45.18 \\ 9.82 & 45.18 & 0 \end{bmatrix} \text{MPa} \qquad \underline{\text{ANS}}$$

2.2 $\sigma_{x'} = 40\cos^2(-30) + (-30)\sin^2(-30) + 2\tau_{xy}\sin(-30)\cos(-30)$ ①

$\sigma_{y'} = 40\sin^2(-30) + (-30)\cos^2(-30) - 2\tau_{xy}\sin(-30)\cos(-30)$ ②

$\tau_{x'y'} = -20 = -(40+30)\sin(-30)\cos(-30) + \tau_{xy}\left(\cos^2(-30) - \sin^2(-30)\right)$ ③

From ① $-20 = 30.311 + 0.5\tau_{xy} \Rightarrow \tau_{xy} = \underline{-100.6\text{ MPa}}$

Sub τ_{xy} into ① & ②

$\sigma_{x'} = \underline{109.6\text{ MPa}}$ $\quad\sigma_{y'} = \underline{-99.6\text{ MPa}}$

2.3 $\sigma_{x'} = 30 = -40\cos^2(135) + \sigma\sin^2(135) + 2(-20)\sin 135\cos 135$ ①

$\tau = -(-40-\sigma)\sin 135\cos 135 + (-20)(\cos^2 135 - \sin^2 135)$ ②

From ① $30 = -20 + 0.5\sigma + 20 \Rightarrow \sigma = \underline{60\text{ MPa}}$

From ② $\tau = -(-40-60)(-\tfrac{1}{2}) - 20(0.5-0.5) = \underline{-50\text{ MPa}}$

2.4 Using Eqs. (2.1-9) & (2.1-10) let $\frac{\sigma_x+\sigma_y}{2} = A$, $\frac{\sigma_x-\sigma_y}{2} = B$, $\tau_{xy} = C$

$\sigma = A + B\cos 2\theta + C\sin 2\theta$, $\quad\tau = -B\sin 2\theta + C\cos 2\theta$

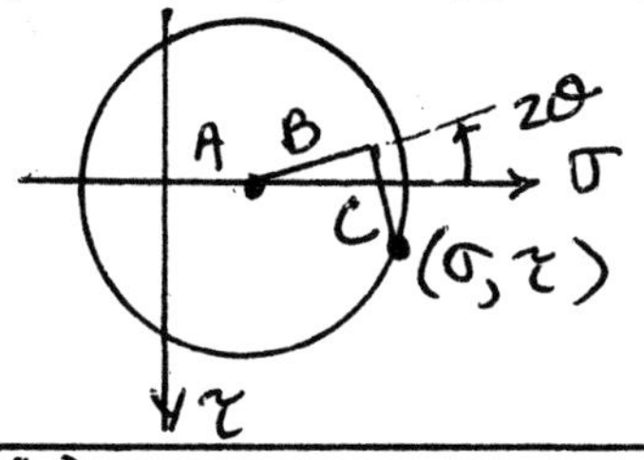

lines B & C move ccw 2θ

2.5 (a) $\sigma_{x'} = \frac{1}{36}\left[2(0)^2 + (-2)(18) + 3(18) + 2(-1)(0)(3\sqrt{2}) + 2(0)(3\sqrt{2})(-3\sqrt{2}) + 2(1)(-3\sqrt{2})(0)\right] = \underline{0.5\text{ kpsi}}$

$\tau_{x'y'} = \frac{1}{36}\Big\{2(0)(-2\sqrt{3}) + (-2)(3\sqrt{2})(2\sqrt{3}) + 3(-3\sqrt{2})(2\sqrt{3}) + (-1)\left[0(2\sqrt{3}) + 3\sqrt{2}(-2\sqrt{3})\right]$
$+\,0[\quad] + (1)\left[0(2\sqrt{3}) + (-3\sqrt{2})(-2\sqrt{3})\right]\Big\} = \underline{-1.225\text{ kpsi}}$

$\tau_{z'x'} = \frac{1}{36}\Big\{2(0)(2\sqrt{6}) + (-2)(3\sqrt{2})(\sqrt{6}) + 3(-3\sqrt{2})(\sqrt{6}) + (-1)\left[0(\sqrt{6}) + 3\sqrt{2}(2\sqrt{6})\right]$
$+\,0[\quad] + (1)\left[0(\sqrt{6}) + (-3\sqrt{2})(2\sqrt{6})\right]\Big\} = \underline{-2.598\text{ kpsi}}$

$\sigma_{y'} = \frac{1}{36}\left[2(12) + (-2)(12) + 3(12) + 2(-1)(-2\sqrt{3})(2\sqrt{3}) + 0 + 2(1)(2\sqrt{3})(-2\sqrt{3})\right] = \underline{1.000\text{ kpsi}}$

(continued)

$$\tau_{y'z'} = \frac{1}{36}\Big\{2(-2\sqrt{3})(2\sqrt{6}) + (-2)(2\sqrt{3})(\sqrt{6}) + 3(2\sqrt{3})(\sqrt{6}) + (-1)\big[(-2\sqrt{3})(\sqrt{6}) + 2\sqrt{3}(2\sqrt{6})\big]$$
$$+ 0[\quad] + (1)\big[(-2\sqrt{3})(\sqrt{6}) + 2\sqrt{3}(2\sqrt{6})\big]\Big\} = \underline{-0.707 \text{ kpsi}}$$

$$\sigma_{z'} = \frac{1}{36}\left[2(24) + (-2)(6) + 3(6) + 2(-1)(2\sqrt{6})(\sqrt{6}) + 0 + 2(1)(\sqrt{6})(2\sqrt{6})\right] = \underline{1.500 \text{ kpsi}}$$

$$[\sigma'] = \begin{bmatrix} 0.5 & -1.225 & -2.598 \\ -1.225 & 1 & -0.707 \\ -2.598 & -0.707 & 1.5 \end{bmatrix} \text{kpsi} \qquad \underline{\text{ANS}}$$

(b) $\sigma = \sigma_{x'} = \underline{0.500 \text{ kpsi}}$ (same eq. used in part (a))

$$\tau = \Big\{\frac{1}{36}\left[2(0) + (-1)(3\sqrt{2}) + (1)(-3\sqrt{2})\right]^2 + \frac{1}{36}\left[(-1)(0) + (-2)(3\sqrt{2}) + (0)(-3\sqrt{2})\right]^2$$
$$+ \frac{1}{36}\left[(1)(0) + (0)(3\sqrt{2}) + 3(-3\sqrt{2})\right]^2 - 0.5^2\Big\}^{1/2} = \underline{2.872 \text{ kpsi}}$$

(c) Check magnitude of $\tau_{x'y'}$ & $\tau_{z'x'}$ from part (a)

$$(\tau_{x'})_{net} = \sqrt{(-1.225)^2 + (-2.598)^2} = \underline{2.872 \text{ kpsi}} \quad \text{agrees with part (b)}$$

2.6 $\sigma_{x'} = \frac{1}{81}\left[25(1)^2 + 0 + (-20)(4)^2 + 2(10)(1)(-8) + 0 + 2(15)(4)(1)\right] = \underline{-4.136 \text{ MPa}}$

$$\tau_{x'y'} = \frac{1}{81}\Big\{25(1)(4) + 0 + (-20)(4)(7) + 10\big[(1)(4) + (-8)(4)\big] + 0$$
$$+ 15\big[(1)(7) + 4(4)\big]\Big\} = \underline{-4.877 \text{ MPa}}$$

$$\tau_{z'x'} = \frac{1}{81}\Big\{25(1)(-8) + 0 + (-20)(4)(4) + 10\big[(1)(1) + (-8)(-8)\big] + 0$$
$$+ 15\big[(1)(4) + (4)(-8)\big]\Big\} = \underline{-3.580 \text{ MPa}}$$

$$\sigma_{y'} = \frac{1}{81}\left[25(4)^2 + 0 + (-20)(7)^2 + 2(10)(4)(4) + 0 + 2(15)(7)(4)\right] = \underline{7.160 \text{ MPa}}$$

$$\tau_{y'z'} = \frac{1}{81}\Big\{25(4)(-8) + 0 + (-20)(7)(4) + 10\big[(4)(1) + 4(-8)\big] + 0$$
$$+ 15\big[4(4) + 7(-8)\big]\Big\} = \underline{-27.654 \text{ MPa}}$$

$$\sigma_{z'} = \frac{1}{81}\left[25(8)^2 + 0 + (-20)(4)^2 + 2(10)(-8)(1) + 0 + 2(15)(4)(-8)\right] = \underline{1.975 \text{ MPa}}$$

$$[\sigma'] = \begin{bmatrix} -4.136 & -4.877 & -3.580 \\ -4.877 & 7.160 & -27.654 \\ -3.580 & -27.654 & 1.975 \end{bmatrix} \text{MPa} \qquad \underline{\text{ANS}}$$

(continued)

(b) $I_1 = 25 + 0 + (-20) = 5 \qquad I_1' = -4.136 + 7.160 + 1.975 = 4.999$ (round-off)

$I_2 = 25(0) + 0(-20) + (-20)(25) - 0^2 - 15^2 - 10^2 = -825$

$I_2' = (-4.136)(7.16) + (7.16)(1.975) + (1.975)(-4.136) - (-27.654)^2 - (-3.58)^2 - (-4.877)^2$

$= -825$

$I_3 = 25(0)(-20) + 2(0)(15)(10) - 25(0)^2 - 0(15)^2 - (-20)(10)^2 = 2000$

$I_3' = (-4.136)(7.16)(1.975) + 2(-27.654)(-3.58)(-4.877) - (-4.136)(-27.654)^2$

$-7.16(-3.58)^2 - (1.975)(-4.877)^2 = 2000$ <u>All check</u>

(c) $\sigma = \sigma_{x'} = \underline{-4.136\ \text{MPa}}$ from part (a)

$$\tau = \left\{ \frac{1}{81} \left\{ [25(1) + 10(-8) + 15(4)]^2 + [10(1) + 0 + 0]^2 + [15(1) + 0 + (-20)4]^2 \right\} - (-4.136)^2 \right\}^{1/2}$$

$= \underline{6.050\ \text{MPa}}$

$$t_x = \frac{1}{(6.05)(9)} [(25 + 4.136)(1) + 10(-8) + 15(4)] = 0.1678$$

$$t_y = \frac{1}{6.05(9)} [(10)(1) + (0 + 4.136)(-8) + 0(4)] = -0.4240$$

$$t_y = \frac{1}{6.05(9)} [15(1) + (0)(-8) + (-20 + 4.136)(4)] = -0.8900$$

As a vector

$\bar{\tau} = 6.05 [0.1678\bar{i} - 0.424\bar{j} - 0.89\bar{k}] = 1.015\bar{i} - 2.565\bar{j} - 5.385\bar{k}$ MPa

(d) One can check magnitudes (minimal check) from part (a)

$(\tau_{x'})_{net} = \sqrt{\tau_{x'y'}^2 + \tau_{z'x'}^2} = \sqrt{(-4.877)^2 + (-3.58)^2} = \underline{6.050\ \text{MPa}}$ checks

Can also check direction by transforming the shears to xyz coordinates according to

$$\{V\}_{xyz} = [T]^T \{V\}_{x'y'z'} \quad \text{where } [T]^T = \frac{1}{9}\begin{bmatrix} 1 & 4 & -8 \\ -8 & 4 & 1 \\ 4 & 7 & 4 \end{bmatrix}$$

$$\bar{\tau}_{x'y'} = -\frac{4.877}{9}(4\bar{i} + 4\bar{j} + 7\bar{k}) = -2.168\bar{i} - 2.168\bar{j} - 3.793\bar{k} \quad ①$$

$$\bar{\tau}_{z'x'} = \frac{-3.580}{9}(-8\bar{i} + \bar{j} + 4\bar{k}) = 3.182\bar{i} - 0.398\bar{j} - 1.591\bar{k} \quad ②$$

① + ② $\Rightarrow (\bar{\tau}_{x'})_{net} = 1.014\bar{i} - 2.566\bar{j} - 5.384k$ MPa

agrees with part (c) with some round-off error

2.7 (a)

$$\sigma_{x'} = \frac{1}{(11)^2}\left[25(2)^2+0+(-20)(9)+2(10)(2)(6)+0+2(15)(9)(2)\right] = \underline{-6.116 \text{ MPa}}$$

$$\tau_{x'y'} = \frac{1}{(11)^2}\left\{25(2)(6)+0+(-20)(9)(-6)+10\left[2(7)+6(6)\right]+0+15\left[2(-6)+9(6)\right]\right\}$$
$$= \underline{20.744 \text{ MPa}}$$

$$\tau_{z'x'} = \frac{1}{(11)^2}\left\{25(2)(9)+0+(-20)(9)(2)+10\left[2(-6)+6(9)\right]+0+15\left[2(2)+9(9)\right]\right\}$$
$$= \underline{14.752 \text{ MPa}}$$

$$\sigma_{y'} = \frac{1}{(11)^2}\left[25(6)^2+0+(-20)(-6)^2+2(10)(6)(7)+0+2(15)(-6)(6)\right] = \underline{-0.496 \text{ MPa}}$$

$$\tau_{y'z'} = \frac{1}{(11)^2}\left\{25(6)(9)+0+(-20)(-6)(2)+10\left[6(-6)+7(9)\right]+0+15\left[6(2)+(-6)(9)\right]\right\}$$
$$= \underline{10.165 \text{ MPa}}$$

$$\sigma_{z'} = \frac{1}{(11)^2}\left[25(9)^2+0+(-20)(2)^2+2(10)(9)(-6)+0+2(15)(2)(9)\right] = \underline{11.612 \text{ MPa}}$$

$$[\sigma'] = \begin{bmatrix} -6.116 & 20.744 & 14.752 \\ 20.744 & -0.496 & 10.165 \\ 14.752 & 10.165 & 11.612 \end{bmatrix} \text{ MPa} \qquad \underline{\text{Ans}}$$

(b) $I_1 = 25+0-20 = 5 \qquad I_1' = -6.116-0.496+11.612 = 5$

$I_2 = 25(0)+0(-20)+(-20)(25)-0^2-15^2-10^2 = -825$

$I_2' = (-6.116)(-0.496)+(-0.496)(11.612)+(11.612)(-6.116)-(10.165)^2-(14.752)^2$
$\quad -(20.744)^2 = -825$

$I_3 = 25(0)(-20)+2(0)(15)(10)-25(0)^2-0(15)^2-(-20)(10)^2 = 2000$

$I_3' = (-6.116)(-0.496)(11.612)+2(10.165)(14.752)(20.744)-(-6.116)(10.165)^2$
$\quad -(-0.496)(14.752)^2-(11.612)(20.744)^2 = 2000$ $\qquad$ All check

(c) $\sigma = \sigma_{x'} = -6.116$ MPa from part (a)

$$\tau = \left\{\left[\frac{1}{(11)^2}\left\{\left[25(2)+10(6)+15(9)\right]^2+\left[10(2)+0+0\right]^2+\left[15(2)+0+(-20)(9)\right]^2\right\}-(-6.116)^2\right]\right\}^{\frac{1}{2}}$$
$$= \underline{25.454 \text{ MPa}}$$

(continued)

$$t_x = \frac{1}{25.454(11)}\left[(25+6.116)(2)+10(6)+15(9)\right] = 0.919$$

$$t_y = \frac{1}{25.454(11)}\left[10(2)+(0+6.116)(6)+0(9)\right] = 0.202$$

$$t_z = \frac{1}{25.454(11)}\left[15(2)+0(6)+(-20+6.116)(9)\right] = -0.339$$

Expressing τ as a vector

$$\bar{\tau} = 25.454\left(0.919\bar{i}+0.202\bar{j}-0.339\bar{k}\right) = 23.39\bar{i}+5.14\bar{j}-8.63\bar{k} \text{ (MPa)}$$

(d) check magnitude (minimal check)

$$(\tau_{x'})_{net} = \sqrt{\tau_{x'y'}^2+\tau_{z'x'}^2} = \sqrt{20.744^2+14.752^2} = 25.455 \text{ MPa} \quad \underline{\text{checks}}$$

checking direction transform $x'y'z'$ vectors to xyz using transformation

$$\{V\}_{xyz} = [T]^T\{V\}_{x'y'z'} \quad \text{where } [T]^T = \frac{1}{11}\begin{bmatrix} 2 & 6 & 9 \\ 6 & 7 & -6 \\ 9 & -6 & 2 \end{bmatrix}$$

$$\bar{\tau}_{x'y'} = \frac{20.744}{11}\left(6\bar{i}+7\bar{j}-6\bar{k}\right) = 11.315\bar{i}+13.201\bar{j}-11.315\bar{k} \quad (1)$$

$$\bar{\tau}_{z'x'} = \frac{20.744}{11}\left(9\bar{i}-6\bar{j}+2\bar{k}\right) = 12.070\bar{i}-8.047\bar{j}+2.682\bar{k} \quad (2)$$

$$(\bar{\tau}_{x'})_{net} = \bar{\tau}_{x'y'}+\bar{\tau}_{z'x'} = 23.39\bar{i}+5.14\bar{j}-8.63\bar{k} \text{ MPa} \quad \underline{\text{checks}}$$

2.8 $n_x = n_y = n_z = \pm\frac{1}{\sqrt{3}}$ $\quad \sigma = \sigma_{ave} = \frac{1}{3}(\sigma_1+\sigma_2+\sigma_3)$

Let $\sigma_x = \sigma_1$, $\sigma_y = \sigma_2$, $\sigma_z = \sigma_3$ $\quad \tau_{xy} = \tau_{yz} = \tau_{zx} = 0$

Eq (2.1-20)

$$\tau = \left[\frac{\sigma_1^2}{3}+\frac{\sigma_2^2}{3}+\frac{\sigma_3^2}{3}-\left(\frac{\sigma_1+\sigma_2+\sigma_3}{3}\right)^2\right]^{1/2} = \frac{1}{3}\left\{2\sigma_1^2+2\sigma_2^2+2\sigma_3^2-2\sigma_1\sigma_2-2\sigma_2\sigma_3-2\sigma_3\sigma_1\right\}^{1/2}$$

$$= \frac{1}{3}\left[(\sigma_1^2-2\sigma_1\sigma_2+\sigma_2^2)+(\sigma_2^2-2\sigma_2\sigma_3+\sigma_3^2)+(\sigma_3^2-2\sigma_3\sigma_1+\sigma_1^2)\right]^{1/2}$$

$$= \frac{1}{3}\left[(\sigma_1-\sigma_2)^2+(\sigma_2-\sigma_3)^2+(\sigma_3-\sigma_1)^2\right]^{1/2} \quad \text{Q.E.D.}$$

2.9 (a) $\sigma = \frac{1}{(11)^2}\left[6(6)^2 + (-3)(6)^2 + 3(7)^2 + 2(4)(6)(6) + 0 + 0\right] = \underline{4.488\ \text{kpsi}}$

$\tau = \left\{\frac{1}{(11)^2}\left\{\left[6(6)+4(6)\right]^2 + \left[4(6)+(-3)(6)\right]^2 + \left[3(7)\right]^2\right\} - (4.488)^2\right\} = \underline{3.681\ \text{kpsi}}$

(b) $t_x = \frac{1}{3.681(11)}\left[(6-4.488)(6) + 4(6)\right] = \underline{0.817}$

$t_y = \frac{1}{3.681(11)}\left[4(6) + (-3-4.488)(6)\right] = \underline{-0.517}$

$t_z = \frac{1}{3.681(11)}\left[(3-4.488)7\right] = \underline{-0.257}$

2.10 (a) BCGF

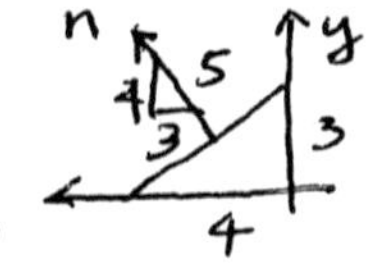

$n_x = 0,\ n_y = \frac{4}{5},\ n_z = \frac{3}{5}$

$\sigma = \frac{1}{(5)^2}\left[-15(0)^2 + 10(4)^2 + 0(3)^2 + 0 + 2(-20)(4)(3) + 0\right] = \underline{-12.8\ \text{MPa}}$

$\tau = \left\{\frac{1}{(5)^2}\left\{\left[0+0+(-25)(3)\right]^2 + \left[0+10(4)+(-20)(3)\right]^2 + \left[0+(-20)(4+0\right]^2\right\} - (-12.8)^2\right\}$

$= \underline{18.25\ \text{MPa}}$

(b) ABEF

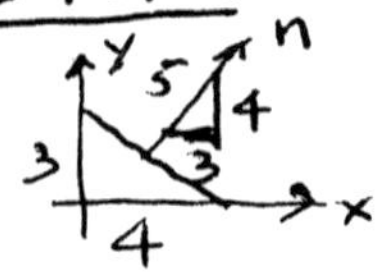

$n_x = \frac{3}{5},\ n_y = \frac{4}{5},\ n_z = 0$

$\sigma = \frac{1}{(5)^2}\left[-15(3)^2 + 10(4)^2 + 0 + 0 + \ \ + 0\right] = \underline{1.0\ \text{MPa}}$

$\tau = \left\{\frac{1}{(5)^2}\left\{\left[-15(3)+0+0\right]^2 + \left[0+10(4)+0\right]^2 + \left[-25(3)+(-20)(4)+0\right]^2\right\} - (1)^2\right\}^{1/2}$

$= \underline{33.24\ \text{MPa}}$

(c) BGE $A_x = \frac{1}{2}(6)(8) = 24,\ A_y = \frac{1}{2}(6)(6) = 18,\ A_z = \frac{1}{2}(6)(8) = 24$

$A_o = \sqrt{A_x^2 + A_y^2 + A_z^2} = 38.419,\ n_x = \frac{A_x}{A_o} = 0.6247,\ n_y = \frac{A_y}{A_o} = 0.4685,\ n_z = 0.6247$

$\sigma = \left[-15(0.6247)^2 + 10(0.4685)^2 + 0 + 0 + 2(-20)(0.4685)(0.6247) + 2(-25)(0.6247)^2\right]$

$= \underline{-34.88\ \text{MPa}}$

$\tau = \left\{\left[-15(0.6247) + 0 + (-25)(0.6247)\right]^2 + \left[0 + 10(0.4685) + (-20)(0.6247)\right]^2 + \left[(-25)(0.6247) + (-20)(0.4685) + 0\right]^2 - (-34.88)^2\right\}^{\frac{1}{2}} = \underline{9.65\ \text{MPa}}$

2.11 $x + ay + bz + c = 0$ $(1,0,0)$ $1 + c = 0$ $\therefore c = -1$

$(0,2,0)$ $2a + c = 0$ $a = -\frac{c}{2} = \frac{1}{2}$, $(0,0,-1)$ $-b + c = 0 \therefore b = c = -1$

$\therefore x + \frac{1}{2}y - z - 1 = 0$ or, $2x + y - 2z = 2$

$d_x = 2$, $d_y = 1$, $d_z = -2$ $\sqrt{2^2 + 1^2 + (-2)^2} = 3$

$n_x = \frac{2}{3}$, $n_y = \frac{1}{3}$, $n_z = -\frac{2}{3}$

(a) $\sigma = \frac{1}{(3)^2}\left[-10(2)^2 + 10(1)^2 + 40(-2)^2 + 2(20)(2)(1) + 2(-20)(1)(-2) + 2(30)(-2)(2)\right]$

$= \underline{5.556 \text{ MPa}}$

$\tau = \left\{\frac{1}{(3)^2}\left\{\left[-10(2) + 20(1) + 30(-2)\right]^2 + \left[20(2) + 10(1) - 20(-2)\right]^2 + \left[30(2) - 20(1) + 40(-2)\right]^2\right\} - 5.556^2\right\}^{1/2} = \underline{38.04 \text{ MPa}}$

(b) $t_x = \frac{1}{38.04(3)}\left[(-10 - 5.556)(2) + (20)(1) + 30(-2)\right] = \underline{-0.6232}$

$t_y = \frac{1}{38.04(3)}\left[20(2) + (10 - 5.556)(1) - 20(-2)\right] = \underline{0.7400}$

$t_z = \frac{1}{38.04(3)}\left[30(2) - 20(1) + (40 - 5.556)(-2)\right] = \underline{-0.2532}$

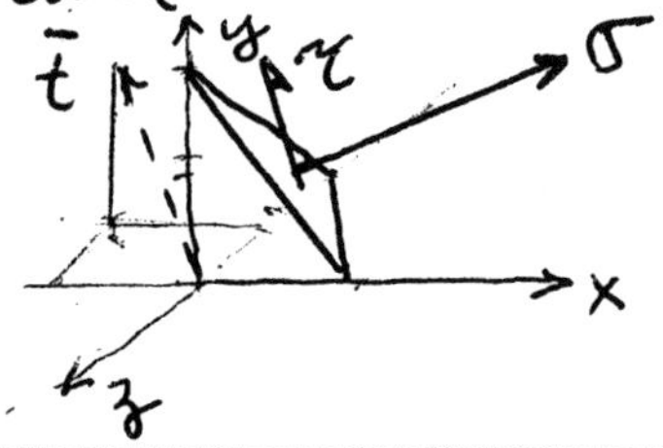

2.12 $d_x = 1$, $d_y = 2$, $d_z = -2$ $\sqrt{1^2 + 2^2 + (-2)^2} = 3$

$n_x = \frac{1}{3}$, $n_y = \frac{2}{3}$, $d_y = -\frac{2}{3}$

$\sigma = \frac{1}{(3)^2}\left[3(1)^2 + 1(2)^2 - 1(-2)^2 + 0 + 2(2)(2)(-2) + 2(-2)(-2)(1)\right] = \underline{-0.556 \text{ kpsi}}$

$\tau = \left\{\frac{1}{(3)^2}\left\{\left[3(1) + 0 + (-2)(-2)\right]^2 + \left[0 + 1(2) + 2(-2)\right]^2 + \left[-2(1) + 2(2) + (-1)(-2)\right]^2\right\} - (-0.556)^2\right\}^{\frac{1}{2}}$

$= \underline{2.712 \text{ kpsi}}$ $t_x = \frac{1}{2.712(3)}\left[(3 + 0.556)(1) + 0 + (-2)(-2)\right] = \underline{0.929}$

$t_y = \frac{1}{2.712(3)}\left[0 + (1 + 0.556)(2) + 2(-2)\right] = \underline{-0.109}$

$t_z = \frac{1}{2.712(3)}\left[-2(1) + 2(2) + (-1 + 0.556)(-2)\right] = \underline{0.355}$

2.13 $F = x^2 + (y-2)^2 + z^2 - 6$

$d_x = \dfrac{\partial F}{\partial x} = 2x, \quad d_y = \dfrac{\partial F}{\partial y} = 2(y-2), \quad d_z = \dfrac{\partial F}{\partial z} = 2z$

At $(1,1,-2)$ $\quad d_x = 2(1) = 2, \quad d_y = 2(1-2) = -2, \quad d_z = 2(-2) = -4$

$\sqrt{2^2 + (-2)^2 + (-4)^2} = \sqrt{24} \quad \therefore n_x = \dfrac{2}{\sqrt{24}}, \; n_y = \dfrac{-2}{\sqrt{24}}, \; n_z = \dfrac{-4}{\sqrt{24}}$

$$\sigma = \frac{1}{24}\left[20(2)^2 + 0 + 10(-4)^2 + 2(35)(2)(-2) + 2(-15)(-2)(-4) + 2(25)(-4)(2)\right]$$

$$= \underline{-28.33 \text{ MPa}}$$

$$\tau = \left\{\frac{1}{24}\left\{\left[20(2)+35(-2)+25(-4)\right]^2 + \left[35(2)+0+(-15)(-4)\right]^2 + \left[(25)(2)+(-15)(-2)+10(-4)\right]^2\right\} - (-28.33)^2\right\}^{\frac{1}{2}} = \underline{32.17 \text{ MPa}}$$

2.14 (a)

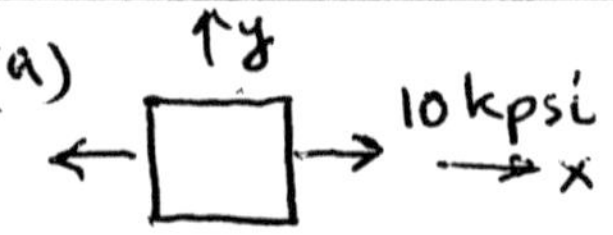

(b) $\sigma_{x'} = 10\cos^2 30 = 7.5$ kpsi

$\sigma_{y'} = 10\sin^2 30 = 2.5$ kpsi

$\tau_{x'y'} = -10\sin 30\cos 30 = -4.33$ kpsi

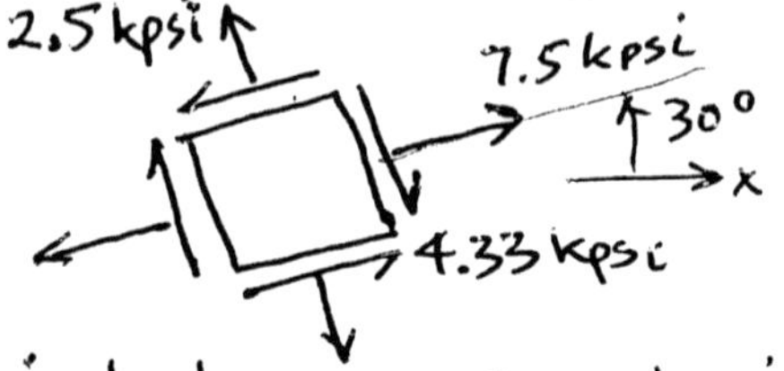

(c) State given are principal stresses. $\sigma_1 = 10$ kpsi, $\sigma_2 = \sigma_3 = 0$

(d)

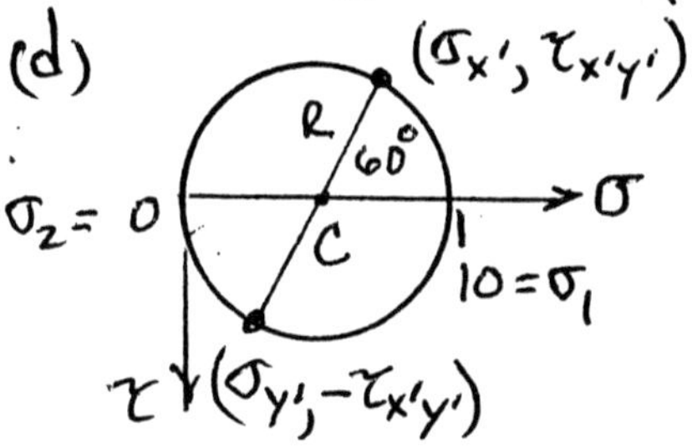

$C = \dfrac{10}{2} = 5$ kpsi $\quad R = 5$ kpsi

$\sigma_{x'} = 5 + 5\cos 60 = 7.5$ kpsi

$\tau_{x'y'} = -5\sin 60 = -4.33$ kpsi

$\sigma_{y'} = 5 - 5\cos 60 = 2.5$ kpsi

(e) $\tau_{max} = \dfrac{\sigma_1 - \sigma_3}{2} = \dfrac{10-0}{2} = \underline{5 \text{ kpsi}}$

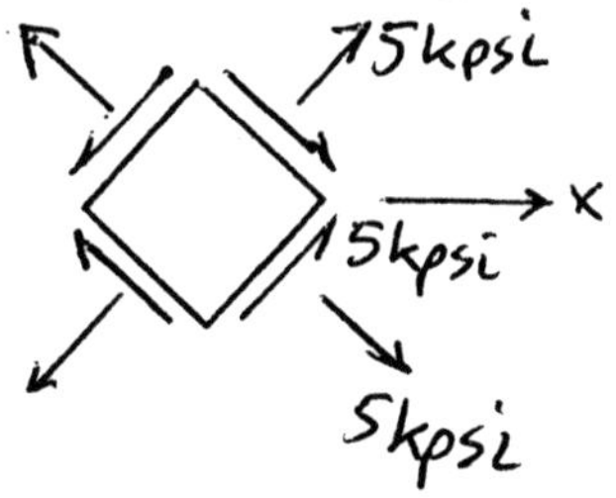

Rotate these surfaces about the x axis. All elements are the same

2.15 (a)

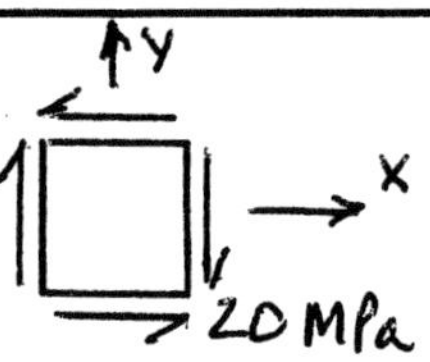

(b)

$$\sigma_{x'} = 2(-20)\cos(-20)\sin(-20) = 12.86\text{ MPa}$$

$$\sigma_{y'} = -2(-20)\sin(-20)\cos(-20) = -12.86\text{ MPa}$$

$$\tau_{x'y'} = (-20)\left[\cos^2(-20) - \sin^2(-20)\right] = -15.32\text{ MPa}$$

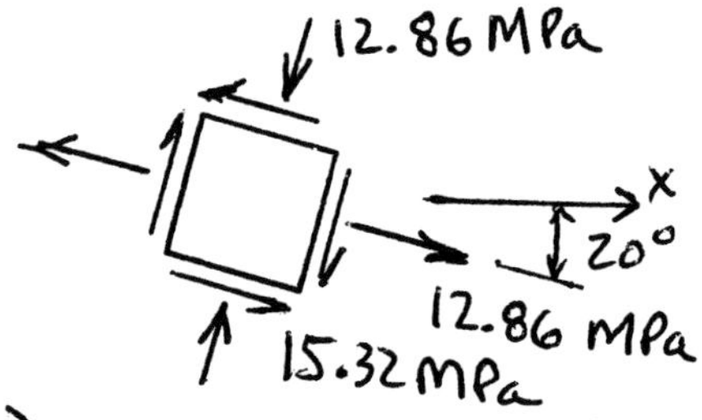

(c) $\sigma_p = \pm\sqrt{20^2} = \pm 20\text{ MPa}$

For $\sigma_p = 20$ $\quad (0-20)\cos\theta - 20\sin\theta = 0 \quad \theta = \tan^{-1}\left(\frac{20}{-20}\right) = -45^\circ$

For $\sigma_p = -20$ $\quad \theta = \tan^{-1}\frac{-20}{-20} = 45^\circ$

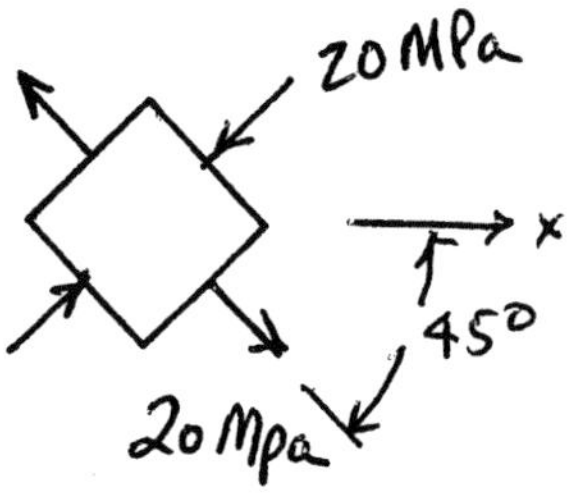

$\sigma_1 = 20\text{ MPa}$, $\sigma_2 = 0$, $\sigma_3 = -20\text{ MPa}$

(d)

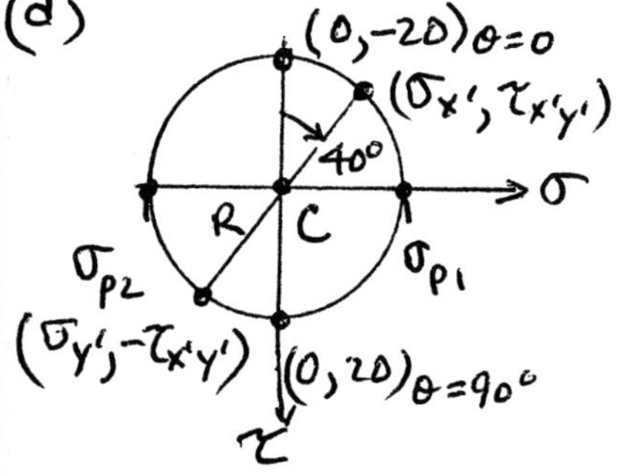

$C = 0$, $R = 20\text{ MPa}$

$\sigma_{x'} = 20\sin 40 = \underline{12.86\text{ MPa}}$

$\tau_{x'y'} = -20\cos 40 = \underline{-15.32\text{ MPa}}$

$\sigma_{y'} = -20\sin 40 = \underline{-12.86\text{ MPa}}$

$\sigma_{p1} = R = \underline{20\text{ MPa}}$, $\sigma_{p2} = -R = \underline{-20\text{ MPa}}$

(e) $\tau_{max} = \frac{\sigma_1 - \sigma_3}{2} = \frac{20+20}{2} = \underline{20\text{ MPa}}$

Given element in part a is ~~mot~~ shear state.

2.16 (a)

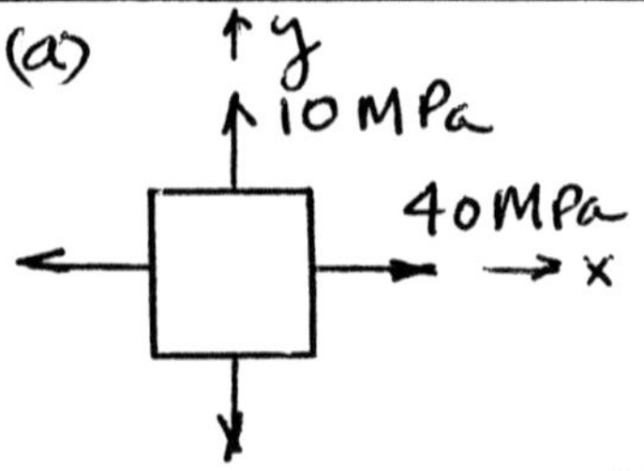

(b) $\sigma_{x'} = 40\cos^2 15 + 10\sin^2 15 = 37.99\ MPa$

$\sigma_{y'} = 40\sin^2 15 + 10\cos^2 15 = 12.01\ MPa$

$\tau_{x'y'} = -(40-10)\sin 15 \cos 15 = -7.5\ MPa$

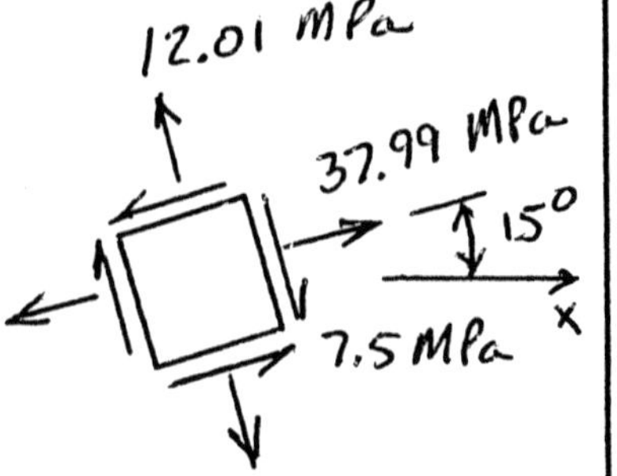

(c) Given state are principal stresses. $\sigma_1 = 40\ MPa$, $\sigma_2 = 10\ MPa$, $\sigma_3 = 0$

(d)

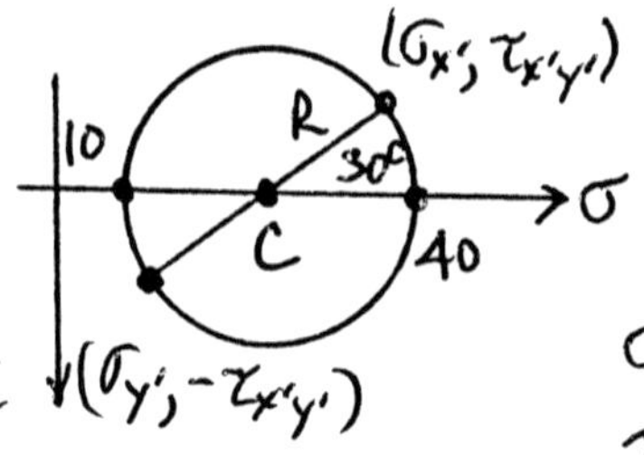

$C = \frac{10+40}{2} = 25\ MPa$

$R = \frac{40-10}{2} = 15\ MPa$

$\sigma_{x'} = 25 + 15\cos 30 = 37.99\ MPa$

$\tau_{x'y'} = -15\sin 30 = -7.5\ MPa$

$\sigma_{p1} = 40$, $\sigma_{p2} = 10\ MPa$ $\quad$ $\sigma_{y'} = 25 - 15\cos 30 = 12.01\ MPa$

(e) $\tau_{max} = \frac{\sigma_1 - \sigma_3}{2} = \frac{40-0}{2} = 20\ MPa$ $\quad$ $\sigma_{Ave} = \frac{40+0}{2} = 20\ MPa$

View down the y axis

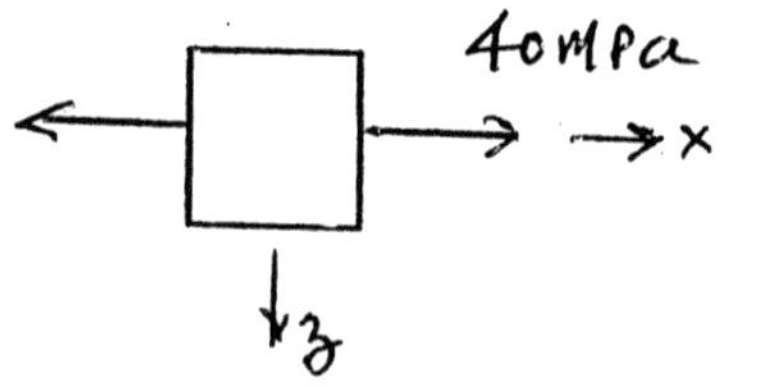

Rotate ±45°

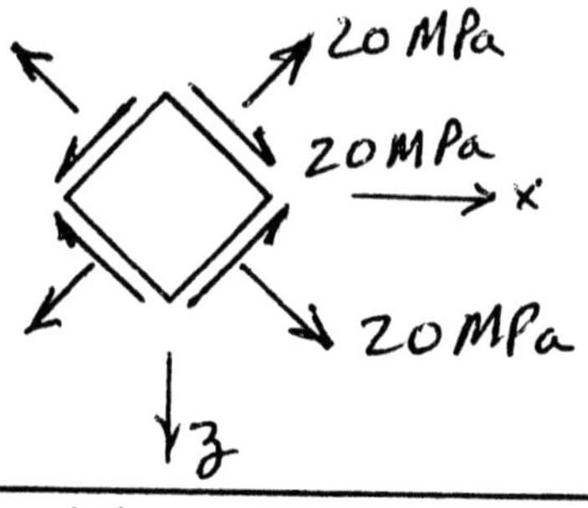

2.17 (a)

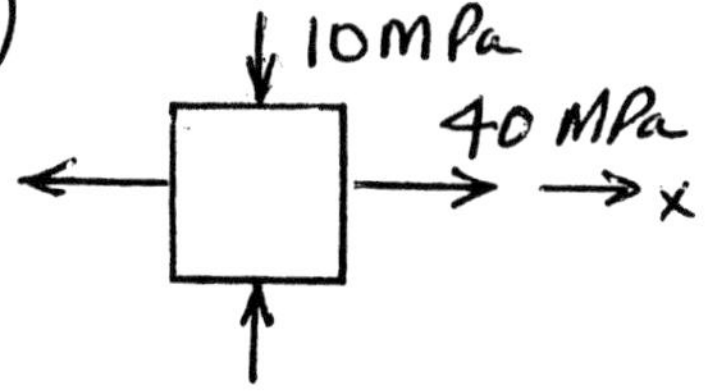

(Continued)

(b)

$\sigma_{x'} = 40\cos^2(-15) + (-10)\sin^2(-15) = 36.65$ MPa

$\sigma_{y'} = 40\sin^2(-15) + (-10)\cos^2(-15) = -6.65$ MPa

$\tau_{x'y'} = -(40+10)\sin(-15)\cos(-15) = 12.5$ MPa

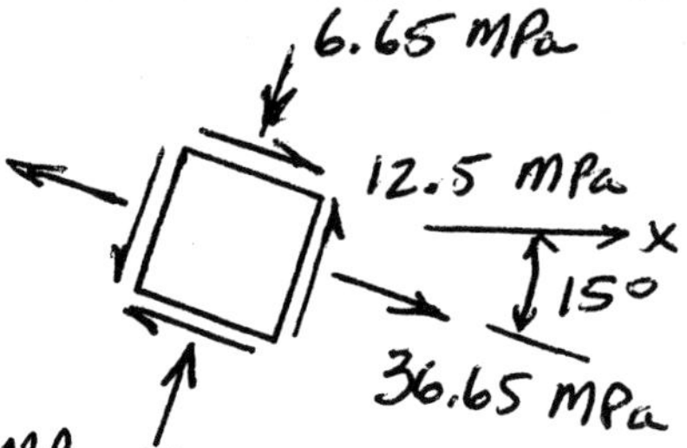

(c) State given are principal stresses. $\sigma_1 = 40$ MPa, $\sigma_2 = 0$, $\sigma_3 = -10$ MPa

(d)

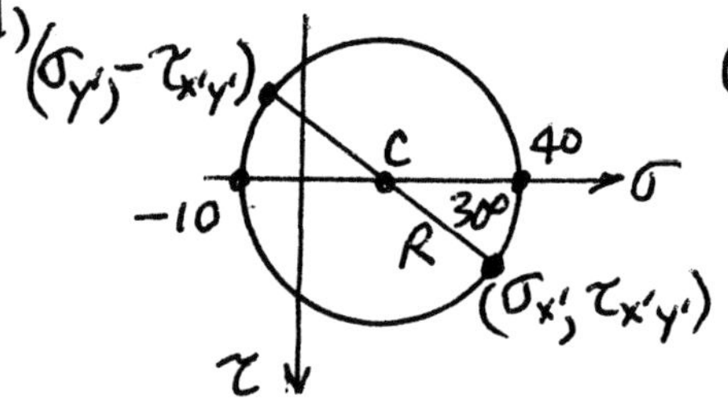

$C = \frac{40-10}{2} = 15$ MPa

$R = \frac{40+10}{2} = 25$ MPa

$\sigma_{x'} = 15 + 25\cos 30 = 36.65$ MPa

$\tau_{x'y'} = 25\sin 30 = 12.5$ MPa

$\sigma_{y'} = 15 - 25\cos 30 = -6.65$ MPa

$\sigma_{p1} = 40$, $\sigma_{p2} = -10$ MPa

(e) $\tau_{max} = \frac{\sigma_1 - \sigma_3}{2} = \frac{40+10}{2} = \underline{25 \text{ MPa}}$ rotate ±45° from x axis

$\sigma_{ave} = \frac{40-10}{2} = 15$ MPa

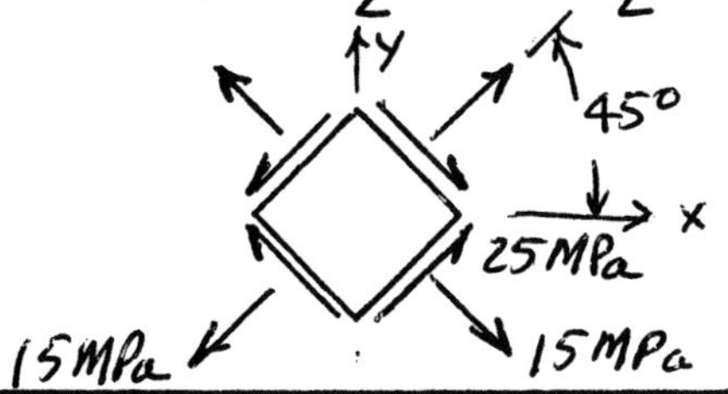

2.18 (a)

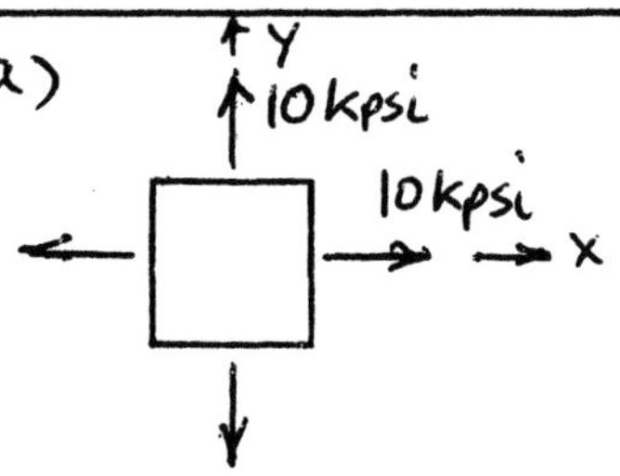

(b)

$\sigma_{x'} = 10\cos^2 60 + 10\sin^2 60 = 10$ kpsi

$\sigma_{y'} = 10\sin^2 60 + 10\cos^2 60 = 10$ kpsi

$\tau_{x'y'} = -(10-10)\sin 60\cos 60 = 0$

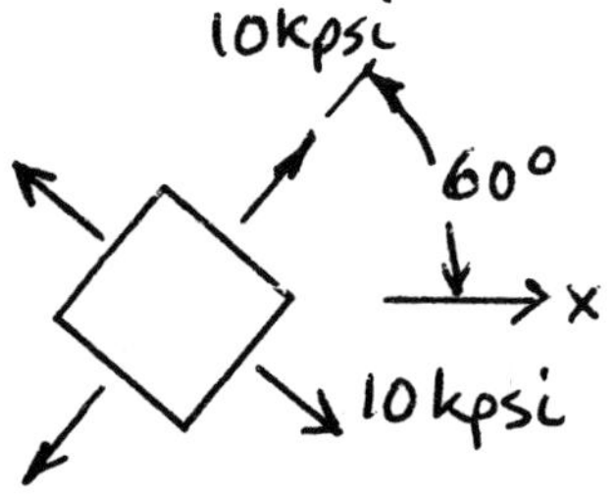

(c) Given state are principal stresses

$\sigma_1 = \sigma_2 = 10$ kpsi, $\sigma_3 = 0$

(d) Mohr's circle is a point

All surfaces ⊥ to xy plane contain $\sigma = 10$ kpsi, $\tau = 0$

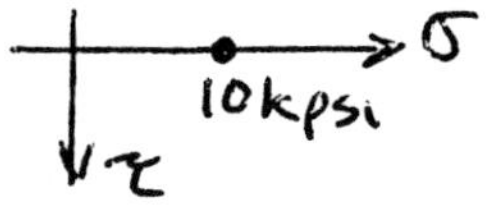

(continued)

(e) $\tau_{max} = \frac{\sigma_1 - \sigma_3}{2} = \frac{10-0}{2} = 5$ kpsi

Look down either x or y axis $\quad \sigma_{ave} = \frac{10+0}{2} = 5$ kpsi

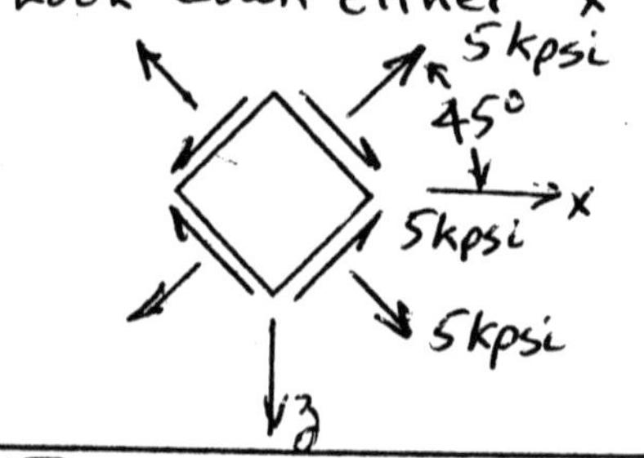

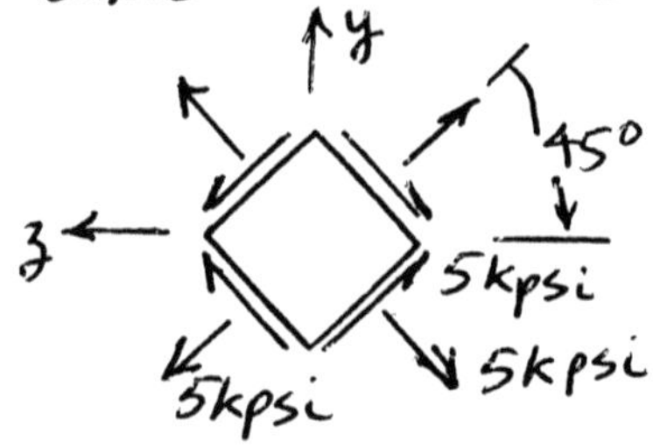

2.19 (a)

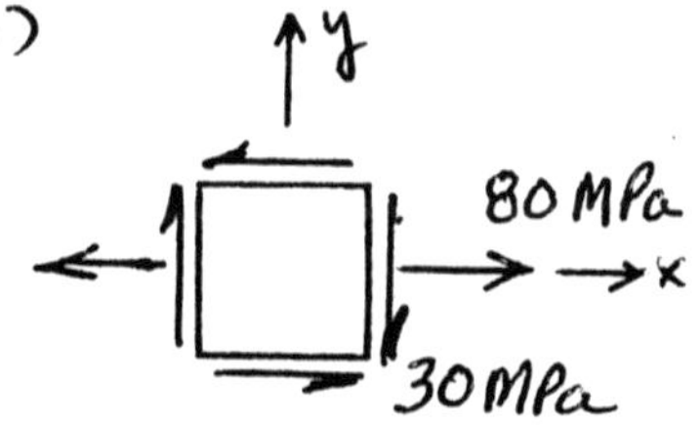

(b)

$\sigma_{x'} = 80\cos^2(-30) + 2(-30)\sin(-30)\cos(-30) = 85.98$ MPa

$\sigma_{y'} = 80\sin^2(-30) - 2(-30)\sin(-30)\cos(-30) = -5.98$ MPa

$\tau_{x'y'} = -80\sin(-30)\cos(-30) + (-30)[\cos^2(-30) - \sin^2(-30)] = 19.64$ MPa

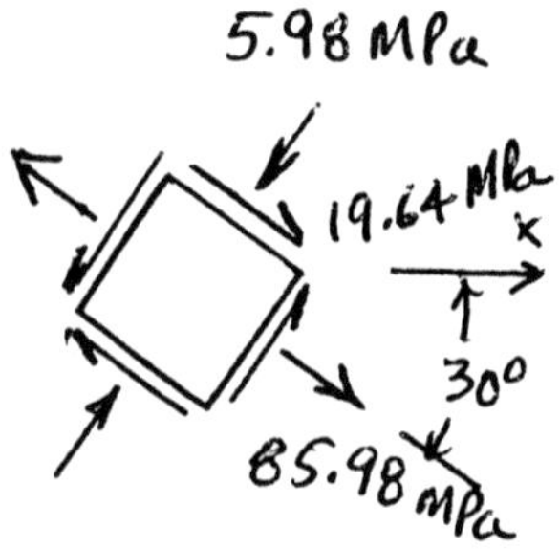

(c) $\sigma_p = \frac{80}{2} \pm \sqrt{\left(\frac{80}{2}\right)^2 + (-30)^2} = 90, -10$ MPa $\quad \therefore \underline{\sigma_1 = 90\text{ MPa}, \sigma_2 = 0, \sigma_3 = -10\text{ MPa}}$

$\sigma_{p1} = 90$ MPa, Eq. (2.1-28a)

$(80-90)\cos\theta_{p1} + (-30)\sin\theta_{p1} = 0 \Rightarrow \theta_{p1} = \tan^{-1}\left(\frac{10}{-30}\right) = -18.43°$

$\sigma_{p2} = -10$ MPa

$(80+10)\cos\theta_{p2} + (-30)\sin\theta_{p2} = 0 \Rightarrow \theta_{p2} = \tan^{-1}\left(\frac{-90}{-30}\right) = 71.57°$

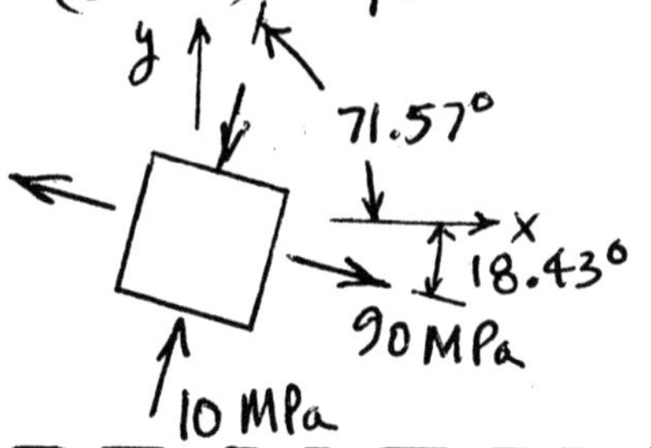

(d)

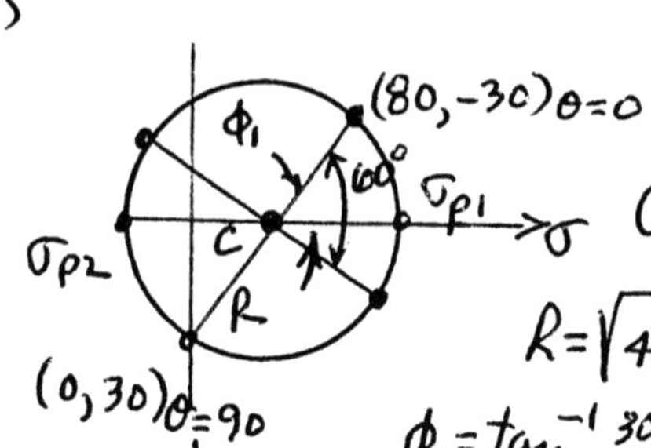

$C = \frac{80}{2} = 40$ MPa

$R = \sqrt{40^2 + 30^2} = 50$ MPa

$\phi_1 = \tan^{-1}\frac{30}{40} = 36.87°$

$\sigma_{x'} = 40 + 50\cos(60 - 36.87) = \underline{85.98\text{ MPa}}$

$\tau_{x'y'} = 50\sin(60 - 36.87) = \underline{19.64\text{ MPa}}$

$\sigma_{y'} = 40 - 50\cos(60 - 36.87) = \underline{-5.98\text{ MPa}}$

$\sigma_{p1} = C + R = \underline{90\text{ MPa}}$

$\sigma_{p2} = C - R = \underline{-10\text{ MPa}}$

$2\theta_{p1} = 60 - \phi_1 = 60 - 36.87 \Rightarrow \theta_{p1} = \underline{18.43°}$ cw from x axis

(continued)

(e) $\tau_{max} = \dfrac{\sigma_1 - \sigma_3}{2} = \dfrac{90+10}{2} = \underline{50\ MPa}$ $\sigma_{ave} = \dfrac{90-10}{2} = 40\ MPa$

$\theta_s = \dfrac{90 - \phi_s}{2} = 26.57^\circ$ ccw to τ_{min}

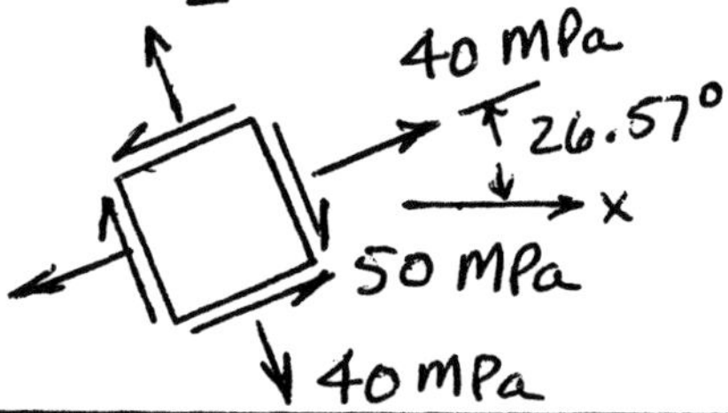

2.20

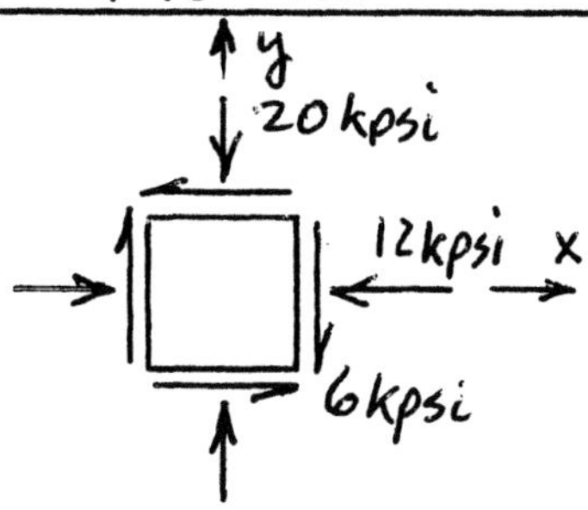

(b)

$\sigma_{x'} = -12\cos^2 25 - 20\sin^2 25 + 2(-6)\cos 25 \sin 25 = -18.03\ kpsi$

$\sigma_{y'} = -12\sin^2 25 - 20\cos^2 25 - 2(-6)\sin 25 \cos 25 = -13.97\ kpsi$

$\tau_{x'y'} = -(-12+20)\sin 25\cos 25 + (-6)(\cos^2 25 - \sin^2 25) = -6.92\ kpsi$

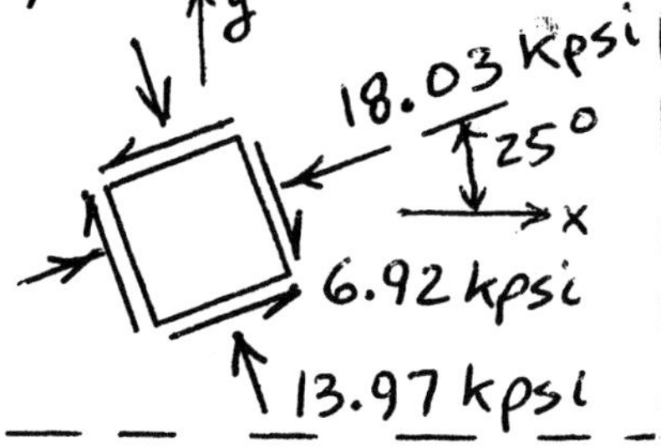

(c) $\sigma_p = \dfrac{-12-20}{2} \pm \sqrt{\left(\dfrac{-12+20}{2}\right)^2 + (-6)^2} = -8.79, -23.21\ kpsi$

$\sigma_1 = 0,\ \sigma_2 = -8.79\ kpsi,\ \sigma_3 = -23.21\ kpsi$

$\sigma_{p1} = -8.79$ Eq. (2.1-28a)

$(-12+8.79)\cos\theta_1 - 6\sin\theta_1 = 0 \Rightarrow \theta_1 = -28.15^\circ$

$\sigma_{p2} = -23.21,\ (-12+23.21)\cos\theta_2 - 6\sin\theta_2 = 0 \Rightarrow \theta_2 = 61.85^\circ$

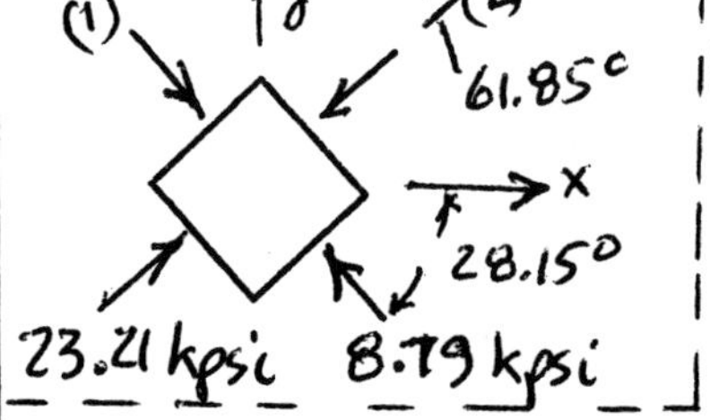

(d)

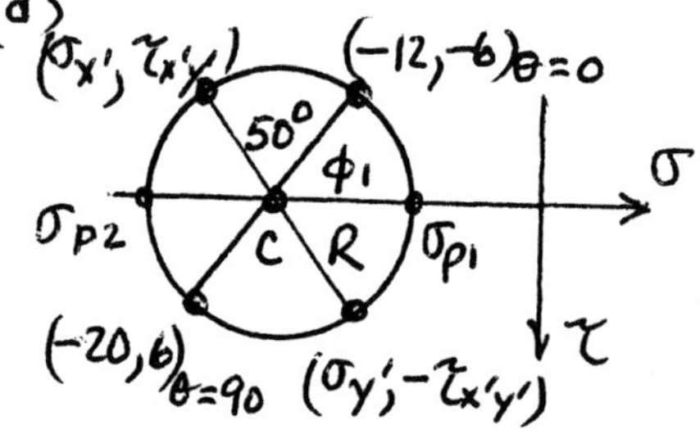

$C = \dfrac{-12-20}{2} = -16\ kpsi$

$R = \sqrt{\left(\dfrac{-12+20}{2}\right)^2 + 6^2} = \sqrt{52}$

$\phi_1 = \tan^{-1}\dfrac{6}{4} = 56.31^\circ$

$\sigma_{x'} = -16 - \sqrt{52}\cos(180-50-56.31) = \underline{-18.03\ kpsi}$

$\tau_{x'y'} = -\sqrt{52}\sin(180-50-56.31) = \underline{-6.92\ kpsi}$

$\sigma_{y'} = -16 + \sqrt{52}\cos(180-50-56.31) = \underline{-13.97\ kpsi}$

$\sigma_{p1,p2} = C \pm R$

$\sigma_{p1} = \underline{-8.79\ kpsi}$

$\sigma_{p2} = \underline{-23.21\ kpsi}$

(Continued)

(d) (continued) $\theta_1 = \frac{\phi_1}{2} = 28.15^\circ$ cw from x axis

(e) $\tau_{max} = \frac{\sigma_1 - \sigma_3}{2} = \frac{0 + 23.21}{2} = \underline{11.61 \text{ kpsi}}$

For surface, view down the axis(1) containing −8.79 kpsi in part (c). Rotate ±45° from this. $\sigma_{Ave} = -11.61$ kpsi

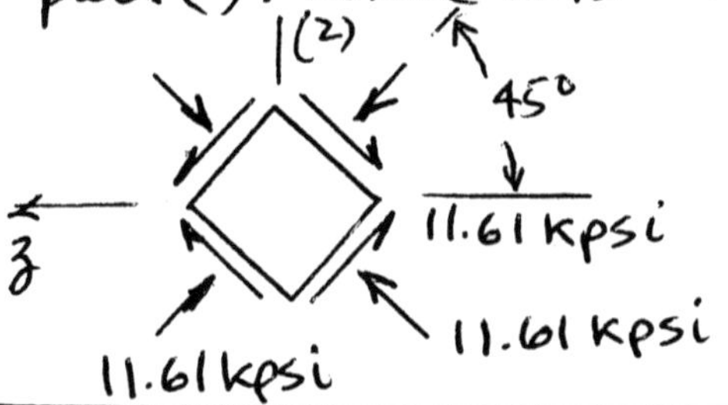

2.21 (a) Eq. (2.1−25)

$$\sigma^3 - (-2+6-4)\sigma^2 + [-2(6)+6(-4)+(-4)(-2)-2^2-(-5)^2-3^2]\sigma$$
$$-[-2(6)(-4)+2(2)(-5)(3)-(-2)(2)^2-6(-5)^2-(-4)(3)^2] = 0$$

Reduces to $\sigma^3 - 66\sigma + 118 = 0$ Using root crossing method, or Programmable calculator, or MathCad or the like, the roots are

$\underline{\sigma = 7.012,\ 1.890,\ -8.903 \text{ kpsi}}$

For $\sigma_1 = 7.012$ kpsi Eqs. (2.1−23) yield

$$\left.\begin{aligned} -9.012n_x + 3n_y - 5n_z &= 0 \\ 3n_x - 1.012n_y + 2n_z &= 0 \\ -5n_x + 2n_y - 11.012n_z &= 0 \end{aligned}\right\}$$

Use first two eqs. and let $n_x = 1$

$3n_y - 5n_z = 9.012$ ①

$1.012n_y - 2n_z = 3$ ②

Solve ① & ② simutaneously to yield $n_y = 3.223$, $n_z = 0.1312$

Normalize $n_x, n_y, \& n_z$: $\sqrt{1^2 + 3.223^2 + 0.1312^2} = 3.3769$

$(n_x)_1 = \frac{1}{3.3769} = \underline{0.2961}$, $(n_y)_2 = \frac{3.223}{3.3769} = \underline{0.9544}$, $(n_z)_1 = \frac{0.1312}{3.3769} = \underline{0.0389}$

In a similar fashion, for $\sigma_2 = 1.890$ kpsi

$(n_x)_2 = \underline{0.7143}$ $(n_y)_2 = \underline{-0.1943}$, $(n_z)_2 = \underline{-0.6723}$

For $\sigma_3 = -8.903$ kpsi: $(n_x)_3 = \underline{0.6341}$, $(n_y)_3 = \underline{-0.2269}$, $(n_z)_3 = \underline{0.7392}$

Check dot products to see if zero.

(b) $\tau_{max} = \frac{\sigma_1 - \sigma_3}{2} = \frac{7.012 + 8.903}{2} = \underline{7.96 \text{ kpsi}}$

2.22 $x + ay + bz - c = 0$

$(0,0,0)$ $\quad 0+0+0-c=0 \Rightarrow c=0$

$(2,-1,3)$ $\quad 2-a+3b=0 \Rightarrow a-3b=2$ ①

$(-2,0,1)$ $\quad -2+b=0 \Rightarrow b=2$ $\quad$ from① $a=3(2)+2=8$

Thus, $\quad x+8y+2z=0 \quad \therefore d_x=1,\ d_y=8,\ d_z=2$

$$\sqrt{1^2+8^2+2^2}=\sqrt{69} \qquad n_x=\frac{1}{\sqrt{69}},\ n_y=\frac{8}{\sqrt{69}},\ n_z=\frac{2}{\sqrt{69}}$$

Eq.s (2.1-18) & (2.1-20)

$$\sigma=\frac{1}{69}\left[-2(1)^2+6(8)^2-4(2)^2+2(3)(1)(8)+2(2)(8)(2)+2(-5)(1)(2)\right]=\underline{6.638\text{ kpsi}}$$

$$\tau=\left\{\frac{1}{69}\left\{\left[-2(1)+3(8)+(-5)(2)\right]^2+\left[3(1)+6(8)+2(2)\right]^2+\left[-5(1)+2(8)-4(2)\right]^2\right\}-6.638^2\right\}^{1/2}$$

$$=\underline{1.414\text{ kpsi}}$$

$$t_x=\frac{1}{1.414\sqrt{69}}\left[(-2-6.638)(1)+3(8)+(-5)(2)\right]=\underline{0.457}$$

$$t_y=\frac{1}{1.414\sqrt{69}}\left[3(1)+(6-6.638)(8)+2(2)\right]=\underline{0.161}$$

$$t_z=\frac{1}{1.414\sqrt{69}}\left[-5(1)+2(8)+(-4-6.638)(2)\right]=\underline{-0.875}$$

2.23 (a) Eq. (2.1-25)

$$\sigma^3-(10+0+10)\sigma^2+\left[0+0+10(10)-(-10\sqrt{2})^2-0-20^2\right]\sigma$$
$$-\left[0+0-10(-10\sqrt{2})^2-0-10(20)^2\right]=0$$

Yields $\quad \sigma^3-20\sigma^2-500\sigma+6000=0$ ①

Sub $\sigma=30$ MPa into ① $\quad 30^3-20(30^2)-500(30)+6000=0$

Thus $\sigma=30$ MPa is a principal stress.

(b) One can solve ① for all roots. Here, using the root 30 to eq ① with synthetic division the quadratic eqn for the remaining two roots emerges

30	1	−20	−500	6000	← coefficients of ①
		30	300	−6000	
	1	10	−200		← coeffients of quadratic eq.

$\therefore \quad \sigma^2+10\sigma-200=0 \Rightarrow (\sigma-10)(\sigma+20)=0$

$\sigma=10,-20$ MPa other roots

$\therefore \quad \sigma_1=30\text{ MPa},\ \sigma_2=10\text{ MPa},\ \sigma_3=-20\text{ MPa}$

(continued)

(c) For $\sigma = 30$ MPa, Eq. (2.1-23) yield

$$\left.\begin{array}{l}(10-30)n_x + 20n_y = 0 \\ 20n_x + (0-30)n_y - 10\sqrt{2}n_z = 0 \\ -10\sqrt{2}n_y + (10-30)n_z = 0\end{array}\right\} \text{Use 1st and 3rd eqs} \quad \begin{array}{l}-n_x + n_y = 0 \quad ② \\ \sqrt{2}n_y + 2n_z = 0 \quad ③\end{array}$$

From ② & ③ $n_x = n_y$, $n_z = -\frac{\sqrt{2}}{2} n_y$ sub this into Eq. (2.1-24)

$$n_y^2 + n_y^2 + \frac{n_y^2}{2} = 1 \Rightarrow \underline{n_y = \sqrt{\tfrac{2}{5}} = n_x}, \quad \underline{n_z = -\tfrac{1}{\sqrt{5}}}$$

(d) n_x, n_y, n_z are directional cosines for plane. Thus eq for plane is $\sqrt{\frac{2}{5}}\,x + \sqrt{\frac{2}{5}}\,y - \frac{1}{\sqrt{5}}\,z = \text{constant}$ or,

$2x + 2y - \sqrt{2}z = \text{constant}$ sub in point A (1,0,0)

$2 + 0 + 0 = \text{constant}$ ∴ $2x + 2y - \sqrt{2}z = 2$

pt B: $0 + 2b + 0 = 2 \Rightarrow b = 1$

pt C: $0 + 0 - \sqrt{2}(c) = 2 \Rightarrow c = -\sqrt{2}$

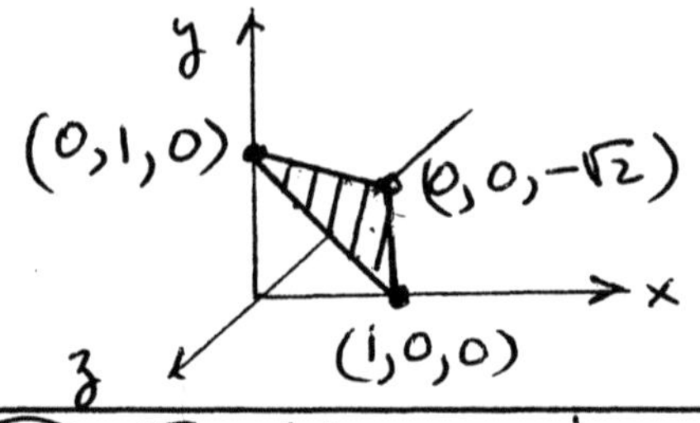

(2.24) Treating n_x & n_y as independent, from Eq. (2.1-18)

$$\frac{\partial\sigma}{\partial n_x} = 2\sigma_x n_x + 2\sigma_z n_z \frac{\partial n_z}{\partial n_x} + 2\tau_{xy}n_y + 2\tau_{yz}n_y\frac{\partial n_z}{\partial n_x} + 2\tau_{zx}n_z + 2\tau_{zx}n_x\frac{\partial n_z}{\partial n_x} \quad ①$$

Likewise, from Eq (2.1-24) take partial wrt n_x

$$2n_x + 2n_z\frac{\partial n_z}{\partial x} = 0 \Rightarrow \frac{\partial n_z}{\partial n_x} = -\frac{n_x}{n_z}$$ sub this into ① giving

$$\frac{\partial\sigma}{\partial n_x} = 2\left[\sigma_x n_x - \sigma_z n_x + \tau_{xy}n_y - \tau_{yz}\frac{n_x n_y}{n_z} + \tau_{zx}n_z - \tau_{zx}\frac{n_x^2}{n_z}\right]$$

From Eq. (2.1-19a) $\sigma_x n_x + \tau_{xy}n_y + \tau_{zx}n_z = -F_x/A_0$ Thus,

$$\frac{\partial\sigma}{\partial n_x} = 2\left[-\frac{F_x}{A_0} - \frac{n_x}{n_z}\left(\sigma_z n_z + \tau_{yz}n_y + \tau_{zx}n_x\right)\right] \quad \text{Term in ( ) is } -\frac{F_z}{A_0}$$

$$= 2\left[-\frac{F_x}{A_0} + \frac{n_x}{n_z}\frac{F_z}{A_0}\right] \quad \text{From Eq (2.1-22)} \begin{array}{l}F_x = -F_x' = -\sigma_p A_0 n_x \\ F_z = -F_z' = -\sigma_p A_0 n_z\end{array}$$

$$\therefore \frac{\partial\sigma}{\partial n_x} = 2\left[\sigma_p n_x + \frac{n_x}{n_z}(-\sigma_p n_z)\right] = 0 \quad \text{Similarly, } \frac{\partial\sigma}{\partial n_y} = 0 \qquad \text{QED.}$$

2.25 (a) Eq. (2.1-25)

$$\sigma^3 - (1+4+4)\sigma^2 + [1(4)+4(4)+4(1)-(-4)^2-(-2)^2-(2)^2]\sigma$$
$$-[1(4)(4)+2(-4)(-2)(2)-1(4)^2-4(-2)^2-4(2)^2] = 0 \text{ gives}$$

$\sigma^3 - 9\sigma^2 = 0 \Rightarrow \underline{\sigma = 9, 0, 0 \text{ kpsi}}$, $\sigma_1 = 9$ kpsi, $\sigma_2 = \sigma_3 = 0$ kpsi

For $\sigma = 9$kpsi, Eqs. (2.1-23) give

$$\begin{rcases} (1-9)n_x + 2n_y - 2n_z = 0 \\ 2n_x + (4-9)n_y - 4n_z = 0 \\ -2n_x - 4n_y + (4-9)n_z = 0 \end{rcases} \text{Arbitrarily use 1st \& 3rd eqs with } n_x = 1 \qquad \begin{rcases} n_y - n_z = 4 \\ 4n_y + 5n_z = -2 \end{rcases} \quad \begin{aligned} n_y &= 2 \\ n_z &= -2 \end{aligned}$$

Normalize $\sqrt{1^2+2^2+(-2)^2} = 3 \quad \therefore \underline{n_x = \frac{1}{3}, n_y = \frac{2}{3}, n_z = -\frac{2}{3}}$

For $\sigma = 0$ (repeated root) Eqs (2.1-23) yield

$$\begin{rcases} n_x + 2n_y - 2n_z = 0 \\ 2n_x + 4n_y - 4n_z = 0 \\ -2n_x - 4n_y + 4n_z = 0 \end{rcases}$$

All three eqs are the same. Can only use one

$n_x + 2n_y - 2n_z = 0$ Substituting any two n's will yield a surface ⊥ to the surfaces containing $\sigma = 9$ kpsi. Thus an infinite set of surfaces exist containing $\sigma = 0$. For example, let $n_x = n_y = 1$

$1 + 2 - 2n_z = 0 \Rightarrow n_z = 1.5 \quad \sqrt{1^2+1^2+1.5^2} = 2.062$

$n_x = \frac{1}{2.062} = 0.4851 = n_y$, $n_z = 0.7276$. If ⊥ to surface containing $\sigma = 9$, the dot products of normals will be zero

$$(0.4851)\left(\frac{1}{3}\right) + (0.4851)\left(\frac{2}{3}\right) + (0.7276)\left(-\frac{2}{3}\right) = 0$$

(b) $\tau_{max} = \frac{\sigma_1 - \sigma_3}{2} = \frac{9-0}{2} = \underline{4.5 \text{ kpsi}}$

2.26 (a) Eq (2.1-25)

$$\sigma^3 - (20+20+20)\sigma^2 + [3(20)^2 - 3(10)^2]\sigma - [20^3 + 2(10^3) - 3(20)(10^2)] = 0$$

yields $\sigma^3 - 60\sigma^2 + 900\sigma - 4000 = 0$ The roots are:

$\underline{\sigma = 40, 10, 10 \text{ MPa}}$: $\sigma_1 = 40$ MPa, $\sigma_2 = \sigma_3 = 10$ MPa

For $\sigma = 40$ MPa, Eq (2.1-23) give

$$\begin{rcases} (20-40)n_x + 10n_y + 10n_z = 0 \\ 10n_x + (20-40)n_y + 10n_z = 0 \\ 10n_x + 10n_y + (20-40)n_z = 0 \end{rcases} \text{Arbitrarily use first two eqs with } n_x = 1 \qquad \begin{rcases} n_y + n_z = 2 \\ 2n_y - n_z = 1 \end{rcases} \quad \begin{aligned} n_y &= 1 \\ n_z &= 1 \end{aligned}$$

(Continued)

Normalize $\sqrt{1^2+1^2+1^2} = \sqrt{3}$ $\therefore n_x = n_y = n_z = 1/\sqrt{3}$

For $\sigma = 10$ MPa (repeated) Eqs (2.1-23) give

$$\left.\begin{array}{l}(20-10)n_x + 10n_y + 10n_z = 0 \\ 10n_x + (20-10)n_y + 10n_z = 0 \\ 10n_x + 10n_y + (20-10)n_z = 0\end{array}\right\} \Rightarrow$$

All three equations are Identical and reduce to

$$n_x + n_y + n_z = 0 \quad (1)$$

Substituting any two values of n into eq (1) will yield a surface ⊥ to surfaces containing $\sigma = 40$ MPa. Thus an infinite set of surfaces exist containing $\sigma = 10$ MPa. For example, let $n_x = n_y = 1$. Sub into (1) yields $n_z = -2$. Normalizing $\sqrt{1^2+1^2+2^2} = \sqrt{6}$

$n_x = n_y = \frac{1}{\sqrt{6}}$, $n_z = -\frac{2}{\sqrt{6}}$. If this surface is ⊥ to surface containing $\sigma = 40$ MPa the dot product of normals is zero

$$\left(\frac{1}{\sqrt{6}}\right)\left(\frac{1}{\sqrt{3}}\right) + \frac{1}{\sqrt{6}}\left(\frac{1}{\sqrt{3}}\right) + \left(-\frac{2}{\sqrt{6}}\right)\left(\frac{1}{\sqrt{3}}\right) = 0$$

(b) $\tau_{max} = \frac{\sigma_1 - \sigma_3}{2} = \frac{40-10}{2} = \underline{15 \text{ MPa}}$

2.27 (a) Eq. (2.1-25)

$$\sigma^3 - (30+10)\sigma^2 + \left[(30)(10) - (-10)^2 - (-20)^2 - (20)^2\right]\sigma$$
$$-\left[2(-10)(-20)(20) - 30(-10)^2 - 10(-20)^2\right] = 0 \text{ yields}$$

$$\sigma^3 - 40\sigma^2 - 600\sigma - 1000 = 0 \quad \text{The roots are:}$$

$\sigma = 51.93, -1.93, -10$ MPa $\Rightarrow \sigma_1 = 51.93$ MPa, $\sigma_2 = -1.93$ MPa, $\sigma_3 = -10$ MPa

For $\sigma = 51.93$, Eq. (2.1-23) give

$$\left.\begin{array}{l}(30-51.93)n_x + 20n_y - 20n_z = 0 \\ 20n_x + (10-51.93)n_y - 10n_z = 0 \\ -20n_x - 10n_y + (0-51.93)n_z = 0\end{array}\right\} \begin{array}{l}\text{Arbitrarily} \\ \text{Using first} \\ \text{two eqs} \\ \text{with} \\ n_x = 1\end{array} \left\{\begin{array}{l}n_y - n_z = 1.0965 \\ 4.193n_y + n_z = 2\end{array}\right\} \begin{array}{l}n_y = 0.596 \\ n_z = -0.5\end{array}$$

Normalize $\sqrt{1^2 + 0.596^2 + (-0.5)^2} = 1.2671$

$(n_x)_1 = \frac{1}{1.2671} = \underline{0.789}$, $(n_y)_1 = \frac{0.596}{1.2671} = \underline{0.471}$, $(n_z)_1 = \frac{-0.5}{1.2671} = \underline{-0.395}$

Similarly, for $\sigma = -1.93$, $(n_x)_2 = \underline{0.421}$, $(n_y)_2 = \underline{-0.882}$, $(n_z) = \underline{-0.210}$

and for $\sigma = -10$, $(n_x)_3 = \underline{0.447}$, $(n_y)_3 = \underline{0}$, $(n_z)_3 = \underline{0.894}$

(b) $\tau_{max} = \frac{1}{2}(\sigma_1 - \sigma_3) = \frac{1}{2}(51.93 + 10) = 30.96$ MPa **ANS**

2.28 (a) Eq. (2.1-30)

$$\sigma_p = \frac{(-15-25)}{2} \pm \sqrt{\left(\frac{-15+25}{2}\right)^2 + (-12)^2} = \underline{-7,\ -33 \text{ kpsi}}$$

For $\sigma_{p1} = -7$ kpsi, Eq (2.1-28a)

$$(-15+7)\cos\theta_1 + (-12)\sin\theta_1 = 0 \Rightarrow \theta_1 = \tan^{-1}\left(\frac{8}{-12}\right) = -33.7^\circ$$

For $\sigma_{p2} = -33$ kpsi $\quad \theta_2 = \tan^{-1}\left(\frac{-18}{-12}\right) = 56.3^\circ$

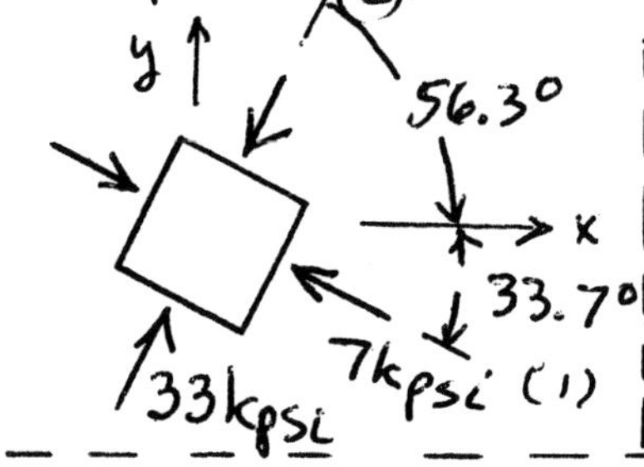

(b)

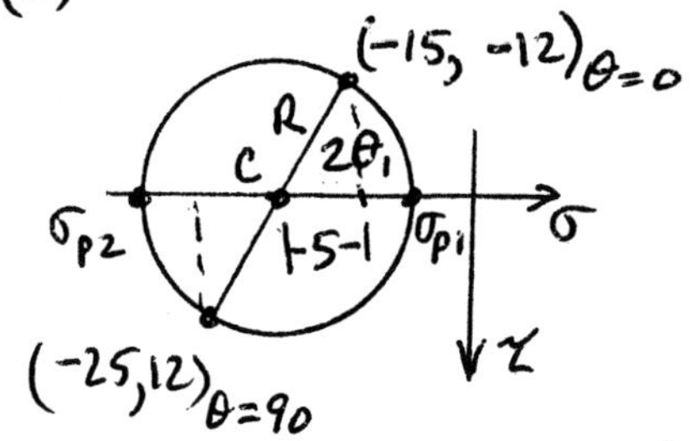

$C = \dfrac{-15-25}{2} = -20$

$R = \sqrt{\left(\dfrac{-15+25}{2}\right)^2 + 12^2} = 13$

$\sigma_{p1} = C + R = -20 + 13 = \underline{7 \text{ kpsi}}$, $\sigma_{p2} = C - R = \underline{-33 \text{ kpsi}}$

$\theta_1 = \frac{1}{2}\tan^{-1}\frac{12}{5} = 33.7^\circ$ cw from x axis to σ_{p1}

(c) $\sigma_1 = 0$, $\sigma_3 = -33$ kpsi

$$\tau_{max} = \frac{\sigma_1 - \sigma_3}{2} = \underline{16.5 \text{ kpsi}}$$

View down the (1) axis containing $\sigma = -7$ kpsi. Then rotate $\pm 45^\circ$ to max shear and $\sigma_{ave} = -16.5$ kpsi

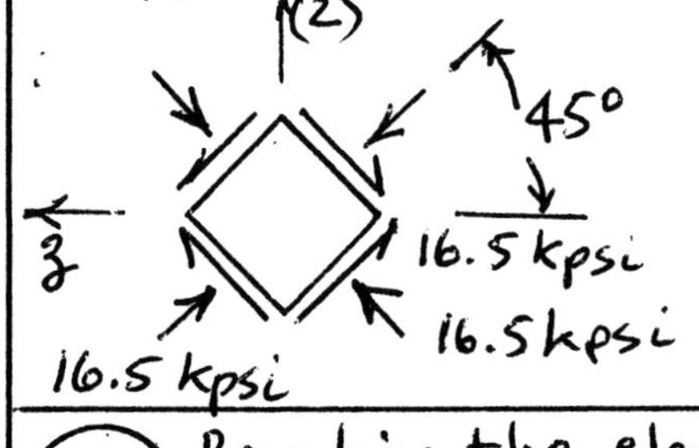

2.29 Repeating the element from Prob. 2.28 solution, part a using (1),(2),(3) axes for $\sigma_1 = 0$, $\sigma_2 = -7$ kpsi, $\sigma_3 = -33$ kpsi resp.

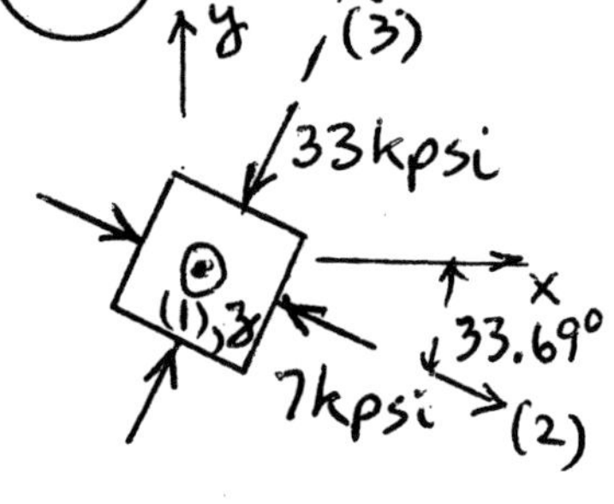

$$\begin{Bmatrix}1\\2\\3\end{Bmatrix} = \begin{bmatrix}0 & 0 & 1\\ \cos 33.69 & -\sin 33.69 & 0\\ \sin 33.69 & \cos 33.69 & 0\end{bmatrix}\begin{Bmatrix}x\\y\\z\end{Bmatrix}$$

$$= \begin{bmatrix}0 & 0 & 1\\ 0.8321 & -0.5547 & 0\\ 0.5547 & 0.8321 & 0\end{bmatrix}\begin{Bmatrix}x\\y\\z\end{Bmatrix} \quad (1)$$

(Continued)

Viewing down (2) axis. x' & y' normals for τ_{max}

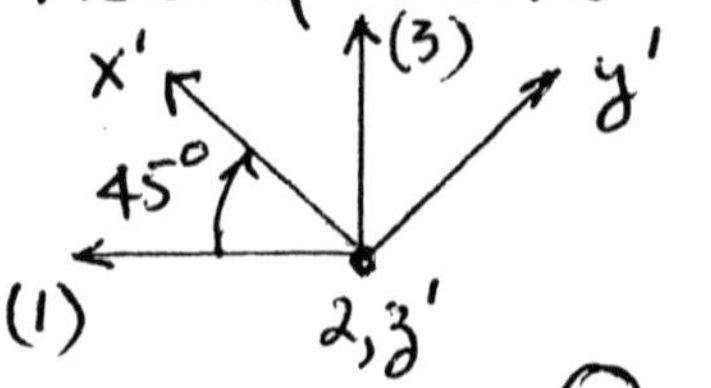

$$\begin{Bmatrix} x' \\ y' \\ z' \end{Bmatrix} = \begin{bmatrix} \cos 45 & 0 & \sin 45 \\ -\sin 45 & 0 & \cos 45 \\ 0 & 1 & 0 \end{bmatrix} \begin{Bmatrix} 1 \\ 2 \\ 3 \end{Bmatrix} = \begin{bmatrix} \sqrt{2}/2 & 0 & \sqrt{2}/2 \\ -\sqrt{2}/2 & 0 & \sqrt{2}/2 \\ 0 & 1 & 0 \end{bmatrix} \begin{Bmatrix} 1 \\ 2 \\ 3 \end{Bmatrix}$$

Sub in eq ① and multiply matrices

$$\begin{Bmatrix} x' \\ y' \\ z' \end{Bmatrix} = \begin{bmatrix} 0.3922 & 0.5883 & 0.7071 \\ 0.3922 & 0.5883 & -0.7071 \\ 0.8321 & -0.5547 & 0 \end{bmatrix} \begin{Bmatrix} x \\ y \\ z \end{Bmatrix}$$

Directional cosines for x' & y' axes are

$\{n\} = \{0.3922 \quad 0.5883 \quad 0.7071\}$ & $\{0.3922 \quad 0.5883 \quad -0.7071\}$ **ANS**

(2.30) Eqs. (2.2-4)

$$\varepsilon_{x'} = \frac{1}{15^2}\left[2(14)^2 + 0 + (-4)(2)^2 + (-1)(14)(-5) + 1(-5)(2) + 2(2)(14)\right](10^{-4}) = \underline{2.187(10^{-4})}$$

$$\varepsilon_{y'} = \frac{1}{15^2}\left[2(2)^2 + 0 + (-4)(11)^2 + (-1)(2)(10) + 1(10)(11) + 2(11)(2)\right](10^{-4}) = \underline{-1.52(10^{-4})}$$

$$\varepsilon_{z'} = \frac{1}{15^2}\left[2(-5)^2 + 0 + (-4)(10)^2 + (-1)(-5)(-10) + 1(-10)(10) + 2(10)(-5)\right](10^{-4}) = \underline{-2.667(10^{-4})}$$

$$\gamma_{x'y'} = \frac{1}{15^2}\Big\{2(14)(2) + 0 + (-4)(2)(11) + \left(-\tfrac{1}{2}\right)\left[(14)(10) + (-5)(2)\right]$$
$$+ \left(\tfrac{1}{2}\right)\left[-5(11) + 2(10)\right] + (1)\left[14(11) + 2(2)\right]\Big\}(10^{-4}) = \underline{0.1933(10^{-4})}$$

$$\gamma_{y'z'} = \frac{1}{15^2}\Big\{2(2)(-5) + 0 + (-4)(11)(10) + \left(-\tfrac{1}{2}\right)\left[2(-10) + 10(-5)\right]$$
$$+ \left(\tfrac{1}{2}\right)\left[10(10) + 11(-10)\right] + (1)\left[2(10) + 11(-5)\right]\Big\}(10^{-4}) = \underline{-2.067(10^{-4})}$$

$$\gamma_{z'x'} = \frac{1}{15^2}\Big\{2(14)(-5) + 0 + (-4)(2)(10) + \left(-\tfrac{1}{2}\right)\left[14(-10) + (-5)(-5)\right]$$
$$+ \tfrac{1}{2}\left[(-5)(10) + 2(-10)\right] + (1)\left[14(10) + 2(-5)\right]\Big\}(10^{-4}) = -0.3(10^{-4})$$

$$[\varepsilon'] = \begin{bmatrix} 2.187 & 0.193 & -0.3 \\ 0.193 & -1.52 & -2.067 \\ -0.3 & -2.067 & -2.667 \end{bmatrix}(10^{-4}) \qquad \underline{\text{ANS}}$$

2.31

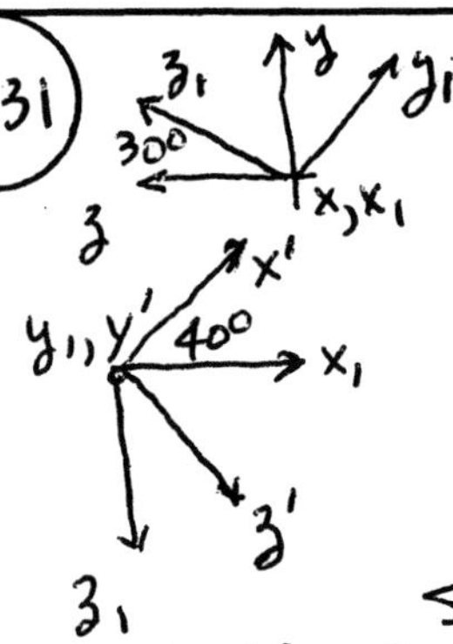

$$\begin{Bmatrix} x_1 \\ y_1 \\ z_1 \end{Bmatrix} = \begin{bmatrix} 1 & 0 & 0 \\ 0 & \cos 30 & -\sin 30 \\ 0 & \sin 30 & \cos 30 \end{bmatrix} \begin{Bmatrix} x \\ y \\ z \end{Bmatrix} = \begin{bmatrix} 1 & 0 & 0 \\ 0 & \sqrt{3}/2 & -1/2 \\ 0 & 1/2 & \sqrt{3}/2 \end{bmatrix} \begin{Bmatrix} x \\ y \\ z \end{Bmatrix} \quad (1)$$

$$\begin{Bmatrix} x' \\ y' \\ z' \end{Bmatrix} = \begin{bmatrix} \cos 40 & 0 & -\sin 40 \\ 0 & 1 & 0 \\ \sin 40 & 0 & \cos 40 \end{bmatrix} \begin{Bmatrix} x_1 \\ y_1 \\ z_1 \end{Bmatrix} = \begin{bmatrix} 0.7660 & 0 & -0.6428 \\ 0 & 1 & 0 \\ 0.6428 & 0 & 0.7660 \end{bmatrix} \begin{Bmatrix} x_1 \\ y_1 \\ z_1 \end{Bmatrix}$$

Sub in (1) & perform matrix multiplication

$$\begin{Bmatrix} x' \\ y' \\ z' \end{Bmatrix} = \begin{bmatrix} 0.7660 & -0.3214 & -0.5567 \\ 0 & 0.8660 & -0.5 \\ 0.6428 & 0.3830 & 0.6634 \end{bmatrix} \begin{Bmatrix} x \\ y \\ z \end{Bmatrix}$$

Eqs. (2.2-4)

$$\varepsilon_{x'} = \left[-4(0.766)^2 + 3(-0.3214)^2 + (-1)(-0.5567)^2 + 0 + (-2)(-0.3214)(-0.5567) + 2(-0.5567)(0.7660)\right](10^{-4}) = \underline{-3.558(10^{-4})}$$

$$\varepsilon_{y'} = \left[-4(0)^2 + 3(0.866)^2 + (-1)(-0.5)^2 + 0 + (-2)(0.866)(-0.5) + 0\right](10^{-4}) = \underline{2.866(10^{-4})}$$

$$\varepsilon_{z'} = \left[-4(0.6428)^2 + 3(0.383)^2 + (-1)(0.6634)^2 + 0 + (-2)(0.383)(0.6634) + 2(0.6634)(0.6428)\right](10^{-4}) = \underline{-1.308(10^{-4})}$$

$$\gamma_{x'y'} = \left\{0 + 3(-0.3214)(0.866) + (-1)(-0.5567)(-0.5) + 0 + (-1)\left[(-0.3214)(-0.5) + (0.5567)(0.866)\right] + (1)\left[0.766(-0.5)\right]\right\}(10^{-4}) = \underline{-1.175(10^{-4})}$$

$$\gamma_{y'z'} = \left\{0 + 3(0.866)(0.383) + (-1)(-0.5)(0.6634) + 0 + (-1)\left[0.866(0.6634) + (-0.5)(0.383)\right] + (1)\left[-0.5(0.6428)\right]\right\}(10^{-4}) = \underline{0.6223(10^{-4})}$$

$$\gamma_{z'x'} = \left\{-4(0.766)(0.6428) + 3(-0.3214)(0.383) + (-1)(-0.5567)(0.6634) + 0 + (-1)\left[(-0.3214)(0.6634) + (-0.5567)(0.383)\right] + (1)\left[0.766(0.6634) + (-0.5567)(0.6428)\right]\right\}(10^{-4})$$

$$= \underline{-1.393(10^{-4})}$$

$$[\varepsilon'] = \begin{bmatrix} -3.558 & -1.175 & -1.393 \\ -1.175 & 2.866 & 0.6223 \\ -1.393 & 0.6223 & -1.308 \end{bmatrix}(10^{-4}) \quad \underline{\text{ANS}}$$

2.32 Writing Eq.(2.1-25) for strain in units of (10^{-4})

$$\epsilon^3-(-4+3-1)\epsilon^2+\left[-4(3)+3(-1)+(-1)(-4)-(-1)^2-(1)^2-0\right]\epsilon$$
$$-\left[-4(3)(-1)+0-(-4)(-1)^2-3(1)^2-0\right]=0 \quad \text{reduces to}$$

$$\epsilon^3+2\epsilon^2-13\epsilon-13=0 \qquad \text{The roots are:}$$

$$\epsilon = 3.2436,\ -0.9289,\ -4.3147$$

$\therefore\ \epsilon_1 = 3.2436(10^{-4})$, $\epsilon_2 = -0.9289(10^{-4})$, $\epsilon_3 = -4.3147(10^{-4})$

For $\epsilon_1 = 3.2436(10^{-4})$ Eqs. (2.1-23) for strain are

$$(-4-3.2436)n_x+(1)n_z=0$$
$$(3-3.2436)n_y+(-1)n_z=0$$
$$(1)n_x+(-1)n_y+(-1-3.2436)n_z=0$$

Arbitrarily select first two eqs and let $n_z = 1$

$$n_x = \frac{1}{7.2436} = 0.1381 \qquad n_y = -\frac{1}{0.2436} = -4.1051$$

Normalize $\sqrt{0.1381^2+(-4.1051)^2+1^2} = 4.2274$ Thus,

$$(n_x)_1 = \frac{0.1381}{4.2274} = 0.03266,\ (n_y)_1 = -\frac{4.1051}{4.2274} = -0.9711,\ (n_z)_1 = \frac{1}{4.2274} = 0.2366$$

Similarly, for $\epsilon_2 = -0.9289(10^{-4})$,

$(n_x)_2 = 0.3009$, $(n_y)_2 = 0.2352$, $(n_z)_2 = 0.9242$

and for $\epsilon_3 = -4.3147(10^{-4})$

$(n_x)_3 = -0.9531$, $(n_y)_3 = 0.0410$, $(n_z)_3 = 0.2999$

2.33 (a) $\epsilon_{x'} = 500\cos^2 30 = 375\mu$, $\epsilon_{y'} = 500\sin^2 30 = 125\mu$

$$\frac{\gamma_{x'y'}}{2} = -500\sin 30\cos 30 \Rightarrow \gamma_{x'y'} = -433\mu$$

(b) Given strains are the principal strains

(c)

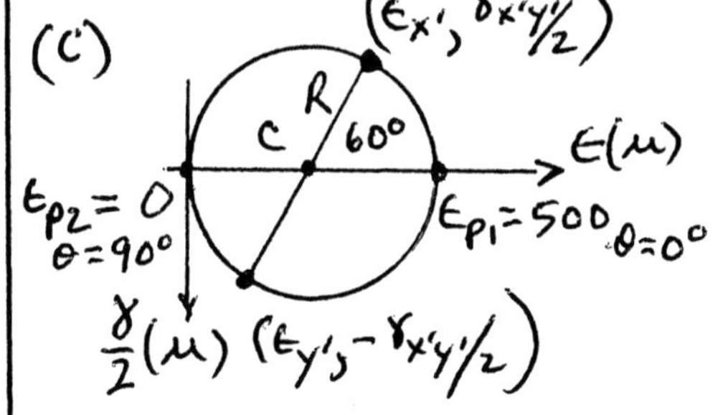

$$C = \frac{500}{2} = 250\mu$$
$$R = \frac{500}{2} = 250\mu$$
$$\epsilon_{x'} = 250+250\cos 60 = 375\mu$$
$$\frac{\gamma_{x'y'}}{2} = -250\sin 60 \Rightarrow \gamma_{x'y'} = -433\mu$$
$$\epsilon_{y'} = 250-250\cos 60 = 125\mu$$

2.34 (a) $\epsilon_{x'} = 2\left(-\frac{200}{2}\right)\cos(-20)\sin(-20) = \underline{64.3\mu}$

$\epsilon_{y'} = -2\left(-\frac{200}{2}\right)\sin(-20)\cos(-20) = \underline{-64.3\mu}$

$\frac{\gamma_{x'y'}}{2} = \left(-\frac{200}{2}\right)\left[\cos^2(-20) - \sin^2(-20)\right] = -76.6\mu$

$\therefore\quad \gamma_{x'y'} = \underline{-153.2\mu}$

(b) $\epsilon_p = \pm\sqrt{\left(-\frac{200}{2}\right)^2} = \underline{\pm 100\mu}$

For $\epsilon_{p1} = 100\mu$, Eq(2.1-28a) for strain is

$-100\cos\theta_1 + \left(-\frac{200}{2}\right)\sin\theta_1 = 0 \Rightarrow \theta_1 = \tan^{-1}\frac{-100}{100} = \underline{-45^\circ}$

For $\epsilon_{p2} = -100\mu$ $\quad \theta_2 = \tan^{-1}\frac{-100}{-100} = \underline{45^\circ}$

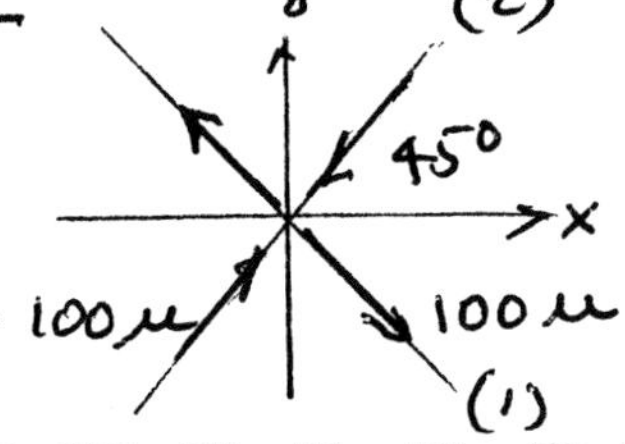

(c)

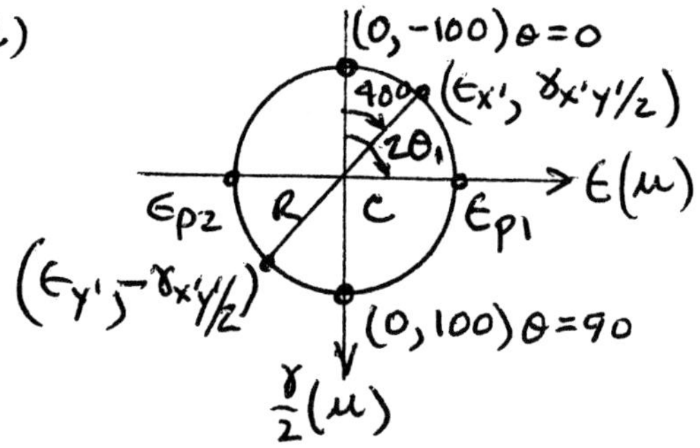

$C = 0$

$R = 100\mu$

$\epsilon_p = \pm 100\mu$

$2\theta_1 = 90^\circ \quad \theta_1 = 45^\circ$ cw to ϵ_{p1}

$\epsilon_{x'} = 100\sin 40 = \underline{64.3\mu} \quad \epsilon_{y'} = -100\sin 40 = \underline{-64.3\mu}$

$\frac{\gamma_{x'y'}}{2} = -100\cos 40 \Rightarrow \gamma_{x'y'} = \underline{-153.2\mu}$

2.35 (a) $\epsilon_{x'} = 400\cos^2 15 + 100\sin^2 15 = \underline{379.9\mu}$

$\epsilon_{y'} = 400\sin^2 15 + 100\cos^2 15 = \underline{120.1\mu}$

$\frac{\gamma_{x'y'}}{2} = -(400-100)\sin 15\cos 15 \Rightarrow \gamma_{x'y'} = \underline{-150\mu}$

(b) Given strains are principal strains

(c)

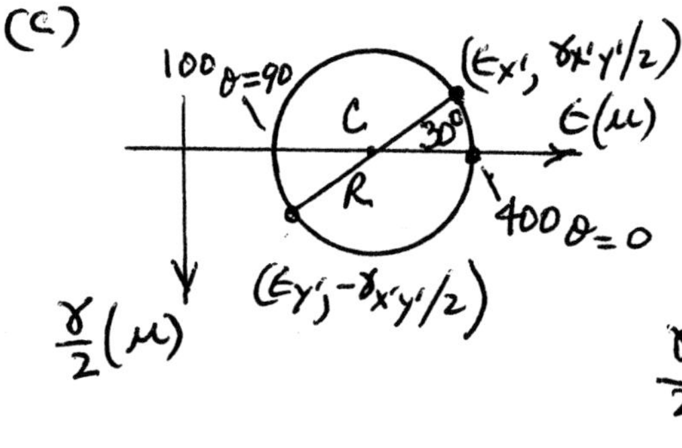

$C = \frac{400+100}{2} = 250\mu$

$R = \frac{400-100}{2} = 150\mu$

$\epsilon_{x'} = 250 + 150\cos 30 = \underline{379.9\mu}$

$\epsilon_{y'} = 250 - 150\cos 30 = \underline{120.1\mu}$

$\frac{\gamma_{x'y'}}{2} = -150\sin 30 = -75$

$\gamma_{x'y'} = \underline{-150\mu}$

2.36 (a) $\epsilon_{x'} = 400\cos^2(-15) + (-100)\sin^2(-15) = \underline{366.5\mu}$

$\epsilon_{y'} = 400\sin^2(-15) + (-100)\cos^2(-15) = \underline{-66.5\mu}$

$\frac{\gamma_{x'y'}}{2} = -(400+100)\sin(-15)\cos(-15) \Rightarrow \gamma_{x'y'} = \underline{250\mu}$

(b) Given strains are principal strains

(c)

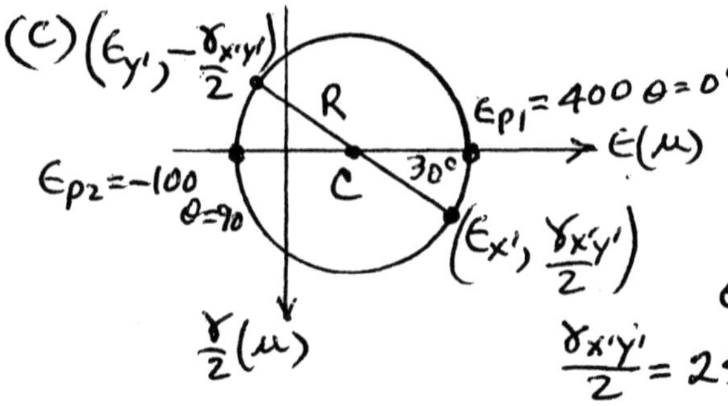

$C = \frac{400-100}{2} = 150\mu$

$R = \frac{400+100}{2} = 250\mu$

$\epsilon_{x'} = 150 + 250\cos 30 = \underline{366.5\mu}$

$\epsilon_{y'} = 150 - 250\cos 30 = \underline{-66.5\mu}$

$\frac{\gamma_{x'y'}}{2} = 250\sin 30 \Rightarrow \gamma_{x'y'} = \underline{250\mu}$

2.37 (a) $\epsilon_{x'} = 100\cos^2 60 + 100\sin^2 60 = \underline{100\mu}$

$\epsilon_{y'} = 100\sin^2 60 + 100\cos^2 60 = \underline{100\mu}$

$\frac{\gamma_{x'y'}}{2} = -(100-100)\sin 60\cos 60 \Rightarrow \gamma_{x'y'} = \underline{0}$

(b) Given strains are principal

(c) Mohr's circle is a point at $(100, 0)\mu$. All strains in xy plane are the same regardless of θ.

2.38 (a) $\epsilon_{x'} = 800\cos^2(-30) + 2\left(-\frac{300}{2}\right)\cos(-30)\sin(-30) = \underline{729.9\mu}$

$\epsilon_{y'} = 800\sin^2(-30) - 2\left(-\frac{300}{2}\right)\sin(-30)\cos(-30) = \underline{70.1\mu}$

$\frac{\gamma_{x'y'}}{2} = -800\sin(-30)\cos(-30) + \left(-\frac{300}{2}\right)\left[\cos^2(-30) - \sin^2(-30)\right]$, $\gamma_{xy} = \underline{542.8\mu}$

(b) $\epsilon_p = \frac{800}{2} \pm \sqrt{\left(\frac{800}{2}\right)^2 + \left(-\frac{300}{2}\right)^2} = \underline{827.2\mu, -27.2\mu}$

For $\epsilon_{p1} = 827.2\mu$, Eq. (2.1-28a) for strain is

$(800 - 827.2)\cos\theta_1 + \left(-\frac{300}{2}\right)\sin\theta_1 = 0 \Rightarrow \theta_1 = \tan^{-1}\frac{27.2}{-150} = \underline{-10.3°}$

For $\epsilon_{p2} = -27.2\mu$ $\quad \theta_2 = \tan^{-1}\frac{827.2}{150} = \underline{79.7°}$

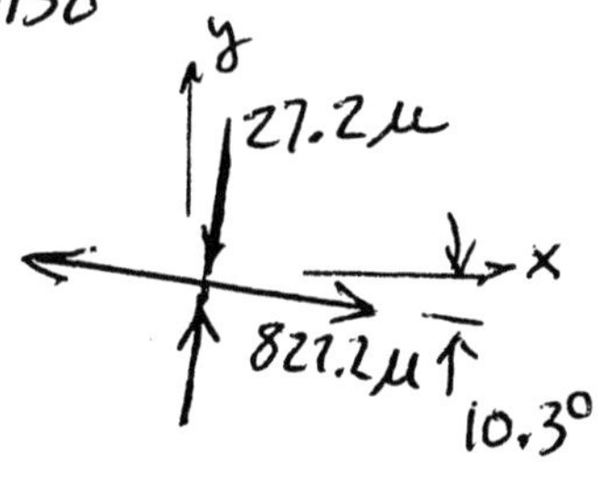

(Continued)

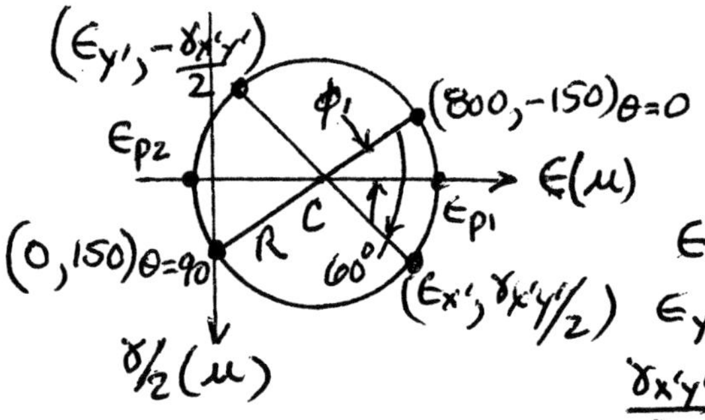

$C = \frac{800}{2} = 400\mu$,

$R = \sqrt{400^2 + 150^2} = 427.2\mu$

$\phi_1 = \tan^{-1}\left(\frac{150}{400}\right) = 20.56°$

$\epsilon_{x'} = 400 + 427.2\cos(60 - 20.56) = \underline{729.9\mu}$

$\epsilon_{y'} = 400 - 427.2\cos(60 - 20.56) = \underline{70.1\mu}$

$\frac{\gamma_{x'y'}}{2} = 427.2\sin(60 - 20.56) \Rightarrow \gamma_{x'y'} = \underline{542.8\mu}$

$\epsilon_{p1} = 400 + 427.2 = \underline{827.2\mu}$ $\quad \epsilon_{p2} = 400 - 427.2 = \underline{-27.2\mu}$

$\theta_1 = \frac{20.56}{2} = 10.3°$ cw from x axis to ϵ_{p1}

2.39 (a) $\epsilon_{x'} = -120\cos^2 25 - 200\sin^2 25 + 2\left(-\frac{60}{2}\right)\sin 25\cos 25 = \underline{-157.3\mu}$

$\epsilon_{y'} = -120\sin^2 25 - 200\cos^2 25 - 2\left(-\frac{60}{2}\right)\sin 25\cos 25 = \underline{-162.7\mu}$

$\frac{\gamma_{x'y'}}{2} = -(-120 + 200)\sin 25\cos 25 + \left(-\frac{60}{2}\right)\left[\cos^2 25 - \sin^2 25\right] \therefore \gamma_{x'y'} = \underline{-99.9\mu}$

(b) $\epsilon_p = \frac{-120 - 200}{2} \pm \sqrt{\left(\frac{-120 + 200}{2}\right)^2 + \left(-\frac{60}{2}\right)^2} = -110\mu, -210\mu$

For $\epsilon_{p1} = -110\mu$, Eq(2.1-28a) for strain is

$(-120 + 110)\cos\theta_1 + \left(-\frac{60}{2}\right)\sin\theta_1 = 0 \Rightarrow \theta_1 = \tan^{-1}\left(\frac{10}{-30}\right) = \underline{-18.4°}$

For $\epsilon_{p2} = -210\mu$, $\theta_2 = \tan^{-1}\left(\frac{-90}{-30}\right) = \underline{71.6°}$

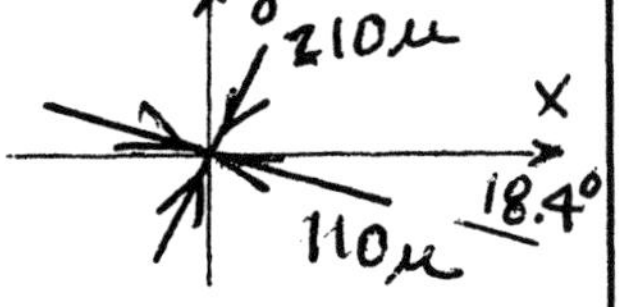

(c)

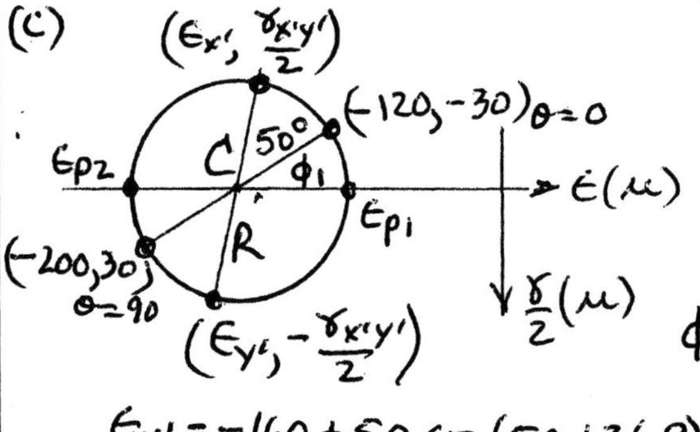

$C = \frac{-120 - 200}{2} = -160\mu$.

$R = \sqrt{\left(\frac{200 - 120}{2}\right)^2 + (30)^2} = 50\mu$

$\phi_1 = \tan^{-1}\left(\frac{30}{40}\right) = 36.9°$

$\epsilon_{x'} = -160 + 50\cos(50 + 36.9) = \underline{-157.3\mu}$

$\epsilon_{y'} = -160 - 50\cos(50 + 36.9) = \underline{-162.7\mu}$

$\frac{\gamma_{x'y'}}{2} = -50\sin(50 + 36.9) \Rightarrow \gamma_{x'y'} = \underline{-99.9\mu}$

$\epsilon_{p1} = -160 + 50 = \underline{-110\mu}$ $\quad \epsilon_{p2} = -160 - 50 = \underline{-210\mu}$

$\theta_1 = \frac{36.9}{2} = 18.4°$ cw from x axis to ϵ_{p1}

(2.40) $\sigma_x = \frac{w}{10I_z}(5x^2+2c^2)y - \frac{w}{3I_z}y^3$

with $\bar{F}_x = 0$

$$\frac{\partial \tau_{xy}}{\partial y} = -\frac{\partial \sigma_x}{\partial x} = -\frac{wxy}{I_z} \quad \text{integrate wrt } y$$

$$\tau_{xy} = -\frac{wxy^2}{2I_z} + f(x) \quad \text{But } \tau_{xy}=0 \text{ at } y=c$$

$$0 = -\frac{wxc^2}{2I_z} + f(x) \Rightarrow f(x) = \frac{wxc^2}{2I_z} \quad \text{sub into } \tau_{xy}$$

$$\therefore \tau_{xy} = \frac{wx}{2I_z}(c^2-y^2) \quad \underline{\text{ANS}}$$

(2.41)

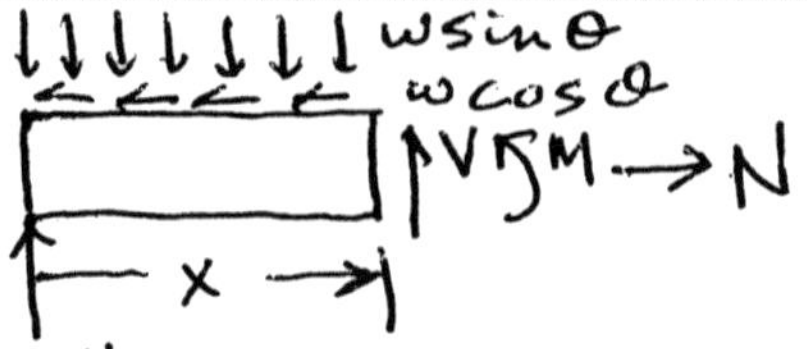

$R = \frac{wL}{2}\sin\theta$

$N = wx\cos\theta$

$M = \frac{wL}{2}\sin\theta\, x - \frac{wx^2}{2}\sin\theta - (wx\cos\theta)(c)$

(a) $\sigma_x = -\frac{My}{I_z} + \frac{N}{A} = -\frac{wLxy}{2I_z}\sin\theta + \frac{wx^2y}{2I_z}\sin\theta + \frac{wxcy}{I_z}\cos\theta + \frac{wx\cos\theta}{A}$

$$= -\frac{wxy(L-x)}{2I_z}\sin\theta + \frac{wc}{I_z}xy\cos\theta + \frac{w}{A}x\cos\theta \quad ①$$

$A = 2bc$. Thus, the term in ①

$$\frac{w}{A}x\cos\theta = \frac{w}{2bc}x\cos\theta = \frac{wxc^2}{3(\frac{2}{3}bc^3)}\cos\theta = \frac{wxc^2}{3I_z}\cos\theta \quad \text{sub into } ①$$

$$\sigma_x = -\frac{wxy(L-x)}{2I_z}\sin\theta + \frac{wc}{3I_z}x(3y+c)\cos\theta \quad \underline{\text{ANS}}$$

(b) $\frac{\partial \tau_{xy}}{\partial y} = -\frac{\partial \sigma_x}{\partial x} = \frac{wy(L-2x)}{2I_z}\sin\theta - \frac{wc(3y+c)}{3I_z}\cos\theta$

$$\tau_{xy} = \frac{wy^2(L-2x)}{4I_z}\sin\theta - \frac{wcy}{6I_z}(3y+2c)\cos\theta + f(x)$$

(Continued)

B.C. at $y=c$ $\tau_{xy} = -\frac{w}{b}\cos\theta$. But $b = \frac{3}{2}\frac{I_z}{c^3}$ $\therefore \tau_{xy} = -\frac{2}{3}\frac{wc^3}{I_z}\cos\theta$

$$-\frac{2}{3}\frac{wc^3}{I_z}\cos\theta = \frac{wc^2(L-2x)}{4I_z}\sin\theta - \frac{wc^2}{6I_z}(3c+2c)\cos\theta + f(x)$$

$$f(x) = \frac{wc^3}{6I_z}\cos\theta - \frac{wc^2}{4I_z}(L-2x)\sin\theta$$

$$\tau_{xy} = -\frac{w}{4I_z}(c^2-y^2)(L-2x)\sin\theta - \frac{wc}{6I_z}(3y^2+2cy-c^2)\cos\theta$$ **ANS**

$$\frac{\partial\sigma_y}{\partial y} = -\frac{\partial\tau_{xy}}{\partial x} = -\frac{w}{2I_z}(c^2-y^2)\sin\theta$$

$$\sigma_y = -\frac{wy}{2I_z}\left(c^2 - \frac{y^2}{3}\right)\sin\theta + g(x)$$

BC. at $y=c$, $\sigma_y = -\frac{w}{b}\sin\theta = -\frac{2}{3}\frac{wc^3}{I_z}\sin\theta$

$$-\frac{2}{3}\frac{wc^3}{I_z}\sin\theta = -\frac{wc}{2I_z}\left(c^2-\frac{c^2}{3}\right)\sin\theta + g(x)$$

$$g(x) = -\frac{wc^3}{3I_z}\sin\theta$$ sub back into σ_y

$$\sigma_y = -\frac{w}{6I_z}(3c^2y - y^3 + 2c^3)\sin\theta$$ **ANS**

(c) At $y=-c$

$$\tau_{xy} = -\frac{w}{4I_z}(c^2-c^2)(L-2x)\sin\theta - \frac{wc}{6I_z}(3c^2-2c^2-c^2) = 0$$

$$\sigma_y = -\frac{w}{6I_z}(-3c^3+c^3+2c^3)\sin\theta = 0$$

$\therefore$ BC's at $y=-c$ are satisfied

2.42 (a)

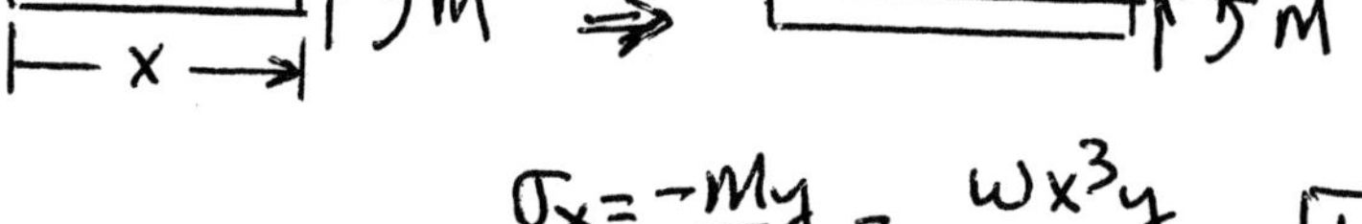

$$M = -\frac{1}{2}\frac{w}{L}x^2\left(\frac{x}{3}\right) = -\frac{wx^3}{6L}$$

$$\sigma_x = -\frac{My}{I_z} = \frac{wx^3y}{6I_zL}$$ **ANS**

(Continued)

with $\bar{F}_x = \bar{F}_y = 0$

$$\frac{\partial \tau_{xy}}{\partial y} = -\frac{\partial \sigma_x}{\partial x} = -\frac{w x^2 y}{2I_z L} \Rightarrow \tau_{xy} = -\frac{w x^2 y^2}{4I_z L} + f(x)$$

B.C. $\tau_{xy} = 0$ at $y = c$ yields $f(x) = \frac{w x^2 c^2}{4I_z L}$

Thus, $\tau_{xy} = \frac{w x^2 (c^2 - y^2)}{4I_z L}$ **ANS**

$$\frac{\partial \sigma_y}{\partial y} = -\frac{\partial \tau_{xy}}{\partial x} = -\frac{w x}{2I_z L}(c^2 - y^2)$$

$$\sigma_y = -\frac{w x y}{6I_z L}(3c^2 - y^2) + g(x)$$

B.C. $\sigma_y = 0$ at $y = -c$

$$0 = -\frac{w x(-c)}{6I_z L}(3c^2 - c^2) + g(x) \Rightarrow g(x) = -\frac{w x c^3}{3I_z L}$$

$$\sigma_y = -\frac{w x}{6I_z L}(3c^2 y - y^3 + 2c^3)$$ **ANS** Note: check BC at $y = c$

(c) $\epsilon_x = \frac{1}{E}\left[\frac{w x^3 y}{6I_z L} + \nu \frac{w x}{6I_z L}(3c^2 y - y^3 + 2c^3)\right] = \frac{w}{6EI_z L}\left[x^3 y + \nu x(3c^2 y - y^3 + 2c^3)\right]$

$$\epsilon_y = \frac{w}{6EI_z}\left[x(3c^2 y - y^3 + 2c^3) - \nu x^3 y\right]$$

$$\gamma_{xy} = \frac{2(1+\nu)}{E}\frac{w x^2}{4I_z L}(c^2 - y^2)$$

$$\frac{\partial^2 \gamma_{xy}}{\partial x \partial y} = -\frac{2(1+\nu)}{EI_z L} w x y, \quad \frac{\partial^2 \epsilon_x}{\partial y^2} = -\frac{\nu w x y}{EI_z L}, \quad \frac{\partial^2 \epsilon_y}{\partial x^2} = -\frac{\nu w x y}{EI_z L}$$

Sub into Eq. (2.5-2a). Results in

$(1+\nu)xy = \nu xy$ $\therefore$ Incompatible

Is compatible at $x = 0$ and/or $y = 0$

2.43 (a) $\epsilon_x = \frac{\partial u}{\partial x} = 3ax^2 - by^2$, $\epsilon_y = \frac{\partial v}{\partial y} = cx^2 - 3dy^2$

$\gamma_{xy} = \frac{\partial v}{\partial x} + \frac{\partial u}{\partial y} = 2cxy - 2bxy = 2(c-b)xy$

$\sigma_x = \frac{E}{1-\nu^2}\left[3ax^2 - by^2 + \nu(cx^2 - 3dy^2)\right] = \frac{E}{1-\nu^2}\left[(3a+\nu c)x^2 - (b+3\nu d)y^2\right]$ ANS

$\sigma_y = \frac{E}{1-\nu^2}\left[cx^2 - 3dy^2 + \nu(3ax^2 - by^2)\right] = \frac{E}{1-\nu^2}\left[(c+3\nu a)x^2 - (3d+\nu b)y^2\right]$ ANS

$\tau_{xy} = \frac{E}{2(1+\nu)} 2(c-b)xy = \frac{E}{1+\nu}(c-b)xy$ ANS

(b) $\Theta = \frac{1}{2}\left(\frac{\partial v}{\partial x} - \frac{\partial u}{\partial y}\right) = \frac{1}{2}(2cxy + 2bxy) = (b+c)xy$

$\Theta(2,1) = 2(b+c)$ ANS

(c) $\frac{\partial^2 \gamma_{xy}}{\partial x \partial y} = 2(c-b)$, $\frac{\partial^2 \epsilon_x}{\partial y^2} = -2b$, $\frac{\partial^2 \epsilon_y}{\partial x^2} = 2c$

Eq.(2.5-2a)

$2(c-b) \stackrel{?}{=} -2b + 2c = 2(c-b)$ displacement field is compatible

2.44 $\frac{\partial \sigma_x}{\partial x} + \frac{\partial \tau_{xy}}{\partial y} = 2bxy - c + 2fxy + gx^2 = 0$

$\therefore$ $c = g = 0$ & $b = -f$

$\frac{\partial \tau_{xy}}{\partial x} + \frac{\partial \sigma_y}{\partial y} = fy^2 + 2gxy + 3dy^2 = 0$

$\therefore$ $g = 0$ and $d = -f/3$

$\epsilon_x = \frac{1}{E}\left[ay^3 + bx^2y - cx - \nu(dy^3 - e)\right]$ $\quad \frac{\partial^2 \epsilon_x}{\partial y^2} = \frac{1}{E}(6ay - 6\nu dy)$

$\epsilon_y = \frac{1}{E}\left[dy^3 - e - \nu(ay^3 + bx^2y - cx)\right]$ $\quad \frac{\partial^2 \epsilon_y}{\partial x^2} = -\frac{2\nu by}{E}$

$\gamma_{xy} = \frac{2(1+\nu)}{E}(fxy^2 + gx^2y - h)$ $\quad \frac{\partial^2 \gamma_{xy}}{\partial x \partial y} = \frac{2(1+\nu)}{E}(2fy + 2gx)$

Eq(2.5-2a)

$\frac{2(1+\nu)}{E}(2fy + 2gx) = \frac{1}{E}(6ay - 6\nu dy - 2\nu by)$

$g = 0$, $4f = 6a$, $4f = -6d - 2b = -6(-f/3) - 2(-f) = 4f$

$\therefore a = \frac{2}{3}f$ $\quad$ e & h are arbitrary

(2.45) $\epsilon_x = \frac{\partial u}{\partial x} = (60x - 30x^2 y)(10^{-3}) = [60(0.05) - 30(0.05)^2(0.02)](10^{-3})$

$= 2.999(10^{-3})$

$\epsilon_y = \frac{\partial v}{\partial y} = (60xy^2 + 10y)(10^{-3}) = [60(0.05)(0.02)^2 + 10(0.02)]10^{-3} = 0.2012(10^{-3})$

$\gamma_{xy} = \frac{\partial v}{\partial x} + \frac{\partial u}{\partial y} = (30x^2 + 20y^3 - 10x^3 + 60y^2)(10^{-3})$

$= [30(0.05)^2 + 20(0.02)^3 - 10(0.05)^3 + 60(0.02)^2](10^{-3}) = 0.0979(10^{-3})$

$\sigma_x = \frac{210(10^9)}{1-0.3^2}[2.999 + 0.3(0.2012)](10^{-3}) = 706(10^6)\,N/m^2 = \underline{706\ MPa}$

$\sigma_y = \frac{210(10^9)}{1-0.3^2}[0.2012 + 0.3(2.999)](10^{-3}) = 254(10^6)\,N/m^2 = \underline{254\ MPa}$

$\tau_{xy} = \frac{210(10^9)}{2(1+0.3)}(0.0979)(10^{-3}) = 7.91(10^6)\,MN/m^2 = \underline{7.91\ MPa}$

$\Theta = \frac{1}{2}\left(\frac{\partial v}{\partial x} - \frac{\partial u}{\partial y}\right) = \frac{1}{2}[30x^2 + 20y^3 + 10x^3 - 60y^2](10^{-3})$

$= \frac{1}{2}[30(0.05)^2 + 20(0.02)^3 + 10(0.05)^3 - 60(0.02)^2](10^{-3}) = \underline{0.0524(10^{-3})\ rad}$

$\frac{\partial^2 \gamma_{xy}}{\partial x \partial y} = 0 \quad \frac{\partial^2 \epsilon_x}{\partial y^2} = 0 \quad \frac{\partial^2 \epsilon_y}{\partial x^2} = 0$ ∴ Eq (2.5-2a) is satisfied; field is compatible

(2.46) $\gamma_{xy} = \frac{2(1+\nu)}{E}\tau_{xy} = -\frac{(1+\nu)}{EI_z}P(c^2 - y^2)$ ①

From Example 2.5-1,

$\gamma_{xy} = \frac{\nu P}{2EI_z}y^2 + \frac{dg(x)}{dx} + \frac{P}{2EI_z}x(2L - x) + \frac{df(y)}{dy}$ ② Set ① = ②

separating functions of x and y yields,

$\frac{dg}{dx} = -\frac{Px}{2EI_z}(2L - x) + C_1, \quad \frac{df}{dy} = -\frac{\nu P y^2}{2EI_z} - \frac{(1+\nu)}{EI_z}P(c^2 - y^2) - C_1$

$g = -\frac{Px^2}{6EI_z}(3L - x) + C_1 x + C_2$

$f = -\frac{\nu P y^3}{6EI_z} - \frac{(1+\nu)Py}{3EI_z}(3c^2 - y^2) - C_1 y + C_3$

(continued)

Sub f & g back into u & v eqs. Let $u = v = \theta = 0$ at $x = y = 0$

This results in $C_2 = C_3 = 0$ and $C_1 = -\frac{(1+\nu)}{2EI_z} Pc^2$

Sub back into u & v and simplify

$$u = \frac{P}{6EI_z}\left[3xy(2L-x) - 3(1+\nu)c^2y + (2+\nu)y^3\right]$$

$$v = -\frac{P}{6EI_z}\left[x^2(3L-x) + 3(1+\nu)c^2x + 3\nu y^2(L-x)\right]$$

2.47 $\epsilon_x = \frac{\partial u}{\partial x} = \frac{w}{EI_z}\left[\frac{(5x^2+2c^2)y}{10} - \frac{y^3}{3} + \frac{\nu}{6}(2c^3+3c^2y-y^3)\right]$

$$\epsilon_y = \frac{\partial v}{\partial y} = \frac{w}{EI_z}\left\{-\frac{1}{6}(2c^3+3c^2y-y^3) - \nu\left[\left(\frac{5x^2+2c^2}{10}\right)y - \frac{y^3}{3}\right]\right\}$$

$$\gamma_{xy} = \frac{\partial v}{\partial x} + \frac{\partial u}{\partial y} = \frac{(1+\nu)}{EI_z}wx(c^2-y^2)$$

Integrate ϵ_x & ϵ_y

$$u = \frac{w}{EI_z}\left[\left(\frac{5}{3}x^3+2c^2x\right)\frac{y}{10} - \frac{xy^3}{3} + \frac{\nu}{6}(2c^3+3c^2y-y^3)x + f(y)\right]$$

$$v = \frac{w}{EI_z}\left\{-\left(2c^3+\frac{3}{2}c^2y-\frac{1}{4}y^3\right)\frac{y}{6} - \nu\left[\left(\frac{5x^2+2c^2}{10}\right)\frac{y^2}{2} - \frac{y^4}{12}\right] + g(x)\right\}$$

Sub u & v into γ_{xy} and simplify

$$\frac{x^3}{6} - \frac{1}{10}(8+5\nu)c^2x + \frac{\partial g}{\partial x} + \frac{\partial f}{\partial y} = 0$$

Set $\partial f/\partial y$ on other side of eq and set $= C_1$. Integrate

$$g = -\frac{x^4}{24} + \frac{1}{20}(8+5\nu)c^2x^2 + C_1x + C_2$$

$$f = -C_1y + C_3$$

Sub f & g back into u & v eqs.

At $(x,y) = (L,0)$ $u = 0$

$$0 = \frac{w}{EI_z}\left[\frac{\nu}{6}(2c^3)L + C_3\right] \Rightarrow C_3 = -\frac{\nu}{3}c^3L$$

At $(x,y) = (L,0)$ $v = 0$

$$0 = \frac{w}{EI_z}\left[-\frac{L^4}{24} + \frac{(8+5\nu)}{20}c^2L^2 + C_1L + C_2\right]$$

(continued)

Thus, $C_1 L + C_2 - \frac{L^4}{24} + \frac{(8+5\nu)}{20} c^2 L^2 = 0$ ①

Rotation:

$$\theta = \frac{1}{2}\left(\frac{\partial v}{\partial x} - \frac{\partial u}{\partial y}\right) = \frac{1}{2}\left[-\frac{x^3}{6} + \frac{(8+5\nu)}{10} c^2 x + C_1 - \left(\frac{5}{3}x^3 + 2c^2 x\right)\frac{1}{10} + xy^2 - \frac{\nu}{6}(3c^2 - 3y^2)x + C_1\right]$$

At $(x,y) = (L,0)$ $\theta = 0$

$$2C_1 - \frac{L^3}{6} + \frac{(8+5\nu)}{10} c^2 L - \frac{5}{30} L^3 - \frac{1}{5} c^2 L - \frac{\nu}{2} c^2 L = 0$$

$C_1 = \frac{(5L^2 - 9c^2)L}{30}$ sub into ① and solve for C_2

$C_2 = -\frac{L^4}{8} - \frac{(2+5\nu)}{20} c^2 L^2$ sub constants into u & v and simplify

$$u = \frac{w}{30EI_z}\left[5x^3 y - 5(2+\nu)xy^3 + 3(2+5\nu)c^2 xy + 10\nu c^3 x - (5L^2 - 9c^2)Ly - 10\nu c^3 L\right]$$

$$v = \frac{w}{120EI_z}\left[-5x^4 + 6(8+5\nu)c^2 x^2 + 4(5L^2 - 9c^2)Lx - 15L^4 - 6(2+5\nu)c^2 L^2\right]$$

(2.48) $\sigma_x = -\frac{My}{I_z}$ $\sigma_y = \tau_{xy} = 0$

$\epsilon_x = \frac{\partial u}{\partial x} = -\frac{My}{EI_z}$, $\epsilon_y = \frac{\partial v}{\partial y} = \frac{\nu My}{EI_z}$, $\gamma_{xy} = \frac{\partial v}{\partial x} + \frac{\partial u}{\partial y} = 0$

$u = -\frac{Mxy}{EI_z} + f(y)$, $v = \frac{\nu My^2}{2EI_z} + g(x)$ sub into γ_{xy} eq.

$\frac{\partial g}{\partial x} - \frac{Mx}{EI_z} + \frac{\partial f}{\partial y} = 0 \Rightarrow \frac{\partial g}{\partial x} = \frac{Mx}{EI_z} + C_1$ $\frac{\partial f}{\partial y} = -C_1$

$g = \frac{Mx^2}{2EI_z} + C_1 x + C_2$ $f = -C_1 y + C_3$

$\therefore$ $u = -\frac{Mxy}{EI_z} - C_1 y + C_3$, $v = \frac{\nu My^2}{2EI_z} + \frac{Mx^2}{2EI_z} + C_1 x + C_2$

(continued)

B.C.s $u = v = 0$ at $x = 0$ yields $C_2 = C_3 = 0$

$$\theta_{xy} = \frac{1}{2}\left(\frac{\partial v}{\partial x} - \frac{\partial u}{\partial y}\right) = \frac{1}{2}\left(\frac{Mx}{EI_z} + C_1 + \frac{Mx}{EI_z} + C_1\right)$$

$\theta_{xy} = 0$ at $x = 0$ yields $C_1 = 0$. Thus

$$u = -\frac{Mxy}{EI_z} \quad \boxed{\text{ANS}} \qquad v = \frac{M}{2EI_z}\left(x^2 + \nu y^2\right) \quad \boxed{\text{ANS}}$$

For $y = 0$ $\quad v = \dfrac{Mx^2}{2EI_z} \quad \theta = \dfrac{Mx}{EI_z}$ $\quad$ checks with Appendix C

(2.49)

$$\frac{\partial^2 \epsilon_x}{\partial y \partial z} = \frac{\partial^3 u}{\partial x \partial y \partial z}$$

$$\frac{\partial^2 \gamma_{yz}}{\partial x^2} = \frac{\partial^3 w}{\partial y \partial x^2} + \frac{\partial^3 v}{\partial z \partial x^2}$$

$$\frac{\partial^2 \gamma_{zx}}{\partial x \partial y} = \frac{\partial^3 w}{\partial x^2 \partial y} + \frac{\partial^3 u}{\partial x \partial y \partial z}$$

$$\frac{\partial^2 \gamma_{xy}}{\partial x \partial z} = \frac{\partial^3 v}{\partial x^2 \partial z} + \frac{\partial^3 u}{\partial x \partial y \partial z}$$

Sub into Eq (2.5-2d)

$$2\frac{\partial^3 u}{\partial x \partial y \partial z} = -\frac{\partial^3 w}{\partial y \partial x^2} - \frac{\partial^3 v}{\partial z \partial x^2} + \frac{\partial^3 w}{\partial x^2 \partial y} + \frac{\partial^3 u}{\partial x \partial y \partial z} + \frac{\partial^3 v}{\partial x^2 \partial z} + \frac{\partial^3 u}{\partial x \partial y \partial z}$$

which can be seen to be satisfied.

The same will apply for Eqs (2.5-2e) & (2.5-2f)

(3.1) $\sigma_2 = \frac{E_2}{E_1}\sigma_1 = \frac{210}{105}\sigma_1 = 2\sigma_1$

$\Sigma \sigma A = 1000\,N$, $\sigma_2(10)(12)(10^{-6}) + \sigma_1[10(12)+20(12)](10^{-6}) = 1000$

($\sigma_2 = 2\sigma_1$) $6(10^{-4})\sigma_1 = 1000$

$\sigma_1 = 1.667(10^6)\,N/m^2 = \underline{1.667\ MPa}$ $\sigma_2 = 2\sigma_1 = \underline{3.333\ MPa}$

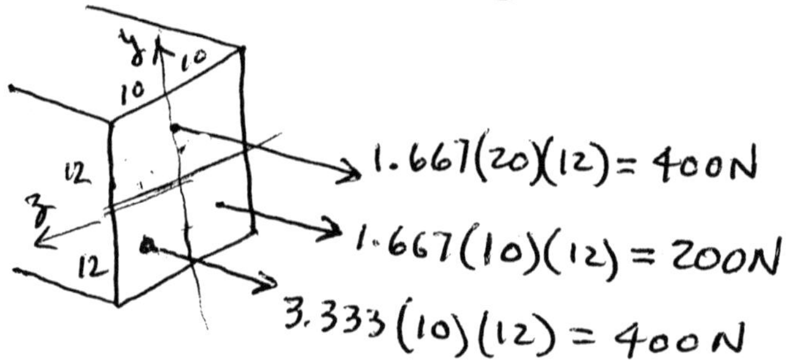

$\Sigma M_y = 0$

$(5-z)400 - 400z - 200(5+z) = 0$

$\underline{z = 1\,mm}$

$\Sigma M_z = 0$

$(6-y)(400) - 600(6+y) = 0$

$\underline{y = -1.2\,mm}$

(3.2) $\int_0^L f\,dx = F \Rightarrow \int_0^L kx^2 dx = F \Rightarrow \frac{1}{3}kL^3 = F \Rightarrow k = \frac{3F}{L^3}$

(b) 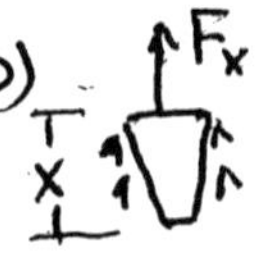

$F_x = -\int f\,dx = -\int \frac{3F}{L^3}x^2 dx = -\frac{Fx^3}{L^3} + C_1$ At $x=0$ $F_x = 0$ $\therefore C_1 = 0$

$F_x = -\frac{Fx^3}{L^3}$ $\sigma_x = \frac{F_x}{A} = -\frac{Fx^3}{AL^3}$ **ANS**

(a) $\epsilon_x = \frac{\sigma_x}{E} = -\frac{Fx^3}{EAL^3} = \frac{du}{dx}$

$u = \int \epsilon_x dx = \frac{-Fx^4}{4EAL^3} + C_2$ Arbitrarily, let $u=0$ at $x=0$ $\therefore C_2 = 0$

$u = -\frac{Fx^4}{4EAL^3}$ **ANS**

(3.3) 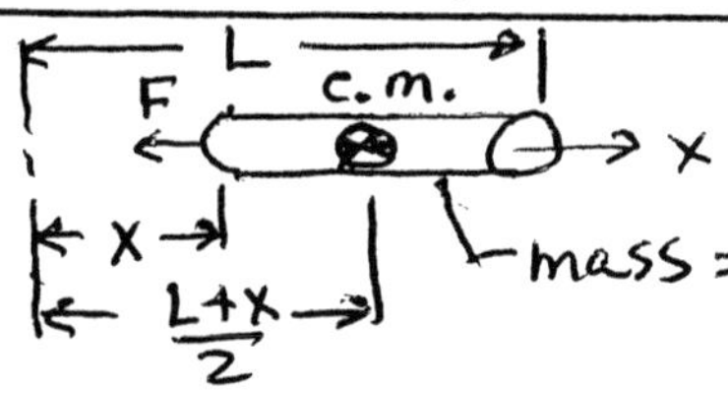

(a) Accel. of cm is $a = -\omega^2\left(\frac{L+x}{2}\right)$

$\Sigma F_x = ma \Rightarrow -F = \rho A(L-x)\left[-\omega^2\left(\frac{L+x}{2}\right)\right]$

$\therefore F = \frac{\rho\omega^2 A}{2}(L^2 - x^2)$ **ANS**

(Continued)

(b) $\epsilon_x = \dfrac{\sigma_x}{E} = \dfrac{du}{dx}$

$$u = \int \frac{\rho\omega^2}{2E}(L^2 - x^2)\,dx = \frac{\rho\omega^2}{6E}x(3L^2 - x^2) + C_1$$

$u = 0$ at $x = 0$ $\therefore$ $C_1 = 0$

$$u = \frac{\rho\omega^2}{6E}x(3L^2 - x^2)$$ **ANS**

(c)

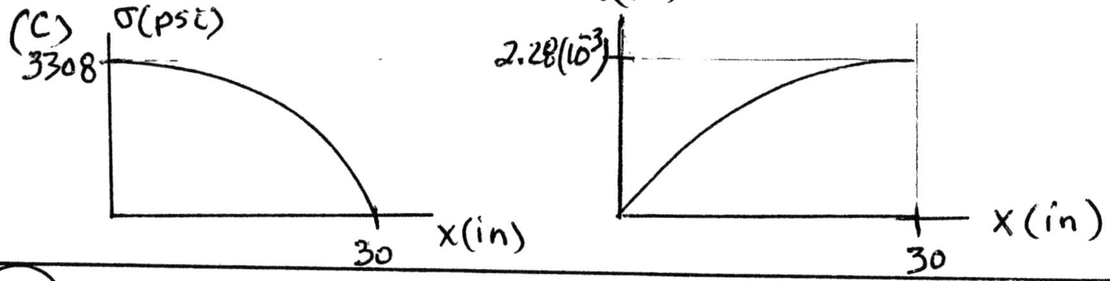

(3.4)

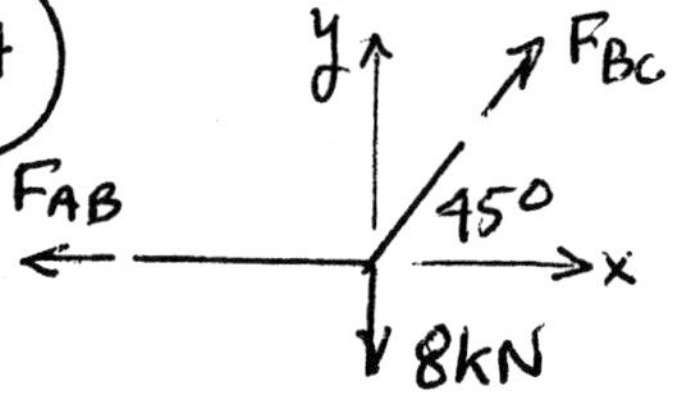

$\Sigma F_y = 0$, $-8 + F_{BC}\sin 45 = 0$, $F_{BC} = 11.313\text{ kN}$

$\Sigma F_x = 0$, $-F_{AB} + 11.313\cos 45 = 0$ $F_{AB} = 8\text{ kN}$

$$\sigma_{AB} = \frac{8(10^3)}{125(10^{-6})} = 64(10^6)\,N/m^2 = 64\,MPa$$

$$\sigma_{BC} = \frac{11.313(10^3)}{125(10^{-6})} = 90.5(10^6)\,N/m^2 = 90.5\,MPa$$

$$\delta_{AB} = \left(\frac{FL}{AE}\right)_{AB} = \frac{8(10^3)(500)}{125(10^{-6})(200)(10^9)} = 0.160\text{ mm}$$

$$\delta_{BC} = \frac{11.313(10^3)(750)}{125(10^{-6})(200)(10^9)} = 0.3394\text{ mm}$$

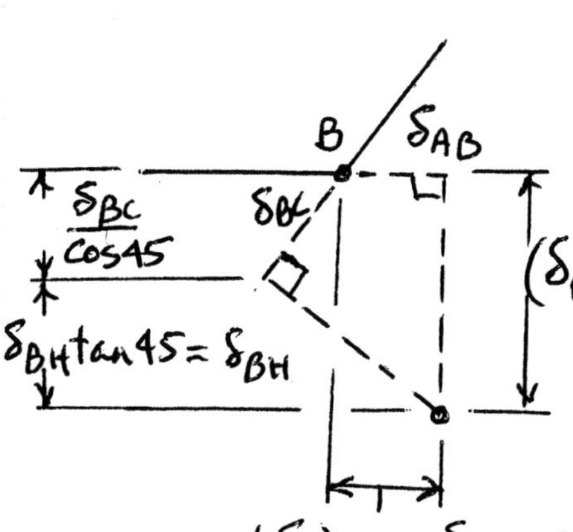

$$(\delta_B)_V = \delta_{BH} + \frac{\delta_{BC}}{\cos 45} = 0.160 + \frac{0.3394}{\cos 45} = 0.640\text{ mm}$$ **ANS**

$(\delta_B)_H = \delta_{AB} = 0.160\text{ mm}$ **ANS**

3.5 (a) AB: $J = \frac{\pi}{32}(1^4 - 0.5^4) = 0.09204 \text{ in}^4$

$$(\tau_{max})_{AB} = \frac{500(0.5)}{0.09204} = 2716 \text{ psi} = \underline{2.72 \text{ kpsi}}$$

BC: $$(\tau_{max})_{BC} = \frac{16(500)}{\pi(1^3)} = 2546 \text{ psi} = \underline{2.546 \text{ kpsi}}$$

CD & DE: $$(\tau_{max})_{CD,DE} = \frac{16(500)}{\pi(0.5^3)} = 20{,}370 \text{ psi} = \underline{20.4 \text{ kpsi}}$$

(b) For section CD let x start at D and go towards C

$\therefore\ r = 0.125(x+2),\quad J = \frac{\pi r^4}{2} = 3.835(10^{-4})(x+2)^4$

$$d\theta = \frac{2(1+\nu)T\,dx}{EJ} = \frac{2(1.29)(500)\,dx}{29(10^6)(3.835)(10^{-4})(x+2)^4} = \frac{0.1160\,dx}{(x+2)^4}$$

$$\theta_{C/D} = \int_0^2 \frac{0.1160\,dx}{(x+2)^4} = -\frac{1}{3}\frac{0.1160}{(x+2)^3}\Big|_0^2 = -\frac{0.116}{3}\left[\frac{1}{4^3} - \frac{1}{2^3}\right] = 0.004229 \text{ rad}$$

$$\theta_{B/A} = \frac{2(1.29)(500)(4)}{29(10^6)(0.09204)} = 0.001933 \text{ rad}$$

$$\theta_{C/B} = \frac{2(1.29)(500)(1)}{29(10^6)(\frac{\pi}{32})(1^4)} = 4.53(10^{-4}) \text{ rad.}$$

$$\theta_{E/D} = \frac{2(1.29)(500)(3)}{29(10^6)(\frac{\pi}{32})(0.5^4)} = 0.021749 \text{ rad.}$$

$$\theta_{E/A} = \Sigma\theta = \underline{0.02836 \text{ rad} = 1.63^\circ}$$

3.6 From the solution of Prob. 1.4

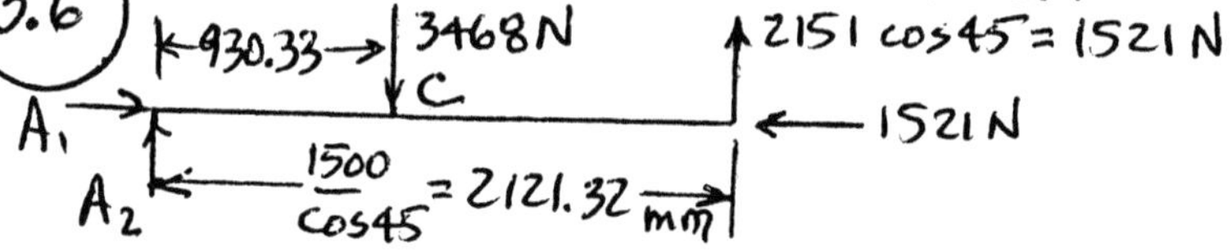

$M_C = 1521(2121.32 - 930.33)(10^{-3}) = 1812$ N-m

Max bending stress: $$\sigma_{max} = \frac{6M}{bh^2} = \frac{6(1812)}{(25)(50^2)(10^{-3})^3} = 173.9(10^6) \text{ N/m}^2 = \underline{174 \text{ MPa}}$$

If include axial stress $-\frac{1521}{(25)(50)} = -1.2$ MPa

$\sigma_{max} = 173.9 - 1.2 = 172.7$ MPa, $\sigma_{min} = -173.9 - 1.2 = \underline{-175.1 \text{ MPa}}$

3.7 $I = \frac{1}{12}\left[100^4 - 84^4\right] = 4.184(10^6)\ mm^4$

(a) $M_{max} = \frac{1}{2}wL^2 + PL = \frac{1}{2}6500(1.5)^2 + 4000(1.5) = 13,313\ N\cdot m$

$$\sigma_{max} = \frac{M_{max}c}{I} = \frac{(13,313)\left(\frac{0.1}{2}\right)}{4.184(10^6)(10^{-3})^4} = 159.1(10^6)N/m^2 = \underline{159\ MPa}$$

(b)

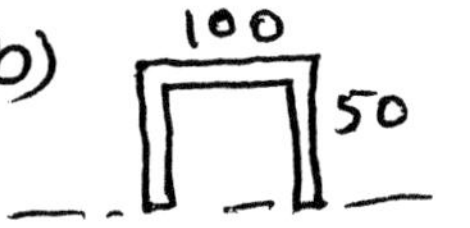

$Q = \Sigma \bar{y}A = \frac{1}{2}(50)(50)(100) - \frac{1}{2}(42)(42)(84) = 50.9(10^3)\ mm^3$

$V_{max} = wL + P = 6500(1.5) + 4000 = 13,750\ N$

$$\tau_{max\,V} = \frac{VQ}{Ib} = \frac{(13,750)(50.9)(10^3)(10^{-3})^3}{4.184(10^6)(10^{-3})^4(2)(0.008)} = 10.5(10^6)N/m^2 = \underline{10.5\ MPa}$$

(c) $\tau_{max} = \frac{\sigma_{max}}{2} = \frac{159.1}{2} = \underline{79.6\ MPa}$

3.8

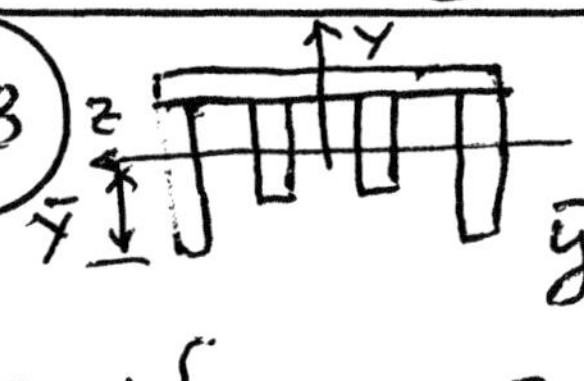

$A = 4(60) + 2\left[36(4) + 21(4)\right] = 696\ mm^2$

$$\bar{y} = \frac{38(4)(60) + 2\left[18(36)(4) + \left(15+\frac{21}{2}\right)(21)(4)\right]}{696} = 26.707\ mm$$

$$I_z = \frac{1}{3}\left\{2(4)(26.707)^3 + 2(4)(26.707-15)^3 - \left[60-4(4)\right](40-4-26.707)^3\right\} = 9.028(10^4)\ mm^4$$

Outer walls

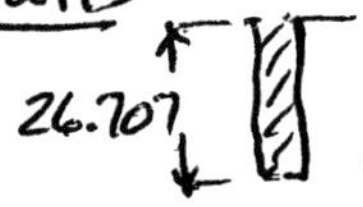

$Q = -\bar{y}A = \frac{(26.707)^2}{2}(4) = 1426.5\ mm^3$

$$\tau = \frac{VQ}{It} = \frac{2000(1426.5)(10^{-3})^3}{9.028(10^4)(10^{-3})^4(0.004)} = 7.90(10^6)N/m^2 = \underline{7.90\ MPa}$$

Inner walls

$Q = \frac{(11.707)^2}{2}(4) = 274.1\ mm^3$

$$\tau = \frac{2000(274.1)(10^{-3})^3}{9.028(10^4)(10^{-3})^4(0.004)} = 1.52(10^6)N/m^2 = \underline{1.52\ MPa}$$

3.9 I about the base of a rectangle is $1/3\,bh^3$

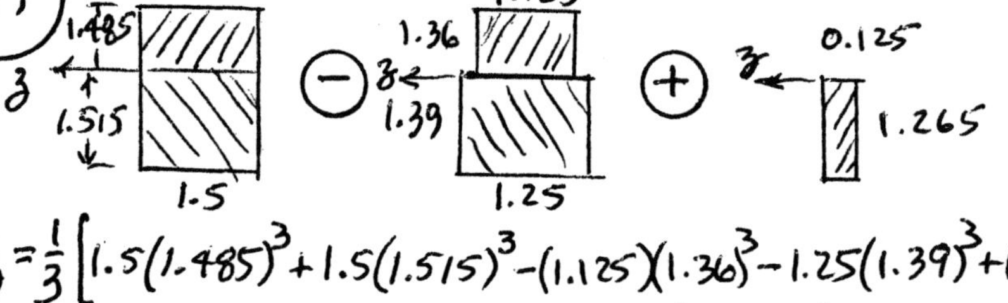

$$I_z = \frac{1}{3}\left[1.5(1.485)^3 + 1.5(1.515)^3 - (1.125)(1.36)^3 - 1.25(1.39)^3 + (0.125)(1.265)^3\right] = 1.398\ in^4$$

(Continued)

(a) Inner wall

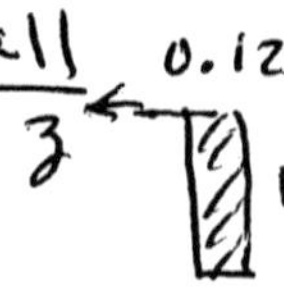

$Q = \bar{y}A = \frac{(1.265)^2}{2}(0.125) = 0.1\ in^3$

$\tau = \frac{VQ}{It} = \frac{500(0.1)}{(1.398)(0.125)} = \underline{286\ psi}$

Outer walls

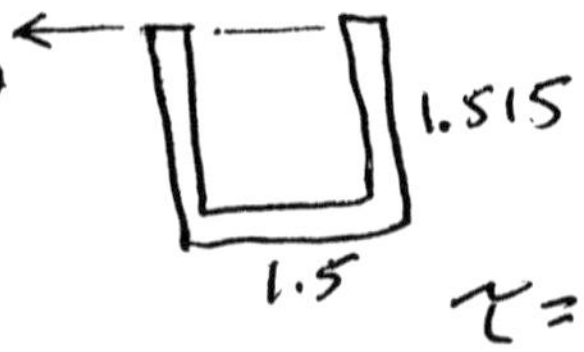

$Q = \Sigma \bar{y}A = \frac{(1.515)^2}{2}(1.5) - \frac{(1.515-0.125)^2}{2}(1.25) = 0.5139\ in^3$

$b = 2(0.125) = 0.25$

$\tau = \frac{500(0.5139)}{1.398(0.25)} = \underline{735\ psi}$

(b) Bending stress $\sigma_{max} = \frac{Mc}{I} = \frac{2000(1.515)}{1.398} = 2167\ psi$

max shear from bending stress is $\sigma_{max}/2 = 1084\ psi$

$\therefore\ \underline{\tau_{max} = 1.08\ kpsi}$ ANS

3.10

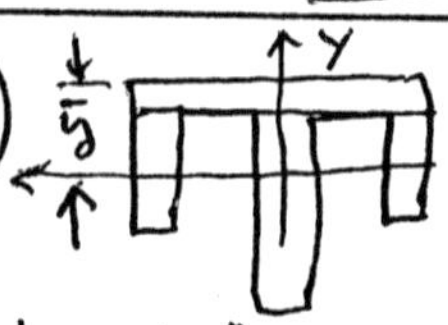

$$\bar{y} = \frac{2.5(45)(5) + 2\frac{(50^2-5^2)}{2}(5) + \frac{(60^2-5^2)(5)}{2}}{45(5) + 2(45)(5) + 55(5)} = 23.026\ mm$$

$$I_z = \frac{1}{12}45(5)^3 + 45(5)(23.026-2.5)^2 + 2\left[\frac{1}{12}5(45)^3 + 5(45)\left(\frac{55}{2} - 23.026\right)^2\right]$$
$$+ \frac{1}{12}5(55)^3 + 5(55)\left(\frac{65}{2} - 23.026\right)^2 = 274.2(10^3)\ mm^4$$

(a) $(M_z)_{wall} = -2000(0.5) = -1000\ N\text{-}m$

$$\sigma_{max} = -\frac{(-1000)(23.026)(10^{-3})}{274.2(10^3)(10^{-3})^4} = 83.98(10^6)N/m^2 = \underline{84.0\ MPa}$$

$$\sigma_{min} = -\frac{(-1000)(23.026-60)(10^{-3})}{274.2(10^3)(10^{-3})^4} = -134.8(10^6)N/m^2 = \underline{-134.8\ MPa}$$

(b) Outer walls

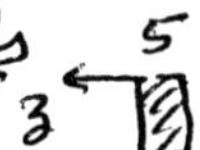

$Q = \frac{(50-23.026)^2}{2}(5) = 1819\ mm^3$

$$\tau_i = \frac{VQ}{It} = \frac{2000(1819)(10^{-3})^3}{274.2(10^3)(10^{-3})^4(0.005)} = 2.65(10^6)N/m^2 = \underline{2.65\ MPa}$$

Inner wall $Q = \frac{(60-23.026)^2}{2}(5) = 3418\ mm^3$

$\tau_o = \frac{3418}{1819}2.65 = \underline{4.99\ MPa}$

(c) $\delta_{max} = FL^3/3EI = 2000(0.5)^3 / [(3)(69)(10^6)274.2(10^3)(10^{-3})^4] = 4.40(10^{-3})m$

$4.40\ mm$ ANS

3.11 (a)

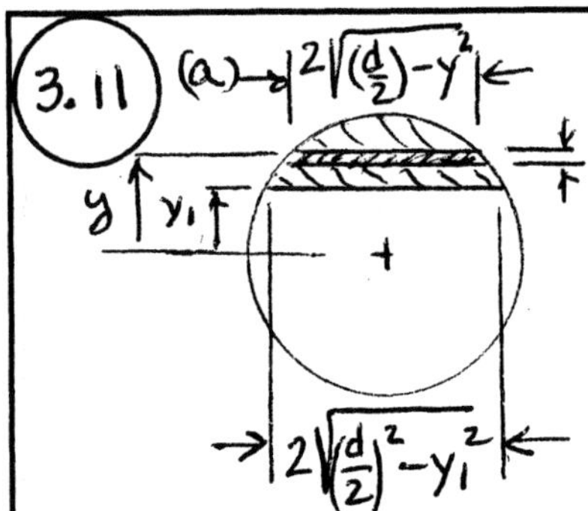

$$Q=\int_{y_1}^{d/2} y\left[2\sqrt{\left(\tfrac{d}{2}\right)^2-y^2}\right]dy=\frac{2}{3}\left[\left(\tfrac{d}{2}\right)^2-y_1^2\right]^{3/2}$$

$$b=2\sqrt{\left(\tfrac{d}{2}\right)^2-y_1^2}$$

$$\tau=\frac{VQ}{Ib}=\frac{V_y\frac{2}{3}\left[\left(\frac{d}{2}\right)^2-y_1^2\right]^{3/2}}{(I)2\sqrt{\left(\frac{d}{2}\right)^2-y_1^2}}=\frac{V_y}{3I_z}\left[\left(\tfrac{d}{2}\right)^2-y_1^2\right]\quad \boxed{\text{ANS}}$$

(b) $\left(\frac{d}{2}\right)^2-y_1^2$ is max when $y_1=0$. Thus,

$$\tau_{max}=\frac{V_y\left(\frac{d}{2}\right)^2}{3\frac{\pi}{64}d^4}=\frac{4}{3}\frac{V_y}{\frac{\pi}{4}d^2}=\frac{4}{3}\frac{V_y}{A}\quad \boxed{\text{ANS}}$$

(c)

other than at $y=0$, τ is not tangent to boundary which it must be.

3.12

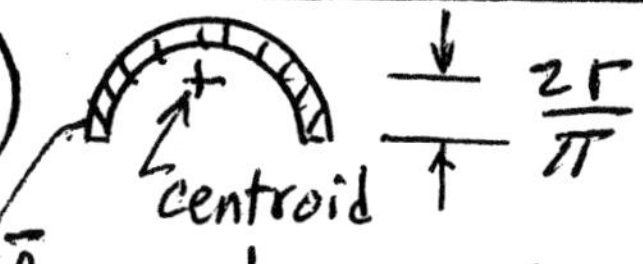

$\bar{A}=\pi r t$

$$Q=\bar{y}\bar{A}=\frac{2r}{\pi}\pi r t=2r^2t$$

$$\tau_{max}=\frac{VQ}{Ib}=\frac{V(2r^2t)}{(\pi r^3t)(2t)}=\frac{V}{\pi r t}$$

total area $A=2\pi r t$ $\quad\therefore\ \tau_{max}=2\frac{V}{A}\qquad k=2\quad \boxed{\text{ANS}}$

3.13 The interface is at $y=0$ where

$$Q=\bar{y}A=\frac{8.25}{2}(11.7)=48.26\ \text{in}^3$$

$$q=\frac{VQ}{I}=\frac{12(10^3)(48.26)}{2\left[146+11.7\left(\frac{8.25}{2}\right)^2\right]}=839.1\ \text{lb/in}$$

$$F_{bolt}=\frac{5(839.1)}{2}=2098\ \text{lb}$$

$$\tau_{avg}=\frac{2098}{\pi(0.5)^2}=10.68(10^3)\ \text{psi}=\underline{10.68\ \text{kpsi}}$$

3.14 $I=\frac{1}{12}(7.5)(3)^3=16.875\ \text{in}^4$

The interface is at $y=0$ where $Q=\frac{(1.5)^2}{2}(7.5)=8.4375\ \text{in}^3$

Eq (3.4-12)

$$\tau_{avg}=\frac{(100)(8.4375)}{16.875}\frac{6}{(2)\frac{\pi}{4}(0.125)^2}=12.22(10^3)\ \text{psi}=\underline{12.2\ \text{kpsi}}$$

3.15

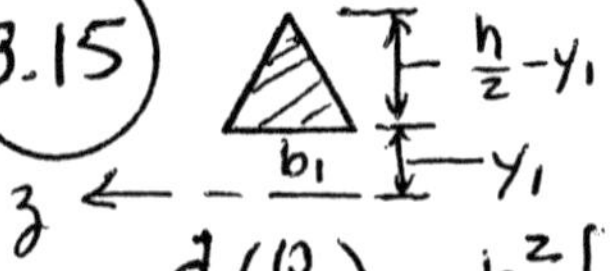

$$\frac{Q}{b_1} = \frac{\bar{y}A}{b_1} = \left[y_1 + \frac{1}{3}\left(\frac{h}{2}-y_1\right)\right]\frac{\left[\frac{1}{2}\left(\frac{h}{2}-y_1\right)b_1\right]}{b_1} = \frac{h^2}{24}\left(1+4\frac{y_1}{h}\right)\left(1-2\frac{y_1}{h}\right)$$

$$\frac{d}{dy_1}\left(\frac{Q}{b_1}\right) = \frac{h^2}{24}\left[\frac{4}{h}\left(1-2\frac{y_1}{h}\right)+\left(-\frac{2}{h}\right)\left(1+4\frac{y_1}{h}\right)\right] = 0$$

reduces to $y_1 = h/8$. Due to symmetry this also applies to $y_1 = -h_1/8$ $\therefore$ τ_{max} at $y = \pm h/8$ **ANS**

3.16

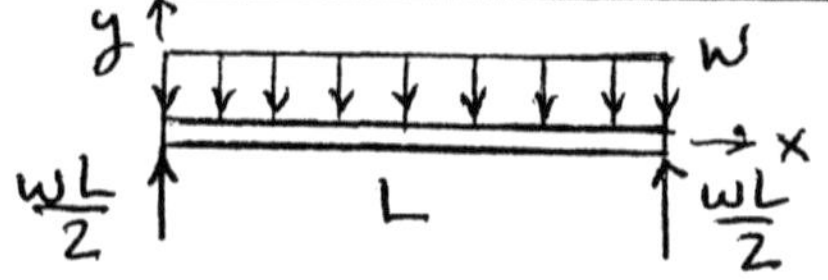

$0 < x < L$ $\quad M = \frac{wL}{2}x - \frac{wx^2}{2}$

$$EIv_c' = \frac{wL}{4}x^2 - \frac{wx^3}{6} + C_1, \quad EIv_c = \frac{wLx^3}{12} - \frac{wx^4}{24} + C_1x + C_2$$

$v_c = 0$ at $x = 0$ $\therefore$ $C_2 = 0$

$v_c = 0$ at $x = L$: $0 = \frac{wL^4}{12} - \frac{wL^4}{24} + C_1L \Rightarrow C_1 = -\frac{wL^3}{24}$

$$\therefore EIv_c = \frac{wLx^3}{12} - \frac{wx^4}{24} - \frac{wL^3x}{24}$$

$$v_c = \frac{wx}{24EI}\left[2Lx^2 - x^3 - L^3\right], \quad \frac{dv_c}{dx} = \frac{w}{24EI}\left[6Lx^2 - 4x^3 - L^3\right]$$

3.17

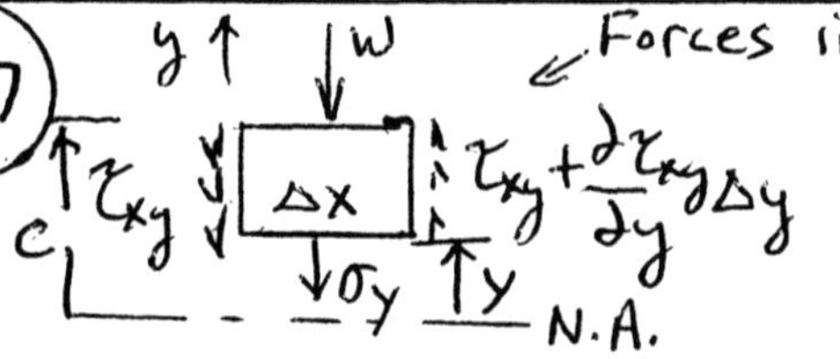

Forces in y direction only

$$\tau_{xy} = \frac{VQ}{I_z b} = \frac{wx}{2I_z}(c^2 - y^2) \quad (1)$$

b = width of beam at y

$$\Sigma F_y = 0 \Rightarrow -w\Delta x - \sigma_y b\Delta x + \int_y^c \frac{\partial \tau_{xy}}{\partial x}\Delta x\, b\,dy = 0$$

$$-w - \sigma_y b + \int_0^c \frac{\partial \tau_{xy}}{\partial x} b\,dy = 0 \quad \text{But from (1)} \quad \frac{\partial \tau_{xy}}{\partial x} = \frac{w}{2I_z}(c^2 - y^2)$$

$$-w - \sigma_y b + \frac{wb}{2I_z}\int_0^c (c^2 - y^2)dy = 0$$

$$-w - \sigma_y b + \frac{wb}{2I_z}\left(\frac{2}{3}c^3 - c^2y + \frac{y^3}{3}\right) = 0$$

But $b = \frac{3}{2}\frac{I_z}{c^3}$

$$\sigma_y = \frac{w}{6I_z}(2c^3 - 3c^2y + 2y^3) - \frac{w}{b}$$

$$= \frac{w}{6I_z}(2c^3 - 3c^2y + 2y^3) - \frac{2wc^3}{3I_z} = \frac{w}{6I_z}(-2c^3 - 3c^2y + 2y^3)$$

$$= -\frac{w}{6I_z}(2c^3 + 3c^2y - y^3)$$ **ANS**

3.18

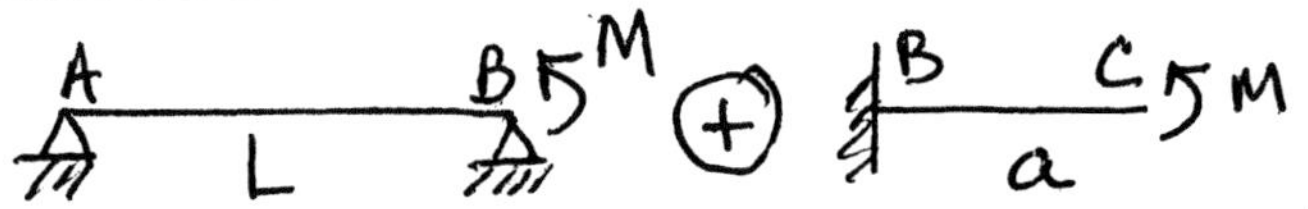

Beam C.7 with $a = L$ | Beam C.4 with $L = a$ $\quad x = x - L$

AB:

$$(v_c)_{AB} = \frac{Mx}{6EIL}(x^2 + 3L^2 - 6L^2 + 2L^2) \qquad \theta_{AB} = \frac{M}{6EIL}(3x^2 + 3L^2 - 6L^2 + 2L^2)$$

$$(v_c)_{AB} = \frac{Mx}{6EIL}(x^2 - L^2) \quad \boxed{\text{ANS}} \qquad \theta_{AB} = \frac{M}{6EIL}(3x^2 - L^2) \quad \underline{\text{ANS}}$$

BC: $$\theta_B = \frac{M}{6EIL}(3L^2 - L^2) = \frac{ML}{3EI}$$

$$(v_c)_{BC} = \frac{M(x-L)^2}{2EI} + \frac{ML}{3EI}(L - x) = \frac{M}{6EI}(3x^2 - 4xL + L^2) \quad \underline{\text{ANS}}$$

$$\theta_{BC} = \frac{M(x-L)}{EI} + \frac{ML}{3EI} = \frac{M}{3EI}(3x - 2L) \quad \underline{\text{ANS}}$$

3.19

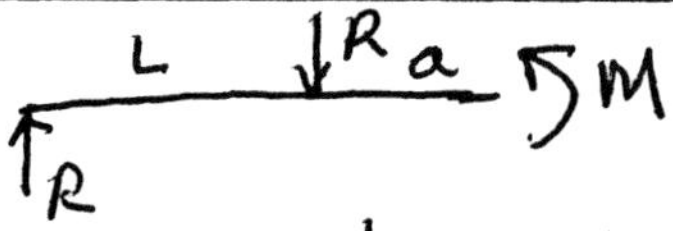

$R = M/L$

$$q(x) = \frac{M}{L}\langle x\rangle^{-1} - \frac{M}{L}\langle x-L\rangle^{-1} - M\langle x-L-a\rangle^{-2}$$

(no need to carry this term as it only terminates V & M curves)

$$V = -\int q\,dx = -\frac{M}{L}\langle x\rangle^{0} + \frac{M}{L}\langle x-L\rangle^{0}$$

$$M = -\int V dx = \frac{M}{L}(x) - \frac{M}{L}\langle x-L\rangle^{1} \qquad \text{where } \langle x\rangle^{1} = x$$

$$EI v_c' = \frac{M}{2L}x^2 - \frac{M}{2L}\langle x-L\rangle^2 + C_1$$

$$EI v_c = \frac{M}{6L}x^3 - \frac{M}{6L}\langle x-L\rangle^3 + C_1 x + C_2$$

$v_c = 0$ at $x = 0$ $\therefore C_2 = 0$

$v_c = 0$ at $x = L$

$$0 = \frac{ML^3}{6L} - 0 + C_1 L \Rightarrow C_1 = -\frac{ML^2}{6L}$$

$$\theta = v_c' = \frac{M}{6EIL}(3x^2 - 3\langle x-L\rangle^2 - L^2) \quad \boxed{\text{ANS}}$$

$$v_c = \frac{M}{6EIL}(x^3 - \langle x-L\rangle^3 - L^2 x)$$

(continued)

Check with Prob 3.18.

AB: $\langle x-L\rangle^n$ doesn't exist

$$(v_c)_{AB} = \frac{M}{6EIL}(x^3 - L^2x) = \frac{Mx}{6EI}(x^2 - L^2) \quad \text{same as Prob. 3.18}$$

$$\theta_{AB} = \frac{M}{6EIL}(3x^2 - L^2) \quad \text{same as Prob. 3.18}$$

BC: $\langle x-L\rangle^n = (x-L)^n$

$$(v_c)_{BC} = \frac{M}{6EIL}\left[x^3 - (x-L)^3 - L^2x\right] = \frac{M}{6EIL}\left[x^3 - x^3 + 3x^2L - 3xL^2 + L^3 - L^2x\right]$$

$$= \frac{M}{6EI}(3x^2 - 4xL + L^2) \quad \text{Same as Prob 3.18}$$

$$\theta_{BC} = \frac{M}{6EIL}\left[3x^2 - 3(x-L)^2 - L^2\right] = \frac{M}{6EIL}(3x^2 - 3x^2 + 6xL - 3L^2 - L^2)$$

$$= \frac{M}{3EI}(3x - 2L) \quad \text{same as Prob 3.18}$$

(3.20)

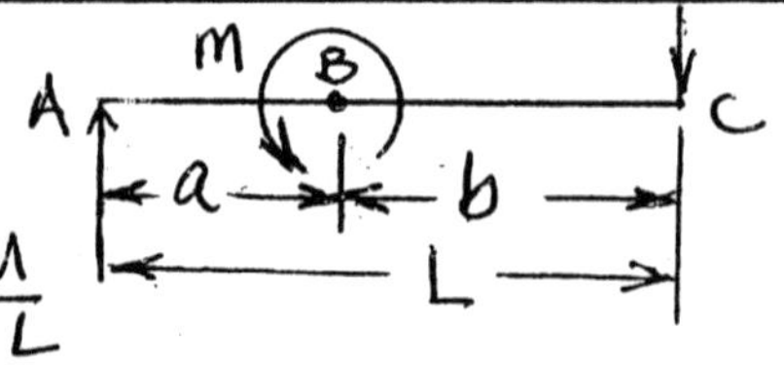

$0 < x < a \quad M_z = \frac{M}{L}x$

$a < x < L \quad M_z = \frac{M}{L}x - M$

$R_A = \frac{M}{L}$

$0 < x < a$

$$EI\frac{dv_c}{dx} = \frac{M}{2L}x^2 + C_1 \quad ①$$

$$EIv_c = \frac{Mx^3}{6L} + C_1x + C_2 \qquad v_c = 0 \text{ at } x = 0 \;\therefore\; C_2 = 0$$

$$EIv_c = \frac{Mx^3}{6L} + C_1x \quad ②$$

$a < x < L$

$$EI\frac{dv_c}{dx} = \frac{M}{2L}x^2 - Mx + C_3 \quad ③$$

$$EIv_c = \frac{M}{6L}x^3 - \frac{M}{2}x^2 + C_3x + C_4 \quad ④$$

$v_c = 0$ at $x = L$

$$0 = \frac{ML^2}{6} - \frac{ML^2}{2} + C_3L + C_4$$

$$C_3L + C_4 = \frac{ML^2}{3} \quad ⑤$$

(Continued)

At $x = a$ ① = ③

$$\frac{Ma^2}{2L} + C_1 = \frac{Ma^2}{2L} - Ma + C_3$$

$$C_1 - C_3 = -Ma \quad ⑥$$

At $x = a$ ② = ④

$$\frac{Ma^3}{6L} + C_1 a = \frac{M}{6L}a^3 - \frac{M}{2}a^2 + C_3 a + C_4$$

$$C_1 a - C_3 a - C_4 = -\frac{M}{2}a^2 \quad ⑦$$

Solving ⑤, ⑥ & ⑦ simultaneously yields

$$C_1 = \frac{M}{6L}(3a^2 + 2L^2 - 6aL), \quad C_3 = \frac{M}{6L}(3a^2 + 2L^2), \quad C_4 = -\frac{M}{2}a^2$$

Substituting constants into ①–④ and simplifying yields

$0 < x < a$

$$v_c = \frac{Mx}{6EIL}(x^2 + 3a^2 - 6aL + 2L^2), \quad \frac{dv_c}{dx} = \frac{M}{6EIL}(3x^2 + 3a^2 - 6aL + 2L^2)$$

$a < x < L$

$$v_c = \frac{M}{6EIL}\left[x^3 - 3Lx^2 + (3a^2 + 2L^2)x - 3a^2 L\right], \quad \frac{dv_c}{dx} = \frac{M}{6EIL}(3x^2 - 6Lx + 3a^2 + 2L^2)$$

3.21 See figure in Prob. 3.20 solution

$$q = \frac{M}{L}\langle x \rangle^{-1} - M\langle x-a \rangle^{-2} - \frac{M}{L}\langle x-L \rangle^{1}$$

— no need to carry term as it only "closes" V_y & M_z curves

$$V_y = -\frac{M}{L}\langle x \rangle^0 + M\langle x-a \rangle^{-1}, \quad M_z = \frac{M}{L}(x) - M\langle x-a \rangle^0$$

where $\langle x \rangle = x$

$$EI v_c' = \frac{M}{2L}x^2 - M\langle x-a \rangle^1 + C_1$$

$$EI v_c = \frac{M}{6L}x^3 - \frac{M}{2}\langle x-a \rangle^2 + C_1 x + C_2$$

$v_c = 0$ at $x = 0$ and $\langle x-a \rangle^2 = 0$. Thus $C_2 = 0$

$v_c = 0$ at $x = L$ and $\langle x-a \rangle^2 = (x-a)^2$

$$0 = \frac{ML^2}{6} - \frac{M}{2}(x-a)^2 + C_1 L \Rightarrow C_1 = \frac{M}{6L}(2L^2 - 6aL + 3a^2)$$

$$v_c = \frac{M}{6EIL}\left[x^3 - 3L\langle x-a \rangle^2 + (2L^2 - 6aL + 3a^2)x\right] \quad \boxed{\text{ANS}}$$

$$\frac{dv_c}{dx} = \frac{M}{6EIL}\left[3x^2 - 6L\langle x-a \rangle^1 + 2L^2 - 6aL + 3a^2\right] \quad \boxed{\text{ANS}}$$

(Continued)

To verify Appendix C equations,

$\underline{0<x<a}$ $\quad \langle x-a\rangle^n = 0$

$$v_c = \frac{M}{6EIL}\left[x^3+(2L^2-6aL+3a^2)x\right] = \frac{Mx}{6EIL}\left(x^2+2L^2-6aL+3a^2\right) \quad ✓$$

$$\frac{dv_c}{dx} = \theta = \frac{M}{6EIL}\left(3x^2+2L^2-6aL+3a^2\right) \quad ✓$$

$\underline{a<x<L}$ $\quad \langle x-a\rangle^n = (x-a)^n$

$$v_c = \frac{M}{6EIL}\left[x^3-3L(x-a)^2+(2L^2-6aL+3a^2)x\right]$$

$$= \frac{M}{6EIL}\left(x^3-3Lx^2+6aLx-3a^2L+2L^2x-6aLx+3a^2x\right)$$

$$= \frac{M}{6EIL}\left(x^3-3Lx^2+(2L^2+3a^2)x-3a^2L\right] \quad ✓$$

$$\frac{dv_c}{dx} = \theta = \frac{M}{6EIL}\left[3x^2-6L(x-a)+2L^2-6aL+3a^2\right]$$

$$= \frac{M}{6EIL}\left(3x^2-6xL+2L^2+3a^2\right) \quad ✓$$

3.22

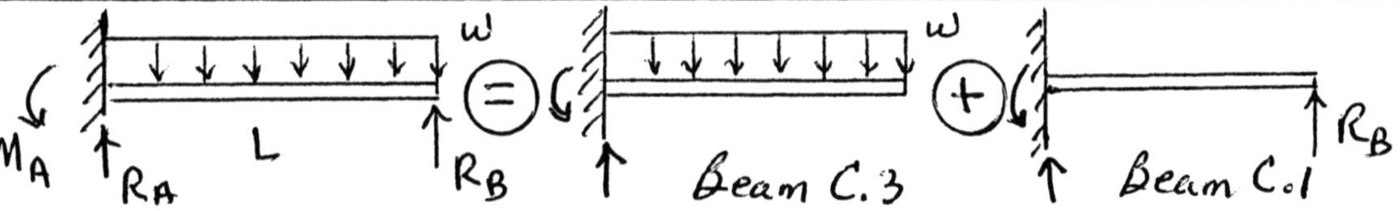

$$(v_c)_B = -\frac{wL^4}{8EI} + \frac{R_BL^3}{3EI} = 0 \Rightarrow R_B = \frac{3}{8}wL \quad ✓$$

$$R_A = wL - R_B = \frac{5}{8}wL \quad ✓$$

$$M_A = \frac{1}{2}wL^2 - R_BL = \frac{1}{8}wL^2 \quad ✓$$

$\underline{0<x<L}$

$$v_c = \frac{wx^2}{24EI}(4Lx-x^2-6L^2) - \frac{R_Bx^2}{6EI}(x-3L) = \frac{wx^2}{48EI}\left(8Lx-2x^2-12L^2-3Lx+9L^2\right)$$

$$= \frac{wx^2}{48EI}\left(-2x^2+5Lx-3L^2\right) = \frac{wx^2}{48EI}(2x-3L)(L-x) \quad ✓$$

$$\theta = \frac{wx}{6EI}(3Lx-x^2-3L^2) - \frac{R_Bx}{2EI}(x-2L) = \frac{wx}{48EI}\left(24Lx-8x^2-24L^2-9Lx+18L^2\right)$$

$$= \frac{wx}{48EI}\left(15Lx-8x^2-6L^2\right) \quad ✓$$

3.23

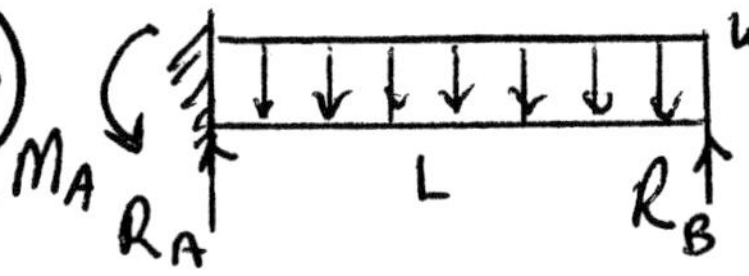

Write eqs in terms of w & R_B only

$R_A = wL - R_B \qquad M_A = \frac{wL^2}{2} - R_B L$

$$q = (wL - R_B)\langle x\rangle^{-1} - \left(\frac{wL^2}{2} - R_B L\right)\langle x\rangle^{-2} - w\langle x\rangle^0 + \underbrace{w\langle x-L\rangle^0 + R_B\langle x-L\rangle^{-1}}$$

can ignore as will not contribute to θ & v_c

$$V_y = -(wL - R_B)\langle x\rangle^0 + \left(\frac{wL^2}{2} - R_B L\right)\langle x\rangle^{-1} + w\langle x\rangle^1$$

$$M_z = (wL - R_B)\langle x\rangle^1 - \left(\frac{wL^2}{2} - R_B L\right)\langle x\rangle^0 - \frac{w}{2}\langle x\rangle^2$$

can drop $\langle\,\rangle$ as functions always exist

$$EI v_c' = \frac{(wL - R_B)}{2}x^2 - \left(\frac{wL^2}{2} - R_B L\right)x - \frac{w}{6}x^3 + C_1 \qquad v_c' = 0 \text{ at } x = 0 \therefore C_1 = 0$$

$$EI v_c = \frac{(wL - R_B)}{6}x^3 - \left(\frac{wL^2}{2} - R_B L\right)\frac{x^2}{2} - \frac{w}{24}x^4 + C_2 \qquad v_c = 0 \text{ at } x = 0 \therefore C_2 = 0$$

$v_c = 0$ at $x = L$

$$0 = (wL - R_B)\frac{L^3}{6} - \left(\frac{wL^2}{2} - R_B L\right)\frac{L^2}{2} - \frac{wL^4}{24} = \frac{L^3}{24}(4wL - 4R_B - 6wL + 12R_B - wL)$$

$$8R_B - 3wL = 0 \Rightarrow R_B = \frac{3}{8}wL \quad \underline{\text{ANS}}$$

$$R_A = wL - R_B = \frac{5}{8}wL, \quad M_A = \frac{wL^2}{2} - R_B L = \frac{1}{8}wL^2 \quad \underline{\text{ANS}}$$

$$EI v_c = \frac{\left(wL - \frac{3}{8}wL\right)}{6}x^3 - \left(\frac{wL^2}{2} - \frac{3}{8}wL^2\right)\frac{x^2}{2} - \frac{w}{24}x^4$$

$$v_c = \frac{wx^2}{48EI}\left(5Lx - 3L^2 - 2x^2\right) = \frac{wx^2}{48EI}(2x - 3L)(L - x) \quad \underline{\text{ANS}}$$

$$EI\theta = \left(wL - \frac{3}{8}wL\right)\frac{x^2}{2} - \left(\frac{wL^2}{2} - \frac{3}{8}wL^2\right)x - \frac{w}{6}x^3$$

$$\theta = \frac{wx}{48EI}\left(15Lx - 8x^2 - 6L^2\right) \quad \underline{\text{ANS}}$$

3.24 (a) Superposition

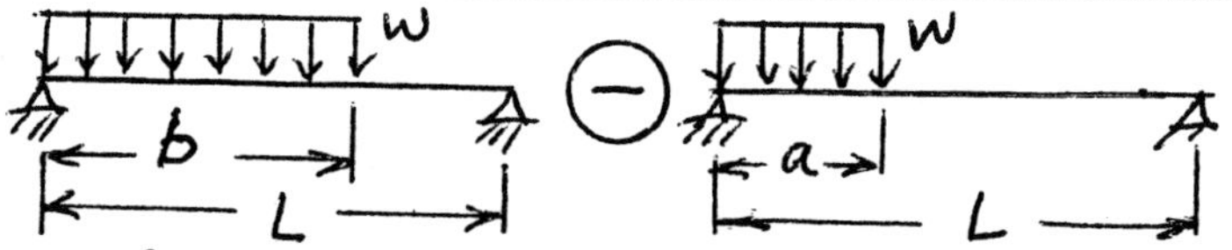

Beam C.10 $\qquad$ Beam C.10

$\underline{0 < x < a}$

$$v_c = \frac{wx}{24EIL}\left\{\left[2bx^2(2L-b) - Lx^3 - b^2(2L-b)^2\right] - \left[2ax^2(2L-a) - Lx^3 - a^2(2L-a)^2\right]\right\}$$

$$= \frac{wx}{24EIL}\left\{2\left[b(2L-b) - a(2L-a)\right]x^2 - b^2(2L-b)^2 + a^2(2L-a)^2\right\} \quad (1)$$

(continued)

$\underline{a < x < b}$

$$v_c = \frac{wx}{24EIL}\left\{\left[2bx^2(2L-b) - Lx^3 - b^2(2L-b)^2\right] - \left[2ax^2(2L-a) - Lx^3 - a^2(2L-a)^2\right]\right\} - \frac{w}{24EI}(x-a)^4$$

$$= \frac{w}{24EIL}\left\{2\left[b(2L-b) - a(2L-a)\right]x^3 - \left[b^2(2L-b)^2 - a^2(2L-a)^2\right]x - L(x-a)^4\right\} \quad (2)$$

$\underline{b < x < L}$ similar to (1) & (2)

$$v_c = \frac{w}{24EIL}\left\{2\left[b(2L-b) - a(2L-a)\right]x^3 - \left[b^2(2L-b)^2 - a^2(2L-a)^2\right]x + L\left[(x-b)^4 - (x-a)^4\right]\right\} \quad (3)$$

(b) Singularity functions. Solve for support reaction

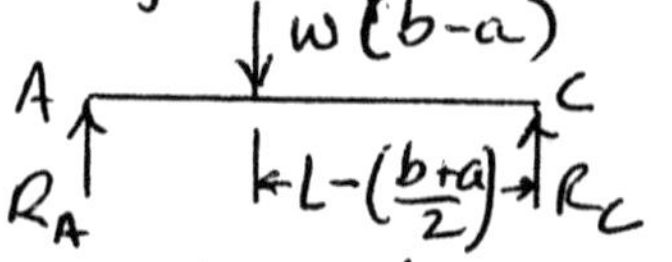

$$R_A = \frac{w(b-a)}{L}\left[L - \frac{b+a}{2}\right] = w(b-a) - \frac{w}{2L}(b^2 - a^2) \quad (4)$$

For actual beam,

$$q = R_A\langle x\rangle^{-1} - w\langle x-a\rangle^0 + w\langle x-b\rangle^0 + R_D\langle x-L\rangle^{-1}$$

← can ignore as it contributes nothing to deflections (the $R_D\langle x-L\rangle^{-1}$ term)

$$V_y = -R_A\langle x\rangle^0 + w\langle x-a\rangle^1 - w\langle x-b\rangle^1$$

$$M_z = R_A\,x - \frac{w}{2}\langle x-a\rangle^2 + \frac{w}{2}\langle x-b\rangle^2 \qquad \text{where } \langle x\rangle = x$$

$$EIv_c' = \frac{R_A}{2}x^2 - \frac{w}{6}\langle x-a\rangle^3 + \frac{w}{6}\langle x-b\rangle^3 + C_1$$

$$EIv_c = \frac{R_A}{6}x^3 - \frac{w}{24}\langle x-a\rangle^4 + \frac{w}{24}\langle x-b\rangle^4 + C_1x + C_2$$

$v_c = 0$ at $x = 0$, and $\langle x-a\rangle^4 = \langle x-b\rangle^4 = 0 \quad \therefore C_2 = 0$

$v_c = 0$ at $x = L$, and $\langle x-a\rangle^4 = (L-a)^4$, $\langle x-b\rangle^4 = (L-b)^4$

$$0 = \frac{R_A L^3}{6} - \frac{w}{24}(L-a)^4 + \frac{w}{24}(L-b)^4 + C_1L$$

$$C_1 = -\frac{R_A L^2}{6} + \frac{w}{24L}(L-a)^4 - \frac{w}{24L}(L-b)^4$$

$$v_c = \frac{w}{24EIL}\left\{2(b-a)(2L-b-a)(x^2-L^2)x - L\langle x-a\rangle^4 + L\langle x-b\rangle^4 + x\left[(L-a)^4 - (L-b)^4\right]\right\}$$

Ans

(continued)

To verify with part (a)

$0<x<a$ $\quad \langle x-a\rangle^4 = \langle x-b\rangle^4 = 0$

$$v_c = \frac{w}{24EIL}\left\{2(b-a)(2L-b-a)(x^2-L^2)x + x\left[(L-a)^4-(L-b)^4\right]\right\}$$

$$= \frac{wx}{24EIL}\left[(4bL-2b^2-4aL+2a^2)x^2 + 4a^2L^2-4a^3L+a^4-4b^2L^2+4b^3L-b^4\right]$$

$$= \frac{wx}{24EIL}\left\{2\left[b(2L-b)-a(2L-a)\right]x^2 - b^2(2L-b)^2 + a^2(2L-a)^2\right] \quad (5) \checkmark$$

$a<x<b$ simply add $-\frac{wL}{24EIL}(x-a)^4$ to (5) yields eq (2) ✓

$b<x<L$ simply add " and $\frac{wL}{24EIL}(x-b)^4$ to (5) yields (3) ✓

(3.25) For reactions

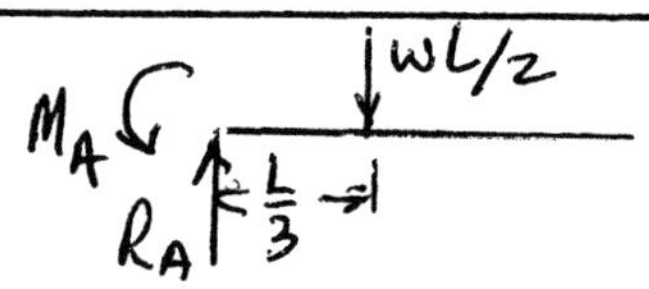

$R_A = wL/2$

$M_A = wL^2/6$

For actual beam

$$q = -\frac{wL^2}{6}\langle x\rangle^{-2} + \frac{wL}{2}\langle x\rangle^{-1} - w\langle x\rangle^0 + \frac{w}{L}\langle x\rangle^1 - \frac{w}{L}\langle x-L\rangle^1$$

(can ignore)

$$V_y = \frac{wL^2}{6}\langle x\rangle^{-1} - \frac{wL}{2}\langle x\rangle^0 + w\langle x\rangle^1 - \frac{w}{2L}\langle x\rangle^2$$

$$M_z = -\frac{wL^2}{6}\langle x\rangle^0 + \frac{wL}{2}\langle x\rangle^1 - \frac{w}{2}\langle x\rangle^2 + \frac{w}{6L}\langle x\rangle^3$$

can ignore $\langle\ \rangle$ brackets as functions always exist

$$EIv_c' = -\frac{wL^2}{6}x + \frac{wLx^2}{4} - \frac{wx^3}{6} + \frac{w}{24L}x^4 + C_1$$

$v_c' = 0$ at $x=0$ $\therefore C_1 = 0$

$$\therefore \theta = v_c' = \frac{wx}{24EIL}\left(x - 4Lx + 6L^2x - 4L^3\right)$$

ANS

$$EIv_c = -\frac{wL^2x^2}{12} + \frac{wLx^3}{12} - \frac{wx^4}{24} + \frac{w}{120L}x^5 + C_2$$

$v_c = 0$ at $x = 0$ $\therefore C_2 = 0$

$$v_c = \frac{wx^2}{120EIL}\left(x^3 - 5Lx^2 + 10L^2x - 10L^3\right)$$

ANS

3.26 (a)

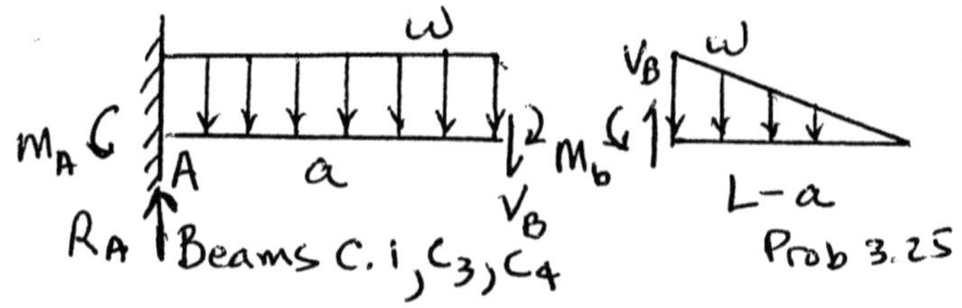

$V_B = \frac{1}{2}w(L-a)$

$M_B = \frac{1}{6}w(L-a)^2$

$0<x<a$

$$v_c = \frac{wx^2}{24EI}(4ax - x^2 - 6a^2) + \frac{\frac{1}{2}w(L-a)x^2}{6EI}(x-3a) - \frac{\frac{1}{6}w(L-a)^2x^2}{2EI}$$

reduces to

$$v_c = -\frac{wx^2}{24EI}\left[x^2 - 2(L+a)x + 2a^2 + 2aL + 2L^2\right] \quad ① \quad \boxed{\text{ANS}}$$

$a<x<L$ Must find θ_B at $x=a$ from beams C.1, C.3, and C4. Deflection from this is $\theta_B(x-a)$. Add this and v_B to the deflection eq. from Prob. 3.25 with $L-a$ for L and $x-a$ for x.

$$\theta_B = \frac{wa}{6EI}(3a^2 - a^2 - 3a^2) + \frac{\frac{1}{2}w(L-a)a}{2EI}(a-2a) - \frac{\frac{1}{6}w(L-a)^2a}{EI}$$

$$= -\frac{wa}{12EI}(a^2 - aL + 2L^2)$$

From ①

$$v_B = -\frac{wa^2}{24EI}\left[a^2 - 2(L+a)a + 2a^2 + 2aL + 2L^2\right] = -\frac{wa^2}{24EI}(a^2 + 2L^2)$$

$$v_c = -\frac{wa^2}{24EI}(a^2+2L^2) - \frac{wa}{12EI}(x-a)(a^2 - aL + 2L^2)$$
$$+ \frac{w}{120EI(L-a)}\left[(x-a)^5 - 5(x-a)^4(L-a) + 10(x-a)^3(L-a)^2 - 10(x-a)^2(L-a)^3\right] \quad ②$$

$\boxed{\text{ANS}}$

(b) Singularity functions. From statics, $R_A = \frac{w}{2}(L+a)$

$M_A = \frac{w}{6}(L^2 + aL + a^2)$

$$q = \frac{w}{2}(L+a)\langle x\rangle^{-1} - \frac{w}{6}(L^2+aL+a^2)\langle x\rangle^{-2} - w\langle x\rangle^0 + \frac{w}{L-a}\langle x-a\rangle^1 - \frac{w}{L-a}\langle x-L\rangle^1$$

(can ignore: $-\frac{w}{L-a}\langle x-L\rangle^1$)

$$V_y = -\frac{w}{2}(L+a)\langle x\rangle^0 + \frac{w}{6}(L^2+aL+a^2)\langle x\rangle^{-1} + w\langle x\rangle^1 - \frac{w}{2(L-a)}\langle x-a\rangle^2$$

$$M_z = \frac{w}{2}(L+a)x - \frac{w}{6}(L^2+aL+a^2)\langle x\rangle^0 - \frac{w}{2}x^2 + \frac{w}{6(L-a)}\langle x-a\rangle^3$$

$$EIv_c' = \frac{w}{4}(L+a)x^2 - \frac{w}{6}(L^2+aL+a^2)x - \frac{w}{6}x^3 + \frac{w}{24(L-a)}\langle x-a\rangle^4 + C_1$$

$v_c' = 0$ at $x=0$ and $\langle x-a\rangle^4 = 0$ ∴ $C_1 = 0$

$$EIv_c = \frac{w}{12}(L+a)x^3 - \frac{w}{12}(L^2+aL+a^2)x^2 - \frac{w}{24}x^4 + \frac{w}{120(L-a)}\langle x-a\rangle^5 + C_2$$

(continued)

$v_c = 0$ at $x = 0$ and $\langle x-a \rangle^5 = 0$ $\therefore C_2 = 0$

Thus,

$$v_c = \frac{w}{120EI(L-a)}\left[10(L^2-a^2)x^3 - 10(L-a)(L^2+aL+a^2)x^2 - 5(L-a)x^4 + \langle x-a \rangle^5\right] \quad \boxed{\text{ANS}} \; ③$$

To verify with part (a)

$\underline{0<x<a}$ $\langle x-a \rangle^5 = 0$ and ③ becomes

$$v_c = \frac{w}{120EI}\left[10(L+a)x^3 - 10(L^2+aL+a^2)x^2 - 5x^4\right]$$

$$= -\frac{wx^2}{24EI}\left[x^2 - 2(L+a)x + 2L^2 + 2aL + 2a^2\right] \quad ✓$$

$\underline{a<x<L}$ Expand eq. ② [except for $(x-a)^5$ term] and simplify

Results in

$$v_c = \frac{w}{24EI}\left[-x^4 + 2(L+a)x^3 - 2(a^2+aL+L^2)x^2\right] + \frac{w(x-a)^5}{120EI(L-a)}$$

$$= \frac{w}{120EI(L-a)}\left[-5(L-a)x^4 + 10(L^2-a^2)x^3 - 10(L-a)(a^2+aL+L^2)x^2 + (x-a)^5\right]$$

which is the same as ③ for $x>a$ ✓

(3.27)

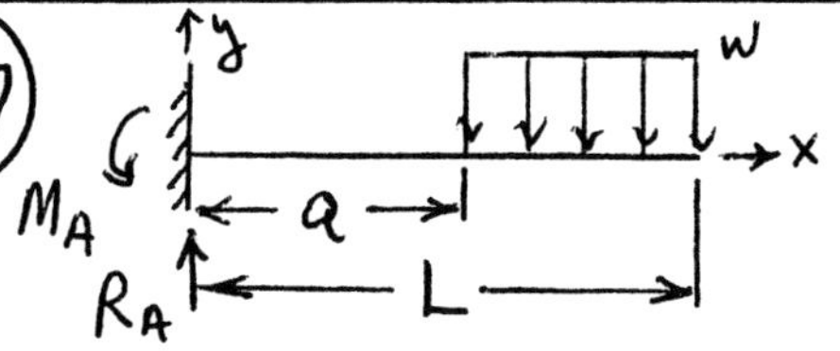

$R_A = w(L-a)$

$M_A = \frac{w}{2}(L^2-a^2)$

$\underline{0<x<a}$

$$M_z = w(L-a)x - \frac{w}{2}(L^2-a^2)$$

$$EIv_c' = \frac{w}{2}(L-a)x^2 - \frac{w}{2}(L^2-a^2)x + C_1 \qquad v_c' = 0 \text{ at } x=0 \;\therefore C_1 = 0$$

$$\therefore \theta = v_c' = \frac{w(L-a)}{2EI}x(x-a-L) \quad ① \; \boxed{\text{ANS}}$$

$$EIv_c = \frac{w}{6}(L-a)x^3 - \frac{w}{4}(L^2-a^2)x^2 + C_2 \qquad v_c = 0 \text{ at } x=0 \;\therefore C_2 = 0$$

$$v_c = \frac{w(L-a)x^2}{12EI}(2x - 3a - 3L) \quad ② \; \boxed{\text{ANS}}$$

$\underline{a<x<L}$

$$M_z = -\frac{w}{2}(L-x)^2 = -\frac{wL^2}{2} + wLx - \frac{wx^2}{2}$$

$$EIv_c' = -\frac{wL^2x}{2} + \frac{wL}{2}x^2 - \frac{wx^3}{6} + C_3 \qquad \text{Equate this to ① at } x=a$$

(Continued)

$$-\frac{wL^2}{2}a + \frac{wL}{2}a^2 - \frac{wa^3}{6} + C_3 = \frac{w}{2}(L-a)a^2 - \frac{w}{2}(L^2-a^2)a$$

$C_3 = wa^3/6$ sub back into v_c' and simplify

$$\therefore \theta = v_c' = -\frac{w}{6EI}(x^3 - 3Lx^2 + 3L^2x - a^3) \quad (3) \quad \boxed{\text{ANS}}$$

$$EIv_c = -\frac{wL^2}{4}x^2 + \frac{wL}{6}x^3 - \frac{wx^4}{24} + \frac{wa^3}{6}x + C_4$$

Equate to (2) at $x = a$

$$-\frac{wL^2}{4}a^2 + \frac{wL}{6}a^3 - \frac{wa^4}{24} + \frac{wa^4}{6} + C_4 = \frac{w}{6}(L-a)a^3 - \frac{w}{4}(L^2-a^2)a^2$$

$C_4 = -\frac{wa^4}{24}$ sub back into v_c and simplify

$$v_c = -\frac{w}{24EI}(x^4 - 4Lx^3 + 6L^2x^2 - 4a^3x + a^4) \quad (4) \quad \boxed{\text{ANS}}$$

(3.28)

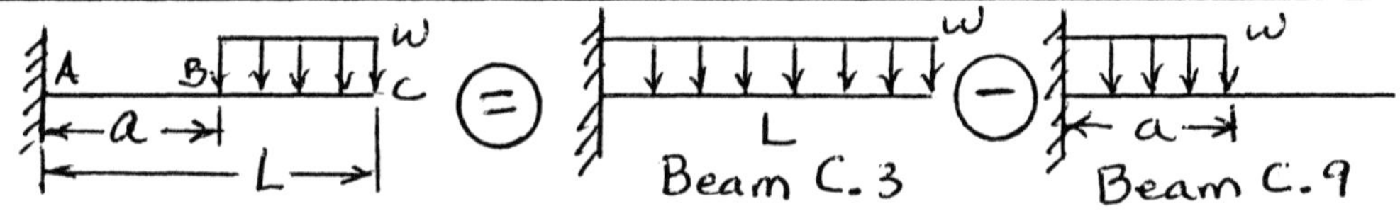

Slope

$$\theta_{AB} = \frac{wx}{6EI}(3Lx - x^2 - 3L^2) - \frac{wx}{6EI}(3ax - x^2 - 3a^2) = -\frac{wx}{2EI}(L-a)(L+a-x) \quad \boxed{\text{ANS}}$$

$$\theta_{BC} = \frac{wx}{6EI}(3Lx - x^2 - 3L^2) - \frac{wa^3}{6EI} = \frac{-w}{6EI}(3L^2x + x^3 - 3Lx^2 - a^3) \quad \boxed{\text{ANS}}$$

Deflection

$$(v_c)_{AB} = \frac{wx^2}{24EI}(4Lx - x^2 - 6L^2) - \frac{wx^2}{24EI}(4ax - x^2 - 6a^2) = -\frac{wx^2(L-a)}{12EI}(3L + 3a - 2x) \quad \boxed{\text{ANS}}$$

$$(v_c)_{BC} = \frac{wx^2}{24EI}(4Lx - x^2 - 6L^2) - \frac{wa^3}{24EI}(a - 4x) = \frac{-w}{24EI}(6L^2x^2 - 4Lx^3 + x^4 + a^4 - 4a^3x) \quad \boxed{\text{ANS}}$$

It is clear that these eqs. agree with Prob 3.27

(3.29) From Prob. 3.27 $R_A = w(L-a)$ $M_A = \frac{w}{2}(L^2-a^2)$

$$q = w(L-a)\langle x\rangle^{-1} - \frac{w}{2}(L^2-a^2)\langle x\rangle^{-2} - w\langle x-a\rangle^0 + w\langle x-L\rangle^0$$

($w\langle x-L\rangle^0$: ignore. No contribution)

$$V_y = -w(L-a)\langle x\rangle^0 + \frac{w}{2}(L^2-a^2)\langle x\rangle^{-1} + w\langle x-a\rangle^1$$

$$M_z = w(L-a)x - \frac{w}{2}(L^2-a^2)\langle x\rangle^0 - \frac{w}{2}\langle x-a\rangle^2$$

(continued)

$$EIv_c' = \frac{w}{2}(L-a)x^2 - \frac{w}{2}(L^2-a^2)x - \frac{w}{6}\langle x-a\rangle^3 + C_1 \qquad \text{at } x=0 \;\; v_c'=0 \;\; \therefore C_1=0$$

$$v_c' = \frac{w}{6EI}\left[3(L-a)(x-L-a)x - \langle x-a\rangle^3\right] \quad (1) \quad \boxed{\text{ANS}}$$

$$EIv_c = \frac{w}{6}(L-a)x^3 - \frac{w}{2}(L^2-a^2)x^2 - \frac{w}{24}\langle x-a\rangle^4 + C_2 \qquad v_c=0 \text{ at } x=0 \;\; \therefore C_2=0$$

$$v_c = \frac{w}{24EI}\left[2(L-a)(2x-3L-3a)x^2 - \langle x-a\rangle^4\right] \quad (2) \quad \boxed{\text{ANS}}$$

Verification:

$0<x<a$ $\quad \langle x-a\rangle^n = 0$

$$(v_c)_{AB} = \frac{wx^2}{12EI}(L-a)(2x-3L-3a) \quad \text{same as Probs. 3.27 \& 3.28}$$

$$\theta_{AB} = (v_c')_{AB} = \frac{wx}{2EI}(L-a)(x-L-a) \quad \text{" " " "}$$

$a<x<L$ $\quad \langle x-a\rangle^n = (x-a)^n$

$$(v_c)_{BC} = \frac{w}{24EI}\left[2x^2(L-a)(2x-3L-3a) - (x-a)^4\right] \quad \text{expand}$$

$$= -\frac{w}{24EI}\left(6L^2x^2 - 4Lx^3 + x^4 + a^4 - 4a^3x\right) \quad \text{same as Probs. 3.27 \& 3.28}$$

$$(v')_{BC} = \theta_{BC} = \frac{w}{6EI}\left[3x(L-a)(x-L-a) - (x-a)^3\right] \quad \text{expand}$$

$$= -\frac{w}{6EI}\left(3L^2x + x^3 - 3Lx^2 - a^3\right) \quad \text{same as Probs. 3.27 \& 3.28}$$

3.30 Equivalent forces for reaction calc.

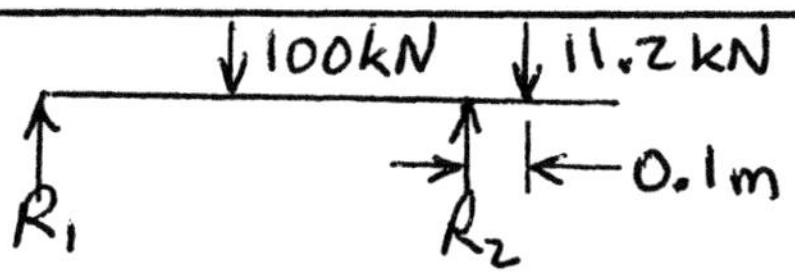

$R_1 = 54.93$ kN
$R_2 = 56.27$ kN

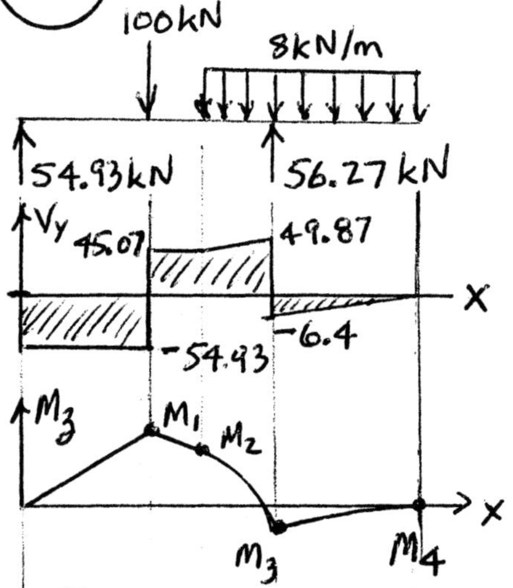

$M_{max} = 43.95$ kN-m

$V_{max} = 54.93$ kN

$M = \int V dx$

$M_1 = 54.93(0.8) = 43.95$ kN-m

$M_2 = 43.95 - (45.07)(0.4) = 25.92$ kN·m

$M_3 = 25.92 - \frac{1}{2}(49.87 + 45.07)(0.6) = -2.56$ kN-m

$M_4 = -2.56 + \frac{1}{2}(6.4)(0.8) = 0$

(continued)

Section

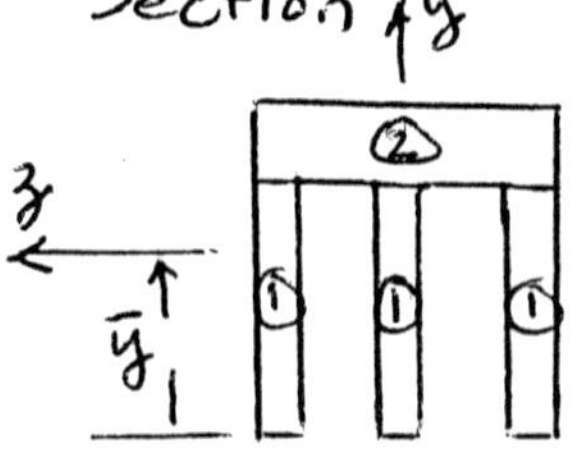

$\bar{y}_{①} = 70\text{mm}$ $A_{①} = 15(140) = 2100\text{ mm}^2$

$\bar{y}_{②} = 160\text{mm}$ $A_{②} = 90(40) = 3600\text{ mm}^2$

$$\bar{y} = \frac{3(70)(2100) + (160)(3600)}{3(2100) + 3600} = 102.7\text{ mm}$$

$$I_z = 3\left[\frac{1}{12}(15)(140)^3 + 15(140)(102.7-70)^2\right] + \frac{1}{12}90(40)^3 + 90(40)(160-102.7)^2$$
$$= 29.33(10^6)\text{ mm}^4$$

(a) max tension occurs under 100kN load at bottom of beam

$$\sigma_{max} = -\frac{(43.95)(10^3)(-102.7)(10^{-3})}{29.33(10^6)(10^{-3})^4} = 153.4(10^6)\text{N/m}^2 = \underline{153.9\text{ MPa}}$$

compression top of beam under 100kN load

$$\sigma_{min} - \frac{43.95(10^3)(180-102.7)(10^{-3})}{29.33(10^6)(10^{-3})^4} = -115.8(10^6)\text{N/m}^2 = \underline{-115.8\text{ MPa}}$$

Max transverse shear occur at neutral axis between left support and 100kN load

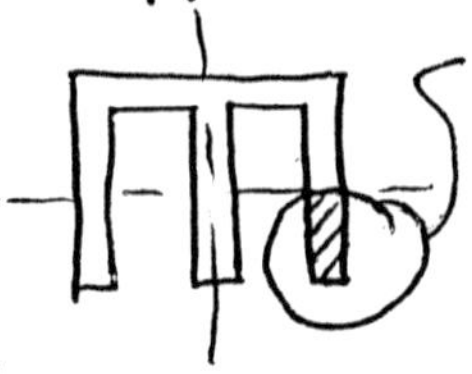

$$Q = -y\bar{A} = \frac{(102.7)^2}{2}(15) = 79.1(10^3)\text{mm}^3 \quad \text{for each web}$$

$$\tau_{max\,V} = \frac{VQ}{Ib} = \frac{54.93(10^3)79.1(10^{-6})}{29.33(10^{-6})15(10^{-3})} = 9.88(10^6)\text{N/m}^2 = \underline{9.88\text{MPa}}$$

(b) $\tau_{max} = \sigma_{max}/2 = \frac{153.9}{2} = \underline{77.0\text{ MPa}}$

(c) For interface

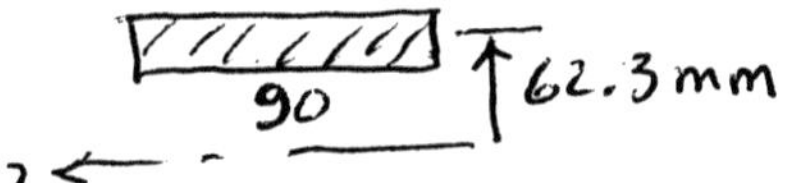

$$Q = 62.3(30)(90) = 168.2\,(10^3)\text{ mm}^3$$

$$q = \frac{VQ}{I} = \frac{54.93(10^3)(168.2)(10^{-6})}{29.33(10^{-6})} = 315(10^3)\text{ N/m}$$

$$F_{bolt} = \frac{1}{2}\left[315(10^3)(0.200)\right] = 31.5(10^3)\text{N}$$

$$\tau_{avg} = \frac{F}{A} = \frac{31.5(10^3)}{\pi/4\,(0.012)^2} = 279(10^6)\text{N/m}^2 = \underline{279\text{ MPa}}$$

(d) $q = \left(54.93\langle x\rangle^{-1} - 100\langle x-0.8\rangle^{-1} - 8\langle x-1.2\rangle^{0} + 56.27\langle x-1.8\rangle^{-1}\right)10^3$

$V_y = \left(-54.93\langle x\rangle^{0} + 100\langle x-0.8\rangle^{0} + 8\langle x-1.2\rangle^{1} - 56.27\langle x-1.8\rangle^{0}\right)10^3$

$M_z = \left(54.93x - 100\langle x-0.8\rangle^{1} - 4\langle x-1.2\rangle^{2} + 56.27\langle x-1.8\rangle^{1}\right)10^3$

(Continued)

$$EI v_c' = \left(\frac{54.93}{2}x^2 - 50\langle x-0.8\rangle^2 - \frac{4}{3}\langle x-1.2\rangle^3 + \frac{56.27}{2}\langle x-1.8\rangle^2 + C_1\right)10^3$$

$$EI v_c = \left(\frac{54.93}{6}x^3 - \frac{50}{3}\langle x-0.8\rangle^3 - \frac{1}{3}\langle x-1.2\rangle^4 + \frac{56.27}{6}\langle x-1.8\rangle^3 + C_1 x + C_2\right)10^3$$

$v_c = 0$ at $x = 0$ $\therefore C_2 = 0$

$v_c = 0$ at $x = 1.8\text{m}$

$$0 = \frac{54.93}{6}(1.8)^3 - \frac{50}{3}(1)^2 - \frac{1}{3}(0.6)^4 + 0 + 1.8C_1 \qquad \therefore C_1 = -20.38$$

Max deflection will occur either near the 100 kN load or at the right end. Near the 100 kN load max deflection occurs where $v_c' = 0$.

$0 < x < 0.8\text{m}$

$$\frac{54.93}{2}x^2 - 20.38 = 0 \Rightarrow x = \pm 0.861\text{m}$$ neither answer is acceptable

$0.8 < x < 1.2\text{ m}$

$$\frac{54.93}{2}x^2 - 50(x-0.8)^2 - 20.38 = 0 \Rightarrow x^2 - 3.55x + 2.324 = 0$$

Solution: $x = 2.684,\ 0.866\text{ m}$ only 0.866m is in range

Sub $x = 0.866\text{m}$ into v_c eq.

$$209(10^9)(29.33)(10^{-6})v_c = \left[\frac{54.93}{6}(0.866)^3 - \frac{50}{3}(0.066)^3 - 20.38(0.866)\right](10^3)$$

$$v_c = -1.910(10^{-3})\text{m} = -1.91\text{ mm}$$

Check right end $(x = 2.6\text{ m})$

$$209(10^9)(29.33)(10^{-6})v_c = \left[\frac{54.93}{6}(2.6)^3 - \frac{50}{3}(1.8)^3 - \frac{1}{3}(1.4)^4 + \frac{56.27}{6}(0.8)^3 - 20.83(2.6)\right](10^3)$$

$$v_c = 2.323(10^{-3})\text{m} = \boxed{2.323\text{ mm} = (v_c)_{max}}$$

3.31

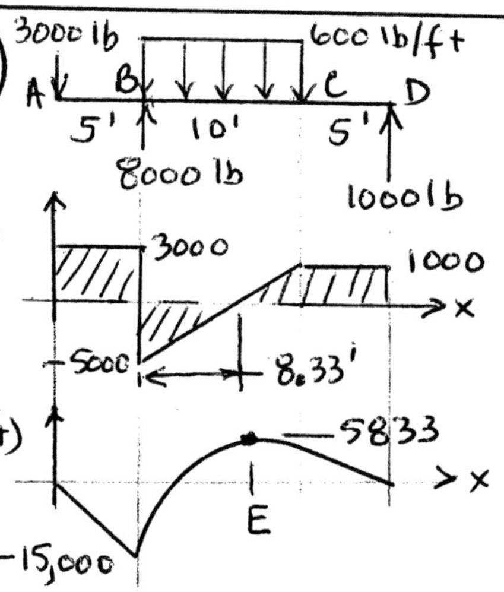

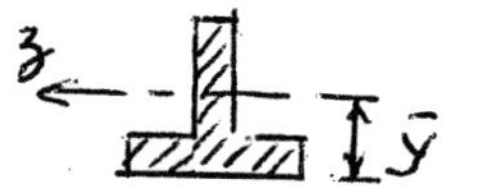

$$\bar{y} = \frac{1(12) + 5(12)}{24} = 3\text{ in}$$

$$I_z = \frac{1}{3}\left[2(5)^3 + 6(3)^3 - 4(1)^3\right] = 136\text{ in}^4$$

(a) B: $x = 5\text{ ft}$

$y = -3\text{in}$ $\quad \sigma = -\frac{(-15000)(12)(-3)}{136} = -3970\text{ psi}$

$y = 5\text{in}$ $\quad \sigma = -\frac{(-15000)(12)(5)}{136} = 6620\text{ psi}$

(Continued)

E: $x = 5 + 8.33 = 13.33$ ft.

$y = -3 \quad \sigma = -\dfrac{5833(12)(-3)}{136} = 1540$ psi

$y = 5 \quad \sigma = -\dfrac{5833(12)(5)}{136} = -2570$ psi

$\sigma_{max} = \underline{6.62 \text{ kpsi}} \qquad \sigma_{min} = \underline{-3.97 \text{ kpsi}}$

(b) $V_{max} = 5000$ lb

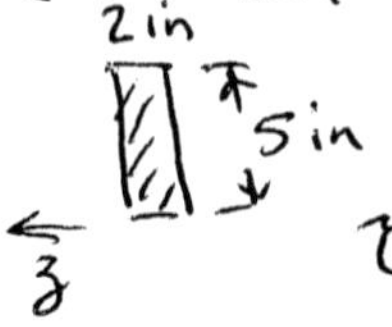

$Q = \dfrac{(5)^2}{2}(2) = 25 \text{ in}^3$

$\tau_{max} = \dfrac{VQ}{Ib} = \dfrac{5000(25)}{(136)(2)} = \underline{460 \text{ psi}}$

3.32

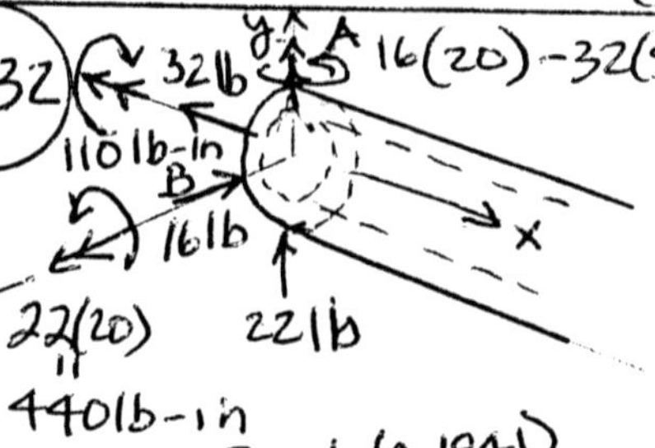

$I = \dfrac{\pi}{8}(0.0625)(1 - 0.0625)^3 = 0.02022 \text{ in}^4$

$J = 2I = 0.04045 \text{ in}^4$

$A = \pi \bar{d} t = \pi(1 - 0.0625)(0.0625) = 0.1841 \text{ in}^2$

$\bar{A} = \frac{1}{2}(0.1841)$, $\bar{y} = \dfrac{2r}{\pi}$

$Q = \bar{y}A = \dfrac{(1 - 0.0625)}{\pi}\left(\dfrac{0.1841}{2}\right) = 0.02747 \text{ in}^3$

Point A:

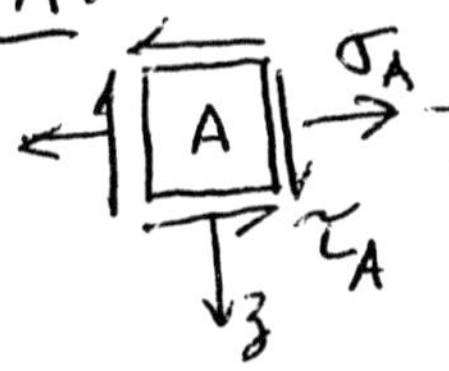

$\sigma_A = \dfrac{M_z c}{I_z} + \dfrac{P}{A} = \dfrac{440(0.5)}{0.02022} + \dfrac{32}{0.1841} = 11{,}050$ psi

$\tau_A = \dfrac{Tc}{J} + \dfrac{V_z Q}{I(2t)} = \dfrac{110(0.5)}{0.04045} + \dfrac{16(0.02747)}{(0.02022)(0.125)} = 1534$ psi

$\tau_{max} = \sqrt{\left(\dfrac{11{,}050}{2}\right)^2 + 1534^2} = 5734 \text{ psi} = \underline{5.73 \text{ kpsi}}$

Point B: ↑y

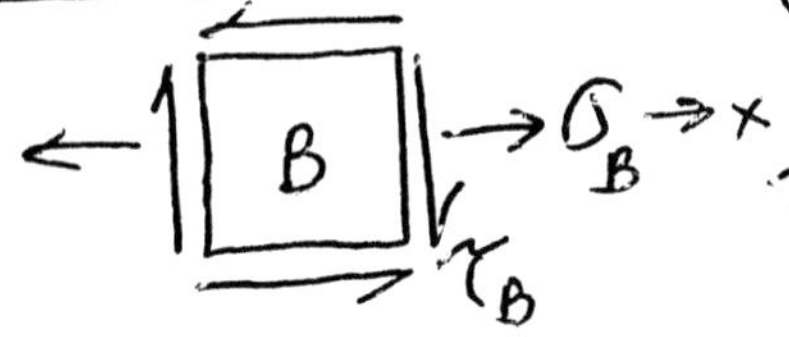

$\sigma_B = -\dfrac{M_y c}{I} + \dfrac{P}{A} = \dfrac{-160(0.5)}{0.02022} + \dfrac{32}{0.1841} = -3783$ psi

$\tau_B = \dfrac{Tc}{J} + \dfrac{V_y Q}{I(2t)} = \dfrac{110(0.5)}{0.04045} + \dfrac{22(0.02747)}{0.02022(0.125)} = 1599$ psi

$\tau_{max} = \sqrt{\left(\dfrac{3783}{2}\right)^2 + 1599^2} = 2477 \text{ psi} = \underline{2.48 \text{ kpsi}}$

Note: Max shear does not occur at point A.

$M_{max} = \sqrt{440^2 + 160^2} = 468$ lb in, $\sigma_{max} \doteq \dfrac{Mc}{I} + \dfrac{P}{A} = 11{,}750$ psi

$\tau_{max} \doteq \sigma_{max}/2 = 5876 \text{ psi} = 5.88 \text{ kpsi}$

3.33

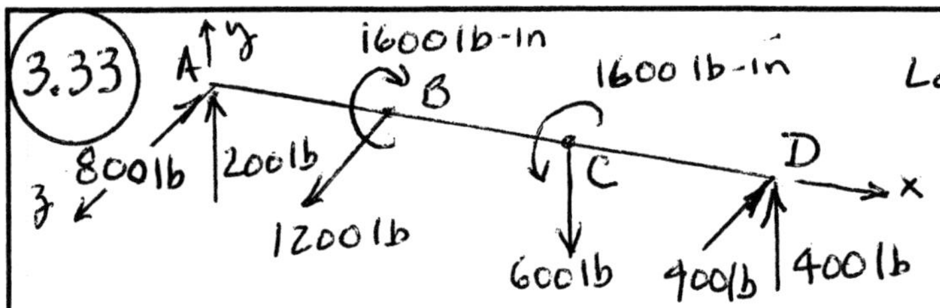

Loading diagram

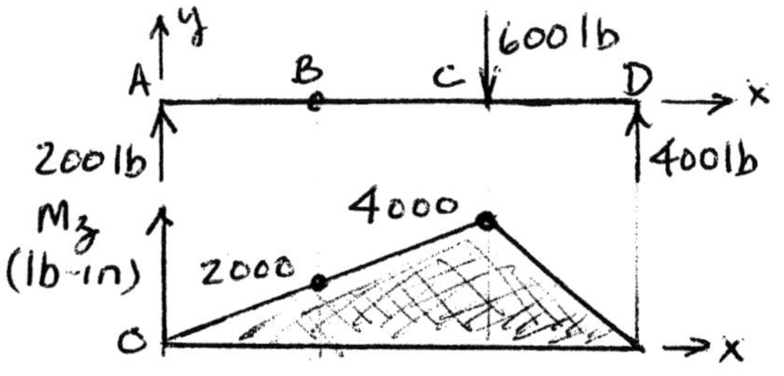

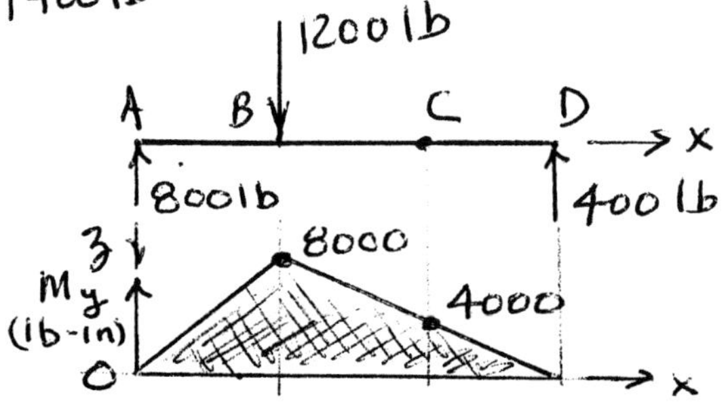

$M_B = \sqrt{2000^2 + 8000^2} = 8246$ lb-in, $M_C = \sqrt{4000^2 + 4000^2} = 5657$ lb-in

$M_{max} = 8246$ lb-in, $\sigma = \dfrac{32M}{\pi d^3} = \dfrac{32(8246)}{\pi(2)^3} = 10{,}500$ psi

Torsion: $\tau = \dfrac{16T}{\pi d^3} = \dfrac{16(1600)}{\pi(2)^3} = 1020$ psi

$\sigma_{max} = \dfrac{10500}{2} + \sqrt{\left(\dfrac{10500}{2}\right)^2 + 1020^2} = 10{,}600 \text{ psi} = \underline{10.6 \text{ kpsi}}$

$\sigma_{min} = \underline{-10.6 \text{ kpsi}}$ (diametrically opposite σ_{max})

$\tau_{max} = \sqrt{\left(\dfrac{10500}{2}\right)^2 + 1020^2} = 5347 \text{ psi} = \underline{5.35 \text{ kpsi}}$

Location: At B ($x = 10^+$)

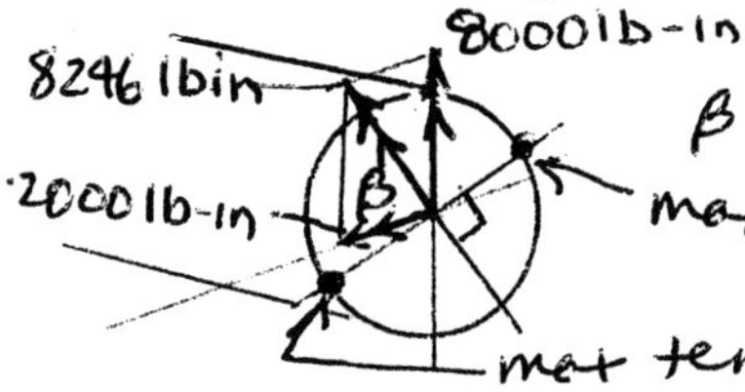

$\beta = \tan^{-1} \dfrac{8000}{2000} = 76°$

(b) Using beam C.5 of Appendix C

$v_B = \dfrac{600(10)(10)}{6(29)(10^6)\frac{\pi}{64}(2)^4(30)}(10^2 + 10^2 - 30^2) = \underline{-0.0102 \text{ in}}$

$w_B = \dfrac{-1200(20)(10)}{6(29)(10^6)\frac{\pi}{64}(2)^4(30)}(10^2 + 20^2 - 30^2) = \underline{0.0234 \text{ in}}$

$(\delta_{net})_B = \sqrt{0.0102^2 + 0.0234^2} = \underline{0.0255 \text{ in}}$

3.34

Loading diagram

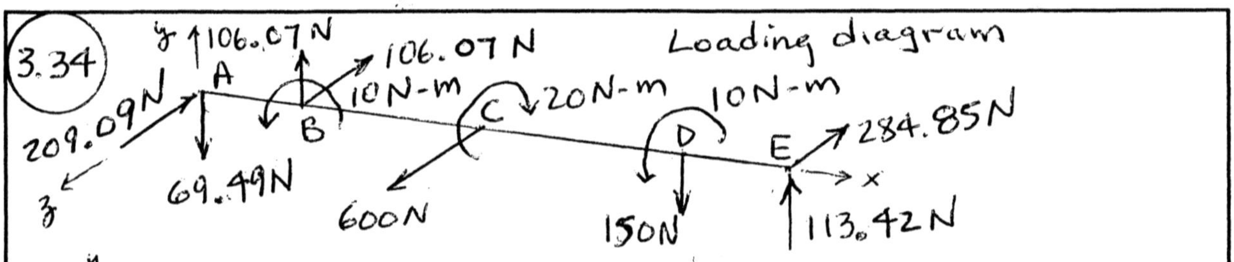

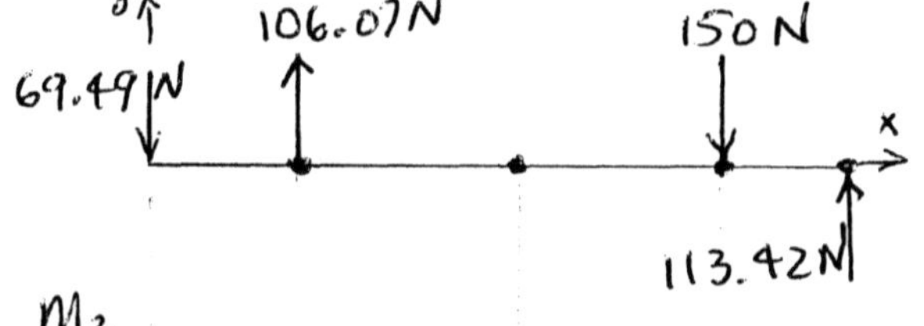

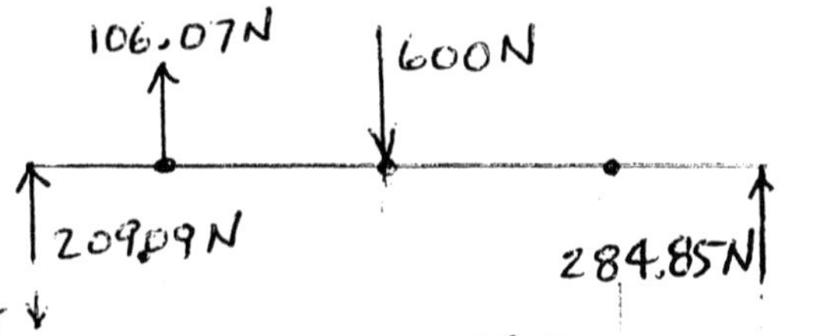

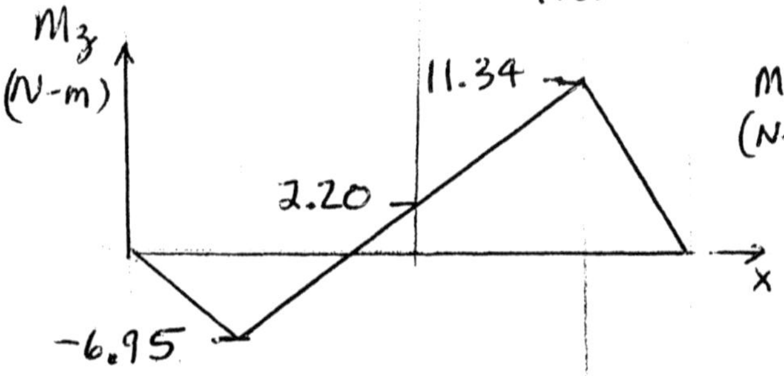

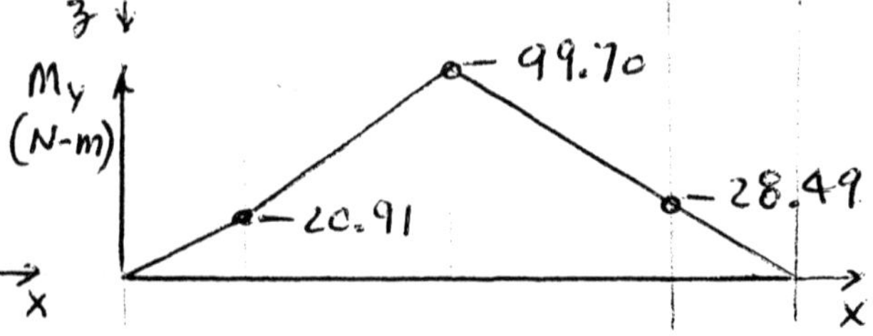

$M_B = \sqrt{(-6.95)^2 + (20.91)^2} = 22.0$ N-m

$M_C = \sqrt{(2.20)^2 + (99.70)^2} = 99.72$ N-m ← obviously the largest

$M_D = \sqrt{(11.34)^2 + (28.49)^2} = 30.66$ N-m

At B

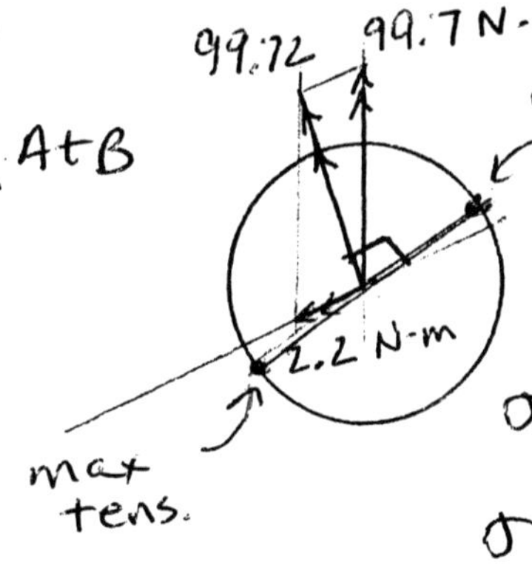

Bending $\sigma = \dfrac{32(99.72)}{\pi(0.030)^3} = 37.6(10^6)\ \text{N/m}^2$

Left or right of point B, $T = 10$ N-m

$\tau = \dfrac{Tc}{J} = \dfrac{16(10)}{\pi(0.03)^3} = 1.89(10^6)\ \text{N/m}^2$

$\sigma_{max} = \dfrac{37.6}{2} + \sqrt{\left(\dfrac{37.6}{2}\right)^2 + 1.89^2} = \underline{37.7 \text{ MPa}}$

$\sigma_{min} = \underline{-37.7 \text{ MPa}}$

$\tau_{max} = \sqrt{\left(\dfrac{37.6}{2}\right)^2 + 1.89^2} = \underline{18.9 \text{ MPa}}$

(b) xy plane

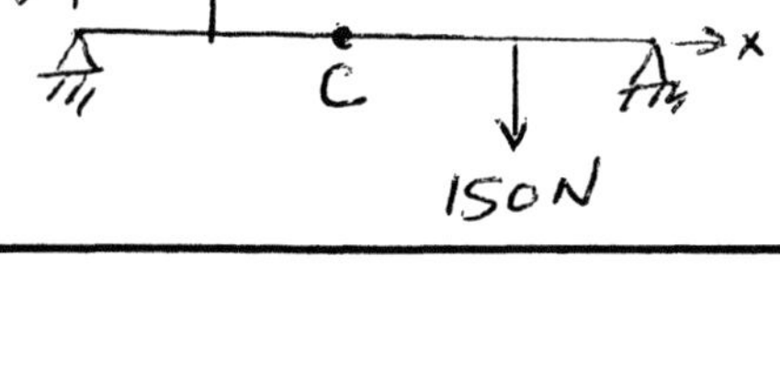

Use beam C.5 for both loads

y_{BC} for 106.07N load

y_{AB} for 150N load

(continued)

$$v_C = \frac{-106.1(0.1)(0.7-0.35)}{6(210)(10^9)\frac{\pi}{64}(0.03)^4(0.7)}\left[(0.35)^2+(0.1)^2-2(0.7)(0.35)\right]$$

$$+\frac{150(0.1)(0.35)}{6(210)(10^9)\frac{\pi}{64}(0.03)^4(0.7)}\left[(0.35)^2+(0.1)^2-(0.7)^2\right]$$

$$= -1.566(10^{-5})\text{ m} = \underline{-0.0157\text{ mm}} \quad \boxed{\text{ANS}}$$

x-z plane

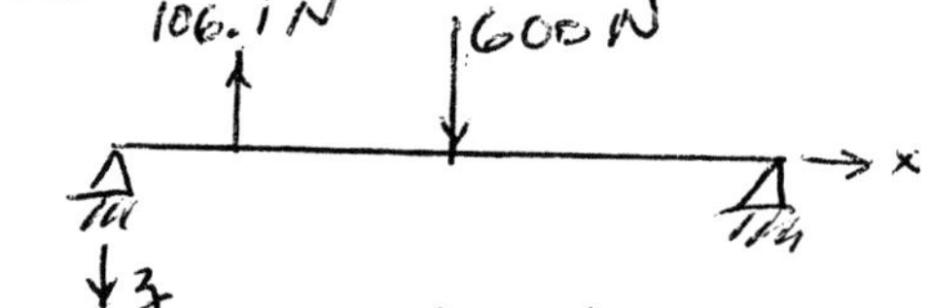

$$w_C = \frac{106.1(0.1)(0.7-0.35)}{6(210)(10^9)\frac{\pi}{64}(0.03)^4(0.7)}\left[(0.35)^2+(0.1)^2-2(0.7)(0.35)\right]$$

$$-\frac{600(0.35)(0.35)}{6(210)(10^9)\frac{\pi}{64}(0.03)^4(0.7)}\left[(0.35)^2+(0.35)^2-(0.7)^2\right]$$

$$= 4.76(10^{-4})\text{m} = \underline{0.476\text{ mm}} \quad \boxed{\text{ANS}}$$

$$(\delta_C)_{net} = \sqrt{0.0157^2+0.476^2} = \underline{0.476\text{mm}}$$

3.35 (a) $J = 2\pi t r^3 \quad r = 19\text{ mm}$

Torsion $\tau_T = \frac{Tc}{J} = \frac{(1500)(0.1)(0.02)}{2\pi(0.002)(0.019)^3} = 34.8(10^6)\text{N/m}^2 = \underline{34.8\text{ MPa}}$

Direct shear: $V = 1500\text{ N}$

$\bar{y} = \frac{2r}{\pi} \quad \bar{A} = \pi r t \quad Q = \frac{2r}{\pi}\pi r t = 2r^2 t \quad I = \pi t r^3$

$$\tau_V = \frac{VQ}{I(2t)} = \frac{(1500)(2)(0.019)^2(0.002)}{\pi(0.002)(0.019)^3 2(0.002)} = 12.56(10^6)\text{N/m}^2 = \underline{12.6\text{ MPa}}$$

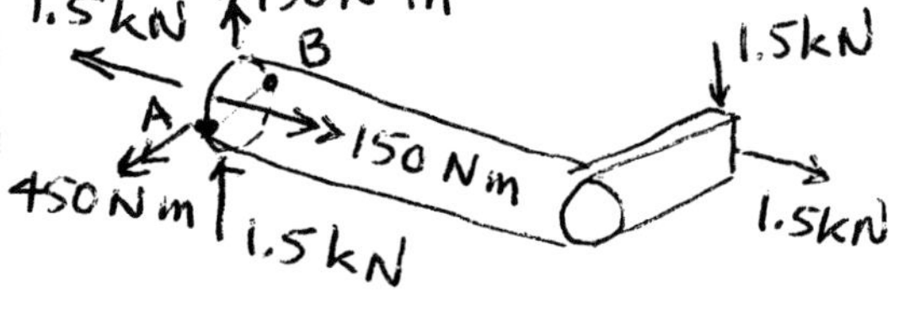

Stresses due to pressure

$$\sigma_\theta = \frac{pr}{t} = \frac{3(10^6)(0.019)}{0.002} = 28.5(10^6)\text{N/m}^2 = \underline{28.5\text{ MPa}}$$

$$\sigma_\ell = \underline{14.25\text{ MPa}}$$

For points A & B, bending and axial stress

$$(\sigma_{b,a})_A = -\frac{Mc}{I}+\frac{P}{A} = -\frac{150(0.02)}{\pi(0.002)(0.019)^3}+\frac{1500}{2\pi(0.019)(0.002)} = -63.3(10^6)\text{ N/m}^2$$

(continued)

$(\sigma_{b,a})_A = \underline{-63.3 \text{ MPa}}$

$$(\sigma_{b,a})_B = \frac{150(0.02)}{\pi(0.002)(0.019)^3} + \frac{1500}{2\pi(0.019)(0.002)} = 75.9(10^6)\text{ N/m}^2 = \underline{75.9 \text{ MPa}}$$

Point A

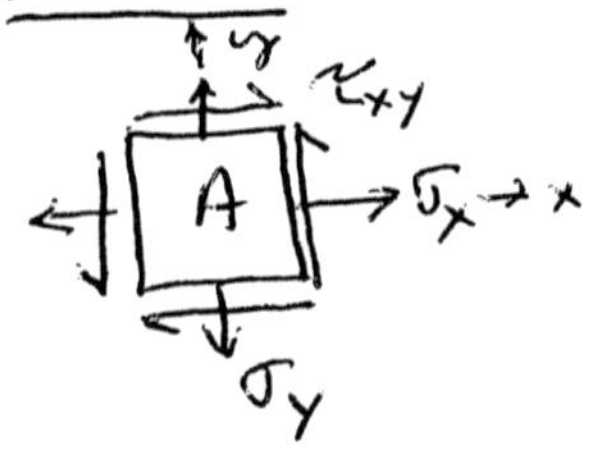

$\sigma_x = 14.25 - 63.3 = -49.1$ MPa

$\sigma_y = 28.5$ MPa

$\tau_{xy} = 34.8 - 12.6 = 22.2$ MPa

ANS

Point B

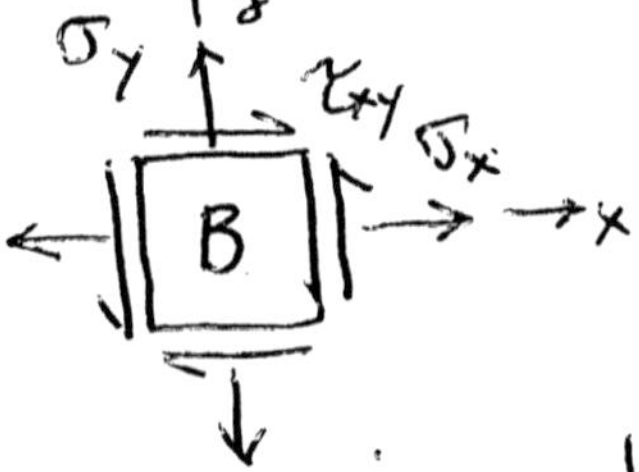

$\sigma_x = 14.25 + 75.9 = 90.1$ MPa

$\sigma_y = 28.5$ MPa

$\tau_{xy} = 34.8 + 12.6 = 47.4$ MPa

ANS

For maximum bending

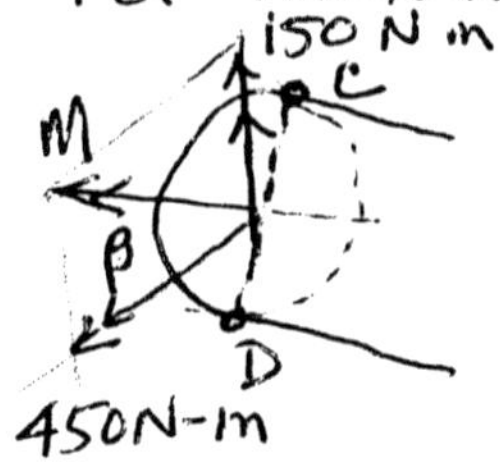

$\beta = \tan^{-1}\frac{150}{450} = 18.4°$

$M = \sqrt{150^2 + 450^2} = 474.3$ N·m

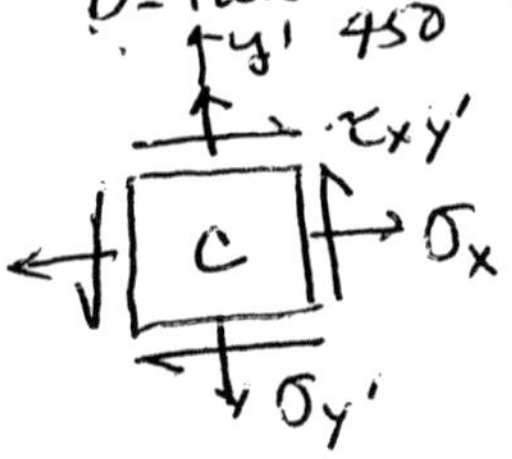

$$\sigma_x = \frac{474.3(0.02)}{\pi(0.002)(0.019)^3} + \frac{1500}{2\pi(0.019)(0.002)} + 14.25(10^6)$$

$$= 240.6(10^6)\text{ N/m}^2 = \underline{240.6 \text{ MPa}}$$

$\sigma_{y'} = \underline{28.5 \text{ MPa}}$

$\tau_{xy'} = \underline{34.8 \text{ MPa}}$

ANS

(b) Point C:

$$\tau_{max} = \sqrt{\left(\frac{240.6 - 28.5}{2}\right)^2 + 34.8^2} = 111.6 \text{ MPa}$$

Point D:

$$\sigma_x = -\frac{474.3(0.02)}{\pi(0.002)(0.019)^3} + \frac{1500}{2\pi(0.019)(0.002)} + 14.25(10^6) = -199.6(10^6)\text{ N/m}^2$$

$$\tau_{max} = \sqrt{\left(\frac{-199.6 - 28.5}{2}\right)^2 + 34.8^2} = \underline{119.2 \text{ MPa}}$$

ANS

(c) Point C:

$$\sigma_{max} = \frac{240.6 + 28.5}{2} + 111.6 = \underline{246.2 \text{ MPa}}$$

ANS

3.36

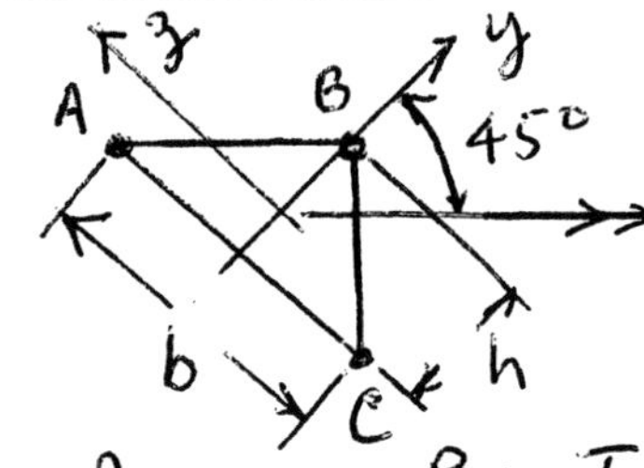

$M = 1000(1.5) = 1500\ N\cdot m$

$b = 90\sqrt{2}\ mm \qquad h = 45\sqrt{2}\ mm$

From Appendix B: $I_z = \frac{1}{36}(90\sqrt{2})(45\sqrt{2})^3 = 911.3(10^3)\ mm^4$

$$I_y = 2\left[\frac{1}{36}45\sqrt{2}(45\sqrt{2})^3 + \frac{1}{2}(45\sqrt{2})^2(15\sqrt{2})^2\right] = 2.734(10^6)\ mm^4$$

$M_y = 1500\cos 45 = 1060.7\ N\text{-}m$, $M_z = -1060.7\ N\text{-}m$

Eq (3.5-1)

Point A:

$$(\sigma_x)_A = \frac{(1060.7)(45\sqrt{2})(10^{-3})}{2.734(10^{-6})} - \frac{(-1060.7)(-\frac{45}{3}\sqrt{2})(10^{-3})}{0.9113(10^{-6})} = 903\ Pa$$

Point B:

$$(\sigma_x)_B = -\frac{(-1060.7)(\frac{2}{3}45\sqrt{2})(10^{-3})}{0.9113(10^{-6})} = 49.4(10^6)\ N/m^2 = 49.4\ MPa$$

Point C:

$$(\sigma_x)_C = \frac{(1060.7)(-45\sqrt{2})(10^{-3})}{2.734(10^{-6})} - \frac{(-1060.7)(-\frac{45}{3}\sqrt{2})(10^{-3})}{0.9113(10^{-6})} = -49.4(10^6)\ N/m^2$$

$= -49.4\ MPa$

Thus, max tension occurs at the wall at point B

$\sigma_{max} = 49.4\ MPa$ ANS

Max compression occurs at the wall at point C

$\sigma_{min} = -49.4\ MPa$ ANS

3.37 cylinder:

$$\sigma_\theta = \frac{pr}{t} = \frac{7(250+6.5)}{12} = 149.6\ MPa$$

$$\sigma_z = \frac{pr}{2t} = 74.8\ MPa$$

Hemi-spherical ends:

$$\sigma_\theta = \sigma_z = \frac{pr}{2t} = 74.8\ MPa$$

(3.38) Consider free expansion of the cylinders
Increase in outer radius of inner cylinder

$$(u_r)_i = \alpha r \Delta T = (23.9)(10^{-6})(240)(100) = 0.5736 \text{ mm}$$

Increase in inner radius of outer cylinder

$$(u_r)_o = (17.3)(10^{-6})(240)(100) = 0.4152 \text{ mm}$$

Radial interference $\delta = 0.5736 - 0.4152 = 0.1584 \text{ mm}$

Pressure develops at interface

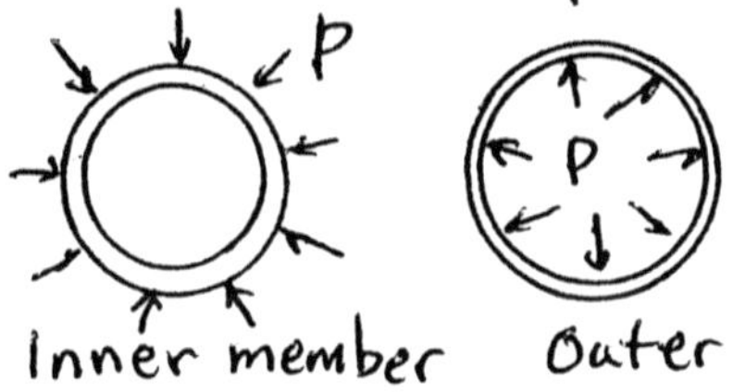

Inner member $\bar{r} = 235$ mm Outer member $\bar{r} = 245$ mm

$$\epsilon_\theta = \frac{u_r}{r} = \frac{\sigma_\theta}{E} = \frac{pr}{2tE}$$

$$\therefore u_r = \frac{pr^2}{2tE} \quad (1)$$

$$(u_r)_i = -\frac{p(235)^2}{2(10)(72)(10^9)} \qquad (u_r)_o = \frac{p(245)^2}{2(10)(190)(10^9)}$$

$$\delta = (u_r)_o - (u_r)_i$$

$$0.1584 = \frac{p}{20(10^9)}\left[\frac{(245)^2}{190} + \frac{(235)^2}{72}\right] \Rightarrow p = 2.925(10^6) \text{ Pa}$$

Inner cylinder:

$$\sigma_\theta = -\frac{pr}{t} = -\frac{2.925(10^6)(235)}{10} = -68.8(10^6) \text{ Pa} = \underline{-68.8 \text{ MPa}}$$

Outer cylinder:

$$\sigma_\theta = \frac{pr}{t} = \frac{2.925(10^6)(245)}{10} = 71.7(10^6) \text{ Pa} = \underline{71.7 \text{ MPa}}$$

(3.39) From Prob. 3.31 $I = 136 \text{ in}^4$

For uniform load use beam C.10 of Appendix C

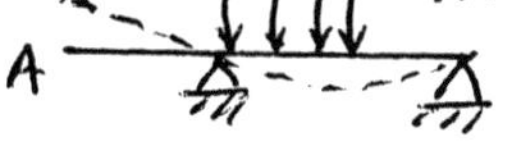

$$\theta_A = \frac{600\{-10^2[2(15)-10]^2\}12^2}{24(30)(10^6)(136)(15)} = -0.00235 \text{ rad}$$

Beam C.8 reversed

$$\theta_A = \frac{-3000[3(20)^2 - 6(20)(20) + 15(45+20)]12^2}{6(30)(10^6)(136)} = 0.00397 \text{ rad}$$

$$(v_c)_A = \frac{-3000(5)^2(20)(12)^3}{3(30)(10^6)(136)} = -0.2118 \text{ in}$$

Superposition: $\theta_A = -0.00235 + 0.00397 = \underline{0.00162 \text{ rad} = 0.093^\circ}$

$(v_c)_A = 0.00235(5)(12) - 0.2118 = \underline{-0.0706 \text{ in}}$

3.40 From the solution of Prob. 1.7

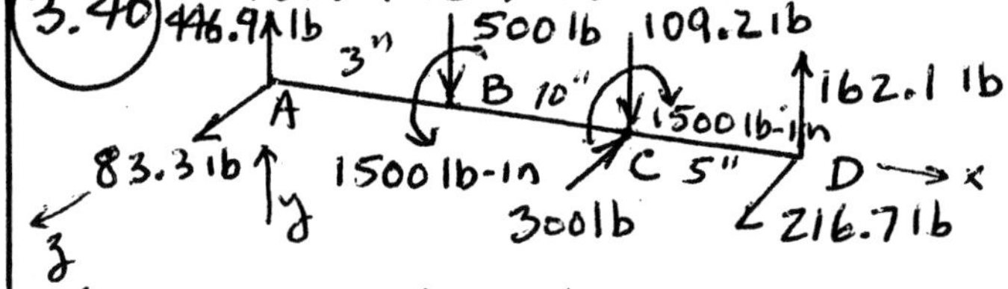

At B: $M_z = 3(446.9) = 1341$ lb-in, $M_y = -83.3(3) = -250$ lb-in

$M_B = \sqrt{1341^2 + (-250)^2} = 1364$ lb-in

At C: $M_z = 162.1(5) = 810$ lb-in, $M_y = -216.7(5) = -1084$ lb-in

$M_C = \sqrt{810^2 + (-1084)^2} = 1353$ lb-in M_B is larger

At $x = 3^+$ in

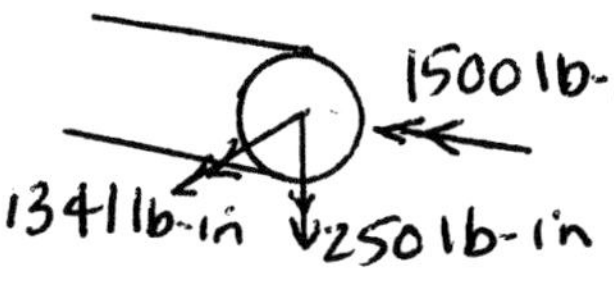

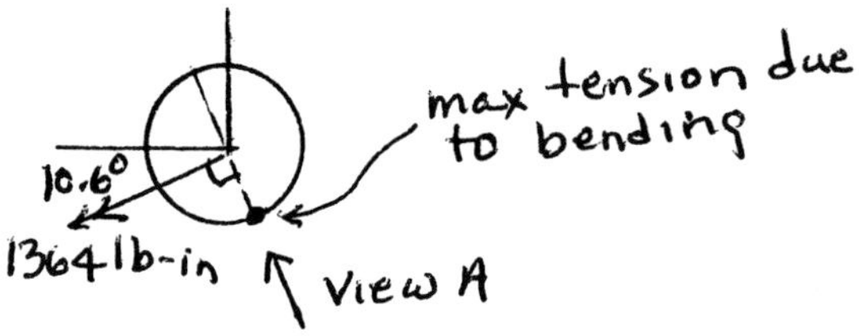

$\tan^{-1}\frac{250}{1341} = 10.6°$

View A

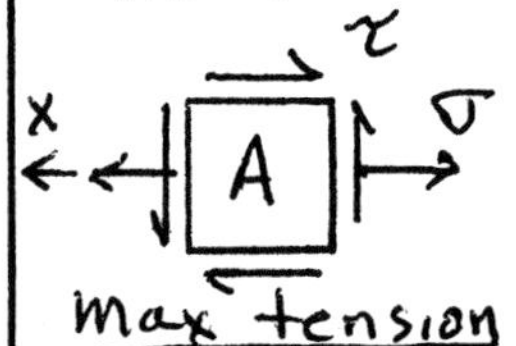

$\sigma = \frac{32(1364)}{\pi(1.5)^3} = 4117$ psi $= 4.117$ kpsi

$\tau = \frac{16(1500)}{\pi(1.5^3)} = 2264$ psi $= 2.264$ kpsi

Max tension

$\sigma_{p1,p2} = \frac{4.117}{2} \pm \sqrt{\left(\frac{4.117}{2}\right)^2 + (2.264)^2} = 5.12, -1.00$ kpsi

$\theta_{p1} = \frac{1}{2}\tan^{-1}\left(\frac{(2)(2.264)}{4.117}\right) = 23.86°$ (1st quad)   **ANS**

Relative to view A

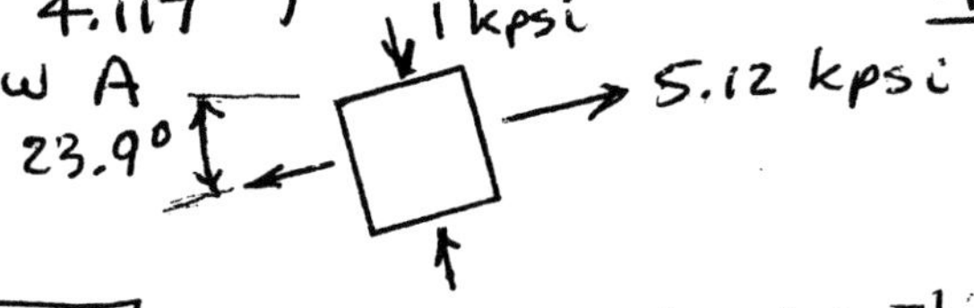

Max shear

$\tau_{max} = \sqrt{\left(\frac{4.117}{2}\right)^2 + (2.264)^2} = 3.06$ kpsi $\theta_s = \frac{1}{2}\tan^{-1}\frac{-4.117}{(2)(2.264)} = -21.14°$

corresponds to τ_{max} ↺
See Prob 3.51

$\sigma_{ave} = \frac{4.117}{2} = 2.06$ kpsi

Relative to view A

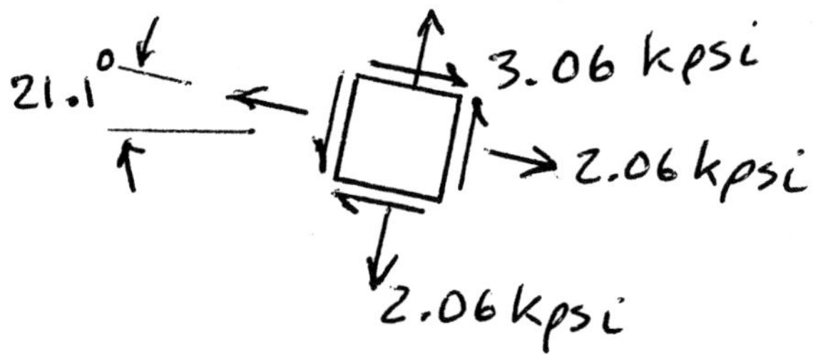

3.41 (a) $\delta_C = \dfrac{1}{200(10^9)}\left[\dfrac{(F+10{,}000)(0.3)}{200(10^{-3})^2} + \dfrac{(10{,}000)(0.5)}{400(10^{-3})^2}\right] = 0.25(10^{-3})\ m$

Solve for F, $\quad F = 15{,}000\ N = \underline{15\ kN}$

(b) Statically indeterminant. First ignore right wall reaction, F_C

$(\delta_C)_1 = \dfrac{1}{200(10^9)}\left[\dfrac{45000(0.3)}{200(10^{-3})^2} + \dfrac{10000(0.5)}{400(10^{-3})^2}\right] = 0.4(10^{-3})\ m$

Next, apply $\leftarrow F_C$ on right end and determine deflection of C due to only F_C.

$(\delta_C)_2 = -\dfrac{F_C}{200(10^9)}\left[\dfrac{0.3}{200(10^{-3})^2} + \dfrac{0.5}{400(10^{-3})^2}\right] = -1.375(10^{-8})F_C$

Now, $(\delta_C)_1 + (\delta_C)_2 = 0.25(10^{-3})\ m$

$0.4(10^{-3}) - 1.375(10^{-8})F_C = 0.25(10^{-3}) \Rightarrow F_C = 10.91(10^3)\ N$

ANS

$F_C = \underline{10.91\ kN} \leftarrow$

$\Sigma F_x = 0 \Rightarrow F_A = 45 - 10.91 = \underline{34.09\ kN} \leftarrow$

(c) For part (a)

$\sigma_{AB} = \dfrac{25(10^3)}{200} = \underline{125\ MPa}$, $\sigma_{BC} = \dfrac{10(10^3)}{400} = \underline{25\ MPa}$ ANS

$\delta_B = \dfrac{25(10^3)(300)}{200(10^9)(200)(10^{-3})^2} = \underline{0.1875\ mm} \rightarrow$

For part (b)

$\sigma_{AB} = \dfrac{34.09(10^3)}{200} = \underline{170\ MPa}$, $\sigma_{BC} = \dfrac{(10-10.91)(10^3)}{400} = \underline{-2.27\ MPa}$

$\delta_B = \dfrac{34.09(10^3)(300)}{200(10^9)(200)(10^{-3})^2} = \underline{0.2557\ mm} \rightarrow$

3.42 Based on the result of Prob. 3.41, C will hit the right wall

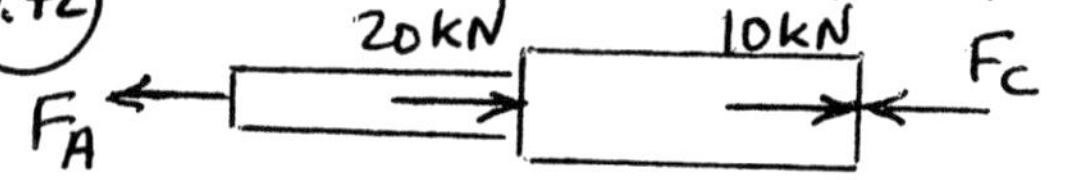

$F_A + F_C = 30\ kN$ ① Step 1

Step 2 $n = 1$ Step 3 Eliminate F_C

$(\delta_C)_1 = \dfrac{30(10^3)(0.3)}{200(10^{-3})^2(200)(10^9)} + \dfrac{10(10^3)(0.5)}{400(10^{-3})^2(200)(10^9)} + 11.7(10^{-6})(0.8)(120)$

$= 1.4107(10^{-3})\ m$

(continued)

Step 4 Determine δ_C due to F_C alone

$$(\delta_C)_2 = \frac{-F_C}{200(10^9)}\left[\frac{0.3}{200(10^{-3})^2}+\frac{0.5}{400(10^{-3})^2}\right] = -1.375(10^{-8})F_C$$

Steps 5 & 6 $1.4107(10^{-3}) - 1.375(10^{-8})F_C = 0.25(10^{-3})$

$F_C = 84.41(10^3)\text{N} = \underline{84.41\text{ kN}}$

Step 7 From ① $F_A = 30 - 84.41 = \underline{-54.41\text{ kN}}$ ∴ compression

$$\delta_B = \alpha L \Delta T + \frac{F_A L_{AB}}{A_{AB}E} = 11.6(10^{-7})(300)(120) + \frac{(-54.41)(10^3)(300)}{200(10^{-3})^2(200)(10^9)} = \underline{1.31(10^{-2})\text{mm}} \rightarrow$$

$$\sigma_{AB} = \frac{-54.41(10^3)}{200} = \underline{-272\text{ MPa}},\quad \sigma_{BC} = \frac{(10-84.41)(10^3)}{400} = \underline{-186\text{ MPa}}$$

(3.43)

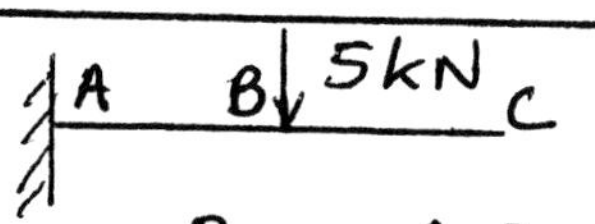

Beam C.2 ⊕

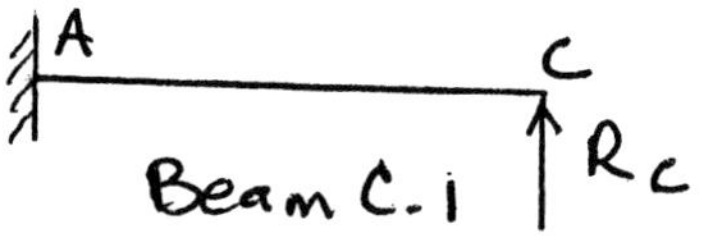

Beam C.1

Can use C.11 directly

$$(v_C)_C = \frac{5(1)^2}{6EI}[1-3(2)] + \frac{R_C(2)^3}{3EI} = 0$$

$R_C = -\left(\frac{3}{8}\right)\left(\frac{5}{6}\right)(-5) = 1.5625\text{ kN} = \underline{1.563\text{ kN}}$

$R_A = 5 - 1.563 = 3.437\text{ kN}\uparrow \quad M_A = 5(1) - 1.5625(2) = 1.875\text{ kNm}$ ↶

$$(v_C)_B = \frac{1}{200(10^3)}\left[\frac{5000(1)^2}{6}(1-3) - \frac{1563(1)^2}{6}(1-6)\right] = -1.823(10^{-3})\text{m} = \underline{-1.823\text{ mm}}$$

$$\sigma_A = \frac{1875}{35(10^{-6})} = 53.6(10^6)\text{N/m}^2 = \underline{53.6\text{ MPa}}$$

$M_B = (1)(1563) = 1563\text{ N-m}, \quad \sigma_B = \frac{1563}{35} = \underline{44.6\text{ MPa}}$

(b)

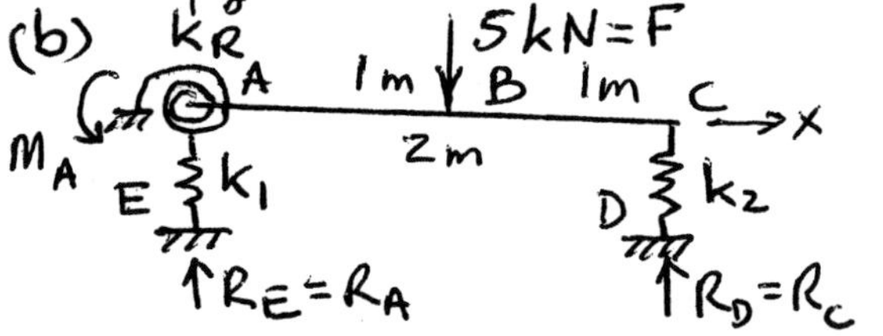

First, ignore R_D

Beam C.2

$$(v_D)_1 = -2\frac{F(1)}{k_R} - \frac{F}{k_1} + \frac{F(1)^2}{6EI}[1-(3)(2)] = -\left(\frac{2}{k_R}+\frac{1}{k_1}+\frac{5}{6EI}\right)F$$

(continued)

Determine v_D due to R_D alone ← Beam C.1

$$(v_D)_2 = 2\frac{(2)R_D}{k_R} + \frac{R_D}{k_1} + \frac{R_D}{k_2} + \frac{R_D(2)^3}{3EI} = \left(\frac{4}{k_R} + \frac{1}{k_1} + \frac{1}{k_2} + \frac{8}{3EI}\right)R_D$$

Now, $(v_D)_1 + (v_D)_2 = 0$. Solving for R_D yields

$$R_D = \frac{\frac{2}{k_R} + \frac{1}{k_1} + \frac{5}{6EI}}{\frac{4}{k_R} + \frac{1}{k_1} + \frac{1}{k_2} + \frac{8}{3EI}} F = \frac{\frac{2}{8(10^6)} + \frac{1}{10(10^6)} + \frac{5}{6(200)(10^3)}}{\frac{4}{8(10^6)} + \frac{1}{10(10^6)} + \frac{1}{5(10^6)} + \frac{8}{3(200)(10^3)}}(5) = \underline{1.598\,kN}$$

$R_A = 5 - 1.598 = 3.402\,kN$, $M_A = 5(1) - 1.598(2) = 1.804\,kN\text{-}m$

$M_C = 1.598(1) = 1.598\,kN\text{-}m$

Deflection of point B can de done similar to point D

$$(v_B)_1 = -\frac{(5000)(1)(1)}{8(10^6)} - \frac{5000}{10(10^6)} - \frac{5000(1)^3}{3(200)(10^3)} = -9.458(10^{-3})\,m$$

$$(v_B)_2 = 1598\left[\frac{(2)(1)}{8(10^6)} + \frac{1}{10(10^6)} - \frac{(1)^2(1-6)}{6(200)(10^3)}\right] = 7.217(10^{-3})\,m$$

$$v_B = (v_B)_1 + (v_{B2}) = -2.241(10^{-3})\,m = \underline{-2.241\,mm}$$

$$\sigma_A = \frac{1804}{35} = \underline{51.5\,MPa}, \quad \sigma_B = \frac{1598}{35} = \underline{45.7\,MPa}$$

(3.44) Step 1 Isolate shaft CD

T_E = torque at E

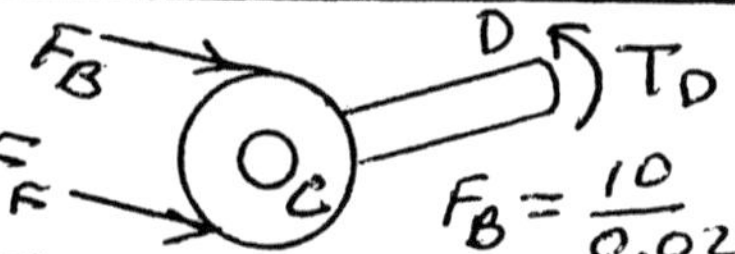

$F_B = \frac{10}{0.025} = 400\,N$

$F_F = \frac{T_E}{0.05} = 20\,T_E$

$\Sigma M_{CD} = 0 \Rightarrow (-400)(0.1) + 20T_E(0.1) + T_D = 0$

$2T_E + T_D = 40$ ①

Step 2 $n = 1$ (1 eq., 2 unknowns)

Step 3 Eliminate T_E and determine θ_E

From ① $(T_D)_1 = 40\,N\text{-}m$

$$(\theta_C)_1 = \frac{2(1+\nu)(T_D)_1 L_{CD}}{EJ} = \frac{1+\nu}{EJ}(2)(40)(0.25) = \frac{20(1+\nu)}{EJ} \quad ②$$

Subsequently, gear F rotates

$$(\theta_F)_1 = -\frac{r_C}{r_F}(\theta_C)_1 = -\frac{100}{50}\left[\frac{20(1+\nu)}{EJ}\right] = -\frac{40(1+\nu)}{EJ} = (\theta_E)_1 \quad ③$$

Step 4 Apply T_E alone. From ①

$T_D = -2T_E$ (continued)

$$(\theta_C)_2 = \frac{(1+\nu)}{EJ}(2)(-2T_E)(0.250) = -\frac{(1+\nu)}{EJ}T_E \quad ④$$

As before, $(\theta_F)_2 = -2(\theta_C)_2 = \frac{2(1+\nu)}{EJ}T_E \quad ⑤$

Now, shaft EF twists,

$$(\theta_{E/F})_2 = \frac{2(1+\nu)T_E(0.5)}{EJ} = \frac{(1+\nu)}{EJ}T_E \quad ⑥$$

Step 5 $\theta_E = ③ + ⑤ + ⑥ = 0$

$$-\frac{40(1+\nu)}{EJ} + \frac{2(1+\nu)}{EJ}T_E + \frac{(1+\nu)}{EJ}T_E = 0 \quad \text{Solve for } T_E$$

$$T_E = \frac{40}{3} = \underline{13.33\ \text{N-m}}$$

Step 6 From ① $T_D = 40 - 2(13.33) = \underline{13.33\ \text{N-m}}$

Stresses

$$\tau_{AB} = \frac{16(10)}{\pi(0.015)^3} = 15.1(10^6)\text{Pa} = \underline{15.1\ \text{MPa}}$$

$$\tau_{CD} = \frac{16(13.33)}{\pi(0.015)^3} = 20.1(10^6)\text{Pa} = \underline{20.1\ \text{MPa}} = \tau_{EF}$$

Gear rotations

$$\theta_F = \frac{2(1+\nu)T_E L_{EF}}{EJ} = \frac{2(1.29)(13.33)(0.5)}{200(10^9)\frac{\pi}{32}(0.015)^4} = \underline{0.0173\text{rad} = 0.99^\circ}$$

same direction as T_A

$$\theta_C = -\frac{50}{100}\theta_F = \underline{-0.496^\circ}, \quad \theta_B = -\frac{100}{25}\theta_C = \underline{1.98^\circ}$$

3.45

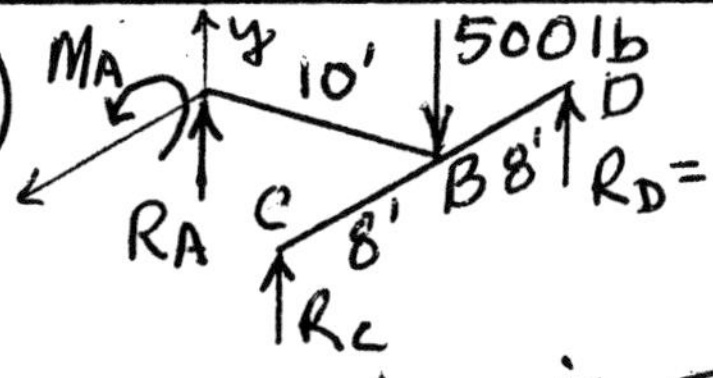

Statics: $R_A + 2R_C = 500 \quad ①$

$M_A + 20R_C = 5000$ ft-lb ②

The geometry is simple here where the six steps outlined in text is unneccessary and more complicated. Separate beams. Solve for v_B for both beams and equate.

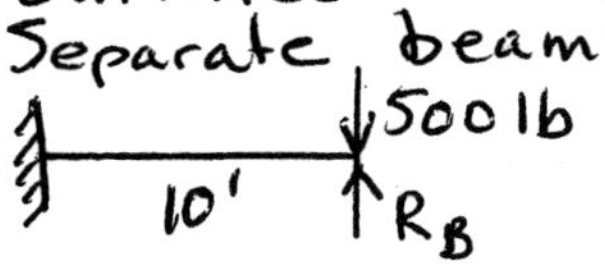

$$(v_B)_1 = -\frac{(500 - R_B)(10^3)}{3EI} \quad ③$$

use consistent units

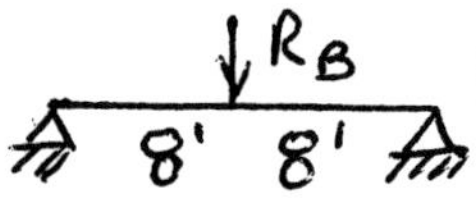

$$(v_B)_2 = -\frac{R_B(16)^3}{48EI} \quad ④$$

(continued)

$③ = ④ \quad -\frac{1000}{3}(500 - R_B) = -\frac{256}{3} R_B$

$$R_B = \frac{1000}{1256}(500) = 398.1 \text{ lb}$$

$R_C = R_D = \frac{R_B}{2} = \underline{199.04 \text{ lb}}$, $R_A = 500 - 2R_C = \underline{101.9 \text{ lb}}$

$M_A = 5000 - 20 R_C = \underline{1019 \text{ lb-ft}}$

(3.46) Step 1

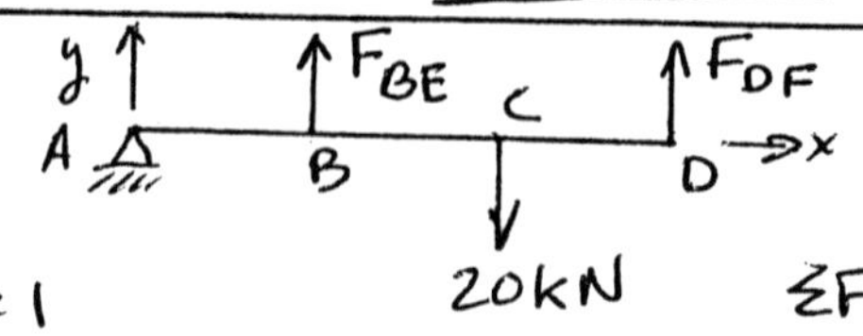

ΣM_A yields

$F_{BE} + 3F_{DF} = 40$ kN ①

Step 2 $n = 1$ (one eq. two unknowns) ΣF_y would only introduce a

Step 3 Release F_{BE} (at point E) and solve for δ_E

With $F_{BE} = 0$, ① gives $F_{DF} = 13.33$ kN

$$(\delta_D)_1 = -\frac{13.33(10^3)(1)}{\pi/4\,(0.012)^2(209)(10^9)} = -5.641(10^{-4})\text{ m}$$

If ABCD is rigid $(\delta_B)_1 = \frac{(\delta_D)_1}{3} = -1.8803(10^{-4})\text{ m} = (\delta_E)_1$

Step 4 Apply F_{BE} to point E alone. From ①

$F_{DF} = -\frac{1}{3} F_{BE}$ (don't worry about compression of a cable as total force will be tension)

$$(\delta_D)_2 = \frac{F_{BE}/3\,(1)}{\frac{\pi}{4}(0.012)^2(209)(10^9)} = 1.4102(10^{-8})\,F_{BE}$$

$$(\delta_B)_2 = \frac{1}{3}(\delta_D)_2 = 4.701(10^{-9})\,F_{BE}$$

$$(\delta_E)_2 = 4.701(10^{-9})F_{BE} + \frac{F_{BE}(1)}{\frac{\pi}{4}(0.012)^2(209)(10^9)} = 4.701(10^{-8})F_{BE}$$

Steps 5 & 6

$$(\delta_E)_1 + (\delta_E)_2 = 0 \Rightarrow -1.8803(10^{-4}) + 4.701(10^{-8})F_{BE} = 0$$

$F_{BE} = 4000\text{ N} = \underline{4 \text{ kN}}$ From ①

$F_{DF} = \frac{1}{3}(40 - 4) = \underline{12 \text{ kN}}$

$\sigma_{BE} = \frac{4000}{\pi/4\,(12)^2} = \underline{35.4 \text{ MPa}}$, $\sigma_{DF} = \frac{12000}{\pi/4\,(12)^2} = \underline{106.1 \text{ MPa}}$

$\delta_B = -\frac{4000(1)(10^3)}{\pi/4\,(0.012)^2(209)(10^9)} = \underline{-0.169 \text{ mm}}$, $\delta_C = 2\delta_B = \underline{-0.338 \text{ mm}}$, $\delta_D = 3\delta_B = \underline{-0.508 \text{ mm}}$

(continued)

(b) From Step 3 of part (a) add elastic deflection to $(\delta_E)_1$

Using beam C.5 from Appendix C

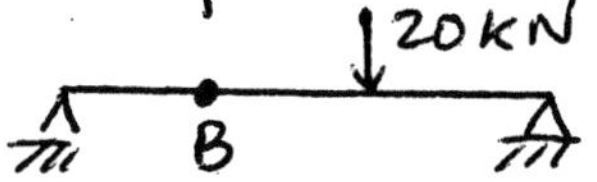

$$(v_B)_1 = \frac{20(10^3)(0.5)(0.5)}{6(209)(10^9)(8)(10^{-7})(1.5)}\left[(0.5)^2 + (0.5)^2 - (1.5)^2\right] = -5.815(10^{-3})\text{ m}$$

$$(\delta_E)_1 = -1.8803(10^{-4}) - 5.815(10^{-3}) = -6.003(10^{-3})\text{ m}$$

Step 4 Add elastic deflection of beam due to F_{BE} at point B

Again, using beam C.5,

$$(v_B)_2 = \frac{-F_{BE}(1)(0.5)}{6(209)(10^9)(8)(10^{-7})(1.5)}\left[(0.5)^2 + (1)^2 - (1.5)^2\right] = 3.323(10^{-7})F_{BE}$$

$$(\delta_E)_2 = 4.701(10^{-8})F_{BE} + 3.323(10^{-7})F_{BE} = 3.793(10^{-7})F_{BE}$$

Steps 5 & 6

$$-6.003(10^{-3}) + 3.793(10^{-7})F_{BE} = 0 \Rightarrow F_{BE} = 15{,}826\text{ N} = \underline{15.83\text{ kN}}$$

Step 7 $F_{DF} = \frac{1}{3}(40 - 15.83) = \underline{8.06\text{ kN}}$

$$\sigma_{BE} = \frac{15.83(10^3)}{\pi/4\,(12)^2} = \underline{140\text{ MPa}}, \quad \sigma_{DF} = \frac{8.06}{\pi/4\,(12)^2} = \underline{71.2\text{ MPa}}$$

$$\delta_B = \frac{-15.83(10^3)(1000)}{\pi/4\,(0.012)^2(209)(10^9)} = \underline{-0.670\text{ mm}}$$

$$\delta_D = \frac{-8.06(10^3)(1000)}{\pi/4\,(0.012)^2(209)(10^9)} = \underline{-0.341\text{ mm}}$$

For point C:

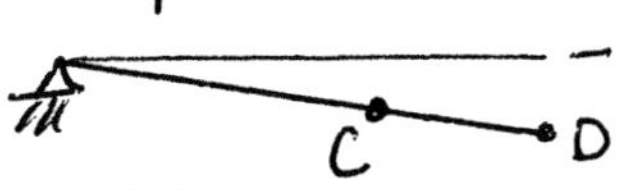

⊕

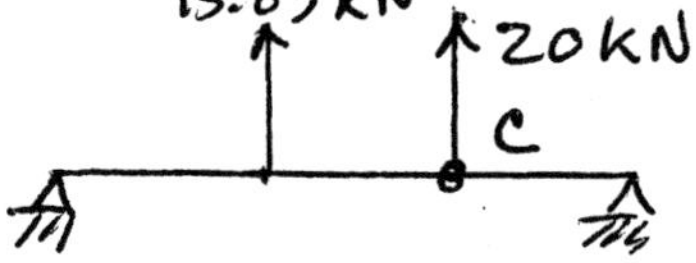

$$(\delta_C)_1 = \frac{2}{3}\delta_D = \frac{2}{3}(-0.341) = -0.227\text{ mm}$$

For beam:

$$(\delta_C)_2 = \frac{-15.83(10^3)(0.5)(1.5-1)}{6(209)(10^9)(8)(10^{-7})(1.5)}\left[(1)^2 + (0.5)^2 - 2(1.5)(1)\right] + \frac{20(10^3)(0.5)(1)}{6(209)(10^9)(8)(10^{-7})(1.5)}\left[(1)^2 + (0.5)^2 - (1.5)^2\right] = -2.043(10^{-3})\text{ m}$$

$$\delta_C = -0.227 - 2.043 = \underline{-2.27\text{ mm}}$$

3.47

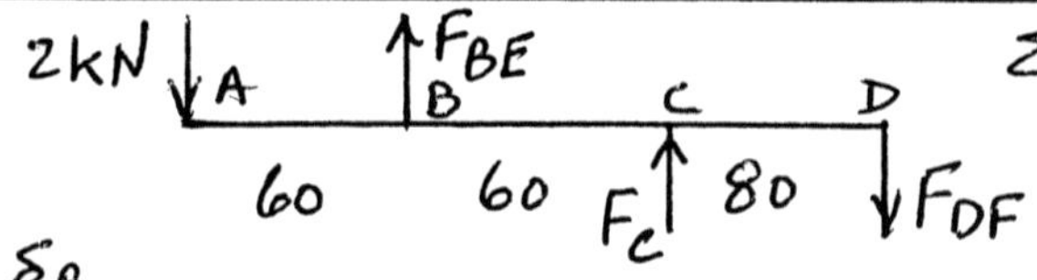

ΣM_C yields

$3F_{BE} + 4F_{DF} = 12$ ①

(a)

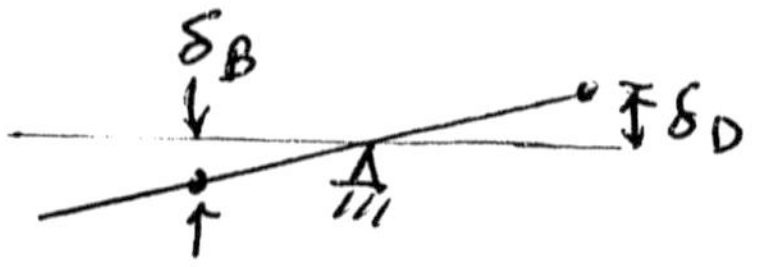

$$\frac{\delta_B}{60} = \frac{\delta_D}{80} \Rightarrow \delta_B = \frac{3}{4}\delta_D \quad ②$$

$$\frac{F_{BE}(50)}{AE} = \frac{3}{4}\,\frac{F_{DF}(60)}{AE} \Rightarrow F_{BE} = \frac{9}{10}F_{DF} \quad ③ \quad \text{Sub into ①}$$

$$3\left(\frac{9}{10}F_{DF}\right) + 4F_{DF} = 12 \Rightarrow \underline{F_{DF} = 1.791\text{ kN}},\ F_{BE} = \frac{9}{10}F_{DF} = \underline{1.612\text{ kN}}$$

$$\sigma_{DF} = \frac{1791}{\frac{\pi}{4}(8)^2} = \underline{35.6\text{ MPa}},\quad \sigma_{BE} = \frac{1612}{\frac{\pi}{4}(8)^2} = \underline{32.1\text{ MPa}}$$

$$\delta_B = \frac{(1612)(50)}{\frac{\pi}{4}(0.008)^2(209)(10^9)} = \underline{7.67(10^{-3})\text{ mm}}\downarrow$$

$$\delta_A = 2\delta_B = \underline{15.34(10^{-3})\text{ mm}}\downarrow \quad \delta_D = \frac{4}{3}\delta_B = \underline{10.23(10^{-3})\text{ mm}}\uparrow$$

(b) Rigid rotation of part (a) is still valid. Considering beam deflection without rotation

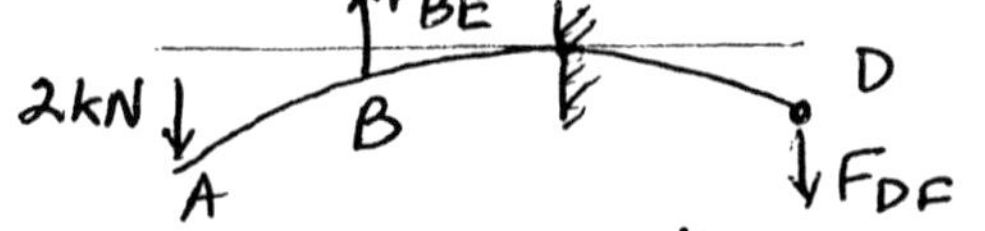

For B superpose beams C.1 & C.2

$$v_B = \frac{2000(0.060)^2\left[0.060 - 3(0.120)\right]}{6(209)(10^9)(200)(10^{-3})^4} + \frac{F_{BE}(0.060)^3}{3(209)(10^9)(200)(10^{-3})^4}$$

$$= 1.7225(10^{-6})F_{BE} - 8.612(10^{-3}) \quad ④$$

$$v_D = \frac{-F_{DF}(0.080)^3}{3(209)(10^9)(200)(10^{-3})^4} = -4.083(10^{-6})F_{DF}$$

The rigid deflection of point B <u>downward</u> is

$$(\delta_B)_R = \delta_{BE} + v_B = \frac{F_{BE}(0.050)}{\frac{\pi}{4}(0.008)^2(209)(10^9)} + 1.7225(10^{-6})F_{BE} - 8.612(10^{-3})$$

$$= 1.7272(10^{-6})F_{BE} - 8.612(10^{-3}) \quad ⑤$$

(Continued)

The rigid deflection of point D upward is

$$(\delta_D)_R = \delta_{DF} - v_D = \frac{F_{DF}(0.060)}{\pi/4(0.008)^2(209)(10^9)} + 4.083(10^{-6})F_{DF} = 4.089(10^{-6})F_{DF} \quad ⑥$$

Substitute ⑤ & ⑥ into ②

$$1.7272(10^{-6})F_{BE} - 8.612(10^{-3}) = \frac{3}{4}(4.089)(10^{-6})F_{DF}$$

or, $F_{BE} = 1.776F_{DF} + 4986$ N or $F_{BE} = 1.776F_{DF} + 4.986$ kN ⑦

Sub into ① $3(1.776F_{DF} + 4.986) + 4F_{DF} = 12$ Solve for F_{DF}

$F_{DF} = \underline{-317\text{ N}}$ Assume bolt is a shoulder bolt and can go into compression (if not, the bolt force would be zero).

Sub F_{DF} into ⑦

$F_{BE} = \underline{4.423\text{ kN}}$

$$\sigma_{BE} = \frac{4423}{\pi/4(8)^2} = \underline{88.0\text{ MPa}}, \quad \sigma_{DF} = \frac{-317}{\frac{\pi}{4}(8)^2} = \underline{-6.31\text{ MPa}}$$

$$\delta_B = \frac{4423(50)}{\pi/4(0.008)^2(209)(10^9)} = 21.05(10^{-3})\text{mm} \downarrow$$

$$\delta_D = \frac{317(60)}{\pi/4(0.008)^2(209)(10^9)} = 1.81(10^{-3})\text{ mm} \downarrow$$

For point A add rigid displacement to elastic. From ⑤

$$(\delta_B)_R = 1.7272(10^{-6})(4423) - 8.612(10^{-3}) = -0.9726(10^{-3})\text{m} = -0.9726\text{mm}$$

$$(\delta_A)_R = \frac{120}{60}(-0.9726) = -1.945\text{ mm} \quad (-\text{ means deflection is } \uparrow)$$

Elastic deflection (similar to v_B)

$$v_A = -\frac{2000(0.12)^3}{3(209)(10^9)(200)(10^{-3})^4} - \frac{4423(0.06)^2[0.06-3(0.12)]}{6(209)(10^9)(200)(10^{-3})^4} = -8.513(10^{-3})\text{ m}$$

$$\delta_A = (\delta_A)_R - v_A = -1.945 + 8.513 = \underline{6.57\text{ mm}} \downarrow$$

3.48 (a) $\sigma_{x'} = 40\cos^2(-20) + 8\sin^2(-20) + 2(-12)\sin(-20)\cos(-20) = \underline{43.97 \text{ MPa}}$

$\sigma_{y'} = 40\sin^2(-20) + 8\cos^2(-20) - 2(-12)\sin(-20)\cos(-20) = \underline{4.03 \text{ MPa}}$

$\tau_{x'y'} = -(40-8)\sin(-20)\cos(-20) + (-12)\left[\cos^2(-20) - \sin^2(-20)\right] = \underline{1.09 \text{ MPa}}$

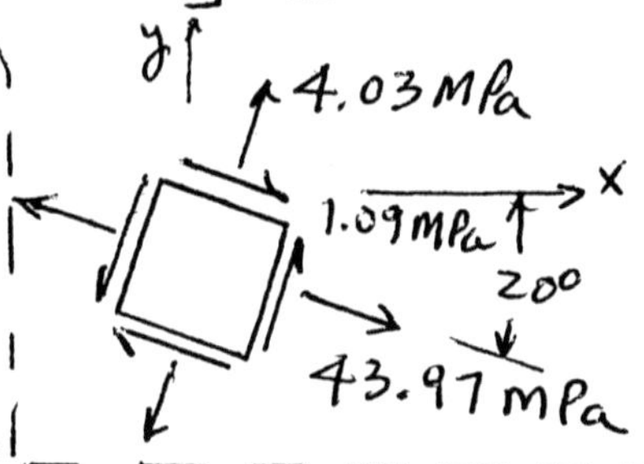

(b) $\theta_p = \frac{1}{2}\tan^{-1}\left[\frac{2(-12)}{40-8}\right] = -18.43°$

$\sigma_{x'} = 40\cos^2(-18.43) + 8\sin^2(-18.43) + 2(-12)\sin(-18.43)\cdot\cos(-18.43)$

$= \underline{44 \text{ MPa}} = \sigma_{p1}$

$\sigma_{y'} = 40\sin^2(-18.43) + 8\cos^2(-18.43) - 2(-12)\sin(-18.43)\cdot\cos(-18.43)$

$= \underline{4 \text{ MPa}} = \sigma_{p2}$

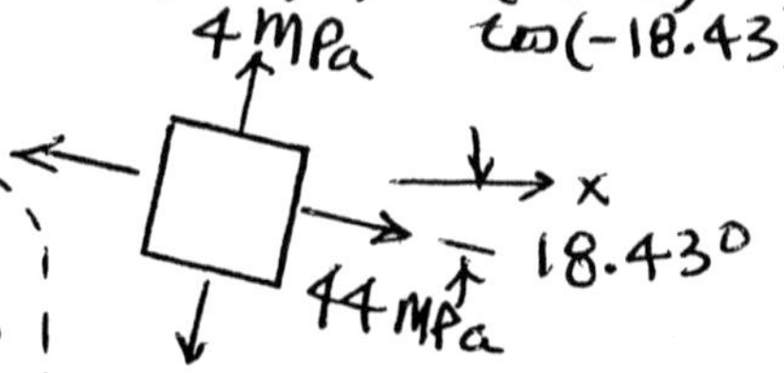

(c)

$\theta_s = \frac{1}{2}\tan^{-1}\left[\frac{8-40}{2(-12)}\right] = 26.57°$

$\sigma_{x'} = 40\cos^2(26.57) + 8\sin^2(26.57) + 2(-12)\sin(26.57)\cos(26.57) = \underline{24 \text{ MPa}}$

$\sigma_{y'} = 40\sin^2(26.57) + 8\cos^2(26.57) - 2(-12)\sin(26.57)\cos(26.57) = \underline{24 \text{ MPa}}$

$\tau_{x'y'} = -(40-8)\sin 26.57\cos 26.57 + (-12)\left[\cos^2 26.57 - \sin^2 26.57\right] = -20 \text{ MPa}$

(d) $\sigma_1 = 44 \text{ MPa}$, $\sigma_2 = 4 \text{ MPa}$, $\sigma_3 = 0$

$\tau_{max} = \frac{\sigma_1 - \sigma_3}{2} = \underline{22 \text{ kpsi}}$

View down axis in part (b) containing the $\sigma_{p2} = 4$ MPa stress

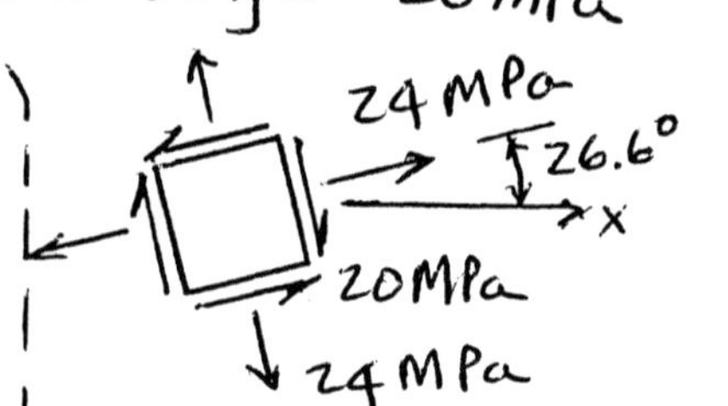

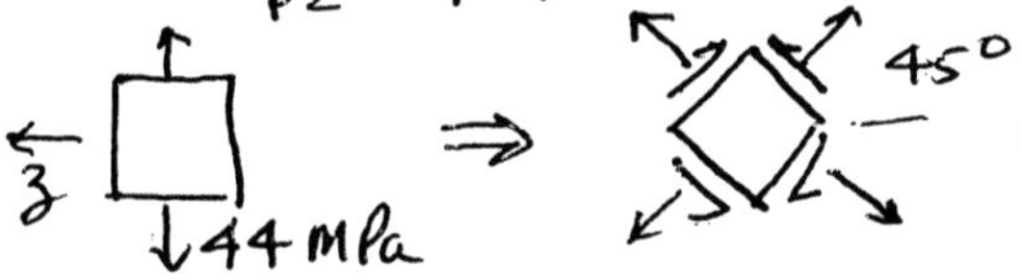

All stresses are 22 kpsi

(e)

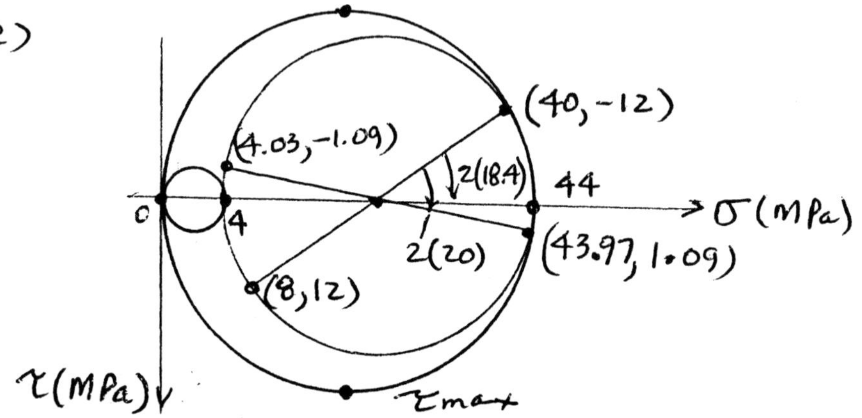

3.49 $\theta_p = \frac{1}{2}\tan^{-1}\left[\frac{2(-12)}{-15+25}\right] = -33.7^\circ$

(a) $\sigma_{x'} = -15\cos^2(-33.7) - 25\sin^2(-33.7) + 2(-12)\sin(-33.7)\cos(-33.7) = \underline{-7\text{ kpsi}}$

$\sigma_{y'} = -15\sin^2(-33.7) - 25\cos^2(-33.7) - 2(-12)\sin(-33.7)\cos(-33.7) = \underline{-33\text{ kpsi}}$

(b)

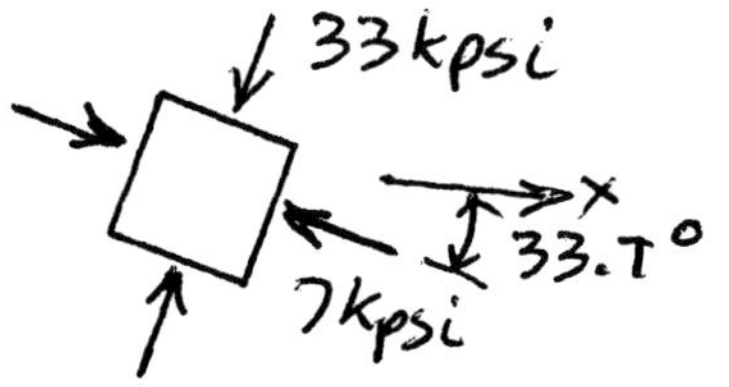

(c) principal stresses (spelling test)

(d) $\tau_{max} = \frac{\sigma_1 - \sigma_3}{2} = \frac{33-0}{2} = \underline{16.5\text{ kpsi}}$

3.50 $\tau_{x'y'} = -(\sigma_x - \sigma_y)\sin\theta\cos\theta + \tau_{xy}(\cos^2\theta - \sin^2\theta)$

$= -\frac{(\sigma_x - \sigma_y)}{2}\sin 2\theta + \tau_{xy}\cos 2\theta$

$\frac{\partial \tau_{x'y'}}{\partial \theta} = -(\sigma_x - \sigma_y)\cos 2\theta_s - 2\tau_{xy}\sin 2\theta_s = 0$

$\tan 2\theta_s = \frac{\sigma_y - \sigma_x}{2\tau_{xy}} \Rightarrow \theta_s = \frac{1}{2}\tan^{-1}\frac{\sigma_y - \sigma_x}{2\tau_{xy}}$ ANS

3.51 From Prob 3.50

$\frac{\partial^2 \tau_{x'y'}}{\partial \theta^2} = 2(\sigma_x - \sigma_y)\sin 2\theta_s - 4\tau_{xy}\cos 2\theta_s$ ①

From the result for θ_s (Prob 3.50)

$\sin 2\theta_s = \frac{\sigma_y - \sigma_x}{\sqrt{(\sigma_y - \sigma_x)^2 + 4\tau_{xy}^2}}$ $\quad \cos 2\theta_s = \frac{2\tau_{xy}}{\sqrt{(\sigma_y - \sigma_x)^2 + 4\tau_{xy}^2}}$

Sub into ①

$\frac{\partial^2 \tau_{x'y'}}{\partial \theta^2} = -2\left[\frac{(\sigma_y - \sigma_x)^2 + 4\tau_{xy}^2}{\sqrt{(\sigma_y - \sigma_x)^2 + 4\tau_{xy}^2}}\right]$ negative, thus when using θ_s eq from Prob 3.51 treating θ_s as single-valued yields τ_{max}.

3.52 At point A

$\sigma_x = \frac{P}{ht}$

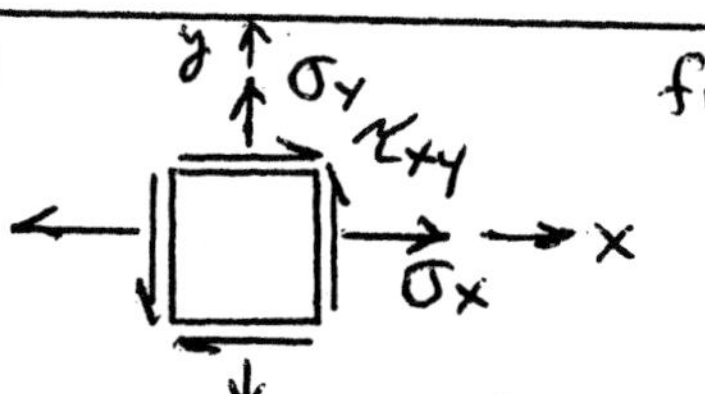

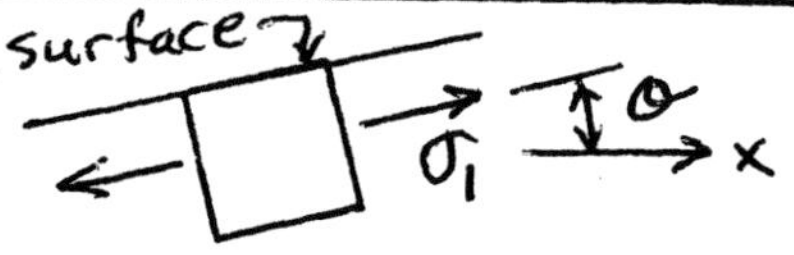

(a) Transform from free surface element to the xy surfaces.

Eq (3.9-1) $\sigma_x = \sigma_1\cos^2(-\theta) \quad = \sigma_1 = \sigma_x/\cos^2\theta$ ANS

(b) Eq (3.9-2) and Eq (3.9-3) $\sigma_y = \sigma_1\sin^2(-\theta) = \frac{\sigma_x}{\cos^2\theta}\sin^2\theta = \sigma_x\tan^2\theta$ ANS

$\tau_{xy} = -\sigma_1\sin(-\theta)\cos(-\theta) = -\frac{\sigma_x}{\cos^2\theta}(-\sin\theta)(\cos\theta) = \sigma_x\tan\theta$ ANS

(3.53) b □h

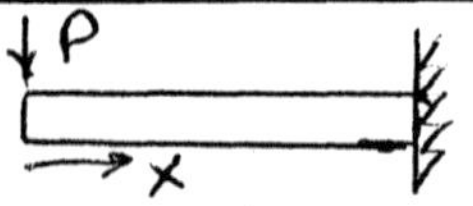

$I = \frac{1}{12}bh^3$, $\sigma = \frac{6M}{bh^2}$, $|M| = Px$

(a) b constant

For bending $\tau_{max} = \frac{\sigma}{2} = \frac{6M}{bh^2} = \frac{6Px}{bh^2} = \tau_{allow}$ (a)

$\therefore h = \sqrt{\frac{3Px}{b\tau_{allow}}}$ (1) However $h \to 0$ as $x \to 0$ and transverse shear stress $\to \infty$

For transverse shear. $\tau_{max} = \frac{3}{2}\frac{V}{A} = \frac{3}{2}\frac{P}{bh}$

$h = \frac{3}{2}\frac{P}{b\tau_{allow}}$ (2) Determine value of x, (x_0), where (1) = (2)

$\frac{3Px_0}{b\tau_{allow}} = \frac{9}{4}\frac{P^2}{b^2\tau_{allow}^2} \Rightarrow x_0 = \frac{3}{4}\frac{P}{b\tau_{allow}}$ (3)

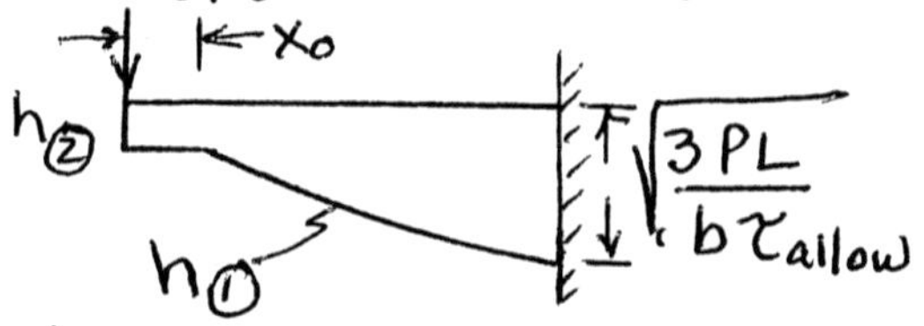

(b) h constant. Bending, from (a) $b = \frac{3Px}{h^2\tau_{allow}}$ (4)

From (2) for trans. shear $b = \frac{3}{2}\frac{P}{h\tau_{allow}}$ (5)

Equate (4) and (5)

$\frac{3Px_0}{h^2\tau_{allow}} = \frac{3}{2}\frac{P}{h\tau_{allow}} \Rightarrow x_0 = h/2$ (6)

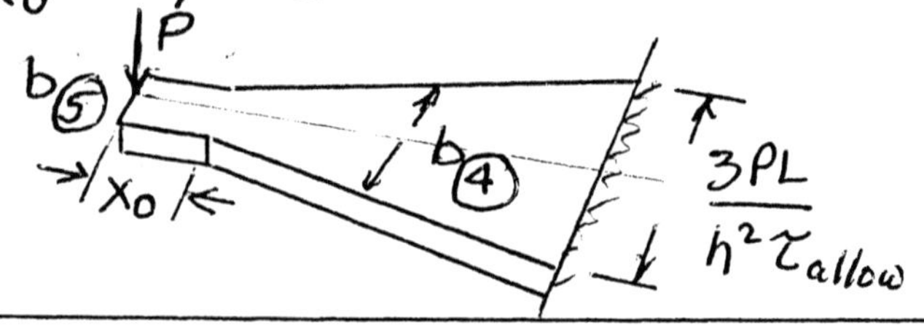

(3.54) Use stress equations replacing σ, τ with $\epsilon, \gamma/2$

$\theta_p = \frac{1}{2}\tan^{-1}\left(\frac{\gamma_{xy}}{\sigma_x - \sigma_y}\right) = \frac{1}{2}\tan^{-1}\left(\frac{-200}{400}\right) = -13.3^\circ$ ↖ to ϵ_{p1}

↑ 4th quad

$\epsilon_{p1,p2} = \frac{300-100}{2} \pm \sqrt{\left(\frac{300+100}{2}\right)^2 + \left(-\frac{200}{2}\right)^2}$

$= 323.6\mu, -123.6\mu$

$\epsilon_{y'} = \epsilon_{p2} = -123.6\mu$ **ANS**

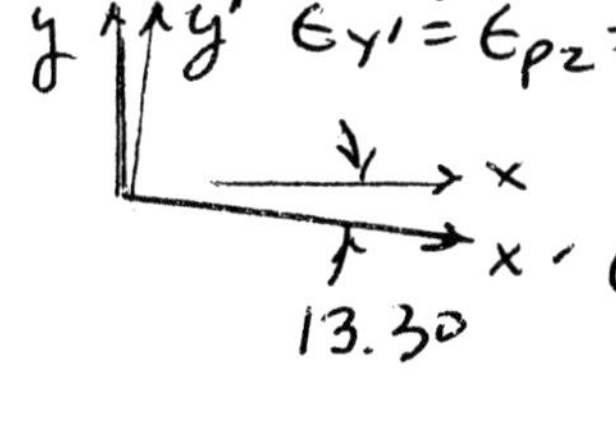

$\epsilon_{x'} = \epsilon_{p1} = 323.6\mu$

3.55

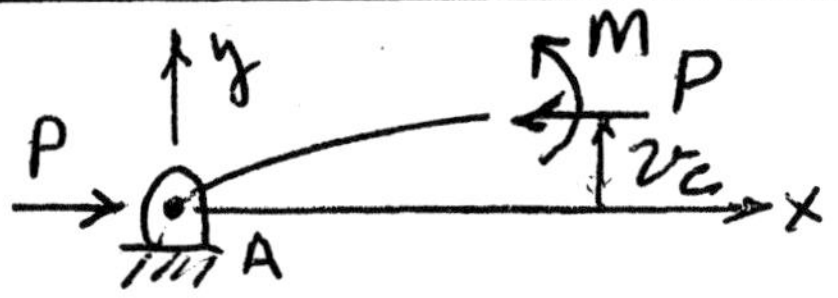

$\Sigma M_A = 0 \quad M + P v_c = 0$

$M = -P v_c = E I_z \dfrac{d^2 v_c}{dx^2}$

$$\frac{d^2 v_c}{dx^2} + \frac{P v_c}{E I_z} = 0 \Rightarrow \frac{d^2 v_c}{dx^2} + k^2 v_c = 0 \quad \text{where } k = \sqrt{\frac{P}{E I_z}}$$

Solution: $v_c = A \cos kx + B \sin kx$

B.C. $v_c = 0$ at $x = 0 \Rightarrow A = 0 \quad \therefore v_c = B \sin kx$ ①

$v_c = 0$ at $x = L \Rightarrow B \sin kL = 0 \Rightarrow \sin kL = 0$

$kL = n\pi \quad n = 1, 2, \ldots \quad P_{cr}$ occurs when $n = 1$

$$k = \frac{\pi}{L} = \sqrt{\frac{P_{cr}}{E I_z}} \Rightarrow P_{cr} = \frac{\pi^2 E I_z}{L^2}$$

Since $P_{cr} = \dfrac{\pi^2 E I_z}{(KL)^2}$ we see that $\underline{K = 1}$ $\underline{\text{ANS}}$

From ① $v_c = B \sin \dfrac{\pi x}{L} \quad (v_c)_{max} = \delta_{max}$ at $x = L/2$

$\therefore B = \delta_{max}$

$v_c = \delta_{max} \sin \dfrac{\pi x}{L}$ $\underline{\text{ANS}}$

3.56

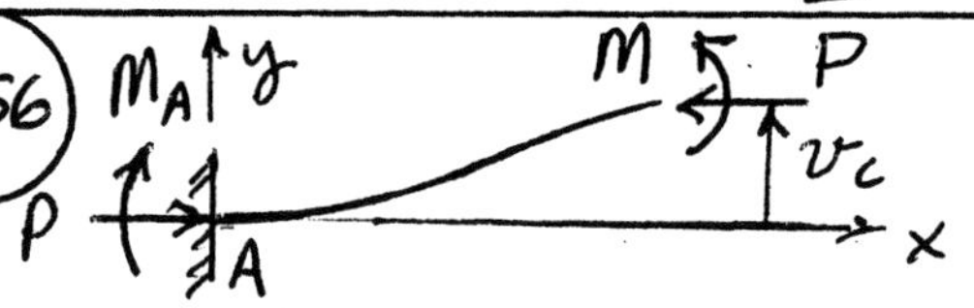

$\Sigma M_A = 0 \quad M - M_A + P v_c = 0$

$M = M_A - P v_c = E I_z \dfrac{d^2 v_c}{dx^2}$

$$\frac{d^2 v_c}{dx^2} + \frac{P}{E I_z} v_c = \frac{M_A}{E I_z} \Rightarrow \frac{d^2 v_c}{dx^2} + k^2 v_c = C \quad \text{where } k = \sqrt{\frac{P}{E I_z}} \text{ and } C = M_A / E I_z$$

Solution: $v_c = \dfrac{C}{k^2} + A \cos kx + B \sin kx$

B.C. $v_c = 0$ at $x = 0 \Rightarrow A = -C/k^2$

$\dfrac{dv_c}{dx} = -Ak \sin kx + Bk \cos kx \qquad \dfrac{dv_c}{dx} = 0$ at $x = 0 \therefore B = 0$

Thus $v_c = \dfrac{C}{k^2}(1 - \cos kx) \quad v_c = 0$ at $x = L$

$\therefore 1 - \cos kL = 0 \Rightarrow \cos kL = 1 \Rightarrow kL = n\pi \quad n = 0, 2, 4 \ldots$ (0 = trivial; 2 = lowest)

$$k_{cr}^2 = \left(\frac{2\pi}{L}\right)^2 = \frac{P_{cr}}{E I_z} \Rightarrow P_{cr} = \frac{\pi^2 E I_z}{(0.5L)^2} \therefore \underline{K = 0.5} \text{ Ans}$$

$v_c = \dfrac{C}{k^2}\left(1 - \cos \dfrac{2\pi x}{L}\right), \; v_c = \delta_{max}$ at $x = \dfrac{L}{2}, \; \therefore \dfrac{C}{k^2} = 0.5\delta_{max}, \; v_c = 0.5\delta_{max}\left(1 - \cos \dfrac{2\pi x}{L}\right)$

(3.57) B' is deflected position

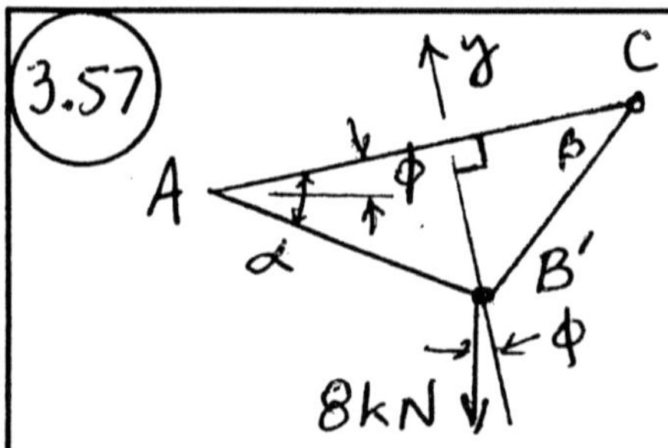

$$L_{AB'} \sin\alpha = L_{B'C} \sin\beta \quad (1)$$

$$L_{AB'} \cos\alpha + L_{B'C} \cos\beta = AC \quad (2)$$

$$AC = \sqrt{(500 + 750\cos 45)^2 + (750 \sin 45)^2} \quad (3)$$

$$\phi = \tan^{-1}\left(\frac{750 \sin 45}{500 + 750 \cos 45}\right) \quad (4)$$

From (1) & (2)

$$\alpha = \cos^{-1}\left(\frac{L_{AB'}^2 - L_{B'C}^2 + AC^2}{2(AC)(L_{AB'})}\right) \quad (5), \qquad \beta = \cos^{-1}\left(\frac{L_{B'C}^2 - L_{AB'}^2 + AC^2}{2(AC)(L_{B'C})}\right) \quad (6)$$

Forces

$$\Sigma F_x = 0 \Rightarrow F_{AB} \cos\alpha - F_{BC} \cos\beta = -8 \sin\phi \quad (7)$$

$$\Sigma F_y = 0 \Rightarrow F_{AB} \sin\alpha + F_{BC} \sin\beta = 8 \cos\phi \quad (8)$$

Solving (7) & (8) simultaneously yields

$$F_{AB} = 8\,\frac{\cos(\beta + \phi)}{\sin(\alpha + \beta)} \quad (9) \qquad F_{BC} = 8\,\frac{\cos(\alpha - \phi)}{\sin(\alpha + \beta)} \quad (10)$$

Stretch of cables

$$\delta_{AB} = \frac{F_{AB}(kN)(10^3)\,L_{AB}(mm)}{E(Pa)\,A(mm^2)(10^{-6})} = \frac{F_{AB}(500)(10^3)}{2(10^9)(125)(10^{-6})} = 2F_{AB} \quad (11)$$

Similarly, $\delta_{BC} = 3F_{BC}$ (12) where forces are in kN deflections in mm

Deflections

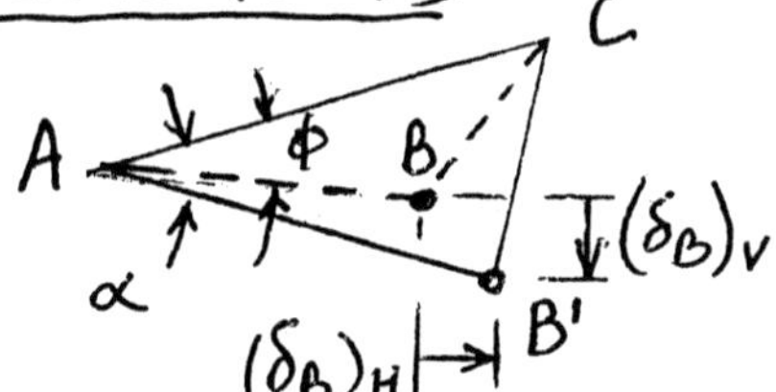

$$(\delta_B)_H = L_{AB'} \cos(\alpha - \phi) - 500 \quad (13)$$

$$(\delta_B)_V = L_{AB'} \sin(\alpha - \phi) \quad (14)$$

Procedure. Start with undeflected static solution for F_{AB} and F_{BC}. Then sub into (11) & (12). $L_{AB'} = L_{AB} + \delta_{AB}$ and $L_{BC'} = L_{BC} + \delta_{BC}$. Sub these into (9) & (10) for new values of F_{AB} and F_{BC}. Repeat until converges. Sub final values of α, $L_{AB'}$ into (13) & (14) for final answer

(continued)

The equations were programed in Excel. Complete convergence occured within 20 iterations. Engineering accuracy was obtained within 5 iterations.

The forces are $F_{AB} = 6.602$ kN, $F_{BC} = 9.856$ kN

giving stresses of $\sigma_{AB} = 52.8$ MPa, $\sigma_{BC} = 78.8$ MPa

The final deflection of pt. B $(\delta_B)_H = 10.69$ mm, $(\delta_B)_V = 50.78$ mm

(b) linear solution (as in Prob. 3.4) is

$F_{AB} = 8$ kN (21% high), $F_{BC} = 11.31$ kN (15% high)

and $(\delta_B)_H = 16$ mm (50% high), $(\delta_B)_V = 63.9$ mm (26% high)

3.58 From eqs. ⑦, ⑧, ①, ②, ⑪, and ⑫ of Prob 3.57 solution

Forces in kN, lengths in mm

$$f_1 = F_{AB}\cos\alpha - F_{BC}\cos\beta + 8\sin\phi = 0 \quad (a)$$

$$f_2 = F_{AB}\sin\alpha + F_{BC}\sin\beta - 8\cos\phi = 0 \quad (b)$$

$$f_3 = (500 + 2F_{AB})\sin\alpha - (750 + 3F_{BC})\sin\beta = 0 \quad (c)$$

$$f_4 = (500 + 2F_{AB})\cos\alpha + (750 + 3F_{BC})\cos\beta - AC = 0 \quad (d)$$

$$\frac{\partial f_1}{\partial \alpha} = -F_{AB}\sin\alpha \qquad \frac{\partial f_1}{\partial \beta} = F_{BC}\sin\beta \qquad \frac{\partial f_1}{\partial F_{AB}} = \cos\alpha$$

$$\frac{\partial f_2}{\partial \alpha} = F_{AB}\cos\alpha \qquad \frac{\partial f_2}{\partial \beta} = F_{BC}\cos\beta \qquad \frac{\partial f_2}{\partial F_{AB}} = \sin\alpha$$

$$\frac{\partial f_3}{\partial \alpha} = (500 + 2F_{AB})\cos\alpha \qquad \frac{\partial f_3}{\partial \beta} = -(750 + 3F_{BC})\cos\beta \qquad \frac{\partial f_3}{\partial F_{AB}} = 2\sin\alpha$$

$$\frac{\partial f_4}{\partial \alpha} = -(500 + 2F_{AB})\sin\alpha \qquad \frac{\partial f_4}{\partial \beta} = -(750 + 3F_{BC})\sin\beta \qquad \frac{\partial f_4}{\partial F_{AB}} = 2\cos\alpha$$

$$\frac{\partial f_1}{\partial F_{BC}} = -\cos\beta$$

$$\frac{\partial f_2}{\partial F_{BC}} = \sin\beta$$

$$\frac{\partial f_3}{\partial F_{BC}} = -3\sin\beta$$

$$\frac{\partial f_4}{\partial F_{BC}} = 3\cos\beta$$

Similar to Example 3.2-5, form $\Delta\alpha$, $\Delta\beta$, ΔF_{AB}, ΔF_{BC} like eqs [e] to [i] of the example. Starting with the linear solution $F_{AB} = 8$ kN, $F_{BC} = 11.313$ kN, $\alpha = \phi = 22.24°$, and $\beta = 17.76°$ the solution converges in 5 iterations to $F_{AB} = 6.601984$ kN, $F_{BC} = 9.855629$ kN

$\alpha = 32.91472°$, $\beta = 20.9605°$

(4.1) plane stress: $\epsilon_x = \frac{1}{E_1}(\sigma_x - \nu_1 \sigma_y)$, plane strain: $\epsilon_x = \frac{1-\nu_2^2}{E_2}\left(\sigma_x - \frac{\nu_2}{1-\nu_2}\sigma_y\right)$

To convert $\frac{1}{E_1} = \frac{1-\nu_2^2}{E_2}$ and $\nu_1 = \frac{\nu_2}{1-\nu_2}$ Solve for E_2, ν_2

$E_2 = E_1(1-\nu_2^2)$ ① $\nu_1(1-\nu_2) = \nu_2$ ②

From ② $\nu_2(1+\nu_1) = \nu_1 \Rightarrow \nu_2 = \frac{\nu_1}{1+\nu_1}$ ③ **ANS**

Sub ③ into ①

$$E_2 = E_1\left[1 - \left(\frac{\nu_1}{1+\nu_1}\right)^2\right] = E_1\left[\frac{1+2\nu_1+\nu_1^2-\nu_1^2}{(1+\nu_1)^2}\right] = E_1\left[\frac{1+2\nu_1}{(1+\nu_1^2)}\right]$$ **ANS**

(4.2) $\sigma_x = \frac{\partial^2 \Phi}{\partial y^2}$, $\sigma_y = \frac{\partial^2 \Phi}{\partial x^2}$, $\tau_{xy} = -\frac{\partial^2 \Phi}{\partial x \partial y}$

Equil. equations with no body forces

$\frac{\partial \sigma_x}{\partial x} + \frac{\partial \tau_{xy}}{\partial y} = 0$ ① and $\frac{\partial \tau_{xy}}{\partial x} + \frac{\partial \sigma_y}{\partial y} = 0$ ② sub in from above

$\frac{\partial^3 \Phi}{\partial x \partial y^2} - \frac{\partial^3 \Phi}{\partial y \partial x \partial y} = 0$ ① $-\frac{\partial^3 \Phi}{\partial x^2 \partial y} + \frac{\partial^3 \Phi}{\partial y \partial x^2} = 0$ ② if Φ continuous is satisfied

(4.3) Sub $\frac{E}{1-\nu^2}$, $\frac{\nu}{1-\nu}$ for E, ν respectively

$$u_r = \left(2 - \frac{\nu}{1-\nu}\right)\frac{pr^2(1-\nu^2)}{2Et} = \left(\frac{2-2\nu-\nu}{1-\nu}\right)\frac{(1-\nu^2)pr^2}{2Et} = (2-3\nu)(1+\nu)\frac{pr^2}{2Et}$$

(4.4) Sub $\frac{E}{1-\nu^2}$, $\frac{\nu}{1-\nu}$ for E, ν respectively

(a) $\epsilon_x = -\frac{(1-\nu^2)M_z y}{EI_z}$, $\epsilon_y = \frac{\nu}{(1-\nu)}\frac{(1-\nu^2)M_z y}{EI_z} = \nu(1+\nu)\frac{M_z y}{EI_z}$

$$\gamma_{xy} = 2\left(1 + \frac{\nu}{1-\nu}\right)\frac{(1-\nu^2)}{E}\frac{V_y Q}{I_z b} = 2\left(\frac{1-\nu+\nu}{1-\nu}\right)\frac{(1-\nu^2)}{E}\frac{V_y Q}{I_z b} = \frac{2(1+\nu)}{E}\frac{V_y Q}{I_z b}$$ no change

$$\frac{d^2 y}{dx^2} = (1-\nu^2)\frac{M}{EI_z}$$

(b) Simply multiply expressions by $(1-\nu^2)$

4.5 $\sigma_x = \frac{\partial^2 \Phi}{\partial y^2} = 2a_3x^2 + 6a_4xy + 12a_5y^2$

$\sigma_y = \frac{\partial^2 \Phi}{\partial x^2} = 12a_1x^2 + 6a_2xy + 2a_3y^2$

$\tau_{xy} = -\frac{\partial^2 \Phi}{\partial x \partial y} = -\left[3a_2x^2 + 4a_3xy + 3a_4y^2\right]$ **ANS**

For compatibility, $\nabla^4 \Phi = 0$

$\frac{\partial^4 \Phi}{\partial x^4} = 24a_1$, $\frac{\partial^4 \Phi}{\partial x^2 \partial y^2} = 4a_3$, $\frac{\partial^4 \Phi}{\partial y^4} = 24a_5$

$\therefore 24a_1 + 2(4a_3) + 24a_5 = 0$ or $3a_1 + a_3 + 3a_5 = 0$ **ANS**

4.6 B.C at $y = \pm c$ $\sigma_y = \tau_{xy} = 0$, at $x = \pm\frac{L}{2}$ $\int_{-c}^{c} \tau_{xy} b\,dy = P$, $\int_{-c}^{c} y\sigma_x b\,dy = 0$

As in Example 4.2-2, $\nabla^4 \Phi = 0$ yields $F = -5G$

$\sigma_x = 6Dy + 6Exy - 30Gx^2y + 20Gy^3$

$\sigma_y = 2B + 2Cy - 10Gy^3$

$\tau_{xy} = -(A + 2Cx + 3Ey^2 - 30Gxy^2)$

$\sigma_y = 0$ at $y = \pm c$

$0 = 2B + 2Cc - 10Gc^3$, $0 = 2B - 2Cc + 10Gc^3$

adding eqs yields $B = 0$, subtracting yields $C = Gc^2$

$\tau_{xy} = 0$ at $x = 0, y = -c$, $\Rightarrow A = -3Ec^2$

Since τ_{xy} is not a function of x, $C = G = 0$ $\therefore \sigma_y = 0$ **ANS**

and $\tau_{xy} = 3E(c^2 - y^2)$ At $x = \frac{L}{2}$, $\int_{-c}^{c} \tau_{xy} b\,dy = P$

$\int_{-c}^{c} 3E(c^2 - y^2) b\,dy = P \Rightarrow E = \frac{P}{4bc^3}$ $\therefore \tau_{xy} = \frac{3P}{4bc^3}(c^2 - y^2)$

Since $I_z = \frac{2}{3}bc^3$ $\tau_{xy} = \frac{P}{2I_z}(c^2 - y^2)$ **ANS**

$\sigma_x = 6Dy + \frac{3P}{2bc^3}xy$, at $x = \frac{L}{2}$ $\int_{-c}^{c} y\left(6Dy + \frac{3P}{2bc^3}\frac{L}{2}y\right)by = 0$

results in (for $x > 0$) $D = -\frac{PL}{8c^3}$ $\therefore \sigma_x = \frac{6Py}{8bc^3}(-L + 2x) = -\frac{Py}{2I_z}(L - 2x)$ **ANS**

For $(x < 0)$ $D = \frac{PL}{8c^3}$ and

$\sigma_x = -\frac{Py}{2I_z}(L + 2x)$ **ANS**

4.7 $\Phi = C_1 + C_2 \ln r + C_3 r^2 + C_4 r^2 \ln r$

$$\sigma_r = \frac{1}{r}\frac{d\Phi}{dr} = \frac{C_2}{r^2} + 2C_3 + 2C_4 \ln r + C_4,$$

$$\sigma_\theta = \frac{d^2\Phi}{dr^2} = -\frac{C_2}{r^2} + 2C_3 + 2C_4 \ln r + 3C_4, \quad \tau_{r\theta} = 0$$

$\sigma_r = -p_i$ at $r = r_i$ and $\sigma_r = -p_o$ at $r = r_o$

$$\frac{C_2}{r_i^2} + 2C_3 + (2\ln r_i + 1)C_4 = -p_i \quad (1) \qquad \frac{C_2}{r_o^2} + 2C_3 + (2\ln r_o + 1)C_4 = -p_o \quad (2)$$

$$\epsilon_r = \frac{\partial u_r}{\partial r} = \frac{1}{E}(\sigma_r - \nu\sigma_\theta) = \frac{1}{E}\left[(1+\nu)\frac{C_2}{r^2} + (1-\nu)(2C_3 + 2C_4 \ln r) + (1-3\nu)C_4\right] \quad (3)$$

$$\epsilon_\theta = \frac{u_r}{r} = \frac{1}{E}(\sigma_\theta - \nu\sigma_r) = \frac{1}{E}\left[-(1+\nu)\frac{C_2}{r^2} + (1-\nu)(2C_3 + 2C_4 \ln r) + (3-\nu)C_4\right]$$

Take derivative

$$\frac{\partial u_r}{\partial r} = \frac{1}{E}\left[(1+\nu)\frac{C_2}{r^2} + (1-\nu)(2C_3 + 2C_4 \ln r + 2C_4)\right] \quad (4) \text{ Equate to } (3)$$

$$\frac{1}{E}\left[(1+\nu)\frac{C_2}{r^2} + (1-\nu)(2C_3 + 2C_4 \ln r) + (1-3\nu)C_4\right] = \frac{1}{E}\left[(1+\nu)\frac{C_2}{r^2} + (1-\nu)(2C_3 + 2C_4 \ln r + 2C_4)\right]$$

yields $(1-3\nu)C_4 = 2(1-\nu)C_4$ for this to be true $C_4 = 0$

Thus, eqs (1) & (2) reduce to

$$\frac{C_2}{r_i^2} + 2C_3 = -p_i, \quad \frac{C_2}{r_o^2} + 2C_3 = p_o \quad \text{solve simultaneously for } C_2 \text{ \& } C_3$$

$$C_2 = \frac{(p_o - p_i) r_i^2 r_o^2}{r_o^2 - r_i^2}, \quad C_3 = \frac{1}{2}\frac{p_i r_i^2 - p_o r_o^2}{r_o^2 - r_i^2} \quad \text{sub into } \sigma_r, \sigma_\theta \text{ eqs.}$$

$$\sigma_r = \frac{1}{r_o^2 - r_i^2}\left[p_i r_i^2 - p_o r_o^2 + \left(\frac{r_i r_o}{r}\right)^2 (p_o - p_i)\right]$$

$$\sigma_\theta = \frac{1}{r_o^2 - r_i^2}\left[p_i r_i^2 - p_o r_o^2 - \left(\frac{r_i r_o}{r}\right)^2 (p_o - p_i)\right]$$

ANS

4.8 Substituting $M = 500$ lb-in, $r_i = 4$, $r_o = 6$, $b = 0.5$ in into Eqs (4.2-16) & (4.2-17) and tabulate results for $4 < r < 6$ in.

$(\sigma_r)_{max}$ and $(\sigma_r)_{min}$ are 0, −153.5 psi respectively

$(\sigma_\theta)_{max}$ and $(\sigma_\theta)_{min}$ are 1324, −1733 psi respectively

Straight beam formulation, $I = \frac{1}{12}(0.5)(2)^3 = 0.333 \text{ in}^4$

$$\sigma = -\frac{My}{I} = -\frac{500(5-r)}{0.333} = -1500(5-r)$$

(continued)

$\sigma_{max,min} = \pm 1500$ psi (13% error)

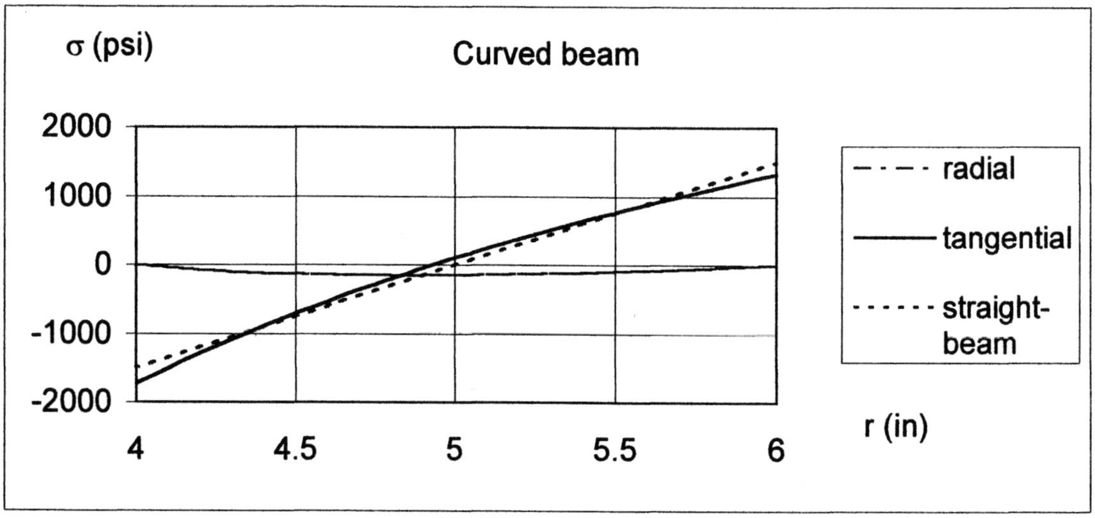

(4.9)

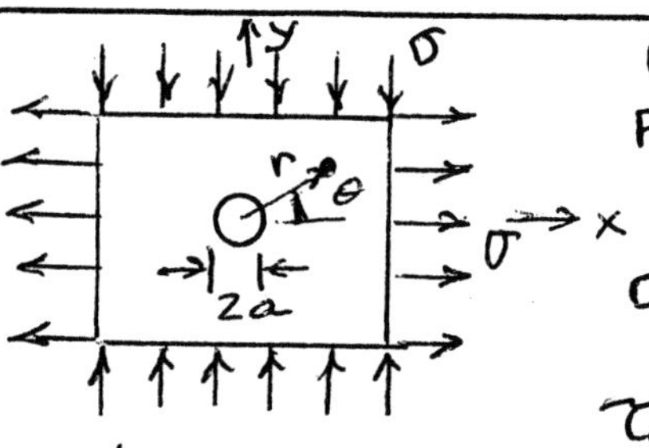

$\sigma_x = \sigma \quad \sigma_y = -\sigma$

For σ_x

$$\sigma_r = \frac{\sigma}{2}\left[1 - \frac{a^2}{r^2} + \left(1 - 4\frac{a^2}{r^2} + 3\frac{a^4}{r^4}\right)\cos 2\theta\right]$$

$$\sigma_\theta = \frac{\sigma}{2}\left[1 + \frac{a^2}{r^2} - \left(1 + 3\frac{a^4}{r^4}\right)\cos 2\theta\right]$$

$$\tau_{r\theta} = -\frac{\sigma}{2}\left(1 + 2\frac{a^2}{r^2} - 3\frac{a^4}{r^4}\right)\sin 2\theta$$

Can covert equations to represent σ_y by substituting $-\sigma$ for σ and $-\cos 2\theta$ for $\cos 2\theta$ and $-\sin 2\theta$ for $\sin 2\theta$ $\left[\begin{matrix}\cos 2(\theta+90) = -\cos 2\theta \\ \sin 2(\theta+90) = -\sin 2\theta\end{matrix}\right]$ Then add to above equations.

$$\therefore \sigma_r = \frac{\sigma}{2}\left[1 - \frac{a^2}{r^2} + \left(1 - 4\frac{a^2}{r^2} + 3\frac{a^4}{r^4}\right)\cos 2\theta\right] - \frac{\sigma}{2}\left[1 - \frac{a^2}{r^2} + \left(1 - 4\frac{a^2}{r^2} + 3\frac{a^4}{r^4}\right)(-\cos 2\theta)\right]$$

$$= \sigma\left(1 - 4\frac{a^2}{r^2} + 3\frac{a^4}{r^4}\right)\cos 2\theta \quad \boxed{\text{ANS}}$$

Likewise $\sigma_\theta = -\sigma\left(1 + 3\frac{a^4}{r^4}\right)\cos 2\theta$ $\boxed{\text{ANS}}$, $\tau_{r\theta} = -\sigma\left(1 + 2\frac{a^2}{r^2} - 3\frac{a^4}{r^4}\right)\sin 2\theta$ $\boxed{\text{ANS}}$

(b) Around hole, $r = a$

$$\frac{\sigma_\theta}{\sigma} = -4\cos\theta$$

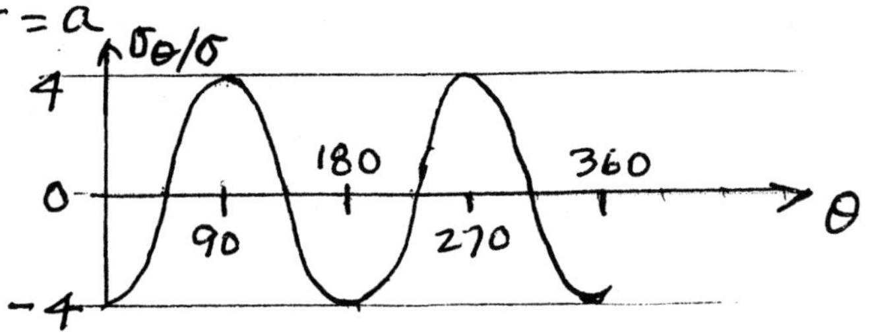

4.10 $\Phi = Cr\theta\sin\theta$

$$\nabla^2\Phi = \frac{\partial^2\Phi}{\partial r^2} + \frac{1}{r}\frac{\partial\Phi}{\partial r} + \frac{1}{r^2}\frac{\partial^2\Phi}{\partial\theta^2} = 0 + \frac{C\theta}{r}\sin\theta - \frac{C\theta}{r}\sin\theta = 0$$

$\therefore \nabla^4\Phi = 0$

4.11 $\Phi = Cr\theta\cos\theta$

$$\sigma_r = \frac{1}{r^2}\frac{\partial^2\Phi}{\partial\theta^2} + \frac{1}{r}\frac{\partial\Phi}{\partial r} = \frac{Cr}{r^2}(-2\sin\theta - \theta\cos\theta) + \frac{C}{r}\theta\cos\theta$$

$$= -2\frac{C}{r}\sin\theta \quad ①$$

$$\sigma_\theta = \frac{\partial^2\Phi}{\partial r^2} = 0,\quad \tau_{r\theta} = \frac{1}{r^2}\frac{\partial\Phi}{\partial\theta} - \frac{1}{r}\frac{\partial^2\Phi}{\partial r\partial\theta} = \frac{C}{r}(\cos\theta - \theta\sin\theta) - \frac{C}{r}(\cos\theta - \theta\sin\theta)$$

$$= 0$$

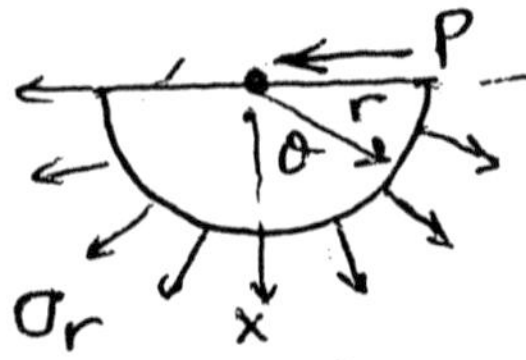

$$\Sigma F_y = 0 \quad -P + \int_{-\pi/2}^{\pi/2} (\sigma_r t r\, d\theta)\sin\theta = 0$$

sub in ①

$$P = -\int_{-\pi/2}^{\pi/2} 2Ct\sin^2\theta\, d\theta = -2Ct\left(\frac{\pi}{2}\right) \quad \therefore C = -\frac{P}{\pi t}$$

$$\therefore \sigma_r = \frac{2P}{\pi t r}\sin\theta,\quad \sigma_\theta = \tau_{r\theta} = 0 \qquad \boxed{\text{ANS}}$$

4.12 Superposition of Eq. (4.2-30) and the result of Prob 4.11

$$\sigma_r = -\frac{2P\sin\phi}{\pi t r}\cos\theta + \frac{2P\cos\phi}{\pi t r}\sin\theta = \frac{2P}{\pi t r}\sin(\theta - \phi) \quad \underline{\text{ANS}}$$

$$\sigma_\theta = \tau_{r\theta} = 0$$

4.13 Based on the derivation of Eq. (4.2-37) it should be clear that $\sigma_r = \sigma_\theta = \tau_{r\theta} = 0$ at $r = R$ for any value of θ except $\pm\pi/2$. However, let $r = R$, $x = R\cos\theta$, $y = R\sin\theta$ and substitute into Eqs. (4.2-37). For example,

$$\sigma_\theta = \sigma_x\sin^2\theta + \sigma_y\cos^2\theta - 2\tau_{xy}\sin\theta\cos\theta$$

$$\sigma_\theta = -\frac{2P}{\pi t}\left[\frac{R(1-\sin\theta)R^2\cos^2\theta\sin^2\theta}{4R^4(1-\sin\theta)^2} + \frac{R(1+\sin\theta)R^2\sin^2\theta\cos^2\theta}{4R^4(1+\sin\theta)^2} + \frac{R^3(1-\sin\theta)^3\cos^2\theta}{4R^4(1-\sin\theta)^2}\right.$$

$$\left. + \frac{R^3(1+\sin\theta)^3\cos^2\theta}{4R^4(1+\sin\theta)^2} + \frac{R^2(1-\sin\theta)^2R\sin\theta\cos^2\theta}{4R^4(1-\sin\theta)^2} - \frac{R^2(1+\sin\theta)^2R\sin\theta\cos^2\theta}{4R^4(1+\sin\theta)^2} - \frac{1}{2R}\right]$$

$$= -\frac{P}{2\pi t R}\frac{(2-2)\cos^4\theta}{\cos^4\theta} = 0 \text{ except for } \theta = \pm\frac{\pi}{2} \text{ where it is indeterminant}$$

$\sigma_r = \tau_{r\theta} = 0$ in a similar manner

4.14 (a) $$\sigma_x - \sigma_y = -\frac{2P}{\pi t}\left[\frac{Rx^2}{(x^2+R^2)^2} + \frac{Rx^2}{(x^2+R^2)^2} - \frac{1}{2R} - \frac{R^3}{(x^2+R^2)^2} - \frac{R^3}{(x^2+R^2)^2} + \frac{1}{2R}\right]$$

$$= \frac{4PR}{\pi t}\frac{(R^2-x^2)}{(R^2+x^2)^2} \quad \text{ANS}$$

(b)

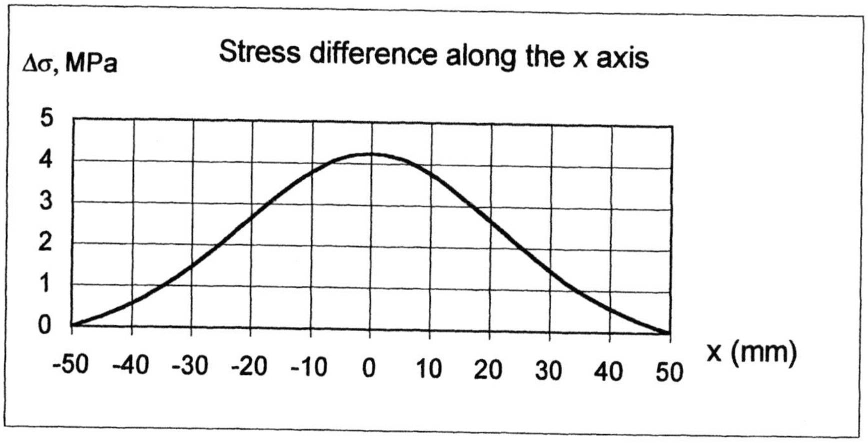

4.15 Equation of ellipse $\left(\frac{y}{a}\right)^2 + \left(\frac{z}{b}\right)^2 = 1$

Let $\bar{y} = \frac{y}{a}$, $\bar{z} = \frac{z}{b}$

$\therefore \Psi = k(\bar{y}^2 + \bar{z}^2 - 1)$ $\quad dy = a\,d\bar{y}$, $dz = b\,d\bar{z}$

Eq (4.3-7) $$T = 2kab\int_{-1}^{1}\left[\int_{-\sqrt{1-\bar{z}^2}}^{\sqrt{1+\bar{z}^2}}(\bar{y}^2+\bar{z}^2-1)\,d\bar{y}\right]d\bar{z} = -\pi kab$$

$\therefore k = -\frac{T}{\pi ab}$ Thus, $\Psi = -\frac{T}{\pi ab}\left[\left(\frac{y}{a}\right)^2 + \left(\frac{z}{b}\right)^2 - 1\right]$ ANS

$$\tau_{xy} = \frac{\partial\Psi}{\partial z} = -\frac{2Tz}{\pi ab^3} \qquad \tau_{zx} = -\frac{\partial\Psi}{\partial y} = \frac{2Ty}{\pi a^3 b}$$

$$(\tau_x)_{net} = \sqrt{\tau_{xy}^2 + \tau_{zx}^2} = \frac{2T}{\pi a^3 b^3}\sqrt{a^4 z^2 + b^4 y^2} \quad \text{ANS}$$

Assuming that τ_{max} occurs at boundary where $z^2 = b^2 - \frac{b^2}{a^2}y^2$ then,

$$(\tau_x)_{boundary} = \frac{2T}{\pi ab^2}\left\{1 - \left[1 - \left(\frac{b}{a}\right)^2\right]\left(\frac{y}{a}\right)^2\right\}^{\frac{1}{2}}$$ If $b > a$ then $\left[1 - \left(\frac{b}{a}\right)^2\right]$ is negative and τ_{max} occurs where y is a maximum. This occurs at minor axis where $y = a$.

Thus $$\tau_{max} = \frac{2T}{\pi ab^2}\left\{1 - \left[1 - \left(\frac{b}{a}\right)^2\right]\right\}^{1/2} = \frac{2T}{\pi a^2 b} \quad \text{ANS}$$ (continued)

For angle of twist, θ', see Eq.(4.3-4)

$$\frac{\partial^2 \Psi}{\partial y^2} = -\frac{2T}{\pi a^3 b}, \quad \frac{\partial^2 \Psi}{\partial z^2} = -\frac{2T}{\pi a b^3}$$

$$\frac{-2T}{\pi a^3 b} - \frac{2T}{\pi a b^3} = -\frac{E}{1+\nu}\theta' \Rightarrow \theta' = \frac{2(1+\nu)(a^2+b^2)T}{\pi E a^3 b^3} \quad \boxed{\text{ANS}}$$

(b) $\tau_{max} = \dfrac{2(400)}{\pi(0.01)^2(0.02)} = 127(10^6)\ Pa = \underline{127\ MPa}$

$$\theta' = \frac{2(1.3)(400)(0.01^2+0.02^2)}{\pi(200)(10^9)(0.01)^3(0.02)^3} = 0.1035\ rad/m = \underline{5.93°/m}$$

(4.16) (a) Equilateral triangle

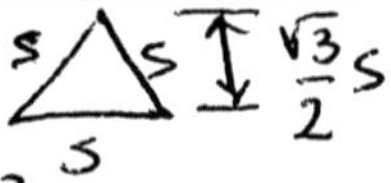

$A = \frac{1}{2}\frac{\sqrt{3}}{2}s^2 = 1$

Eq(n) from Example 4.3-2 $\quad S = 1.5197$

$$\tau_{max} = 20\frac{T}{s^3} = \frac{20(1)}{(1.5197)^3} = \underline{5.699}$$

(b) solid cylinder $\quad A = \pi r^2 = 1 \Rightarrow r = 0.5642$

$$\tau_{max} = \frac{2T}{\pi r^3} = \frac{2(1)}{\pi(0.5642)^3} = \underline{3.545}$$

(c) Thin-walled tube $(t = 0.1r)$

$$A = \pi\left[(1.05r)^2 - (0.95r)^2\right] = 1 \Rightarrow r = 1.262$$

$$\tau_{max} = \frac{2(1)(1.05r)}{\pi\left[(1.05r)^4 - (0.95r)^4\right]} = \underline{0.8302}$$

(d) Solid square $\quad A = s^2 = 1 \Rightarrow s = 1$

Eq(4.3-11) $\quad \tau_{max} = \dfrac{(1)}{(0.208)(1)^3} = \underline{4.808}$

(e) Rectangle with $h/b = 2$

$A = 2a^2 = 1 \quad a = \sqrt{2}/2$

$$\tau_{max} = \frac{1}{(0.246)(\sqrt{2})(\sqrt{2}/2)^2} = \underline{5.749}$$

(f) I-beam: $A = (4a)(0.1a) = 1 \Rightarrow a = 2.5$

Eq (4.3-14) $\quad h = 4a,\ b = 0.1a \quad \tau_{max} = \dfrac{3(1)}{(4)(0.1)^2(2.5)^3} = \underline{4.8}$

(continued)

Divide all solutions by 0.8302 (case c) to normalize

Case	a	b	c	d	e	f
$\tau_{max}/\tau_{case\,c}$	6.865	4.276	1.0	5.791	6.925	5.782

4.17 $\frac{h}{b} = 2$ Table 4.3-1 $k_1 = 0.229$, $k_2 = 0.246$

Eq. (4.3-11) $\tau_{max} = \dfrac{600}{(0.246)(0.05)(0.025)^2} = 73.8(10^6)\,\text{Pa} = \underline{73.8\ \text{MPa}}$

$\dfrac{\theta}{0.5} = \dfrac{2(1.3)(600)}{(0.229)(200)(10^9)(0.05)(0.025)^3} \Rightarrow \theta = \underline{2.18(10^{-2})\,\text{rad} = 1.25^\circ}$

4.18 Using Excel or a math software to calculate series. Here Mathcad was used.

k_1: To obtain six-place accuracy, n was taken to 13

$$\frac{1}{3}\cdot\left[1-\frac{192}{1.5\cdot\pi^5}\cdot\sum_{n=1}^{13}\left((-1)^{n+1}+1\right)\cdot\frac{\tanh\left(1.5\cdot n\cdot\frac{\pi}{2}\right)}{(2\cdot n^5)}\right]=0.195761$$

k: To obtain six-place accuracy, n was taken to 5

$$1-\frac{8}{\pi^2}\cdot\sum_{n=1}^{5}\frac{(-1)^{n+1}+1}{2\cdot n^2\cdot\cosh\left(1.5\cdot n\cdot\frac{\pi}{2}\right)}=0.847562$$

k_2:

$$\frac{0.195761}{0.847562}=0.23097$$

4.19 (a) $\tau = \dfrac{3(200)}{(2+3)(1/4)^2} = 1920\,\text{psi} = \underline{1.92\ \text{kpsi}}$

$\theta' = \dfrac{6(1.3)(200)}{30(10^6)(2+3)(1/4)^3} = 6.66(10^{-4})\,\text{rad/in} = \underline{0.038\ \text{deg/in}}$

(b) $\tau = \dfrac{3(200)}{\pi(7/8)(1/4)^2} = 3490\,\text{psi} = \underline{3.49\ \text{kpsi}}$

(Continued)

$$\theta' = \frac{6(1.3)(200)}{30(10^6)(\pi)(7/8)(1/4)^3} = 1.21(10^{-3})\text{ rad/in} = \underline{0.069\text{ deg/in}}$$

(c) $J_e = \frac{2}{3}(2.5)\left(\frac{1}{8}\right)^3 + \frac{2}{3}\left(\frac{3}{16}\right)^3 = 0.00765\text{ in}^4$

$$\tau_{max} = \frac{200(3/16)}{0.00765} = 4900\text{ psi} = \underline{4.90\text{ kpsi}}$$

$$\theta' = \frac{2(1.3)(200)}{30(10^6)(0.00765)} = 2.27(10^{-3})\text{ rad/in} = \underline{0.13\text{ deg/in}}$$

5.1 $\bar{A} = 50(75) + \pi(25)^2 = 5.713(10^3)\text{ mm}^2 = 5.713(10^{-3})\text{ m}^2$

$\bar{S} = 2(75) + 2\pi(25) = 307.1\text{ mm}$

(a) $$\tau = \frac{2000}{2(5.713)(10^3)(0.003)} = \underline{58.3\text{ MPa}}$$

(b) $$\theta = \frac{(1.29)(2000)(1)(307.1)(10^{-3})}{2(210)(10^9)(5.713)^2(10^{-3})^2(0.003)} = \underline{0.0193\text{ rad} = 1.10^\circ}$$

5.2

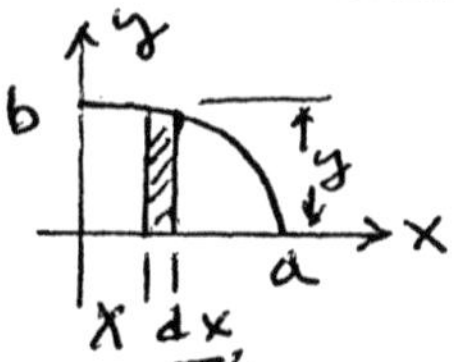

$y = b\sqrt{1 - x^2/a^2} = \frac{b}{a}\sqrt{a^2 - x^2}$

$\bar{A} = 4\int_0^a \frac{b}{a}\sqrt{a^2 - x^2}\,dx$ Let $x = a\sin\theta$, $dx = a\cos\theta\,d\theta$

$$\bar{A} = 4\frac{b}{a}a^2\int_0^{\pi/2}\cos^2\theta\,d\theta = 4\int_0^{\pi/2}\left(\frac{1+\cos 2\theta}{2}\right)d\theta = \pi ab$$

$\bar{A} = \pi(75)(50) = 11.78(10^3)\text{ mm}^2$

The perimeter length, $\bar{S}$, is not as easy and involves elliptic integrals and tables. An easy alternative is to numerically evaluate $\bar{S}$.* $d\bar{S} = \sqrt{dx^2 + dy^2}$. Using a spreadsheet increment x (in small, equal steps) and $y = \frac{b}{a}\sqrt{a^2 - x^2}$. Evaluate dx and dy for each increment and determine $d\bar{S} = \sqrt{dx^2 + dy^2}$ and sum. For a quarter of the ellipse, the result is 99.1504 mm. Thus $\bar{S} = 4(99.1504) = 396.6\text{ mm}$ (continued)

* See also, "Roark' Formulas for Stress and Strain," W. C. Young, 6th Ed., 1989, page 351, Section 13.

$$\bar{S} \approx \pi(a+b)\left[1 + 0.258\left(\frac{a-b}{a+b}\right)^2\right] = \pi(125)\left[1 + 0.258\left(\frac{25}{125}\right)^2\right] = 396.75\text{ mm}$$

$$\tau = \frac{1000}{2(11.78)(10^3)(0.004)} = \underline{10.6\ MPa}$$

$$\theta' = \frac{(1.33)(1000)(396.6)(10^{-3})}{2(70)(10^9)(11.78)^2(10^{-3})^2(0.004)} = \underline{6.79(10^{-3})\ rad/m = 0.389\ deg/m}$$

5.3 $\bar{A} = 25^2 + 4\left(\frac{\pi}{2}\right)(6.25)^2 - 4\left(\frac{\pi}{4}\right)(6.25)^2 = 747.7\ mm^2$

$\bar{S} = 12\left(\frac{1}{4}\right)(2\pi)(6.25) = 117.81\ mm$

$$\tau = \frac{100}{2(747.7)(0.0025)} = \underline{26.7\ MPa}$$

$$\theta = \frac{(1.29)(100)(0.5)(117.81)(10^{-3})}{2(207)(10^9)(747.7)^2(10^{-3})^4(2.5)(10^{-3})}$$

$= \underline{1.313(10^{-2})\ rad = 0.75^\circ}$

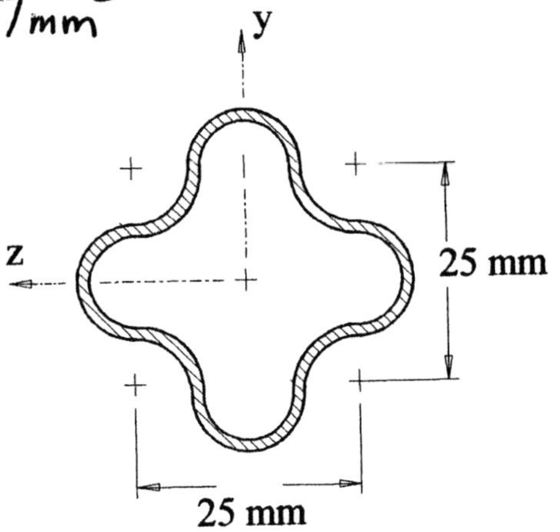

5.4 $\bar{A} = \frac{\pi}{2}(40)^2 = 2.513(10^3)\ mm^2$, $\bar{S} = \pi(40) + 2(40) = 205.66\ mm$

(a) $$\tau = \frac{500}{2(2.513)(10^3)(0.003)} = \underline{33.2\ MPa}$$

(b) $$\theta = \frac{1.29(500)(0.3)(205.66)(10^{-3})}{2(210)(10^9)(2.513)^2(10^{-3})^2(0.003)} = \underline{5.0(10^{-3})\ rad = 0.287^\circ}$$

5.5 From symmetry, the flow in each cell is the same. The flow in the horizontal wall therefore will be zero and we can treat the section as a single cell and ignore the horizontal wall.

$\bar{A} = 60(50) + \pi(25)^2 = 4.963(10^3)\ mm^2$

$\bar{S} = 4(30) + 2\pi(25) = 277.1\ mm$

$$\tau = \frac{2500}{2(4.963)(10^3)(0.005)} = \underline{50.4\ MPa}$$

$$\theta' = \frac{(1.33)(2500)(277.1)(10^{-3})}{2(70)(10^9)(4.963)^2(10^{-3})^2(0.005)} = \underline{0.0534\ rad/m = 3.06^\circ/m}$$

5.6 For an estimate, average $\bar{A}$ and $\bar{S}$

$\bar{A} \approx \frac{\pi}{2}(25^2+20^2) = 1.61(10^3)\,mm^2 \qquad t_{min} = 4\,mm$

$\bar{S} \approx \frac{2\pi}{2}(25+20) = 141.4 \qquad t_{avg} = 5\,mm$

$$\tau \approx \frac{500}{2(1.61)(10^3)(0.004)} = \underline{38.8\ MPa}$$

For steel, let $E = 207$ GPa and $\nu = 0.29$

$$\theta' = \frac{(1.29)(500)(141.4)(10^{-3})}{2(207)(10^9)(1.61)^2(10^{-3})^2(0.005)} = \underline{0.0168\ rad/m = 0.96°/m}$$

5.7 $\bar{A} = \frac{\pi}{2}(20)^2 + 40^2 = 2.228(10^3)\,mm^2$

$\bar{S} = \pi(20) + 3(40) = 182.8\,mm$

$$\tau = \frac{125}{2(2.228)(10^3)(0.003)} = \underline{9.35\ MPa}$$

$$\theta = \frac{(1.3)(125)(0.8)(0.1828)}{2(210)(10^9)(2.228)^2(10^{-3})^2(0.003)} = \underline{3.80(10^{-3})\,rad = 0.22°}$$

5.8 $\bar{A} = \frac{1}{2}(1.5)^2 \sin 60 = 0.9743\,in^2$, $\bar{S} = 3(1.5) = 4.5\,in$

(a) $$\tau = \frac{5000}{2(0.9743)(0.125)} = 20{,}530\ psi = \underline{20.5\ kpsi}$$

(b) $$\theta = \frac{(1.29)(5000)(20)(4.5)}{2(29)(10^6)(0.9743)^2(0.125)} = \underline{0.0844\ rad = 4.83°}$$

5.9 (a) Circular $\bar{S} = 2\pi r \Rightarrow r = \frac{\bar{S}}{2\pi} \Rightarrow \bar{A} = \pi r^2 = \frac{\bar{S}^2}{4\pi}$

$$\tau = \frac{T}{2\bar{A}t} = 2\pi\frac{T}{\bar{S}^2 t} = 6.283\frac{T}{\bar{S}^2 t} \quad \boxed{ANS}$$

$$\theta' = \frac{(1+\nu)T\bar{S}}{2E\left(\frac{\bar{S}^2}{4\pi}\right)^2 t} = 78.96\frac{(1+\nu)T}{E\bar{S}^3 t} \quad \boxed{ANS}$$

(b) square $4l = \bar{S} \Rightarrow l = \bar{S}/4 \Rightarrow \bar{A} = l^2 = \bar{S}^2/16$

$$\tau = \frac{T}{2(\bar{S}^2/16)t} = 8\frac{T}{\bar{S}^2 t} \quad \boxed{ANS}$$

(continued)

$$\theta' = \frac{(1+\nu)T\bar{S}}{2E\left(\frac{\bar{S}^2}{16}\right)^2 t} = 128(1+\nu)\frac{T}{E\bar{S}^2 t} \quad \boxed{\text{ANS}}$$

(c) Equilateral triangle $3l = \bar{S} \Rightarrow l = \frac{\bar{S}}{3} \Rightarrow \bar{A} = \frac{\sqrt{3}}{4}l^2 = \frac{\sqrt{3}}{36}\bar{S}^2$

$$\tau = \frac{T}{2\left(\frac{\sqrt{3}}{36}\bar{S}^2\right)t} = \frac{18}{\sqrt{3}}\frac{T}{\bar{S}^2 t} = 10.39\frac{T}{\bar{S}^2 t} \quad \boxed{\text{ANS}}$$

$$\theta' = \frac{(1+\nu)T\bar{S}}{2E\left(\frac{\sqrt{3}}{36}\bar{S}^2\right)^2 t} = 216\frac{(1+\nu)T}{E\bar{S}^3 t} \quad \boxed{\text{ANS}}$$

Normalize results wrt circular section

	$\frac{\tau\bar{S}^2 t}{2\pi}$	$\frac{\theta' E\bar{S}^3 t}{8\pi^2(1+\nu)T}$
Circular	1	1
Square	1.273	1.621
Triangular	1.654	2.736

(5.10) Due to symmetry, all quarter circular cells are identical

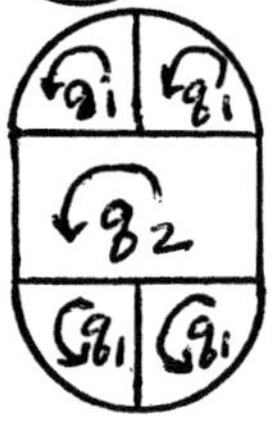

$$T = 2\left[(4)q_1\frac{\pi}{4}(1)^2 + q_2(1)(2)\right] \Rightarrow 5000 = 2\pi q_1 + 4q_2 \quad (1)$$

q_1 cell:
$$\theta = \frac{(1+\nu)L}{E\left(\frac{\pi}{4}\right)(1)^2}\left[\frac{q_1}{0.125}\left(\frac{2\pi(1)}{4}\right) + \frac{(q_1-q_1)(1)}{0.125} + \frac{(q_1-q_2)(1)}{0.25}\right]$$

or, $21.093\,q_1 - 5.093\,q_2 - C = 0$ (2) where $C = \frac{\theta E}{(1+\nu)L}$

q_2 cell:
$$\theta = \frac{(1+\nu)L}{E(1)(2)}\left[\frac{q_2(2)}{0.25} + \frac{4(q_2-q_1)(1)}{0.25}\right]$$

or, $8q_1 - 12q_2 + C = 0$ (3)

solve (1), (2), and (3) simultaneously to obtain

$q_1 = 381.9$ lb/in, $q_2 = 650.1$ lb/in, $C = 4745$ lb/in^3

Stresses

$\tau_1 = \frac{q_1}{0.125} = 3055\,psi = \underline{3.06\,kpsi}$, $\tau_2 = \frac{q_1 - q_1}{0.125} = \underline{0}$

$\tau_3 = \frac{q_2 - q_1}{0.25} = 1073\,psi = \underline{1.07\,kpsi}$

$\tau_3 = q_2/0.25 = 2600\,psi = \underline{2.60\,kpsi}$

$$\theta = \frac{(1+\nu)CL}{E} = \frac{(1.33)(4745)(24)}{10(10^6)} = \underline{0.0152\,rad = 0.87^\circ}$$

5.11 Let q_1 be inner cell and q_2 for outer cells

Length of straight wall $\ell = \sqrt{2}(75)$ mm

$\bar{A}_1 = \ell^2 = 2(75)^2 = 11.25(10^3)\text{ mm}^2 = 11.25(10^{-3})\text{ m}^2$

$A_2 = \frac{\pi}{4}(75)^2 - \frac{1}{2}(75)^2 = 1.605(10^3)\text{ mm}^2 = 1.605(10^{-3})\text{ m}^2$

$T = 2\left[q_1\bar{A}_1 + 4q_2A_2\right] \Rightarrow 20(10^3) = 2\left[11.25q_1 + 4(1.605)q_2\right](10^{-3})$

or, $2.25q_1 + 1.2843q_2 = 20(10^5)$ ①

<u>Cell 1:</u> $\theta = \dfrac{(1.33)(0.5)(4)(q_1 - q_2)(\sqrt{2})(0.075)}{69(10^9)(11.25)(10^{-3})(0.003)}$

or, $1.2115q_1 - 1.2115q_2 = (10^7)\theta$ ②

<u>Cell 2:</u> $\theta = \dfrac{1.33(0.5)}{69(10^9)(1.605)(10^{-3})}\left[\dfrac{q_2\frac{\pi}{2}(0.075)}{0.004} + \dfrac{(q_2 - q_1)(\sqrt{2})(0.075)}{0.003}\right]$

or, $-2.123q_1 + 3.892q_2 = (10^7)\theta$ ③

Set ② = ③ and simplify obtaining

$3.335q_1 - 5.103q_2 = 0$ ④ Solve ① & ④ simultaneously

obtaining $q_1 = 6.474(10^5)\text{ N/m}$, $q_2 = 4.231(10^5)\text{ N/m}$

Straight walls: $\tau_1 = \dfrac{q_1 - q_2}{0.003} = 74.8(10^6)\text{ Pa} = \underline{74.8\text{ MPa}}$

Circular walls: $\tau_2 = \dfrac{q_2}{0.004} = 105.8(10^6)\text{ Pa} = \underline{105.8\text{ MPa}}$

sub q_1 & q_2 into ② or ③ obtaining $\underline{\theta = 0.0271\text{ rad} = 1.56^\circ}$

5.12 Due to symmetry, top and bottom cells are equal

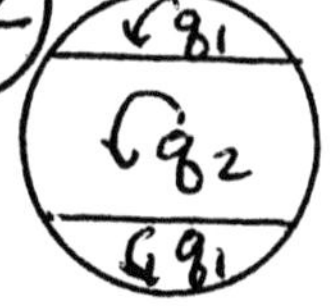

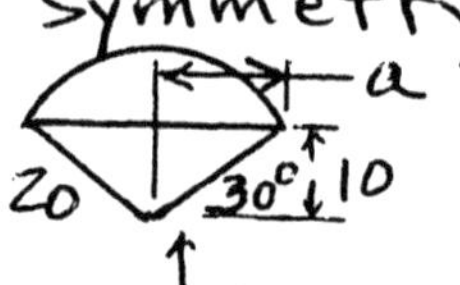

Total Area $= \frac{\pi}{3}(20)^2 = 418.9\text{ mm}^2$

$a = 20\sin 60 = 17.32$ mm

$\text{Area}_\triangledown = 2\left(\frac{1}{2}\right)(17.32)(10) = 173.2\text{ mm}^2$

Cell 1: $\bar{A}_1 = 418.9 - 173.2 = 245.7\text{ mm}^2$

Cell 2: $\bar{A}_2 = \pi(20)^2 - 2(245.7) = 765.3\text{ mm}^2$

Eq. (5.2-8) $150 = 2\left[2(q_1)(245.7) + (q_2)(765.3)\right](10^{-6})$

yields $q_1 + 1.557q_2 = 152.6(10^3)$ ①

(Continued)

Twist equations

Cell 1: $\frac{\theta}{L} = \frac{(1+\nu)}{E(245.7)(10^{-6})}\left[\frac{\frac{2}{3}\pi(20)}{4}q_1 + \frac{2(17.32)}{3}(q_1-q_2)\right]$

or, $8.961q_1 - 4.699q_2 - C = 0$ ② where $C = \frac{E\theta}{(1+\nu)L}(10^{-4})$

Cell 2: $\frac{\theta}{L} = \frac{(1+\nu)}{E(765.3)(10^{-6})}\left[\frac{2(\frac{\pi}{3})20}{4}q_2 + \frac{2(q_2-q_1)(2)(17.32)}{3}\right]$

or, $-3.018q_1 + 4.386q_2 - C = 0$ ③

Solving ①, ②, & ③ simultaneously yields

$q_1 = 4.979(10^4)\ N/m$, $q_2 = 6.565(10^4)N/m$, $C = 1.377(10^5)N/m^3$

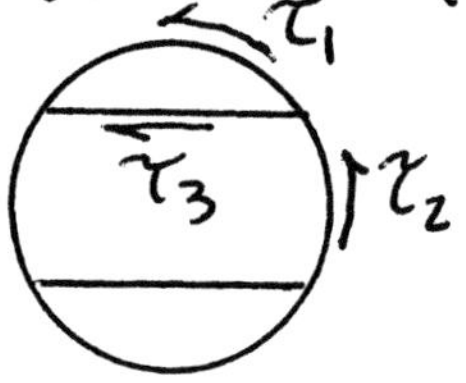

$\tau_1 = \frac{q_1}{4(10^{-3})} = 12.45(10^6)Pa = \underline{12.45\ MPa}$

$\tau_2 = \frac{q_2}{4(10^{-3})} = 16.41(10^6)Pa = \underline{16.41\ MPa}$

$\tau_3 = \frac{q_2-q_1}{3(10^{-3})} = 5.29(10^6)Pa = \underline{5.29\ MPa}$

$\frac{\theta}{L} = \frac{(1+\nu)C}{E(10^{-4})} = \frac{(1.33)(1.377)(10^5)}{70(10^9)(10^{-4})} = \underline{0.0262\ rad/m = 1.5°/m}$

5.13 (a) Because of symmetry, the shear flow in each cell of the proposed design are equal. Thus the flows are cancelled in the inner walls and little or no stress exists. For this reason, the proposed design is equivalent to the current design and the objectives will not be achieved.

(b) Make the walls thicker and with fillets without changing the outer dimensions

5.14 (a) $I_y = \frac{1}{12}(6.25)(0.25)^3 + 2\left[\frac{1}{12}(0.25)(1.875)^3 + (0.25)(1.875)\left(\frac{2+0.125}{2}\right)^2\right]$

$= \underline{1.3411\ in^4}$

$I_z = \frac{1}{12}(0.25)(6.25)^3 + 2\left[\frac{1}{12}(1.875)(0.25)^3 + (0.25)(1.875)(3)^2\right] = \underline{13.5286\ in^4}$

$I_{yz} = 2(0.25)(1.875)(3)\left(\frac{2.125}{2}\right) = \underline{2.9883\ in^4}$

(continued)

(b) $\theta = \frac{1}{2}\tan^{-1}\left[\frac{2(2.9883)}{13.5286-1.3411}\right] = 13.06^\circ$

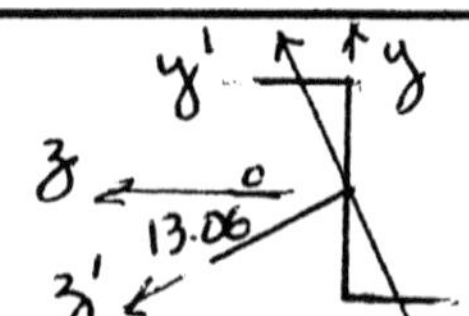

(c) $I_m = I_{y'} = \frac{1}{2}(1.341+13.529)+\frac{1}{2}(1.341-13.529)\cos\left(2(13.06)\right)-2.988\sin\left(2(13.06)\right)$

$= \underline{0.6478 \text{ in}^4}$

$I_n = I_{z'} = \frac{1}{2}(1.341+13.529)-\frac{1}{2}(1.341-13.529)\cos\left(2(13.06)\right)+2.988\sin\left(2(13.06)\right)$

$= \underline{14.2219 \text{ in}^4}$

(d) $M_z = 20$ kip-in, $M_m = 20\sin(13.06) = 4.52$ kip-in,

$M_n = 20\cos(13.06) = 19.48$ kip-in

Neutral plane $\alpha = \tan^{-1}\frac{19.48}{4.52}\frac{0.6478}{14.2219} = 11.11^\circ$

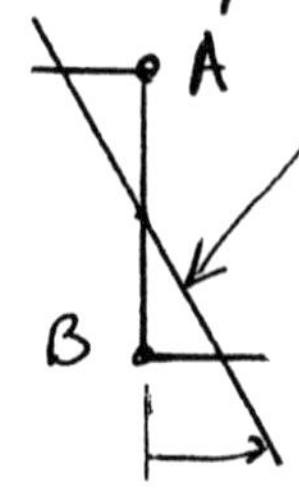

$13.06 + 11.11 = 24.17^\circ$

Point A: $y = 3.125 \quad z = -0.125$

$m = 3.125\cos 13.06 - 0.125\sin 13.06 = 3.016$ in

$n = -3.125\sin 13.06 - 0.125\cos 13.06 = -0.828$ in

$\sigma_A = \frac{4.519(-0.828)}{0.6478} - \frac{19.483(3.016)}{14.2219} = \underline{-9.91 \text{ kpsi}}$

point B is antisymmetric $\therefore$ $\sigma_B = \underline{9.91 \text{ kpsi}}$

(5.15) $I_y = I_{YY} = 0.392 \text{ in}^4$, $I_z = I_{XX} = 1.09 \text{ in}^4$

$I_m = I_{ZZ} = A r_z^2 = 1.19(0.435)^2 = 0.225 \text{ in}^4$

$I_n = I_y + I_z - I_m = 0.392 + 1.09 - 0.225 = 1.257 \text{ in}^4$

$\theta = -\alpha = -\tan^{-1} 0.440 = -23.75^\circ$

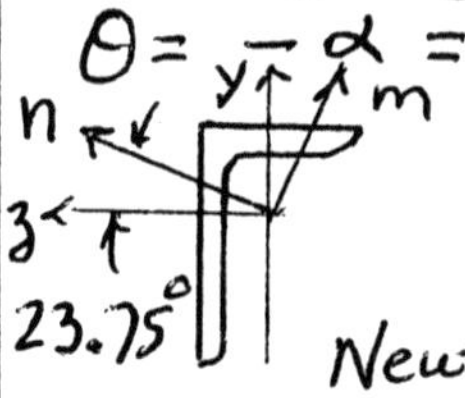

$M_m = -2000\sin 23.75 = -805.5$ lb-in

$M_n = 2000\cos 23.75 = 1831$ lb-in

Neutral plane: $\alpha = \tan^{-1}\left(\frac{1831}{-805.5}\frac{0.225}{1.257}\right) = -22.14^\circ$

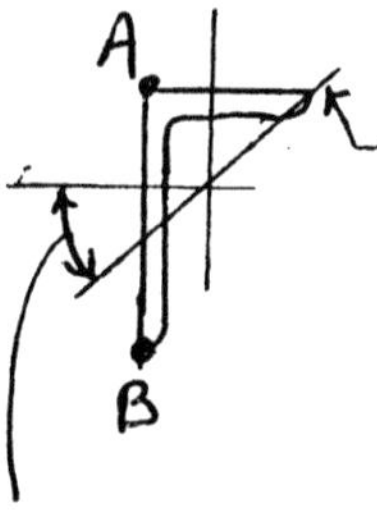

Point A: $y = 0.993$, $z = 0.493$

$m = 0.993\cos(-23.75) + 0.493\sin(-23.75) = 0.7104$ in

$n = -0.993\sin(-23.75) + 0.493\cos(-23.75) = 0.8512$ in

$\sigma_A = \frac{-805.5(0.8512)}{0.225} - \frac{1831(0.7104)}{1.257} = \underline{-4082 \text{ psi} = -4.08 \text{ kpsi}}$

$23.75 + 22.14 = 45.9^\circ$

Point B: $y = -(3-0.993) = -2.007$

$z \approx 0.493 - 0.25 = 0.243$

(continued)

$m = -2.007\cos(-23.75) + 0.243\sin(-23.75) = -1.935 \text{ in}$

$n = -(-2.007)\sin(-23.75) + 0.243\cos(-23.75) = -0.586$

$$\sigma_B = \frac{-805.5(-0.586)}{0.225} - \frac{1831(-1.935)}{1.257} = \underline{4916 \text{ psi} = 4.92 \text{ kpsi}}$$

5.16 (a) $I_y = I_z = \frac{1}{3}\left[(2)(0.833)^3 + (1)(1-0.833)^3 + (1)(2-0.833)^3\right] = \underline{0.9167 \text{ in}^4}$

$I_{yz} = 2(1)(0.833-0.5)(1-0.833) + 1(1)(0.5-0.833)(0.833-1.5) = \underline{0.3333 \text{ in}^4}$

$$\theta = \frac{1}{2}\tan^{-1}\left[\frac{2(0.3333)}{0.9167-0.9167}\right] = 45^\circ \text{ (obvious?)}$$

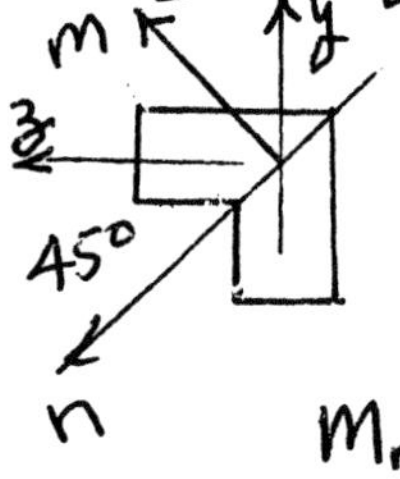

$$I_m = I_{y'} = \frac{0.9167+0.9167}{2} + 0 - 0.3333\sin 90 = 0.5834 \text{ in}^4$$

$I_n = I_{z'} = 0.9167 + 0.3333 = 1.25 \text{ in}^4$

$M_z = -500(20) = -10{,}000 \text{ lb-in}$

$M_m = M_n = -10{,}000\frac{\sqrt{2}}{2} = -5000\sqrt{2} \text{ lb-in}$

Point B: $m = 0.833\cos 45 + 1.167\sin 45 = 1.414 \text{ in}$

$n = -0.833\sin 45 + 1.167\cos 45 = 0.2362 \text{ in}$

$$\sigma_B = \frac{-5000\sqrt{2}(0.2362)}{0.5834} - \frac{(-5000\sqrt{2})(1.414)}{1.25} = 5136 \text{ psi} = \underline{5.14 \text{ kpsi}}$$

Alternatively, using Eq. (5.3-5)

$$\sigma_B = \frac{-(-10{,}000)(0.9167)(0.833) + (-10{,}000)(0.3333)(1.167)}{(0.9167)^2 - (0.3333)^2} = 5136 \text{ psi} = \underline{5.14 \text{ kpsi}}$$

(b) At end of beam, $F_n = -F_m = 500\frac{\sqrt{2}}{2} = 250\sqrt{2}$ lb

$$\delta_n = -\delta_m = \frac{(250\sqrt{2})(20)^3}{3(30)(10^6)(0.9167)} = 0.0343 \text{ in} \quad \left(\frac{FL^3}{3EI}\right)$$

$v = \delta_m\cos 45 - \delta_n\sin 45 = \underline{-0.0485 \text{ in}}$

$w = \delta_m\sin 45 + \delta_n\cos 45 = \underline{0}$

5.17

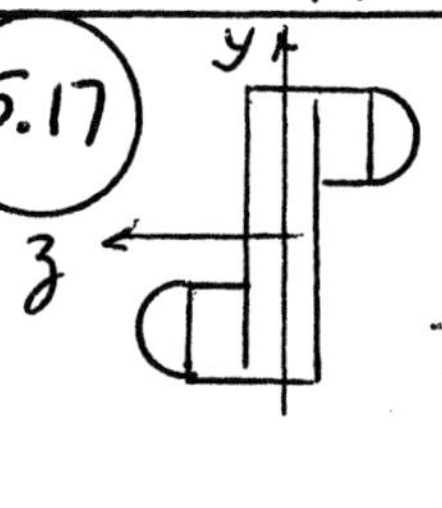

$$I_y = \frac{1}{12}(2)\left(\frac{1}{4}\right)^3 + 2\left[\frac{1}{12}\left(\frac{1}{2}\right)\left(\frac{3}{8}\right)^3 + \left(\frac{1}{2}\right)\left(\frac{3}{8}\right)(0.3125)^2\right]$$
$$+ 2\left[\frac{\pi}{8}\left(\frac{1}{4}\right)^4 - \frac{\pi}{2}\left(\frac{1}{4}\right)^2\left(\frac{4}{3\pi}\cdot\frac{1}{4}\right)^2 + \frac{\pi}{2}\left(\frac{1}{4}\right)^2\left(\frac{4}{3\pi}\cdot\frac{1}{4} + \frac{1}{2}\right)^2\right] = 0.1166 \text{ in}^4$$

$$I_z = \frac{1}{12}\left(\frac{1}{4}\right)(2)^3 + 2\left[\frac{1}{12}\left(\frac{3}{8}\right)\left(\frac{1}{2}\right)^3 + \left(\frac{3}{8}\right)\left(\frac{1}{2}\right)\left(\frac{3}{4}\right)^2\right.$$
$$\left. + 2\left[\frac{\pi}{8}\left(\frac{1}{4}\right)^4 + \frac{\pi}{2}\left(\frac{1}{4}\right)^2\left(\frac{3}{4}\right)^2\right]\right] = 0.4989 \text{ in}^4$$

$$I_{yz} = 2\left[\left(\frac{1}{2}\right)\left(\frac{3}{8}\right)\left(\frac{3}{4}\right)(-0.3125) + \frac{\pi}{2}\left(\frac{1}{4}\right)^2\left(\frac{3}{4}\right)\left(-\frac{4}{3\pi}\cdot\frac{1}{4} - \frac{1}{2}\right)\right] = -0.1771 \text{ in}^4$$

(continued)

$$\theta = \frac{1}{2}\tan^{-1}\left[\frac{2(-0.1771)}{0.4989-0.1166}\right] = -21.41^\circ$$

$$I_m = I_{y'} = \frac{0.4989+0.1166}{2} + \frac{0.1166-0.4989}{2}\cos(-42.82) - (-0.1771)\sin(-42.82)$$

$$= 0.04717 \text{ in}^4$$

$$I_n = I_{z'} = \frac{0.4989+0.1166}{2} - \left(\frac{0.1166-0.4989}{2}\right)\cos(-42.82) + (-0.1771)\sin(-42.82)$$

$$= 0.5683 \text{ in}^4$$

Loading of beam is statically indeterminant. For the load analysis assume deflections are constrained to the xy plane. Assume right reaction is applied subject to the constraint of zero deflection at support.

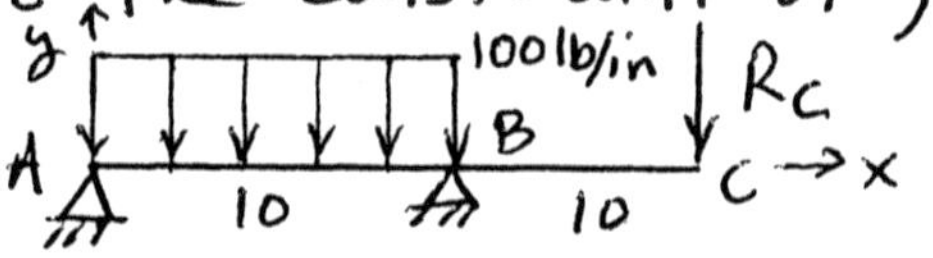

$v_C = 0$

For v_C due to uniform load, find slope at B and multiply by length of 10 in. From beam C.6 of Appendix C

$$\theta_B = \frac{100}{24EI}\left[6(10)(10)^2 - 4(10)^3 - 10^3\right] = 4166.7/EI$$

$$\therefore (v_C)_1 = 10\,\theta_B = 41.667(10^3)/EI$$

For v_C due to R_C see beam C.8

$$(v_C)_2 = \frac{R_C(20-10)}{6EI}\left[(20-10)^2 - 10(60-10)\right] = -666.7R_C/EI$$

$v_C = (v_C)_1 + (v_C)_2 = 0$ yields $R_C = 62.5$ lb

From statics $R_A = 437.5$ lb, $R_B = 625$ lb

Shear and bending moment diagrams

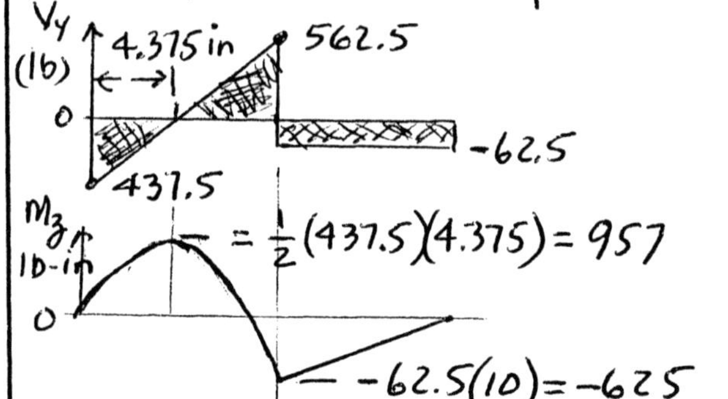

$M_{max} = 957$ lb-in

$M_m = -957 \sin 21.42 = -349$ lb-in

$M_n = 957 \cos 21.42 = 891$ lb-in

Neutral plane:

$$\alpha = \tan^{-1}\left(\frac{891}{-349}\,\frac{0.04717}{0.5683}\right) = -12^\circ$$

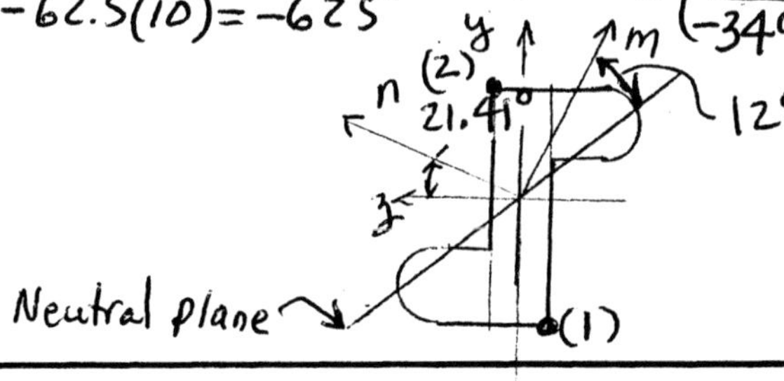

(Continued)

Point(1) $y=-1,\ z=-\frac{1}{8}$

$$m = -1\cos(-21.41) + \left(-\tfrac{1}{8}\right)\sin(-21.41) = -0.885 \text{ in}$$

$$n = -(-1)\sin(-21.41) + \left(-\tfrac{1}{8}\right)\cos(-21.41) = -0.4814 \text{ in}$$

$$\sigma_{max} = -\frac{349(-0.4814)}{0.04717} - \frac{(891)(-0.885)}{0.5683} = 4950 \text{ psi} = \underline{4.95 \text{ kpsi}}$$

Point(2) $y=1,\ z=\frac{1}{8}$ yields $\sigma_{min} = \underline{-4.95 \text{ kpsi}}$

(5.18)

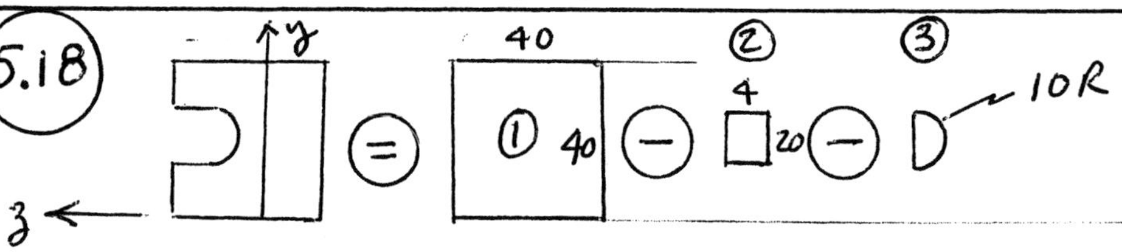

y-axis

$$I_{1y} = \frac{1}{12}40^4 = 213.33(10^3)\text{ mm}^4,\quad I_{2y} = \frac{1}{12}20(4)^3 + 4(20)(18)^2 = 26.03(10^3)\text{ mm}^4$$

$$I_{3y} = \frac{\pi}{8}(10^4) - \frac{\pi}{2}(10)^2\left(\frac{4(10)}{3\pi}\right)^2 + \frac{\pi}{2}(10^2)\left(16 - \frac{4(10)}{3\pi}\right)^2 = 22.81(10^3)\text{ mm}^4$$

$$I_y = 2\left(I_{1y} - I_{2y} - I_{3y}\right) = 329.00(10^3)\text{ mm}^4$$

z-axis

$$I_{1z} = \frac{1}{3}(40)^4 = 853.33(10^3)\text{ mm}^4,\quad I_{2z} = \frac{1}{12}4(20)^3 + 4(20)(20^2) = 34.67(10^3)\text{ mm}^4$$

$$I_{3z} = \frac{\pi}{8}(10)^4 + \frac{\pi}{2}(10)^2(20)^2 = 66.76(10^3)\text{ mm}^4$$

$$I_z = 2\left[I_{1z} - I_{2z} - I_{3z}\right] = 1.5038(10^6)\text{ mm}^4$$

y-z

$$I_{1yz} = 0,\quad I_{2yz} = 20(18)(4)(20) = 28.8(10^3)\text{ mm}^4$$

$$I_{3yz} = (20)\left[16 - \frac{4(10)}{3\pi}\right]\left(\frac{\pi}{2}\right)(10^2) = 36.93(10^3)\text{ mm}^4$$

$$I_{yz} = 2\left(I_{1yz} - I_{2yz} - I_{3yz}\right) = -131.46(10^{-3})\text{ mm}^4$$

$$\theta = \frac{1}{2}\tan^{-1}\left(\frac{(2)(-131.46)}{1503.8 - 329}\right) = -6.31^\circ$$

$$I_m = I_{y'} = \left[\frac{329 + 1503.8}{2} + \frac{329 - 1503.8}{2}\cos(-12.62) - (-131.46)\sin(-12.62)\right](10^3)$$

$$= 314.5(10^3)\text{ mm}^4$$

$$I_n = I_{z'} = \left[\frac{329 + 1503.8}{2} - \frac{329 - 1503.8}{2}\cos(-12.62) + (-131.46)\sin(-12.62)\right](10^3)$$

$$= 1.5183(10^6)\text{ mm}^4$$

(Continued)

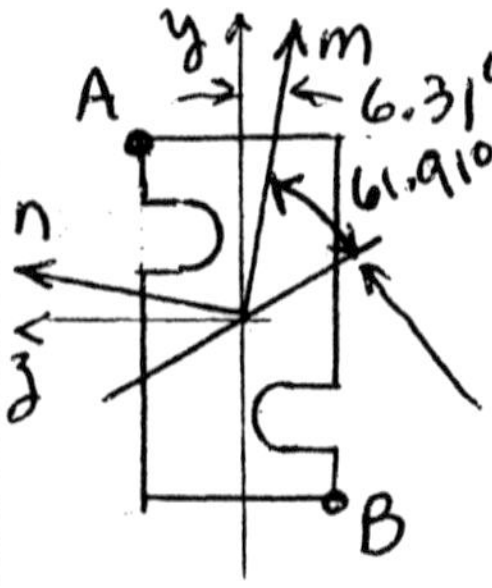

$M_z = -2000(0.75) = -1500 \text{ N·m}$

$M_m = 1500 \sin 6.31 = 164.8 \text{ N·m}$

$M_n = -1500 \cos 6.31 = -1491 \text{ N·m}$

$$\alpha = \tan^{-1}\left(\frac{-1491}{164.8}\,\frac{314.5}{1518.3}\right) = -61.91^\circ$$

Point A: $y = 40 \text{ mm}$, $z = 20 \text{mm}$

$m = 40\cos(-6.31) + 20\sin(-6.31) = 37.56 \text{ mm}$

$n = -40\sin(-6.31) + 20\cos(-6.31) = 24.27 \text{ mm}$

$$\sigma_{max} = \frac{164.8(24.27)(10^{-3})}{314.5(10^{3})(10^{-3})^4} - \frac{(-1491)(37.56)(10^{-3})}{1.5183(10^{6})(10^{-3})^4} = 49.6(10^6) \text{ Pa}$$

$= \underline{49.6 \text{ MPa}}$ At point B: $\sigma_{min} = \underline{-49.6 \text{ MPa}}$

Deflection of right end $\left(\delta = \frac{FL^3}{3EI}\right)$

End force $F_m = -2\cos 6.31 = -1.988 \text{ kN}$

$$\delta_m = \frac{-1.988(10^3)(0.75)^3}{3(210)(10^9)(1.5183)(10^6)(10^{-3})^4} = -8.77(10^{-4}) \text{m} = -0.877 \text{ mm}$$

$F_n = -2\sin 6.31 = -0.2197 \text{ kN}$

$$\delta_n = \frac{-0.2197(10^3)(0.75)^3}{3(210)(10^9)(314.5)(10^3)(10^{-3})^4} = -4.68(10^{-4}) \text{m} = -0.468 \text{ mm}$$

$v = -0.877\cos(-6.31) - (-0.468)\sin(-6.31) = \underline{-0.923 \text{ mm}}$

$w = -0.877\sin(-6.31) - (0.468)\cos(-6.31) = \underline{-0.369 \text{ mm}}$

$\delta_{total} = \sqrt{\delta_m^2 + \delta_n^2} = \sqrt{v^2 + w^2} = \underline{0.994 \text{mm}}$

5.19

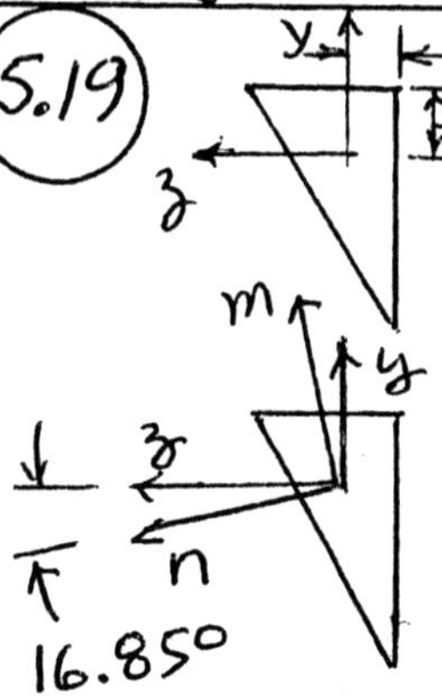

From Appendix B

$I_y = \frac{1}{36}(150)(75)^3 = 1.758(10^6) \text{ mm}^4$

$I_z = \frac{1}{36}(75)(150)^3 = 7.031(10^6) \text{ mm}^4$

$I_{yz} = \frac{1}{72}(75)^2(150)^2 = 1.758(10^6) \text{ mm}^4$

$$\theta = \frac{1}{2}\tan^{-1}\left[\frac{2(1.758)}{7.031 - 1.758}\right] = 16.85^\circ$$

(continued)

$$I_{\substack{m\\n}} = \left[\frac{1}{2}(1.758+7.031) \pm \frac{1}{2}(1.758-7.031)\cos 33.7 \mp 1.758\sin 33.7\right](10^6)$$

$I_m = 1.226(10^6)\ mm^4$, $I_n = 7.563(10^6)\ mm^4$

$M_z = -1000(1.5) = -1500$ N-m, $M_m = -1500 \sin 16.85 = -435$ N·m

$M_n = -1500 \cos 16.85 = -1436$ N·m

Neutral plane $\alpha = \tan^{-1}\left(\frac{-1436}{-435}\ \frac{1.226}{7.563}\right) = 28.15°$

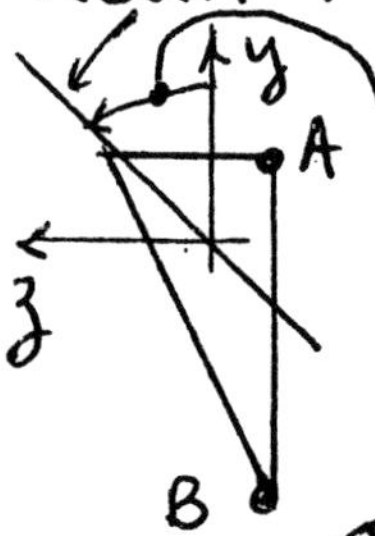

$16.85 + 28.15 = 45°$ Points A & B are farthest away.

Point A: $y = 50$ mm $z = -25$ mm

$m = 50\cos 16.85 - 25 \sin 16.85 = 40.61$ mm

$n = -50\sin 16.85 - 25\cos 16.85 = -38.42$ mm

$$\sigma_A = \frac{(-435)(-38.42)(10^{-3})}{1.226(10^6)} - \frac{(-1436)(40.61)(10^3)}{7.563(10^6)} = \underline{21.34\ MPa}$$

Point B: $y = -100$ mm $z = -25$ mm

$m = -100\cos 16.85 - 25\sin 16.85 = -102.95$ mm

$n = 100 \sin 16.85 - 25\cos 16.85 = 5.056$ mm

$$\sigma_B = \frac{(-435)(5.056)(10^3)}{1.226(10^6)} - \frac{(-1436)(-102.95)(10^3)}{7.563(10^6)} = \underline{-21.34\ MPa}$$

(b) At end, $F_m = -1000\cos 16.85 = -957 N$, $F_n = 1000 \sin 16.85 = 289.8 N$

Using $FL^3/3EI$

$$\delta_m = \frac{-957(1.5)^3}{3(70)(10^9)(7.563)(10^6)(10^{-3})^4} = -2.034(10^{-3})m = -2.034\ mm$$

$$\delta_n = \frac{289.8(1.5)^3}{3(70)(10^9)(1.226)(10^6)(10^{-3})^4} = 3.799(10^{-3})m = 3.799\ mm$$

$v = -2.034\cos 16.85 - 3.799\sin 16.85 = \underline{-3.048\ mm}$

$w = -2.034\sin 16.85 + 3.799\cos 16.85 = \underline{3.046\ mm}$

$\delta_{max} = \sqrt{\delta_m^2 + \delta_n^2} = \sqrt{v^2 + w^2} = \underline{4.31\ mm}$

(5.20) With $\beta = 0$ $I_y = I_z = \frac{1}{12}a^4$, $I_{yz} = 0$

For any value of θ in Eqs. (E.2-3) $I_{y'} = I_{z'} = \frac{1}{12}a^4$, $I_{y'z'} = 0$

For $\beta = 0$, upper left corner coordinates are $(a/2, a/2)$

For β, $y_{max} = a/2 \cos\beta + a/2 \sin\beta = \frac{\sqrt{2}}{2} a \cos(\pi/4 - \beta)$

$\sigma_{max} = \frac{M_z y_{max}}{I} = \frac{12 M_z}{a^4}\frac{\sqrt{2}}{2}\cos\left(\frac{\pi}{4}-\beta\right) = \frac{6\sqrt{2}}{a^3} M_z \cos\left(\frac{\pi}{4} - \beta\right)$ **ANS**

(5.21)

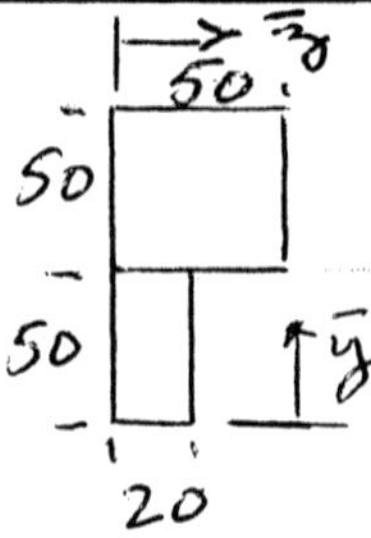

$$A = 20(50) + 50(50) - \tfrac{1}{2}\pi(12)^2 = 3273.8 \text{ mm}^2$$

$$\bar{y} = \frac{25(20)(50) + 75(50)^2 - \left(50 + \frac{4}{3}\cdot\frac{12}{\pi}\right)\left(\frac{1}{2}\right)(\pi)(12)^2}{3273.8} = 61.103 \text{ mm}$$

$$\bar{z} = \frac{10(20)(50) + 25(50)^2 - 32\left(\frac{1}{2}\right)(\pi)(12)^2}{3273.8} = 19.935 \text{ mm}$$

$$I_y = \frac{1}{12}(50)(20)^3 + 20(50)(19.935 - 10)^2 + \frac{1}{12}50^4 + 50^2(25 - 19.935)^2 - \frac{\pi}{8}(12)^4 - \frac{\pi}{2}(12)^2(32 - 19.935)^2 = 675.94(10^3)\text{ mm}^4$$

$$I_z = \frac{1}{12}(20)(50)^3 + 20(50)(61.103 - 25)^2 + \frac{1}{12}50^4 + 50^2(75 - 61.103)^2 - \left[\frac{\pi}{8}(12)^4 - \frac{\pi}{2}(12)^2\left(\frac{4}{3}\,\frac{12}{\pi}\right)^2\right] - \frac{\pi}{2}(12)^2\left(50 + \frac{4}{3}\,\frac{12}{\pi} - 61.103\right)^2 = 2.505(10^6)\text{ mm}^4$$

$$I_{yz} = 20(50)(25 - 61.103)(19.935 - 10) + 50^2(75 - 61.103)(19.935 - 25) - \left[\frac{\pi}{2}(12)^2\left(50 + \frac{4}{3}\,\frac{12}{\pi} - 61.103\right)(19.935 - 35)\right] = -555.13(10^3)\text{ mm}^4$$

$$\theta_p = \frac{1}{2}\tan^{-1}\left[\frac{2(-555.13)}{2505 - 675.94}\right] = -15.63^\circ$$

$$I_{m,n} = \left[\frac{1}{2}(675.94 + 2505) \pm \frac{1}{2}(675.94 - 2505)\cos(-31.26) \mp 555.13\sin(-31.26)\right](10^3)$$

$$I_m = 520.6(10^3)\text{ mm}^4,\quad I_n = 2.660(10^6)\text{ mm}^4$$

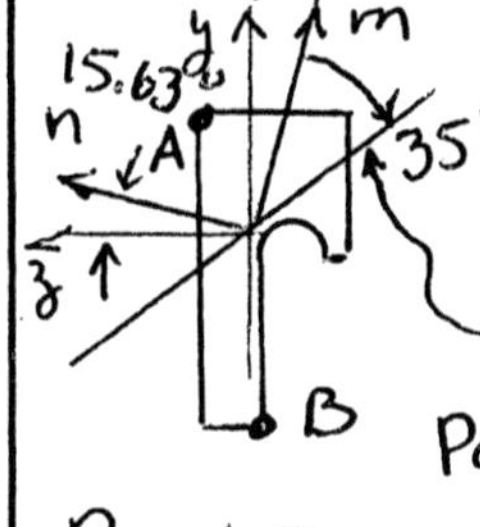

$M_z = -4000(0.5) = -2000$ Nm, $M_m = 2000\sin 15.63 = 538.8$ Nm

$M_n = -2000\cos 15.63 = -1926.1$ Nm

Neutral plane: $\alpha = \tan^{-1}\left(\frac{-1926.1}{538.8}\,\frac{520.6}{2660}\right) = -35^\circ$

Points A & B are farthest from Neutral Plane

Point A: $y = 100 - 61.103 = 38.9$ mm, $z = 19.94$ mm

$m = 38.9\cos(-15.63) + 19.94\sin(-15.63) = 32.09$ mm

$n = -38.9\sin(-15.63) + 19.94\cos(-15.63) = 29.68$ mm

$$\sigma_A = \frac{538.8(29.68)(10^3)}{520.6(10^3)} - \frac{(-1926.1)(32.09)(10^3)}{2.660(10^6)} = \underline{54.0 \text{ MPa}}$$

Point B: $y = -61.103$ mm, $z = 19.935 - 20 = -0.065$ mm

$m = -61.103\cos(-15.63) - 0.065\sin(-15.63) = -58.83$ mm

$n = 61.103\sin(-15.63) - 0.065\cos(-15.63) = -16.52$ mm

$$\sigma_B = \frac{(538.8)(-16.52)(10^3)}{520.6(10^3)} - \frac{(-1926.1)(-58.83)(10^3)}{2.660(10^6)} = \underline{-59.7 \text{ MPa}}$$

(continued)

(b) Force at end $F_n = -4000 \sin 15.63 = -1078\,N$

$F_m = -4000 \cos 15.63 = -3852\ N$

For deflections, $\delta = FL^3/3EI$

$$\delta_m = \frac{-3852(0.5)^3}{3(70)(10^9)(2.660)(10^6)(10^{-3})^4} = -862(10^{-6})\,m = -0.862\ mm$$

$$\delta_n = \frac{-1078(0.5)^3}{3(70)(10^9)(520.6)(10^3)(10^{-3})^4} = -1.233(10^{-3})\,m = -1.233\ mm$$

y, z deflections, transform

$v = -0.862 \cos(-15.63) + 1.233 \sin(-15.63) = \underline{-1.162\ mm}$

$w = -0.862 \sin(-15.63) - 1.233 \cos(-15.63) = \underline{-0.955\ mm}$

$\delta_{max} = \sqrt{\delta_m^2 + \delta_n^2} = \sqrt{v^2 + w^2} = \underline{1.504\ mm}$

(5.22) Ignoring terms of order t^3

$$I_z \approx \frac{1}{12}(2)(24)^3 + 2(12)(2)(12)^2 + 2\left[\frac{1}{12}(2)(12)^3 + 12(2)(18)^2\right] + 2(12)(2)(24)^2$$
$$= 52.99(10^3)\ mm^4$$

AB

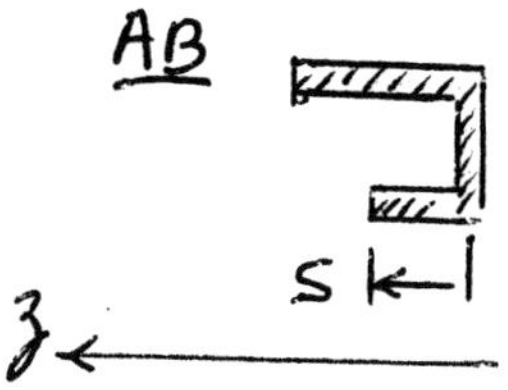

$Q = \Sigma \bar{y} A = 24(2)(12) + 18(2)(12) + 12(2)s$
$= 1008 + 24s \quad mm^3$

$$\tau = \frac{VQ}{It} = \frac{(4000)(1008+24s)}{52.99(10^3)(2)} = \underline{38.04 + 0.9058s\ MPa}$$
where s in mm

Net force: $V_{AB} = \int_0^{12} \tau t\, ds = \int_0^{12} (38.04 + 0.9058s)(2)\,ds = 1043\,N$

$= \underline{1.043\ kN}$

(5.23) Ignoring terms of order t^3

$I_z \approx \frac{1}{12}(0.1)(2)^3 + \frac{1}{12}(0.1)(3)^3 = 0.2917\ in^4$

From Example 3.4-3 $\tau_{xy} = \frac{V_y}{2I_z}\left[\left(\frac{h}{2}\right)^2 - y^2\right]$ for vertical walls

Right wall: $\tau_1 = \frac{V_y}{2(0.2917)}(1.5^2 - y^2)$

Net force $V_1 = 2\int_0^{1.5} \frac{V_y}{(0.5834)}(2.25 - y^2)(0.1)\,dy = 0.7714\,V_y$

(continued)

Although unnecessary check left wall

$\tau_2 = \dfrac{V_y}{2(0.2917)}(1-y^2)$, $V_2 = 2\displaystyle\int_0^1 \frac{V_y}{0.5834}(1-y^2)(0.1)dy = 0.2286\,V_y$

Note that $V_1 + V_2 = V_y$

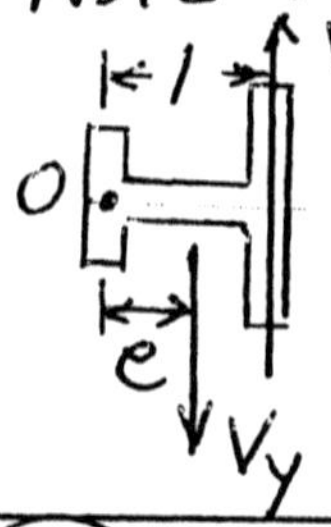

$\Sigma M_0 = 0 \Rightarrow (0.7714\,V_y)(1) - V_y e = 0$

$\underline{e = 0.771 \text{ in}}$

(5.24) $I_z = \dfrac{1}{2}\left[\dfrac{\pi}{8}t(2r)^3\right] = \dfrac{\pi}{2}tr^3$

$A = r\theta t$, $\bar{r} = \dfrac{2r}{\theta}\sin\left(\dfrac{\theta}{2}\right)$

$\bar{y} = \bar{r}\cos\dfrac{\theta}{2} = \dfrac{2r}{\theta}\sin\dfrac{\theta}{2}\cos\dfrac{\theta}{2} = \dfrac{r}{\theta}\sin\theta$

$Q = \bar{y}A = r^2 t\sin\theta$

(a) $\tau = \dfrac{V_y Q}{I_z t} = \dfrac{V_y r^2 t\sin\theta}{\frac{\pi}{2}tr^3 t} = \dfrac{2}{\pi}\dfrac{V_y}{rt}\sin\theta$ ANS

(b) $\Sigma M_0 = 0 \quad V_y e - 2\displaystyle\int_0^{\pi/2} r\tau(t r\,d\theta) = 0$

$V_y e - 2\left(\dfrac{2}{\pi}\dfrac{V_y}{r}\right)tr^2\displaystyle\int_0^{\pi/2}\sin\theta\,d\theta = 0$

$e = \dfrac{4r}{\pi}(-\cos\theta)\Big|_0^{\frac{\pi}{2}} = \dfrac{4}{\pi}r$ ANS

(5.25) From the solution to Prob 5.22 $I_z \simeq 52.99(10^3)\,mm^4$ and $V_{AB} = 1043$ N.

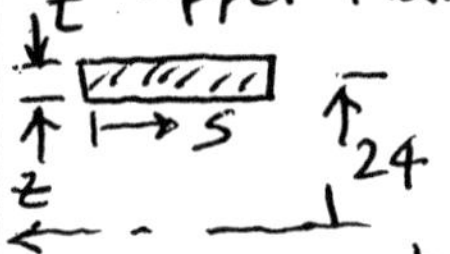

Upper flange: $Q_1 = \bar{y}A = 24st$, $\tau_1 = \dfrac{4000(st)}{52.99(10^3)(2)} = 1.8125$ MPa, s in mm

$V_1 = \displaystyle\int_0^{12}\tau t\,ds = \int_0^{12}(1.8125)(2)ds = 261$ N

Upper vert. wall:

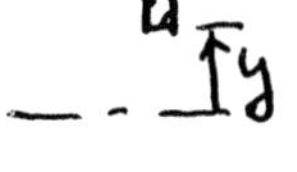

$Q_2 = 24(12)(2) + \dfrac{2}{2}(24^2 - y^2) = 1152 - y^2$

$\tau_2 = \dfrac{4000(1152-y^2)}{52.99(10^3)(2)} = 43.48 - 0.03774\,y^2$ MPa, y in mm

$V_2 = \displaystyle\int_{12}^{24}(43.48 - 0.03774\,y^2)(2)dy = 739$ N

(continued)

Center vert. wall (only for a check)

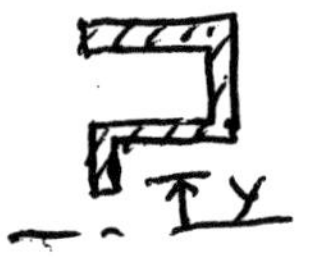

$$Q = 24(2)(2) + \frac{2}{2}(24^2 - 12^2) + 12(12)(2) + \frac{2}{2}(12^2 - y^2)$$
$$= 1440 - y^2$$

$$\tau_3 = \frac{4000(1440-y^2)}{52.99(10^3)(2)} = 54.35 - 0.03774y^2 \text{ MPa, } y \text{ in mm}$$

$$V_3 = 2\int_0^{12}(54.35 - 0.03774y^2)(2)\,dy = 2522 \text{ N}$$

Note $V_3 + 2V_2 = 4000 \text{ N}$ (check of forces)

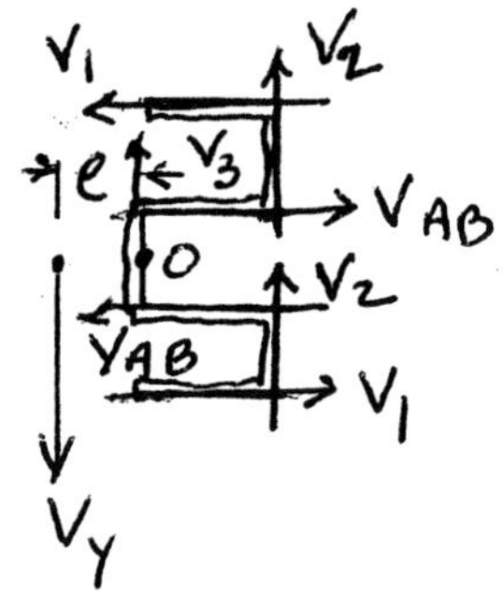

$$\Sigma M_O = 0$$

$$V_y e - V_{AB}(24) + 2(V_2)(12) + V_3(48) = 0$$

$$e = \frac{(1043)(24) - 2(739)(12) - 261(48)}{4000}$$
$$= -1.305 \text{ in}$$ ($\therefore$ shear center is to the right of O)

(5.26) (a)

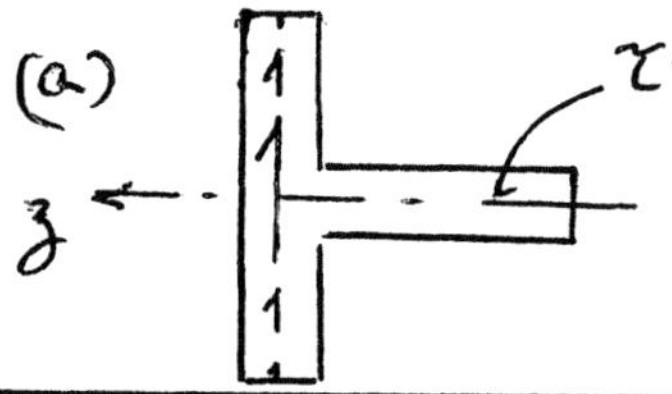

(b) $\underline{e = 0}$ Load must go through web to avoid torsion.

(5.27)

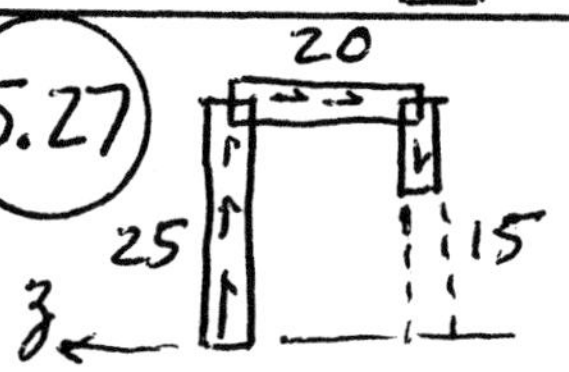

$$I_z \approx 2\left[\frac{2}{3}(4)(25)^3 + 4(20)(25)^2 - \frac{1}{3}(4)(15)^3\right] = 174.3(10^3)\text{ mm}^4$$

Right vert. section

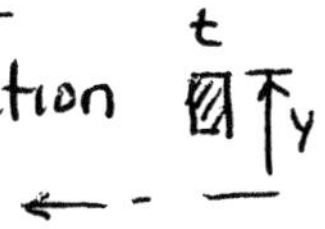

$$Q_1 = \frac{t}{2}(y^2 - 15^2), \quad \tau_1 = \frac{V\frac{t}{2}(y^2-225)}{It} = \frac{V(y^2-225)}{2I}$$

$$V_1 = \int_{15}^{25}\frac{V(y^2-225)}{2(174.3)(10^3)}(4)\,dy = 0.0210\,V$$

Horiz. Section

$$Q_2 = \Sigma\bar{y}A = (15 + \tfrac{10}{2})(10)t + 25st = 25(8+s)t$$

$$\tau_2 = \frac{V[25(8+s)t]}{It} = 25(8+s)\frac{V}{I}$$

z

$$V_2 = \int_0^{20}\tau_2 t\,ds = \int_0^{20}\frac{25(8+s)V}{174.3(10^3)}(4)\,ds = 0.2065\,V$$

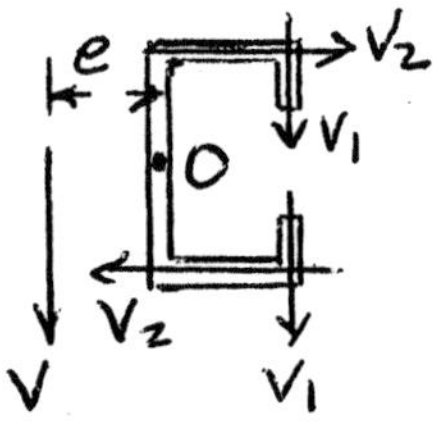

$$\Sigma M_O = 0 \quad Ve - 2V_2(25) - 2V_1(20) = 0$$

$$e = 2(0.2065)(25) + 2(0.0210)(20)$$
$$= \underline{11.17 \text{ mm}}$$

5.28

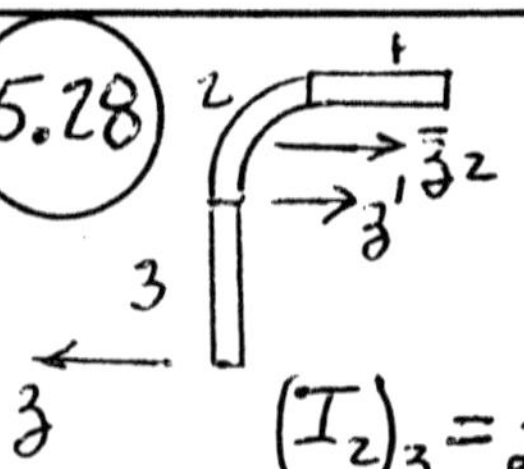

$$(I_1)_z \approx (2)(10)(20)^2 = 8(10^3)\ \text{mm}^4$$

For section 2 $\quad I_{z'} = \frac{\pi}{4} t r^3 = \frac{\pi}{4}(2)(10^3) = 1.571(10^3)\ \text{mm}^4$

$$I_{z_2} = 1.571(10^3) - \frac{1}{4}(2\pi)(10)(2)\left[\frac{2(10)}{\pi}\right]^2 = 297.8\ \text{mm}^4$$

$$(I_2)_z = 297.8 + \frac{1}{4}(2\pi)(10)(2)\left[10 + \frac{2}{\pi}(10)\right]^2 = 8.713(10^3)\ \text{mm}^4$$

$$(I_3)_z = \frac{1}{3}(2)(10^3) = 666.7\ \text{mm}^4$$

$$I_z = 2\left[(I_1)_z + (I_2)_z + (I_3)_z\right] = 34.76(10^3)\ \text{mm}^4$$

Section 1: $\quad Q_1 = 20st, \quad \tau_1 = \dfrac{V(20st)}{34.76(10^3)t} = 5.754(10^{-4})Vs$

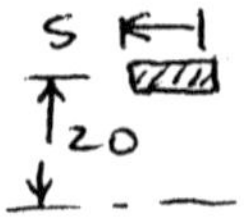

$$V_1 = \int_0^{10} 5.754(10^{-4})Vs(2)\,ds = 5.754(10^{-2})V \quad ①$$

Section 2: (see prob 5.24 solution)

$$Q_2 = \Sigma \bar{y}A = 20(10)t + \left(\frac{r\sin\theta}{\theta} + 10\right) r\theta t$$

$$= (200 + 100\sin\theta + 100\theta)t$$

$$\tau_2 = \frac{V(100)(2+\sin\theta+\theta)t}{34.76(10^3)t} = 2.877(10^{-3})V(2+\sin\theta+\theta) \quad ②$$

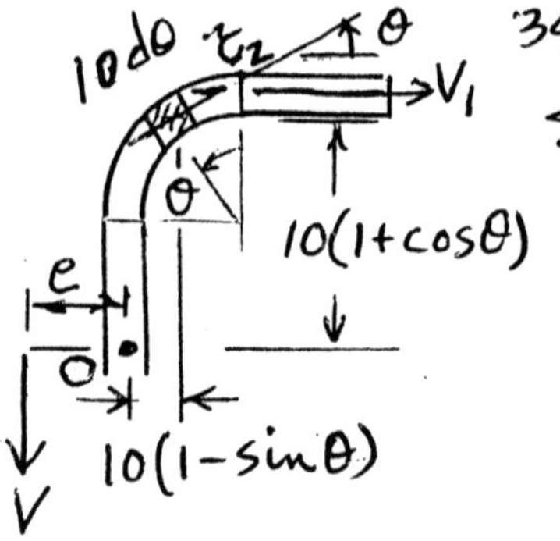

$\Sigma M_O = 0$

$$Ve = 2V_1(20) + 2\int_0^{\pi/2}(\tau_2\cos\theta)\left[10(1+\cos\theta)\right](2)(10\,d\theta)$$

$$-2\int_0^{\pi/2}(\tau_2\sin\theta)\left[10(1-\sin\theta)\right](2)(10\,d\theta)$$

Sub in ① & ②, cancel V

$$e = 2.302 + 1.1508\int_0^{\pi/2}(2+\sin\theta+\theta)(\cos\theta)(1+\cos\theta)\,d\theta$$

$$-1.1508\int_0^{\pi/2}(2+\sin\theta+\theta)(\sin\theta)(1-\sin\theta)\,d\theta$$

$$= 2.302 + 1.1508\int_0^{\pi/2}(2\cos\theta + \sin\theta\cos\theta + \theta\cos\theta + 2\cos^2\theta + \cos^2\theta\sin\theta + \theta\cos^2\theta)\,d\theta$$

$$-1.1508\int_0^{\pi/2}(2\sin\theta - \sin^2\theta + \theta\sin\theta - \sin^3\theta - \theta\sin^2\theta)\,d\theta$$

(Note: The last term in each integral can be combined to yield θ)

Integration yields

$$\underline{e = 7.67\ \text{mm}}$$

5.29 $I_z \approx \frac{1}{12}(0.1)(2.4)^3 + 2\left[(0.1)(1)(1.2)^2 + \frac{1}{12}(0.1)(0.4)^3 + (0.1)(0.4)(1.4)^2\right]$

$= 0.5611 \text{ in}^4$

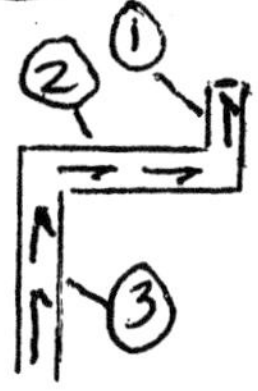

Section ①

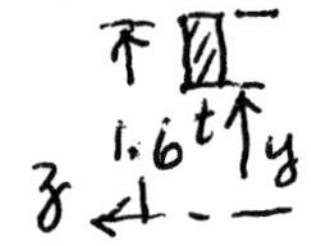

$Q_1 = \frac{t}{2}(1.6^2 - y^2)$, $\tau_1 = \dfrac{V_y \frac{t}{2}(2.56 - y^2)}{0.5611t} = 0.891 V_y (2.56 - y^2)$

$V_1 = \int_{1.2}^{1.6} 0.891 V_y (2.56 - y^2)(0.1)dy = 0.02091 V_y$

Section ②

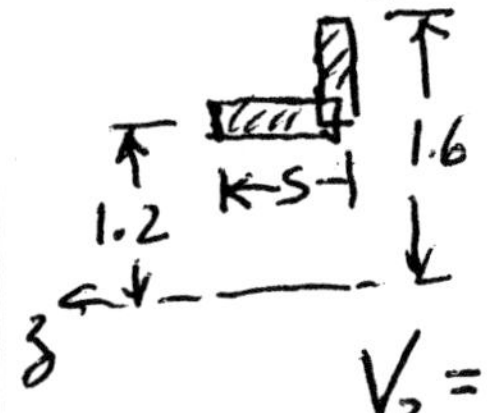

$Q_2 = \Sigma \bar{y} A = \frac{t}{2}(1.6^2 - 1.2^2) + 1.2st = (0.56 + 1.2s)t$

$\tau_2 = \dfrac{V_y t(0.56 + 1.2s)}{0.5611t} = (0.998 + 2.139s)V_y$

$V_2 = \int_0^1 (0.998 + 2.139s)V_y(0.1)ds = 0.2067 V_y$

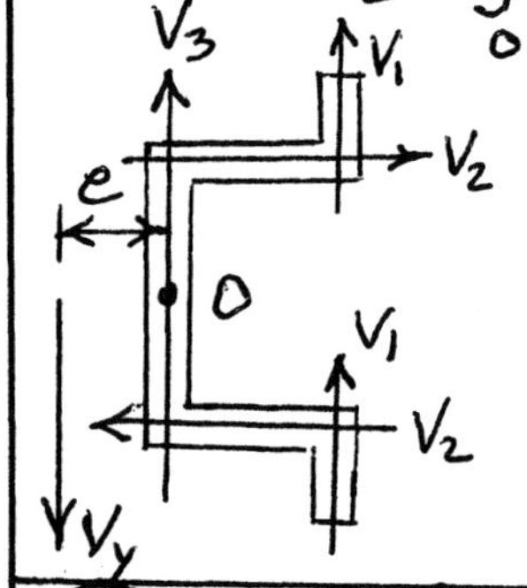

$\Sigma M_0 = 0$

$V_y e + 2V_1(1.0) - V_2(2.4) = 0$

$e = 2.4(0.2067) - 2(0.02091)$

$= 0.454 \text{ in}$

5.30

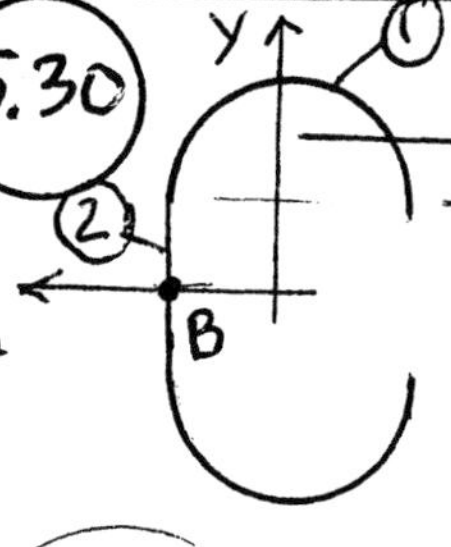

$(I_1)_z = \underbrace{\frac{\pi}{2}(0.1)(0.5)^3 - \pi(0.5)(0.1)\left(\frac{1}{\pi}\right)^2}_{(I_1)_{\bar{z}_1}} + \underbrace{\pi(0.5)(0.1)}_{A}\left(0.5 + \frac{1}{\pi}\right)^2$

$= 0.1089 \text{ in}^4$, $(I_2)_z = \frac{1}{12}(0.1)(1)^3 = 8.33(10^{-3}) \text{ in}^4$

$I_z = (I_2)_z + 2(I_1)_z = 0.2261 \text{ in}^4$

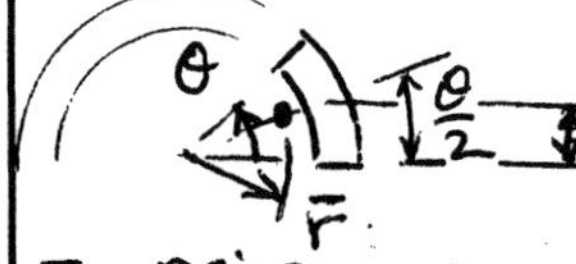

$\bar{r}\sin\frac{\theta}{2} = \frac{2r}{\theta}\sin^2\frac{\theta}{2}$

$\bar{r} = \dfrac{r\sin\theta/2}{\theta/2} = \frac{2r}{\theta}\sin\frac{\theta}{2}$

$Q_1 = \bar{y}A = \left(0.5 + \frac{2r}{\theta}\sin^2\frac{\theta}{2}\right)(r\theta t)$

$r = 0.5$

$Q_1 = 0.25\left(\theta + 2\sin^2\frac{\theta}{2}\right)t$

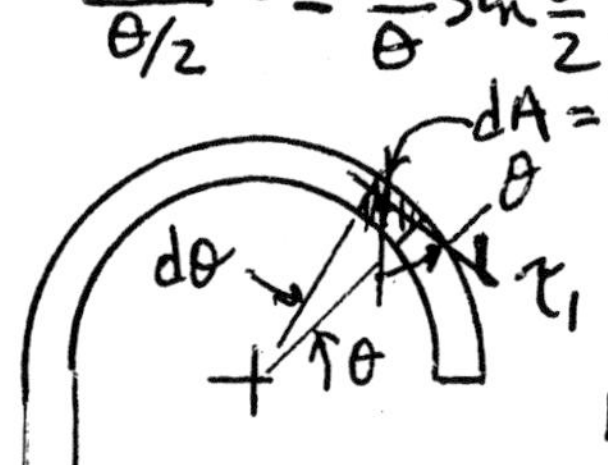

$\tau_1 = \dfrac{V_y(0.25)\left(\theta + 2\sin^2\frac{\theta}{2}\right)t}{0.2261t} = 1.106\left(\theta + 2\sin^2\frac{\theta}{2}\right)V_y$

Determine the moment about point B due to τ_1

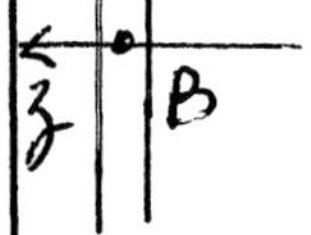

$M_B = \int_0^{\pi} 1.106\left(\theta + 2\sin^2\frac{\theta}{2}\right)V_y(\sin\theta)(0.5 + 0.5\sin\theta)tr\,d\theta$

$+ \int_0^{\pi} 1.106\left(\theta + 2\sin^2\frac{\theta}{2}\right)V_y(\cos\theta)(0.5 + 0.5\cos\theta)tr\,d\theta$

$t = 0.1$, $r = 0.5$ (continued)

$$M_B = 0.0276\,V_y \int_0^{\pi} \Big[\theta\sin\theta + \theta\sin^2\theta + 2\sin\theta\sin^2\tfrac{\theta}{2} + 2\sin^2\theta\sin^2\tfrac{\theta}{2} + \theta\cos\theta + 2\cos\theta\sin^2\tfrac{\theta}{2} + \theta\cos^2\theta + 2\cos^2\theta\sin^2\tfrac{\theta}{2}\Big]d\theta$$

$$= 0.0276\,V_y \int_0^{\pi} \Big[\theta + \theta\sin\theta + \theta\cos\theta + 2\sin\theta\sin^2\tfrac{\theta}{2} + 2\cos\theta\sin^2\tfrac{\theta}{2} + 2\sin^2\tfrac{\theta}{2}\Big]d\theta$$

Use $\sin^2\frac{\theta}{2} = \frac{1-\cos\theta}{2}$ and integrate yielding

$$M_B = 0.2667\,V_y$$

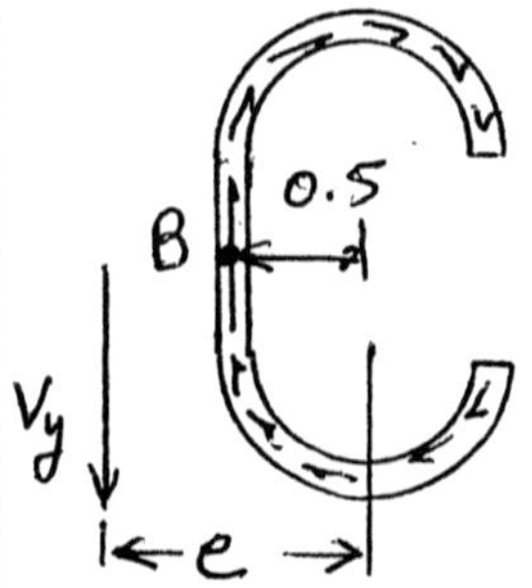

$$\Sigma M_B = 2(0.2667V_y) - V_y(e-0.5) = 0$$

$$\underline{e = 1.033 \text{ in}}$$

(5.31) $I_z \approx \frac{1}{12}(3)(80)^3 + 2(3)(30)(20^2+40^2) = 488(10^3)\,\text{mm}^4$

(a) Shear in upper and lower flanges are the same as the channel in Example 5.4-1 where

$$V_1 = \frac{V_y a b^2 t}{2I_z} = \frac{V_y(40)(30^2)(3)}{2(488)(10^3)} = 0.1107\,V_y$$

For inner flanges, let $a = 20$ mm

$$V_2 = \frac{V_y(20)(30^2)(3)}{2(488)(10^3)} = 0.05533\,V_y$$

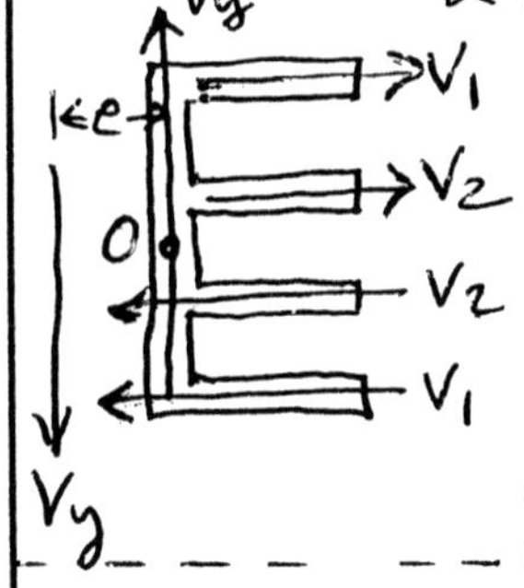

$$\Sigma M_O = 0 \quad V_y e - 2[(0.05533V_y)(20) + (0.1107V_y)(40)] = 0$$

$$\underline{e = 11.07 \text{ mm}}$$

(b)

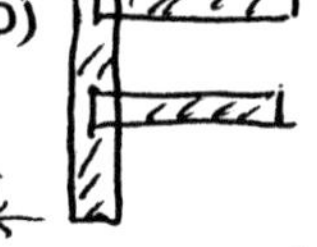

$$Q = \Sigma \bar{y}A = 40(30)(3) + 20(30)(3) + 20(40)(3) = 7.80(10^3)\,\text{mm}^3$$

$$\tau_{max} = \frac{(500)(7.80)(10^3)}{488(10^3)(3)} = \underline{2.66 \text{ MPa}}$$

5.32 (a)

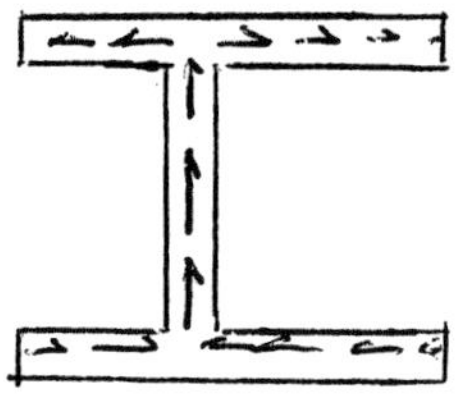

(b) $I_z \approx \frac{1}{12}(4)(200)^3 + 2[150(4)(100)^2] = 14.667(10^6)\,\text{mm}^4$

upper and lower right flange. From Example 5.4-1

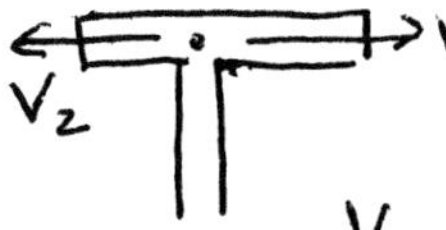

$$V_1 = \frac{V_y(100)(100^2)(4)}{2(14.667)(10^6)} = 0.13636\,V_y$$

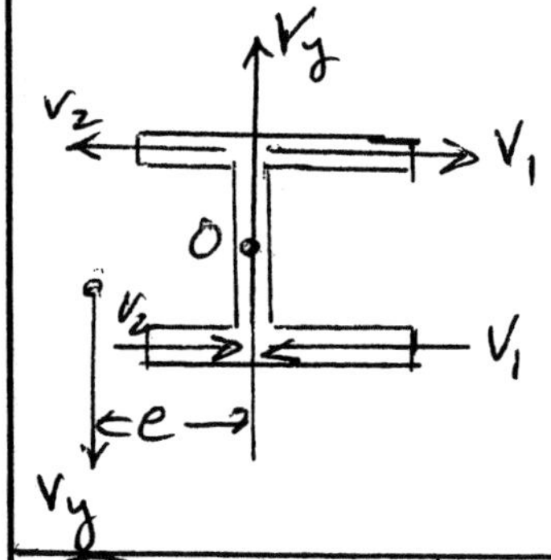

$$V_2 = \frac{V_y(100)(50)^2(4)}{2(14.667)(10^6)} = 0.03409\,V_y$$

$\Sigma M_O = 0$

$$V_y e - V_1(200) + V_2(200) = 0$$

$$e = 0.13636(200) - 0.03409(200)$$

$$= \underline{20.5\text{ mm}}$$

5.33

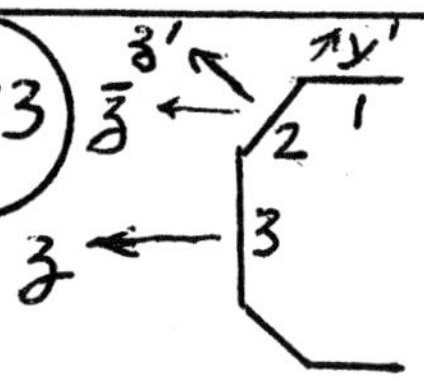

$(I_1)_{\bar{z}} \approx 2(0.1)4^2 = 3.2\,\text{in}^4$, $(I_3)_{\bar{z}} = \frac{1}{12}(0.1)4^3 = 0.533\,\text{in}^4$

$(I_2)_{\bar{z}'} = \frac{1}{12}(0.1)(\sqrt{2}\cdot 2)^3 = 0.1886$, $(I_2)_{y'} \approx 0$, $(I_2)_{y'\bar{z}'} = 0$

Transform, $\theta = 45°$

$$(I_2)_{\bar{z}} = \frac{0.1886}{2} + \frac{0.1886}{2}\cos 90 = 0.0943\,\text{in}^4$$

$$(I_2)_z = 0.0943 + (2\sqrt{2})(0.1)3^2 = 2.640\,\text{in}^4$$

$$I_z = 2(I_1)_z + 2(I_{2z}) + (I_3)_z = 12.21\,\text{in}^4$$

Outer flanges: Same as channel in Example 5.4-1

$$V_1 = \frac{V_y}{2I_z}ab^2t = \frac{V_y}{2(12.21)}(4)(2)^2(0.1) = 0.0655\,V_y$$

Section 2: $Q_2 = \Sigma \bar{y}A = 4(2)t + (4 - \frac{\sqrt{2}}{4}s)st = (8 + 4s - \frac{\sqrt{2}}{4}s^2)t$

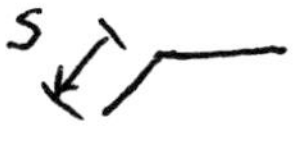

$$\tau_2 = \frac{V_y(8 + 4s - \frac{\sqrt{2}}{4}s^2)t}{12.21t}$$

$$V_2 = \frac{V_y}{12.21}\int_0^{2\sqrt{2}}(8 + 4s - \frac{\sqrt{2}}{4}s^2)(0.1)ds = 0.2944\,V_y$$

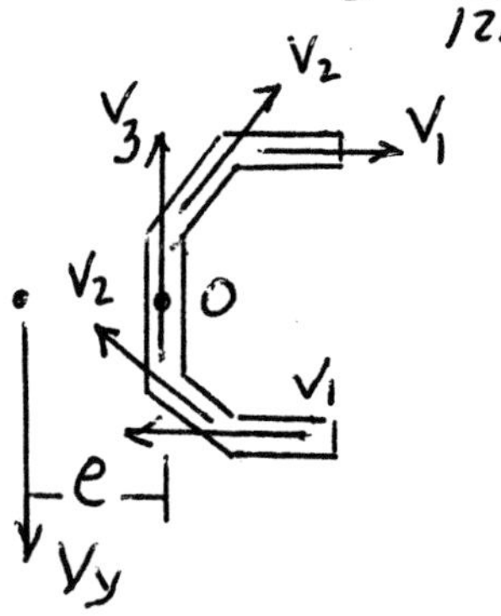

$\Sigma M_O = 0$

$$V_y e - 8V_1 - 4\left(\frac{\sqrt{2}}{2}V_2\right) = 0$$

$$e = 8(0.0655) + 2\sqrt{2}(0.2944)$$

$$= \underline{1.357\text{ in}}$$

5.34 The results for V_y are given in Example 5.4-1

$$I_z \approx \frac{2}{3}(2.5)(50^3) + 2(2.5)(50)(50^2) = 833.3(10^3)\ \text{mm}^4$$

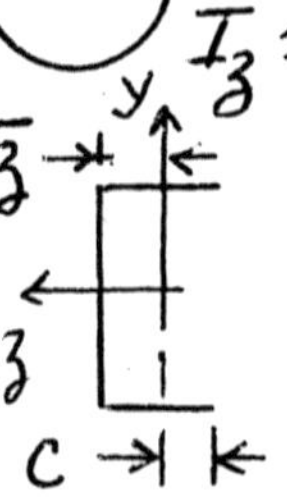

$$\bar{z} \approx \frac{(2)(25)(2.5)(50)}{(2)(2.5)(50) + (2.5)(100)} = 12.5\ \text{mm}$$

$$I_y \approx 2\left[\frac{1}{12}(2.5)(50^3) + (2.5)(50)(25-12.5)^2\right] + (2.5)(100)(12.5)^2 = 130.2(10^3)\ \text{mm}^4$$

Considering only V_z, shear in horizontal flanges is the same as Example 3.4-3. That is, $Q = \frac{t}{2}(c^2 - z^2)$

$$\tau = \frac{2400\,\frac{t}{2}(c^2 - z^2)}{130.2(10^3)\,t} = 9.217(10^{-3})(c^2 - z^2)\ \text{MPa} \quad (c \text{ \& } z \text{ in mm})$$

$$\tau\Big|_{z=0} = 9.217(10^{-3})\left[(-37.5)^2\right] = 12.96\ \text{MPa}, \quad \tau_{z=12.5\,\text{mm}} = 11.52\ \text{MPa}$$

Considering only V_z, shear in vertical web will be linear (same as horizontal flange for V_y). The shear will be zero at $y = 0$ (where $Q = \int z\,dA = 0$), to 11.52 MPa at $y = 50$ mm (to match shear in flange). Thus,

$\tau_1 = 12.96$ MPa

$\tau_2 = 11.52$ MPa

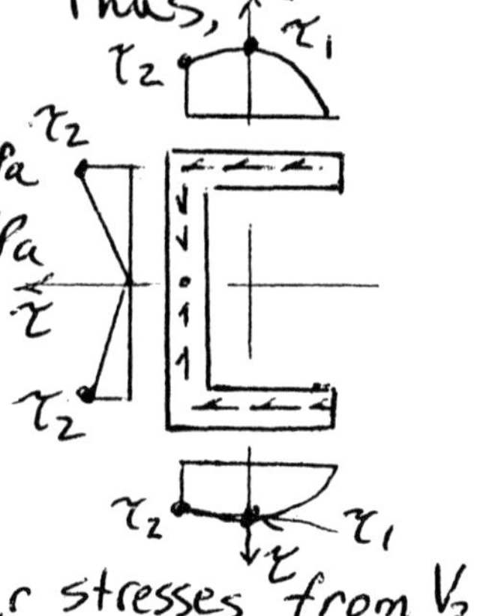

Shear stresses from V_z

Due to V_y [See Example 5.4-1, Eqs. (a) and (b)]

Max shear in flange,

$$\tau_3 = \frac{V_y a b}{I_z} = \frac{1800(50)(50)}{833.3(10^3)} = 5.4\ \text{MPa}$$

Max shear in web (at $y = 0$),

$$\tau_4 = \frac{V_y a(2b + a)}{2 I_z} = \frac{1800(50)(150)}{2(833.3)(10^3)} = 8.1\ \text{MPa}$$

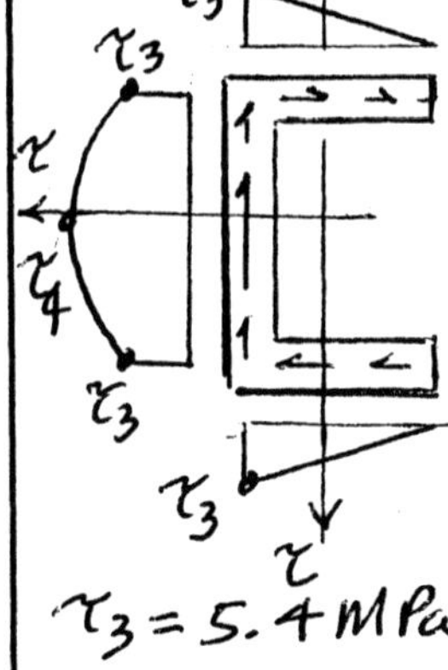

$\tau_3 = 5.4$ MPa, $\tau_4 = 8.1$ MPa

Shear stresses from V_y

Superposition

$\tau_5 = \tau_2 - \tau_3 = 6.12$ MPa

$\tau_6 = \tau_4 = 8.1$ MPa

$\tau_7 = \tau_2 + \tau_3 = 16.92$ MPa

ANS

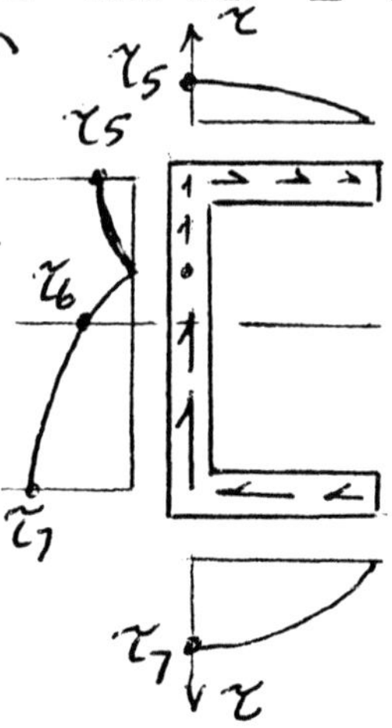

(5.35) $A = 4(50+20+20) = 360 \text{ mm}^2$

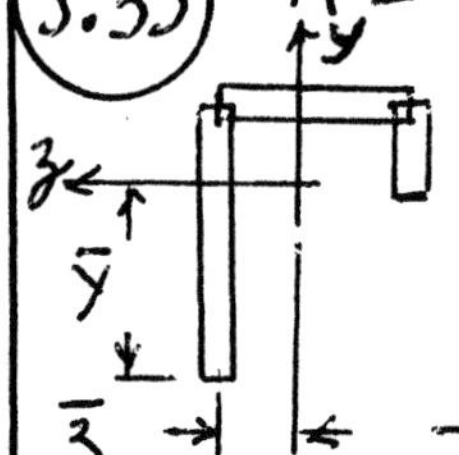

$$\bar{y} = \frac{25(4)(50)+50(4)(20)+(50-10)(4)(20)}{360} = 33.89 \text{ mm}$$

$$\bar{z} = \frac{10(4)(20)+20(4)(20)}{360} = 6.667 \text{ mm}$$

$$I_y = 4(50)(6.667)^2 + \frac{1}{12}(4)(20)^3 + 4(20)(10-6.67)^2 + 4(20)(20-6.667)^2 = 26.67(10^3) \text{ mm}^4$$

$$I_z = \frac{1}{12}(4)(50^3)+(4)(50)(33.89-25)^2 + 4(20)(50-33.89)^2 + \frac{1}{12}(4)(20^3) + 4(20)(40-33.89)^2 = 83.89(10^3) \text{ mm}^4$$

$$I_{yz} = (4)(50)(25-33.89)(6.667) + 4(20)(50-33.89)(6.667-10) + 4(20)(40-33.89)(6.667-20) = -22.67(10^3) \text{ mm}^4$$

$$\theta_p = \frac{1}{2}\tan^{-1}\left[\frac{2(-22.67)}{83.89-26.67}\right] = -19.20^\circ$$

$$I_{m,n} = I_{y',z'} = \left[\frac{26.67+83.89}{2} \pm \frac{26.67-83.89}{2}\cos(-38.4) \mp (-22.67)\sin(-38.4)\right](10^3)$$

$I_m = 18.78(10^3)\text{mm}$, $I_n = 91.78(10^3)\text{mm}$

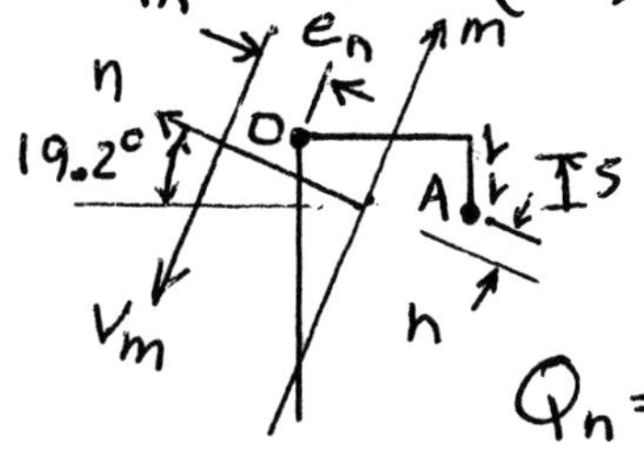

Sum moment about O. Thus, shear in right vertical wall is needed.

point A: $y = -3.89$, $z = 6.667-20 = -13.33$ mm

$h = -3.89\cos(-19.2) - 13.33\sin(-19.2) = 0.711$ mm

$Q_n = \bar{m}A = (0.711 + \frac{s}{2}\cos 19.2)st = (0.711 + 0.472s)st$

$$V_1 = \int \tau t\,ds = \int_0^{20} \frac{V_m(0.711+0.472s)st}{91.78(10^3)t}(4)ds = 0.0611\,V_m$$

$\Sigma M_o = 0 \quad V_m e_n - (0.0611 V_m)(20) = 0, \quad e_n = \underline{1.22 \text{ mm}}$

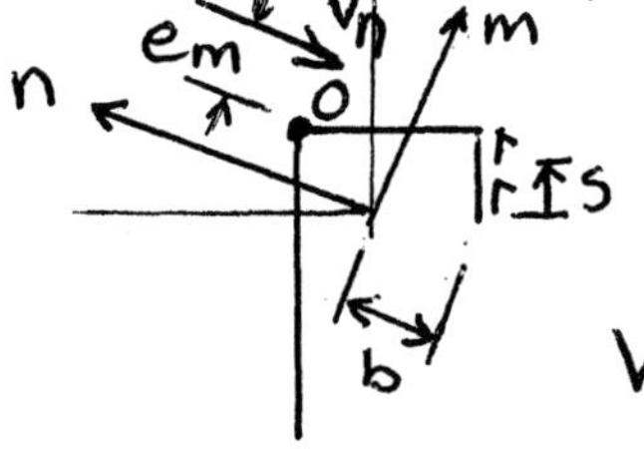

$b = |3.89\sin(-19.2) - 13.33\cos(-19.2)| = 13.87$ mm

$Q_m = (13.87 - \frac{s}{2}\sin 19.2)st = (13.87 - 0.1644s)st$

$$V_2 = \int_0^{20} \frac{V_n(13.87-0.1644s)st}{18.78(10^3)t}(4)ds = 0.4975\,V_n$$

$\Sigma M_o = 0 \quad V_n e_m - 0.4975 V_n(20) = 0, \quad e_m = \underline{9.95 \text{ mm}}$

(continued)

y and z components

$e_y = 9.95\cos(-19.2) - 1.22\sin(-19.2) = \underline{9.80\text{ mm}}$

$e_z = 9.95\sin(-19.2) + 1.22\cos(-19.2) = \underline{-2.12\text{ mm}}$

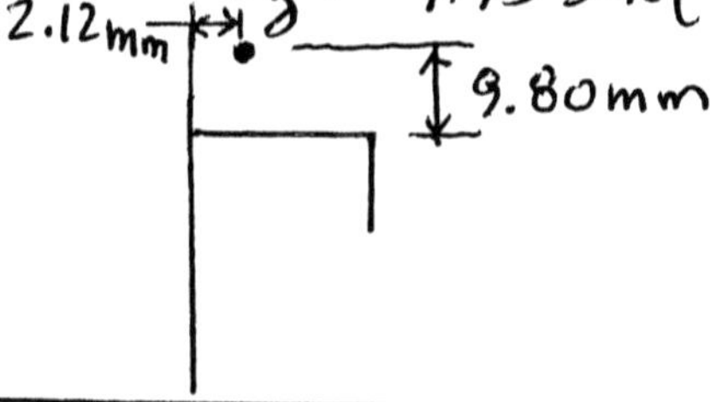

ANS

(5.36)

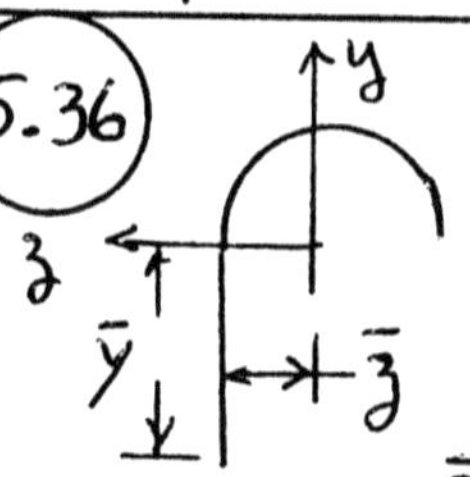

$A = 0.125[1.5 + \pi(0.5)] = 0.3838\text{ in}^2$

$\bar{y} = \dfrac{0.75(0.125)(1.5) + \left[1.5 + \frac{2(0.5)}{\pi}\right](\pi)(0.5)(0.125)}{0.3838}$

$= 1.296\text{ in}$

$\bar{z} = \dfrac{0.5(\pi)(0.5)(0.125)}{0.3838} = 0.2558\text{ in}$

$I_y \approx (0.125)(1.5)(0.2558)^2 + \frac{\pi}{2}(0.125)(0.5)^3 + \pi(0.5)(0.125)(0.5 - 0.2558)^2$

$= 0.04852\text{ in}^4$

$I_z \approx \frac{1}{12}(0.125)(1.5)^3 + (0.125)(1.5)(1.296 - 0.75)^2 + \frac{\pi}{2}(0.125)(0.5)^3$

$+ \pi(0.5)(0.125)\left\{-\left[\frac{2(0.5)}{\pi}\right]^2 + \left[1.5 + \frac{2(0.5)}{\pi} - 1.296\right]^2\right\} = 0.1493\text{ in}^4$

$I_{yz} \approx (0.125)(1.5)(0.75 - 1.296)(0.2558) + \pi(0.5)(0.125)\left[1.5 + \frac{2(0.5)}{\pi} - 1.296\right](0.2558 - 0.5)$

$\approx -0.05123\text{ in}^4$

$\theta_p = \frac{1}{2}\tan^{-1}\left[\dfrac{2(-0.05123)}{0.1493 - 0.04852}\right] = -22.74°$

$I_{m,n} = \dfrac{0.04852 + 0.1493}{2} \pm \dfrac{(0.04852 - 0.1493)}{2}\cos(-45.48) \mp (-0.05123)\sin(-45.48)$

$I_m = 0.02705\text{ in}^4, \quad I_n = 0.1708\text{ in}^4$

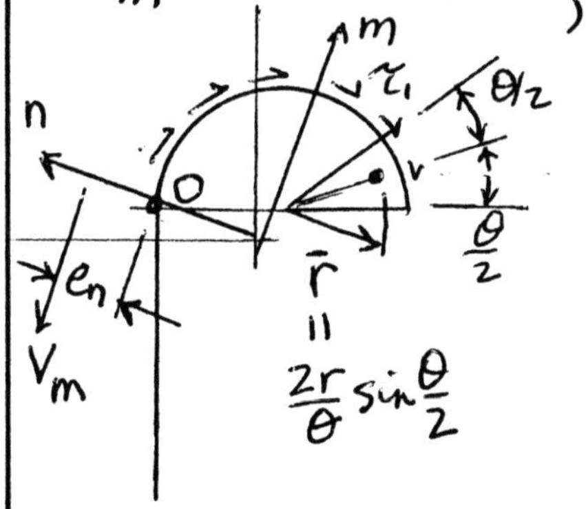

(y, z) location of rθ segment (see Prob. 5.30)

$\bar{y}_c = 1.5 - 1.296 + \frac{2(0.5)}{\theta}\sin^2\frac{\theta}{2} = 0.204 + \frac{1}{\theta}\sin^2\frac{\theta}{2}$

$\bar{z}_c = -\left(0.5 - 0.2558 + \frac{2(0.5)}{\theta}\sin\frac{\theta}{2}\cos\frac{\theta}{2}\right.$

$= -\left(0.2442 + \frac{0.5}{\theta}\sin\theta\right)$

(Continued)

$$\bar{m}_c = \left(0.204 + \frac{1}{\theta}\sin^2\frac{\theta}{2}\right)\cos(-22.74) - \left(0.2442 + \frac{0.5}{\theta}\sin\theta\right)\sin(-22.74)$$

$$= 0.2826 + \frac{0.9221}{\theta}\sin^2\frac{\theta}{2} + \frac{0.1935}{\theta}\sin\theta$$

$$Q_n = \bar{m}_c A = \bar{m}_c r\theta t, \quad \tau_1 = \frac{V_m Q_n}{I_n t} = \frac{V_m}{I_n}\bar{m}_c r\theta$$

Moment about O

$$M_O = \int_0^{\pi}\left[(\tau_1 \sin\theta)(r\sin\theta) r t\, d\theta + (\tau_1\cos\theta)(r + r\cos\theta) r t\, d\theta\right]$$

$$= \frac{V_m r^3 t}{I_n}\int_0^{\pi}\Big(0.2826\,\theta\sin^2\theta + 0.9221\sin^2\theta\sin^2\frac{\theta}{2} + 0.1935\sin^3\theta$$
$$+ 0.2826\,\theta\cos\theta + 0.9221\cos\theta\sin^2\frac{\theta}{2} + 0.1935\sin\theta\cos\theta$$
$$+ 0.2826\,\theta\cos^2\theta + 0.9221\cos^2\theta\sin^2\frac{\theta}{2} + 0.1935\sin\theta\cos^2\theta\Big)d\theta$$

$$= \frac{V_m}{0.1708}(0.5)^3(0.125)(1.941) = 0.178\,V_m$$

$$V_m e_n = 0.178\,V_m \Rightarrow \underline{e_n = 0.178\text{ in}}$$

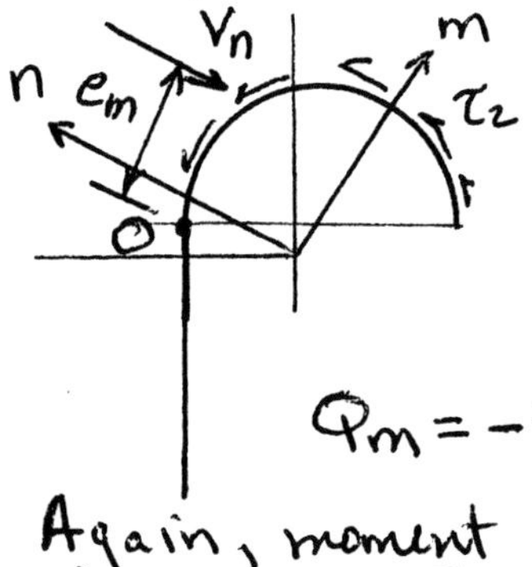

similar to $\bar{m}_c$,

$$\bar{n}_c = -\left(0.204 + \frac{1}{\theta}\sin^2\frac{\theta}{2}\right)\sin(-22.74) - \left(0.2442 + \frac{0.5}{\theta}\sin\theta\right)\cos(-22.74)$$

$$= -0.1462 + \frac{0.387}{\theta}\sin^2\frac{\theta}{2} - \frac{0.461}{\theta}\sin\theta$$

$$Q_m = -\bar{n}_c A = -\bar{n}_c r\theta t, \quad \tau_2 = \frac{V_n Q_m}{I_m t} = -\frac{V_n}{I_m}\bar{n}_c r\theta$$

Again, moment about O

$$M_O = \frac{V_n r^3 t}{I_m}\int_0^{\pi}\Big(0.1462\,\theta\sin^2\theta - 0.387\sin^2\frac{\theta}{2}\sin^2\theta + 0.461\sin^3\theta$$
$$+ 0.1462\,\theta\cos\theta - 0.387\cos\theta\sin^2\frac{\theta}{2} + 0.461\sin\theta\cos\theta$$
$$+ 0.1462\,\theta\cos^2\theta - 0.387\cos^2\theta\sin^2\frac{\theta}{2} + 0.461\sin\theta\cos^2\theta\Big)d\theta$$

$$= \frac{V_n(0.5)^3(0.125)(1.047)}{0.02705} = 0.605\,V_n$$

$$e_m V_n = 0.605\,V_n \Rightarrow \underline{e_m = 0.605\text{ in}}$$

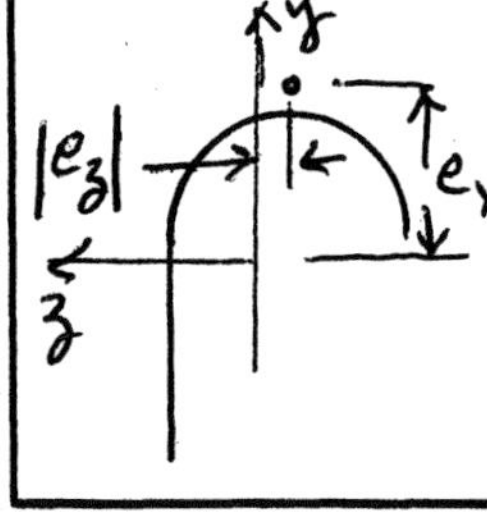

$$e_y = 0.605\cos(-22.74) - 0.178\sin(-22.74) = \underline{0.627\text{ in}}$$

$$e_z = 0.605\sin(-22.74) + 0.178\cos(-22.74) = \underline{-0.070\text{ in}}$$

(5.37) $I_z \approx \frac{1}{12}(2t)(2r)^3 + \frac{\pi}{2}tr^3 = 2.904\,tr^3$

Open cell as shown for transverse shear

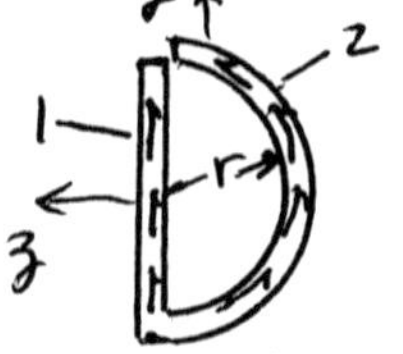

Wall 1:

$$q_{01} = \frac{V_y}{2I_z}(r^2-y^2)2t = \frac{V_y t}{I_z}(r^2-y^2) \quad (1)$$

Wall 2:

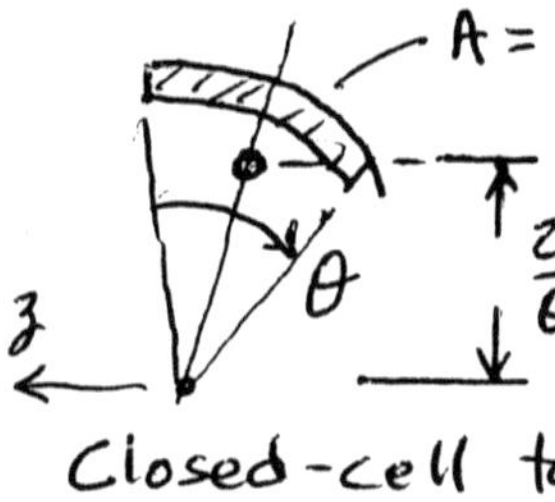

$A = r\theta t$

$\frac{2r}{\theta}\sin\frac{\theta}{2}\cos\frac{\theta}{2} = \frac{r}{\theta}\sin\theta$

$$Q = \bar{y}A = r^2 t\sin\theta$$

$$q_{02} = \frac{V_y r^2 t\sin\theta}{I_z} \quad (2)$$

Closed-cell torsion

For loading through shear center, $\theta = 0$

$$\theta = \frac{(1+\nu)L}{E\bar{A}}\left(\sum_{i=1}^{2}\int_{\text{wall } i}\frac{q_i}{t_i}ds\right) = 0$$

$$\therefore \int_{\text{wall 1}}\frac{q_c - q_{01}}{2t} + \int_{\text{wall 2}}\frac{q_c + q_{02}}{t} = 0$$

$$\left(\frac{2r}{2t} + \frac{\pi r}{t}\right)q_c - 2\frac{V_y t}{2t I_z}\int_0^r (r^2-y^2)dy + 2\frac{V_y r^3 t}{t I_z}\int_0^{\pi/2}\sin\theta\, d\theta = 0$$

Integration and simplification yields

$$q_c = -\frac{4}{3(1+\pi)}\frac{V_y r^2 t}{I_z}$$

Stress distributions

Wall 1: ↑+ $\tau_1 = \frac{q_{01} - q_c}{2t} = \frac{V_y}{2I_z}(r^2-y^2) + \frac{2}{3(1+\pi)}\frac{V_y r^2}{I_z}$

$$= \frac{V_y}{6I_z}\left[\frac{7+3\pi}{1+\pi}r^2 - 3y^2\right] = (0.228r^2 - 0.172y^2)\frac{V_y}{tr^3}$$

ANS

Wall 2: ↻+

$$\tau_2 = \frac{V_y}{I_z}r^2\sin\theta - \frac{4}{3(1+\pi)}\frac{V_y r^2}{I_z} = \frac{V_y r^2}{I_z}\left(\sin\theta - \frac{4}{3(1+\pi)}\right) = (0.344\sin\theta - 0.111)\frac{V_y}{tr}$$

ANS

(continued)

Shear center $\Sigma M_0 = 0$

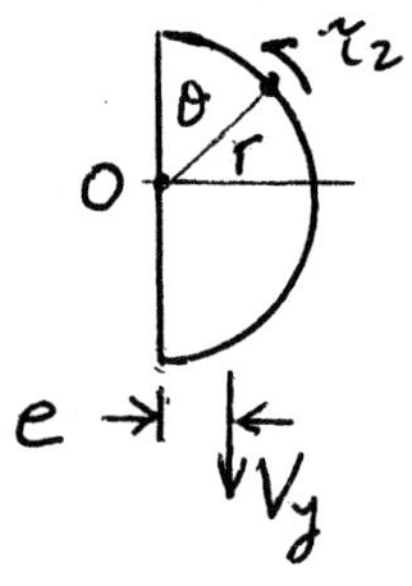

$$-V_y e + 2\int_0^{\pi/2}\left[(\tau_2\cos\theta)(r\cos\theta) + (\tau_2\sin\theta)(r\sin\theta)\right]t\,r\,d\theta = 0$$

$$V_y e = 2tr^2\int_0^{\pi/2}\tau_2\,d\theta = 2tr^2\int_0^{\pi/2}(0.344\sin\theta - 0.111)\frac{V_y}{tr}\,d\theta$$

$$= 2V_y r\left(0.344 - 0.111\frac{\pi}{2}\right) = 0.339\,V_y r$$

$\therefore$ $\underline{e = 0.339\,r}$

(5.38) $I_z = \frac{1}{12}(3)(100)^3 + \frac{1}{12}(6)(100)^3 + 2(6)(50)(50^2) = 2.25(10^6)\text{ mm}^4$

Open cell as shown: Wall 1 & Wall 5 are the same:

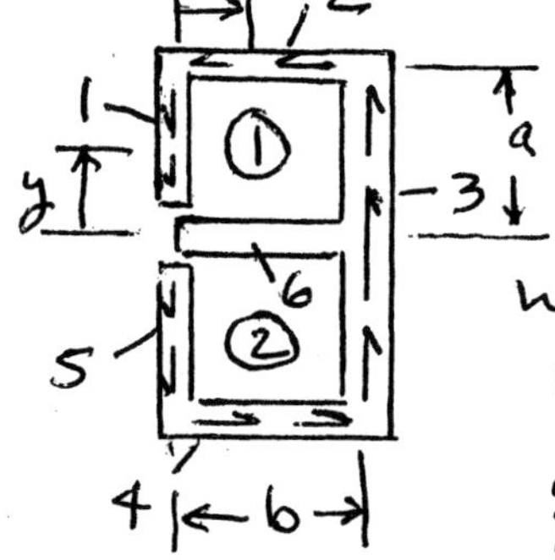

$$Q_1 = \frac{1}{2}ty^2$$

$$q_{01} = \frac{V_y}{I_z}\left(\frac{1}{2}ty^2\right) = \frac{1}{2}\frac{V_y t y^2}{I_z} \quad (1) \quad = q_{05}$$

Wall 2 & wall 4 are the same:

$$Q_2 = \frac{1}{2}ta^2 + as(2t)$$

$$q_{02} = q_{04} = \frac{V_y}{2I_z}ta(a+4s) \quad (2)$$

Wall 3:

$$Q_3 = \frac{1}{2}ta^2 + 2abt + \frac{1}{2}(a^2 - y^2)(2t)$$

$$q_{03} = \frac{V_y t}{2I_z}(3a^2 + 4ab - 2y^2) \quad (3)$$

wall 6: $q_{06} = 0$

Due to symmetry the torsional shear flow in each cell will be the same. Thus $q_{c1} = q_{c2} = q_c$

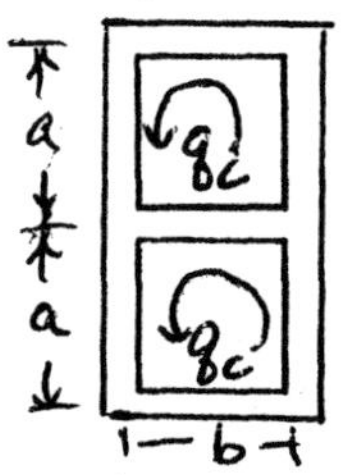

Loading through shear center, $\theta = 0$

For each cell,

$$\theta = \frac{(1+\nu)L}{E\bar{A}_i}\left(\int_{\text{wall}_1}\frac{q_{01}+q_c}{t}ds + \int_{\text{wall}_2}\frac{q_{02}+q_c}{2t}ds + \int_{\text{wall}_3}\frac{q_{03}+q_c}{2t}ds + \int_{\text{wall } 6}\frac{q_c - q_c}{t}ds\right) = 0$$

$\therefore$ ( ) = 0

$$\left(\frac{a}{t} + \frac{b}{2t} + \frac{a}{2t}\right)q_c + \int_0^a \frac{1}{2}\frac{V_y y^2}{I_z}dy + \int_0^b \frac{V_y t a(a+4s)}{2I_z(2t)}ds + \int_0^a \frac{V_y t}{2I_z(2t)}(3a^2 + 4ab - 2y^2)dy = 0$$

yields

$$q_c = -\frac{at}{2(3a+b)}\frac{V_y}{I_z}(3a^2 + 5ab + 2b^2) = -\frac{50(3)(2000)(10)(50)^2}{2(4)(50)2.25(10^6)}$$

$$= -8.333\text{ N/mm}$$

(continued)

Walls 1 & 5 ↑+ (y in mm)

$$\tau_1 = -\frac{(q_{02}+q_c)}{2t} = -\left[\frac{\frac{1}{2}\frac{2000(3)y^2}{2.25(10^6)} - 8.333}{3}\right]$$

$$= 4.44(10^{-4})y^2 - 2.778 \text{ MPa} = \tau_5 \quad \boxed{\text{ANS}}$$

Walls 2 & 4 $\overset{+}{\leftarrow}$ (s in mm)

$$\tau_2 = \frac{q_{02}+q_c}{2t} = \frac{\frac{2000(3)}{2.25(10^6)}\left(\frac{50^2}{2}+100s\right) - 8.33}{6}$$

$$= 4.44(10^{-2})s - 0.833 \text{ MPa} = \tau_4 \quad \boxed{\text{ANS}}$$

Wall 3 ↑+ (y in mm)

$$\tau_3 = \frac{q_{03}+q_c}{2t} = \frac{\frac{2000(3)}{2.25(10^6)}\left[\frac{3}{2}(50)^2 + 2(50)^2 - y^2\right] - 8.333}{6}$$

$$= 2.5 - 4.44(10^{-4})y^2 \text{ MPa} \quad \boxed{\text{ANS}}$$

Wall 6 $\tau_6 = 0$

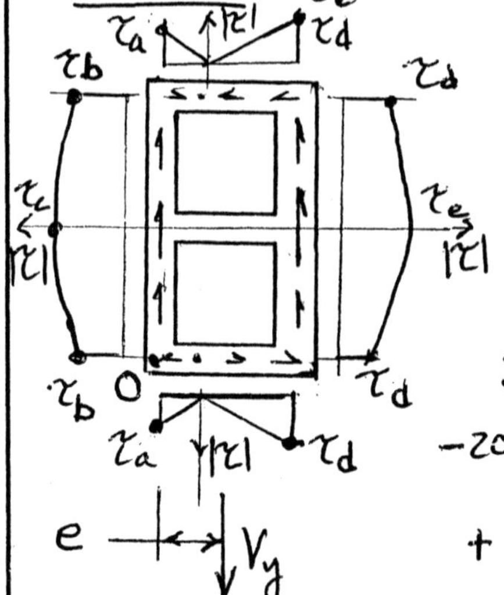

$\tau_a = 0.833$ MPa
$\tau_b = 1.667$ MPa
$\tau_c = 2.778$ MPa
$\tau_d = 1.389$ MPa
$\tau_e = 2.5$ MPa

Note: Consistent with concept of "flow"

Shear center: $\Sigma M_0 = 0$

$$-2000e + 100\int_0^{50}\left[4.44(10^{-2})s - 0.833\right](6)\,ds + 50(2)\int_0^{50}\left[2.5 - 4.44(10^{-4})y^2\right](6)\,dy = 0$$

$$e = \frac{1}{2000}\left\{600\left[2.22(10^{-2})(50^2) - 0.833(50)\right] + 600\left[2.5(50) - \frac{4.44(10^{-4})}{3}(50)^3\right]\right\}$$

$$= 36.1 \text{ mm}$$

(5.39) $I_z \approx 2\left[\frac{1}{12}(3)(100^3) + 3(50)(50^2)\right] = 1.25(10^6)\text{ mm}^4$

Vertical walls: $Q_1 = \Sigma \bar{y}A = 2\left(\frac{50+y}{2}\right)(50-y)(3) + 50(50)(3)$

$= 15(10^3) - 3y^2 \text{ mm}^3$ (y in mm)

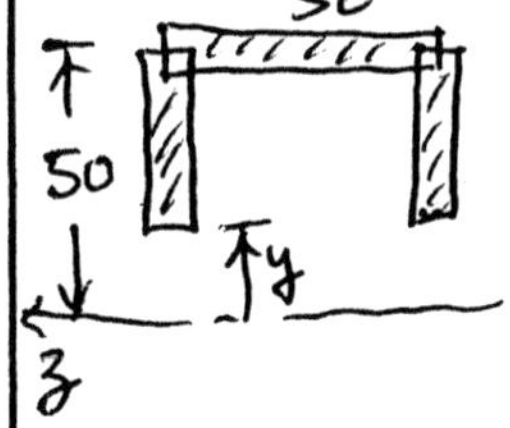

$$\tau_1 = \frac{2000\left[15(10^3) - 3y^2\right]}{1.25(10^6)(2)(3)} = 4\left[1 - 2(10^{-4})y^2\right] \text{ MPa}$$

(continued)

Upper horizontal section:

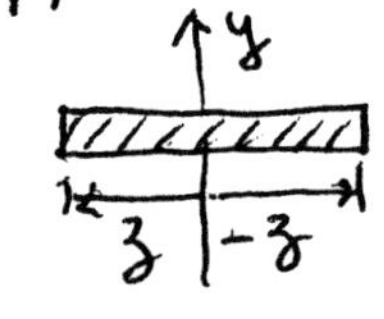

$$Q_z = 50(2z)(3) = 300z \text{ mm}$$

$$\tau_z = \frac{2000(300z)}{1.25(10^6)(2)(3)} = 80(10^{-3})z \text{ MPa}$$

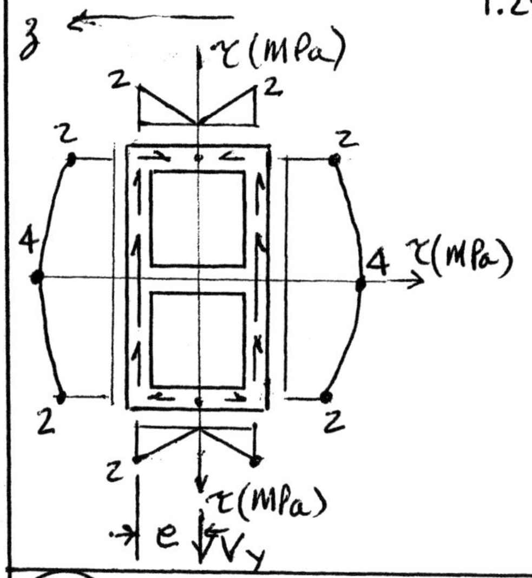

Due to symmetry the shear center is the geometric center

(5.40) $I_z \approx 2\left[\frac{1}{12}(0.125)(4)^3 + (0.125)(2)(2)^2\right] = 3.333 \text{ in}^4$

Open section as shown

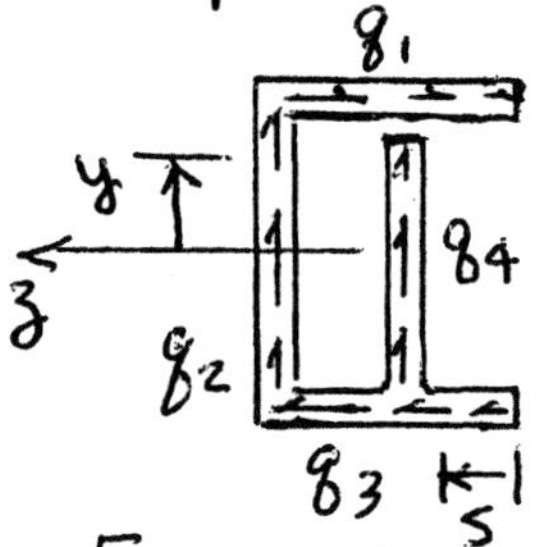

For q_1, q_3 (same as Example 5.4-1 with $a=2$)

$$Q_1 = ast = 0.25s$$

$$q_1 = q_3 = \frac{V_y Q_1}{I_z} = \frac{5000(0.25s)}{3.333} = 375s \text{ lb/in}$$

For q_2 (same as web of Example 5.4-1)

$$Q_2 = abt + \frac{t}{2}(a^2 - y^2) = 2(2)(0.125) + \frac{0.125}{2}(2^2 - y^2) = 0.5 + 0.0625(4 - y^2)$$

$$q_2 = \frac{5000}{3.333}\left[0.5 + 0.0625(4 - y^2)\right] = 1125 - 93.75y^2 \text{ lb/in}$$

For q_4 (same as vertical wall of Example 3.4-3)

$$Q_4 = \frac{t}{2}\left[\left(\frac{h}{2}\right)^2 - y^2\right] = 0.0625(4 - y^2)$$

$$q_4 = \frac{5000}{3.333}\left[0.0625(4 - y^2)\right] = 375 - 93.75y^2 \text{ lb/in}$$

<u>For torsion</u>

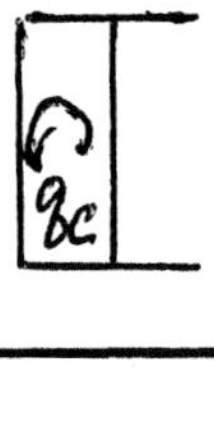

Loading at shear center $\theta = 0$ in single cell

(continued)

$\int \frac{q}{t}\,ds$ for q_c is simply $\frac{q_c S}{t} = \frac{q_c(1+4+1+4)}{0.125} = 80\,q_c$

$\theta = 0$ for cell yields

$$80q_c - 2\int_1^2 \frac{375s}{0.125}ds - 2\int_0^2 \frac{1125-93.75y^2}{0.125}dy + 2\int_0^2 \frac{375-93.75y^2}{0.125}dy = 0$$

Integration yields $q_c = 412.5$ lb/in

Shear Flow Diagram

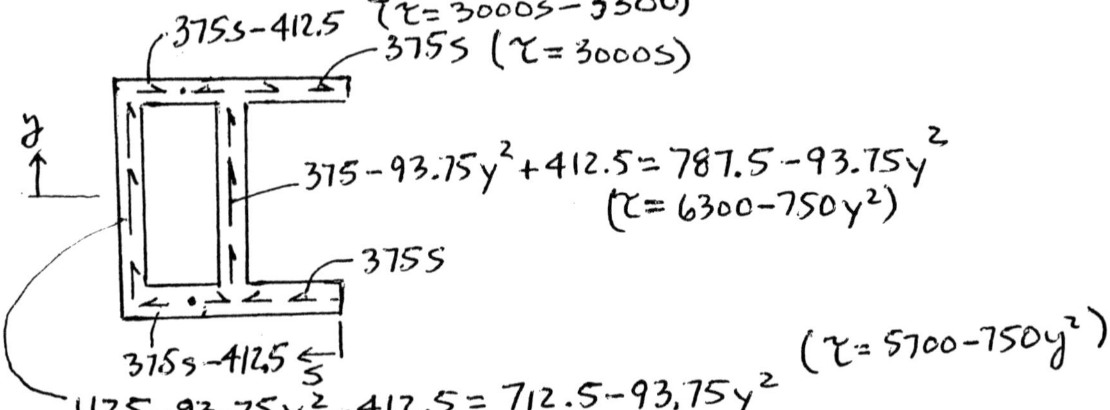

For shear stresses (psi) divide q by 0.125

In flanges $q = 0$ when $375s_0 = 412.5 \Rightarrow s_0 = 1.10$ in

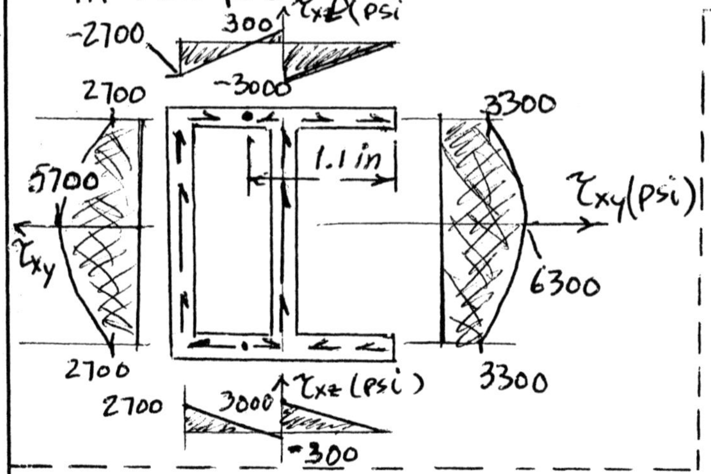

For shear center, sum moments about point O and ignore forces intersecting point O.

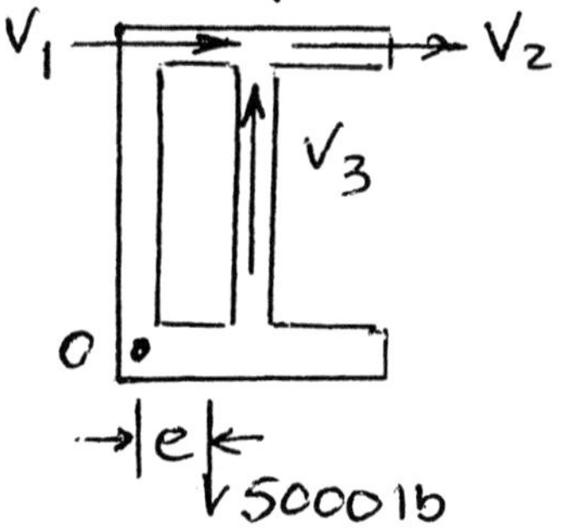

$\Sigma M_0 = 0 \Rightarrow -5000e - 4(V_1+V_2) + 2V_3 = 0$

$e = \frac{2V_3 - 4(V_1+V_2)}{5000}$ ① $V_1 = \int_1^2 (3000s-3300)(0.125)ds = 150$ lb

$V_2 = \int_0^1 (3000s)(0.125)ds = 187.5$ lb, $V_3 = 2\int_0^2 (6300-750y^2)(0.125)dy = 2650$ lb

Sub into ①

$$e = \frac{2(2650) - 4(150+187.5)}{5000} = \underline{0.790 \text{ in}}$$

(5.41) Make equivalent section of aluminum $E_s/E_a = 3$

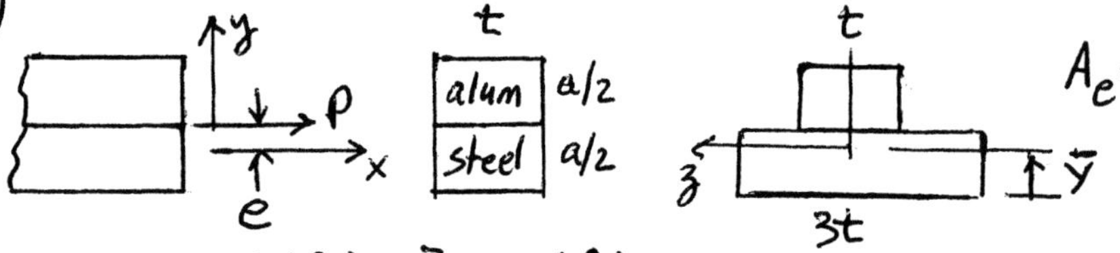

$A_e = 3t\left(\frac{a}{2}\right) + t\left(\frac{a}{2}\right) = 2at$

$$\bar{y} = \frac{\frac{a}{4}(3t)\left(\frac{a}{2}\right) + \frac{3a}{4}(t)\left(\frac{a}{2}\right)}{2at} = \frac{3}{8}a \qquad \therefore\ e = \frac{a}{8}$$

(a) $M_z = -Pe = -\frac{1}{8}Pa$ **ANS**

(b) Stresses

$$(I_z)_e = \frac{1}{12}(3t)\left(\frac{a}{2}\right)^3 + (3t)\left(\frac{a}{2}\right)\left(\frac{a}{4} - \frac{3}{8}a\right)^2 + \frac{1}{12}t\left(\frac{a}{2}\right)^3 + (t)\left(\frac{a}{2}\right)\left(\frac{3}{4}a - \frac{3}{8}a\right)^2$$

$$= \frac{13}{96}ta^3$$

Aluminum: max σ at $y = 5/8\,a$

$$\sigma_A = -\frac{\left(-\frac{Pa}{8}\right)\left(\frac{5a}{8}\right)}{\frac{13}{96}ta^3} + \frac{P}{2at} = \frac{14}{13}\frac{P}{at} \quad \textbf{ANS}$$

Steel: max at $y = a/8$

$$\sigma_s = -(3)\frac{\left(-\frac{Pa}{8}\right)\left(\frac{a}{8}\right)}{\frac{13}{96}ta^3} + 3\frac{P}{2at} = \frac{24}{13}\frac{P}{at} \quad \textbf{ANS}$$

(5.42) Equivalent aluminum $E_s/E_a = 3$

(a)

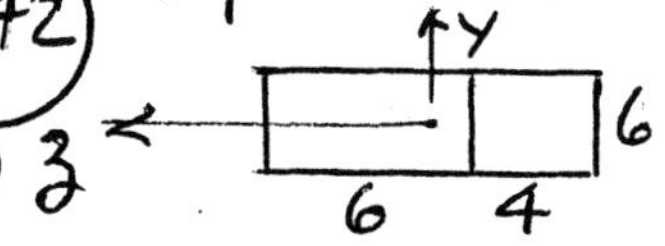

$(I_z)_e = \frac{1}{12}(10)(6)^3 = 180\ in^4$

$$(\sigma_{al})_{max} = \frac{Mc}{I} = \frac{36(5000)(3)}{180} = 3000\ psi = \underline{3\ kpsi}$$

$$(\sigma_{st})_{max} = 3\frac{Mc}{I} = \underline{9\ kpsi}$$

(b) Bending about y axis for equiv. alum. section

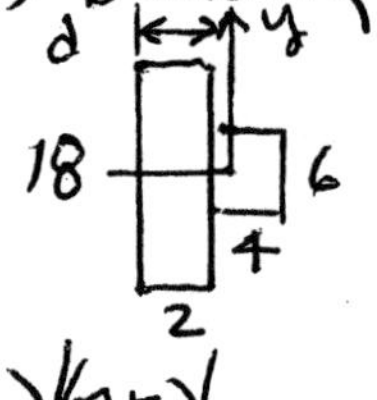

$A_e = 18(2) + 4(6) = 60\ in^2$

$$d = \frac{(1)(2)(18) + 4(4)(6)}{60} = \underline{2.2\ in}$$

(c) $$(v_c)_{max} = \frac{FL^3}{3EI} = \frac{5000(36)^3}{(3)(10)(10^6)(180)} = \underline{0.0432\ in}$$

(5.43) Tube: $I_s = \frac{\pi}{64}(35^4 - 20^4) = 65.81(10^3)\ mm^4$

(a) The second-area moment of a triangle about its base is $I = \frac{1}{36}bh^3 + \frac{1}{2}bh\left(\frac{h}{3}\right)^2 = \frac{1}{12}bh^3$

For the aluminum section:

$$I_A = 2\left(\tfrac{1}{12}\right)(50)(45^3) - \tfrac{\pi}{64}(35^4) = 685.71(10^3)\ mm^4$$

Equivalent section of Aluminum $E_s/E_a = 3$

$$I_e = 685.71(10^3) + 3(65.81)(10^3) = 883.13(10^3)\ mm^4$$

$$\sigma_{Alum} = \frac{Mc_A}{I} = \frac{(1000)(45)(10^3)}{883.13(10^3)} = \underline{50.96\ MPa}$$

$$\sigma_{Steel} = 3\frac{Mc_s}{I} = 3\frac{(1000)(\frac{35}{2})(10^3)}{883.13(10^3)} = \underline{59.45\ MPa}$$

(b) strains

at $y = 45\,mm$ $\quad \epsilon = \frac{50.96}{70(10^3)} = 727.9\mu$

$= \frac{35}{2}$ $\quad \epsilon = \frac{59.45}{210(10^3)} = 283.1\mu$

Stress distribution

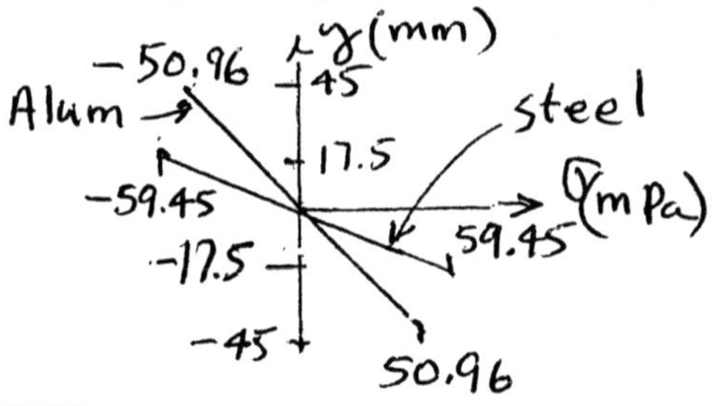

Strain distribution

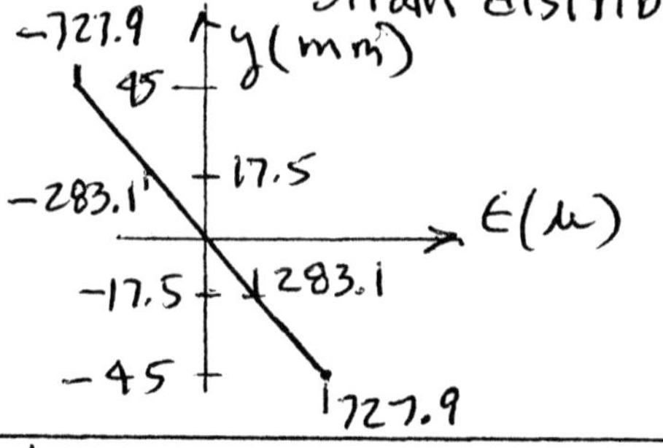

(5.44) Equivalent Section : steel

$$I_e = \frac{\pi}{64}\left[\left(\frac{70}{210}\right)50^4 + \left(\frac{105}{210}\right)(54^4 - 50^4) + (58^4 - 50^4)\right] = 406.3(10^3)\ mm^4$$

<u>Stresses</u>

Steel $\quad \sigma_s = \frac{2(10^3)(29)(10^3)}{406.3(10^3)} = \underline{142.8\ MPa}$

Brass $\quad \sigma_B = \frac{105}{210}\frac{2(10^3)(27)(10^3)}{406.3(10^3)} = \underline{66.5\ MPa}$

Alum. $\quad \sigma_A = \frac{70}{210}\frac{2(10^3)(25)(10^3)}{406.3(10^3)} = \underline{41.0\ MPa}$

(5.45) Equivalent section of aluminum. For plastic

$$b_e = \frac{0.04E_a}{E_a} b = 0.04\, b$$

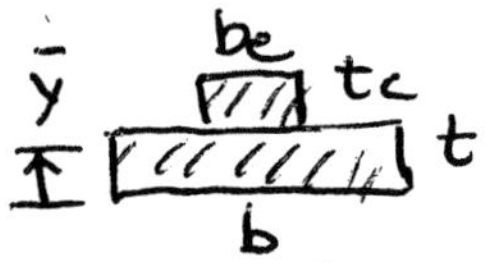

$$\bar{y} = \frac{\frac{t}{2}(bt) + \left(t+\frac{t_c}{2}\right)(t_c)(0.04b)}{b(t) + 0.04bt_c} \qquad \text{Let } \frac{t_c}{t} = r$$

Reduces to $\bar{y} = \dfrac{25+2r+r^2}{50+2r}\, t$ ①

$$I_e = \frac{1}{3}\left[b\bar{y}^3 + (b-0.04b)(t-\bar{y})^3 + (0.04b)(t+t_c-\bar{y})^3\right] \quad \text{Sub in ①}$$

$$= \frac{bt^3}{3}\left[\left(\frac{25+2r+r^2}{50+2r}\right)^3 + 0.96\left(\frac{25-r^2}{50+2r}\right)^3 + 0.04\left(\frac{25+50r+r^2}{50+2r}\right)^3\right]$$

$I_e = K\,bt^3$ ② where $K = \dfrac{1}{3(50+r)^3}\left[(25+2r+r^2)^3 + 0.96(25-r^2)^3 + 0.04(25+50r+r^2)^3\right]$

At midpoint of coating

$$\bar{\epsilon} = \frac{My}{E_a I_e} = \frac{M\left(t+\frac{t_c}{2}-\bar{y}\right)}{E_a K bt^3} = \frac{25+25r}{50+2r}\,\frac{M}{E_a K bt^2}$$

$$C_1 = \frac{\bar{\epsilon}}{\epsilon} = \frac{\frac{25}{2}\frac{(1+r)}{(25+r)}\frac{M}{E_a K bt^2}}{6M/E_a bt^2} = \frac{25(1+r)}{12(25+r)K} \qquad \underline{\text{ANS}}$$

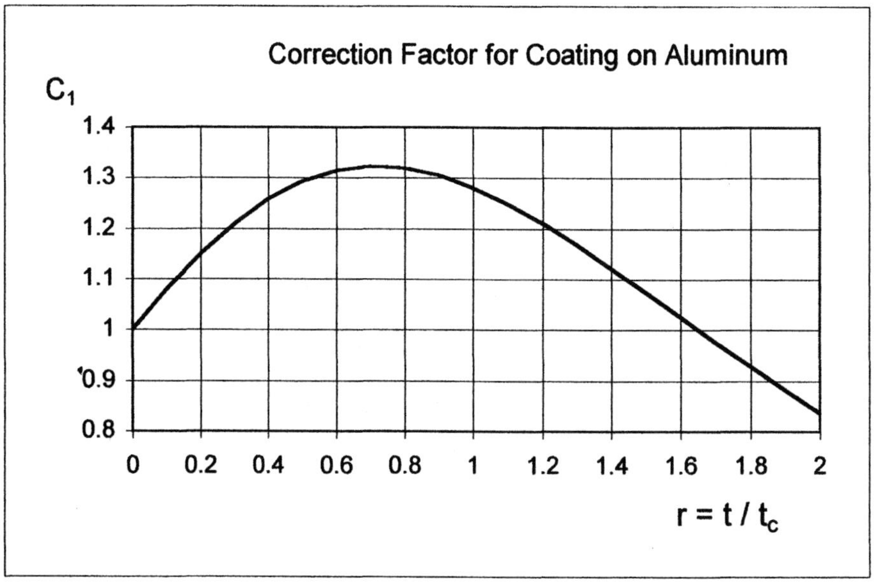

5.46 Equivalent section of steel $E_b/E_s = 0.5$

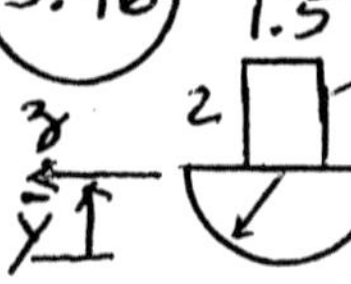

$A_1 = 3\text{ in}^2,\ \bar{y}_1 = 2.5\text{ in}$

$A_2 = \frac{\pi}{2}(1.5^2) = 3.5343\text{ in}^2,\ \bar{y}_2 = 1.5 - \frac{4(1.5)}{3\pi} = 0.86338\text{ in}$

(a)

$$\bar{y} = \frac{2.5(3) + 0.86338(3.5343)}{3 + 3.5343} = 1.615\text{ in}$$

$$(I_z)_e = \frac{1}{12}(1.5)(2)^3 + 3(2.5 - 1.615)^2 + \left(\frac{\pi}{8} - \frac{8}{9\pi}\right)(1.5)^4 + 3.5343(1.615 - 0.8634)^2$$

$$= 5.902\text{ in}^4 \qquad M_{max} = wL^2/2 = \frac{200(20)^2}{2} = 40(10^3)\text{ lb-in}$$

$$(\sigma_{min})_{steel} = \frac{40(10^3)(-1.615)}{5.902} = -10{,}950\text{ psi} = \underline{-10.95\text{ kpsi}}$$

$$(\sigma_{max})_{steel} = \frac{40(10^3)(1.5 - 1.615)}{5.902} = \underline{-779\text{ psi}}$$

No tension in steel

$$(\sigma_{max})_{brass} = \frac{1}{2}\,\frac{40(10^3)(3.5 - 1.615)}{5.902} = 6390\text{ psi} = \underline{6.39\text{ psi}}$$

$$(\sigma_{min})_{brass} = \frac{1}{2}\,\frac{40(10^3)(1.5 - 1.615)}{5.902} = \underline{-390\text{ psi}}$$

$$\text{(b)}\ (v_c)_{x=L} = \frac{-200(20^4)}{8(30)(10^6)(5.902)} = \underline{-0.0226\text{ in}}$$

Redoing the problem in brass the section width doubles that shown above. This gives the same $\bar{y}$ but $(I_z)_e$ doubles. This gives the same calculations for stress since the steel stresses must be multiplied by a factor of 2 from the equiv. stresses. The deflection calculation will be the same since $(I_z)_e$ doubles but E is for brass which is ½ steel E.

5.47 Equivalent Section of Aluminum $E_s/E_a = 210/70 = 3$

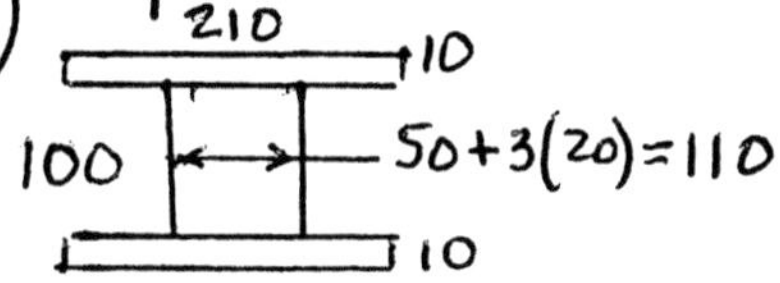

$$I_e = \frac{1}{12}(210)(120^3) - \frac{1}{12}(210 - 110)(100^3)$$

$$= 21.91(10^6)\text{ mm}^4$$

$\sigma_{max/min}$ in aluminum occurs at $y = \pm 50\text{ mm}$

$$\sigma_{max} = \frac{(2F_{max})(50)(10^3)}{21.91(10^6)} = 60\text{ MPa} \Rightarrow F_{max} = 13{,}144\text{ N} = \underline{13.1\text{ kN}}$$

$$\delta_{max} = \frac{13.144(10^3)(2)^3}{3(70)(10^9)(21.91)(10^6)(10^{-3})^4} = 0.0229\text{ m} = \underline{22.9\text{ mm}}$$

5.48 Make equivalent of Steel $\quad E_a/E_s = 1/3$

Outer shell: $I_o = \dfrac{\pi}{64}\left[(d+2t)^4 - d^4\right]$

Inner core (aluminum): $I_i = \dfrac{1}{3}\left(\dfrac{\pi}{64}d^4\right)$

$$I_e = I_o + I_i = \frac{\pi}{64}\left[(d+2t)^4 - d^4 + \frac{d^4}{3}\right] = \frac{\pi}{192}\left[3(d+2t)^4 - 2d^4\right] \quad \boxed{\text{ANS}}$$

5.49

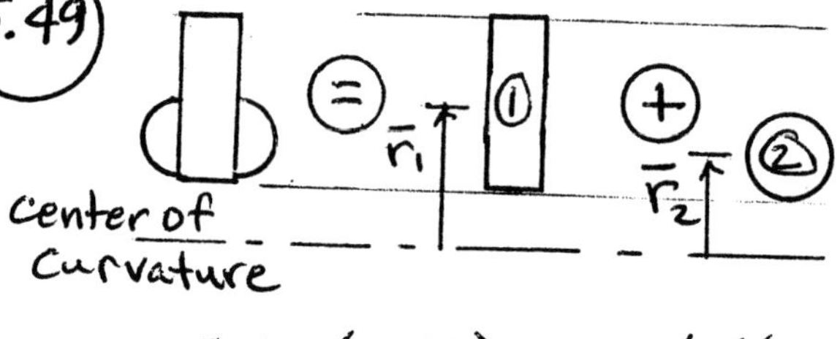

$A_① = (0.5)(2.5) = 1.25\ \text{in}^2$

$\bar r_① = 2 + \dfrac{2.5}{2} = 3.25\ \text{in}$

$A_② = \pi(0.5^2)$

$r_② = 2.5\ \text{in}$

$$r_c = \frac{3.25(1.25) + 2.5(\pi)(0.5^2)}{1.25 + \pi(0.5^2)} = 2.96060\ \text{in}$$

$\leftarrow A = 2.0354\ldots\ \text{in}^2$ (denominator)

$$\int\frac{dA}{r} = \left[\int\frac{dA}{r}\right]_① + \left[\int\frac{dA}{r}\right]_② = 0.5\ln\frac{4.5}{2} + 2\pi\left(2.5 - \sqrt{2.5^2 - 0.5^2}\right) = 0.72283\ldots$$

$$r_n = \frac{2.0354\ldots}{0.72283\ldots} = 2.81587\ \text{in}, \quad e = r_c - r_n = 0.14473\ \text{in}$$

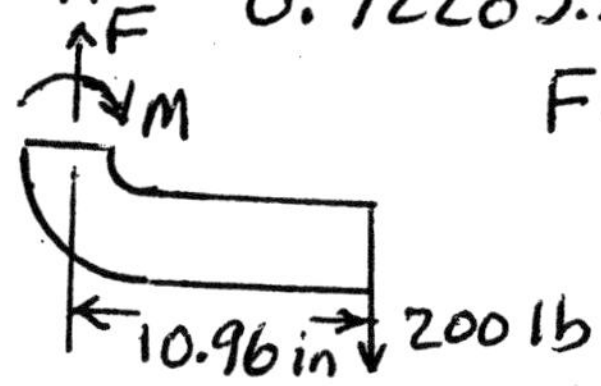

$F = 200\ \text{lb} \quad M = -200(10.96) = -2192\ \text{lb-in}$

$$\sigma_\theta = -\frac{(-2192)(2.816 - r)}{(2.035)(0.14473)(r)} + \frac{200}{2.035}$$

$$= \left(\frac{20.96}{r} - 7.343\right)10^3\ \text{psi} = \frac{20.96}{r} - 7.343\ \text{kpsi} \quad \boxed{\text{ANS}}$$

$\sigma_\theta\big|_{r=2} = \underline{3.13\ \text{kpsi}}, \quad \sigma_\theta\big|_{r=4.5} = \underline{-2.69\ \text{kpsi}}$

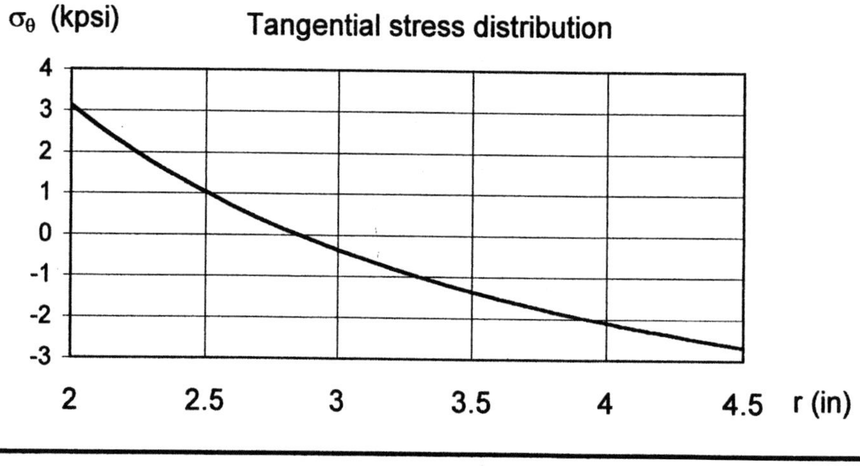

(5.50)

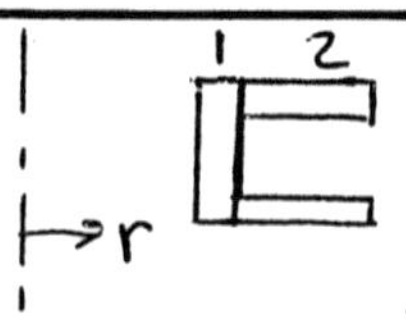

$A = 20(60) + 2(10)(60) = 2400\ \text{mm}^2$

$r_c = \dfrac{(25+10)(20)(60) + 2(25+20+30)(10)(60)}{2400} = 55\ \text{mm}$

$r_n = \dfrac{2400}{60 \ln \frac{45}{25} + 2(10) \ln \frac{105}{45}} = 45.9654\ \text{mm}$

$e = r_c - r_n = 9.035\ \text{mm}$

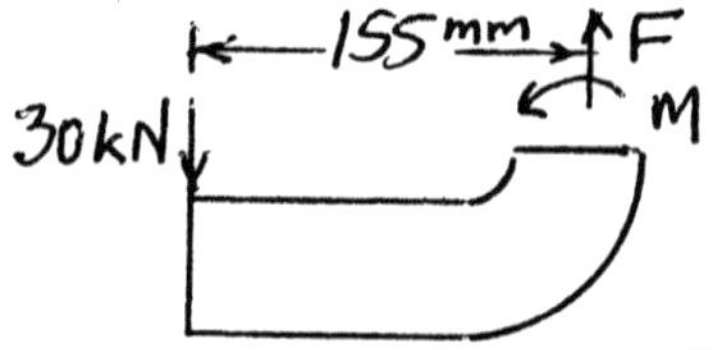

$F = 30\ \text{kN}$

$M = -30(155) = -4650\ \text{N}\cdot\text{m}$

(a) $\sigma_\theta = -\dfrac{(-4650)(45.965-r)(10^3)}{(2400)(9.035)\, r} + \dfrac{30(10^3)}{2400} = \dfrac{9855}{r} - 201.9\ \text{MPa}$ ① ANS (r in mm)

$\sigma_\theta \big|_{r=25\text{mm}} = \underline{192.3\ \text{MPa}}, \quad \sigma_\theta \big|_{r=105\text{mm}} = \underline{-108.0\ \text{MPa}}$

(b) $\underline{25\text{mm} < r < 45\text{mm}}$, $b = 60\text{mm}$, $\int \frac{dA}{r} = 60 \ln \frac{r}{25}$, $A_1 = 60(r-25)$

$\sigma_r = -\dfrac{(-4650)(10^3)\left[45.965(60) \ln \frac{r}{25} - 60(r-25)\right]}{(2400)(60)(r)(9.035)}$

$= \dfrac{9.857(10^3)}{r} \ln \dfrac{r}{25} - 214.44 + \dfrac{5.361(10^3)}{r}\ \text{MPa}$ ② ANS

$45\text{mm} < r < 105\text{mm}$, $b = 20\text{mm}$, $\int \frac{dA}{r} = 60 \ln \frac{45}{25} + 20 \ln \frac{r}{45} = 35.267 + 20 \ln \frac{r}{45}$

$A_1 = 60(20) + 20(r-45) = 300 + 20r$

$\sigma_r = -\dfrac{(-4650)\left[45.965\left(35.267 + 20 \ln \frac{r}{45}\right) - (300+20r)\right]}{2400(20)(r)(9.035)}$

$= \dfrac{9.857(10^3)}{r} \ln \dfrac{r}{45} + \dfrac{1.416(10^4)}{r} - 214.44\ \text{MPa}$ ③ ANS

For $(\sigma_r)_{max}$ take derivative of ② wrt r & set to zero

$\dfrac{d\sigma_r}{dr} = -\dfrac{9.857(10^3)}{r^2} \ln \dfrac{r}{25} + \dfrac{9.857(10^3)}{r^2} - \dfrac{5.361(10^3)}{r^2} = 0$

Reduces to $\ln \frac{r}{25} = 0.4561 \Rightarrow r = 39.45\ \text{mm}$ sub into ②

$(\sigma_r)_{max} = \underline{35.43\ \text{mm}}$ (continued)

at $r = 45\text{mm}$, Eq ③ $(\sigma_r)_{max} = \underline{100\ \text{MPa}}$

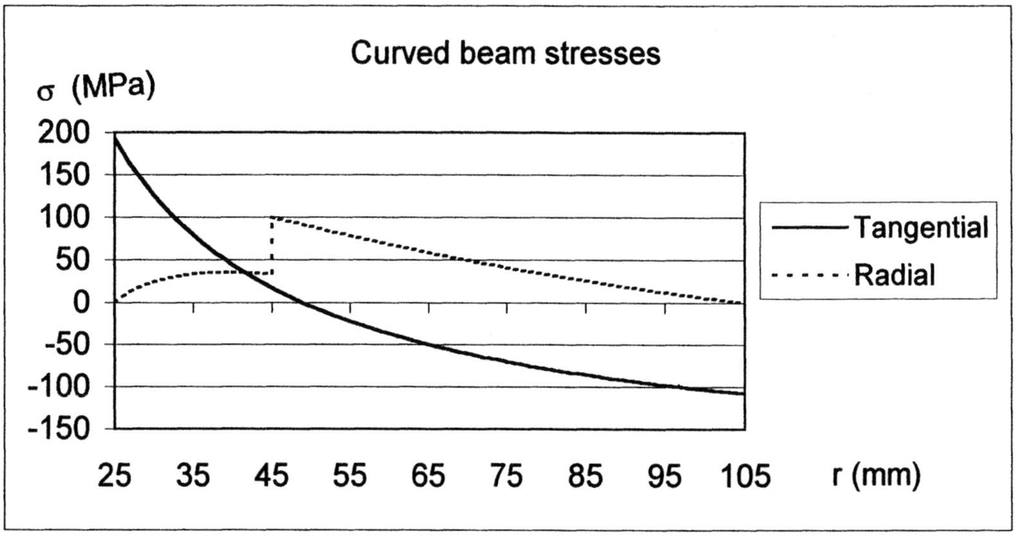

5.51 (a) $A = 3\ \text{in}^2$, $r_c = 3\ \text{in}$, $\int \frac{dA}{r} = 1.5 \ln \frac{4}{2} = 1.5 \ln 2$

$$r_n = \frac{3}{1.5 \ln 2} = 2.88539\ \text{in}, \quad e = r_c - r_n = 0.1146\ \text{in}$$

$$M = -750(8+3) = -8250\ \text{lb·in}, \quad F = 750\ \text{lb}$$

$$\sigma_\theta\Big|_{r=2} = -\frac{(-8250)(2.8854-2)}{(3)(0.1146)(2)} + \frac{750}{3} = 10.87(10^3)\ \text{psi} = \underline{10.9\ \text{kpsi}}$$

$$\sigma_\theta\Big|_{r=4} = -\frac{(-8250)(2.8854-4)}{3(0.1146)(4)} + \frac{750}{3} = -6.44(10^3)\ \text{psi} = \underline{-6.44\ \text{kpsi}}$$

(b)

$$\sigma_r\Big|_{r=3} = -\frac{(-8250)\left[2.8854(1.5)\ln\frac{3}{2} - 1(1.5)\right]}{3(1.5)(3)(0.1146)} = 1359\ \text{psi} = \underline{1.36\ \text{kpsi}}$$

5.52 $A = \frac{1}{2}\left(\frac{1}{4}+\frac{1}{2}\right)(1) = 0.375\ \text{in}^2$

$$\int \frac{dA}{r} = \left[\frac{0.5(4) - (0.25)(3)}{1}\right] \ln\frac{4}{3} - \frac{1}{4} = 0.10960\ \text{in}$$

$$r_n = \frac{0.375}{0.10960} = 3.42145\ \text{in}$$

From Example B. 2-1, Appendix B,

$$r_c = 3 + \frac{1}{3}\left[\frac{0.5 + 2(0.25)}{0.5 + 0.25}\right] = 3.44444\ \text{in}$$

$$e = r_c - r_n = 0.02299\ \text{in}$$

$M = 100(4+3.444) = 744.4$ lb-in, $F = -100$ lb.

$$\sigma_\theta\Big|_{r=3} = -\frac{(744.4)(3.4215-3)}{(0.375)(0.02299)(3)} - \frac{100}{0.375} = -12{,}400\text{ psi} = \underline{-12.4\text{ kpsi}}$$

$$\sigma_\theta\Big|_{r=4} = -\frac{(744.4)(3.4215-4)}{0.375(0.02299)(4)} - \frac{100}{0.375} = 12{,}220\text{ psi} = \underline{12.2\text{ kpsi}}$$

Radial stresses

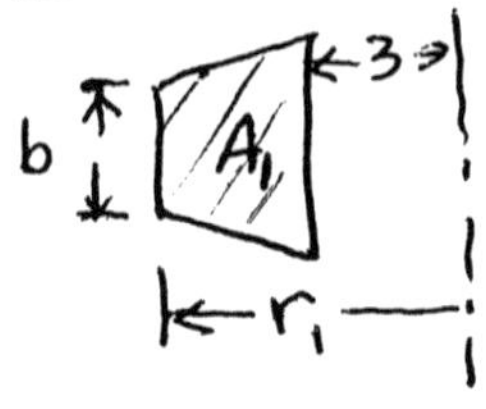

$b = \frac{1}{4}(5-r_1)$, $A_1 = \frac{1}{2}[0.5+0.25(5-r_1)](r_1-3)$

$$A_1 = \frac{1}{8}(7-r_1)(r_1-3)$$

$$\int_3^{r_1}\frac{dA}{r} = \frac{[0.5r_1 - \frac{1}{4}(5-r_1)(3)]}{r_1-3}\ln\frac{r_1}{3} - \frac{1}{2} + \frac{1}{4}(5-r_1)$$

$$= \frac{5r_1-15}{4(r_1-3)}\ln\frac{r_1}{3} + \frac{3-r_1}{4}$$

sub into Eq. (5.6-11) and simplify

$$\sigma_r = \frac{-3.466(10^5)}{r_1(5-r_1)}\left\{0.8554\left[\frac{5r_1-15}{r_1-3}\ln\frac{r_1}{3} + 3 - r_1\right] - \frac{1}{8}(7-r_1)(r_1-3)\right\}$$

At $r_1 = 3.422$, $\sigma_r = -847$ psi. $|\sigma_r|_{max} \approx \underline{847\text{ psi}}$

(5.53) $E_1 = 200$ GPa, $E_2 = 100$ GPa. Make of E_2 mat.

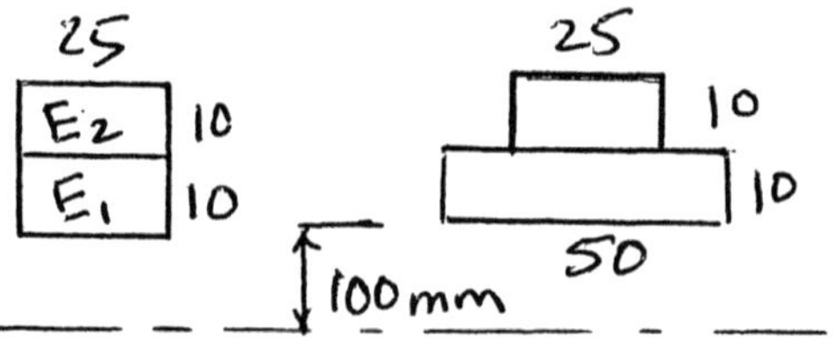

$A_e = 25(10) + 50(10) = 750\text{ mm}^2$

$$(r_c)_e = \frac{105(50)(10) + 115(25)(10)}{750} = 108.33333\text{ mm}$$

$$\left(\int\frac{dA}{r}\right)_e = 50\ln\frac{110}{100} + 25\ln\frac{120}{110} = 6.941\ldots\text{ mm},\quad r_n = \frac{750}{6.941\ldots} = 108.05681\text{ mm}$$

$e = r_c - r_n = 0.2765$ mm

$$\sigma_\theta\Big|_{r=120\text{mm}} = -\frac{50(10^3)(108.06-120)}{(750)(0.2765)(120)} = \underline{24.0\text{ MPa}}$$

$$\sigma_\theta\Big|_{r=100\text{mm}} = -\left(\frac{200}{100}\right)\frac{50(10^3)(108.06-100)}{(750)(0.2765)(100)} = \underline{-38.9\text{ MPa}}$$

Materials are better in present position. This places more rigidity at the inner radius which is necessary for optimizing curved beam sections

5.54 $w = Cxy$

$$M_x = -D\left(\frac{\partial^2 w}{\partial x^2} + \nu\frac{\partial^2 w}{\partial y^2}\right) = 0, \quad M_y = -D\left(\frac{\partial^2 w}{\partial y^2} + \nu\frac{\partial^2 w}{\partial x^2}\right) = 0$$

$$M_{xy} = -D(1-\nu)C = \text{constant} \quad \boxed{\text{ANS}}$$

5.55
$$\int_0^b p(x,y)\sin\frac{m\pi y}{b}\,dy = \int_0^b \sum_m\sum_n C_{mn}\sin\left(\frac{m\pi x}{a}\right)\left(\sin\frac{n\pi y}{b}\right)\left(\sin\frac{m\pi y}{b}\right)dy$$

In general, $\int_0^b \sin\frac{n\pi}{b}y\,\sin\frac{m\pi}{b}y\,dy = 0$ for $m \neq n$

For $m = n$
$$\int_0^b \sin^2\frac{n\pi y}{b}\,dy = \left(\frac{y}{2} - \sin\frac{2n\pi}{b}\right)\Bigg|_0^b = \frac{b}{2}$$

$$\therefore \int_0^b p(x,y)\sin\frac{n\pi y}{b}\,dy = \frac{b}{2}C_{mn}\sin\frac{m\pi x}{a}$$

Next, integrate wrt x, and in a similar fashion
$$\int_0^a\left[\int_0^b p(x,y)\sin\frac{n\pi y}{b}\,dy\right]\sin\frac{m\pi x}{a}\,dx = \frac{a}{2}\left[\frac{b}{2}C_{mn}\right] \quad \text{Thus}$$

$$C_{mn} = \frac{4}{ab}\int_{x=0}^{a}\int_{y=0}^{b} p(x,y)\sin\frac{m\pi x}{a}\sin\frac{n\pi y}{b}\,dx\,dy \quad \boxed{\text{ANS}}$$

5.56 (b)
$$\frac{\partial^4 w}{\partial x^4} = w_0\left(\frac{\pi}{a}\right)^4\sin\left(\frac{\pi x}{a}\right)\sin\left(\frac{\pi y}{b}\right) \quad (1)$$

$$\frac{\partial^4 w}{\partial x^2\partial y^2} = w_0\left(\frac{\pi}{a}\right)^2\left(\frac{\pi}{b}\right)^2\sin\left(\frac{\pi x}{a}\right)\sin\left(\frac{\pi y}{b}\right) \quad (2)$$

$$\frac{\partial^4 w}{\partial y^4} = w_0\left(\frac{\pi}{b}\right)^4\sin\left(\frac{\pi x}{a}\right)\sin\left(\frac{\pi y}{b}\right) \quad (3)$$

Sub eqs (1), (2), (3), and (a) into Eq.(5.7-12) results in

$$\pi^4 w_0\left[\left(\frac{1}{a}\right)^4 + 2\left(\frac{1}{a}\right)^2\left(\frac{1}{b}\right)^2 + \left(\frac{1}{b}\right)^4\right] = \frac{p_0}{D}$$

$$\pi^4 w_0\left[\frac{b^4 + 2a^2b^2 + a^4}{a^4b^4}\right] = \frac{p_0}{D} \Rightarrow w_0 = \frac{p_0}{D\pi^4}\left(\frac{a^2b^2}{a^2+b^2}\right)^2$$

$$\therefore w(x,y) = \frac{p_0}{D\pi^4}\left(\frac{a^2b^2}{a^2+b^2}\right)^2\sin\frac{\pi x}{a}\sin\frac{\pi y}{b} \quad (4) \quad \boxed{\text{ANS}}$$

(continued)

(a) Simple supports

(1) $w=0$ at $x=0, a$ for any y Eq ④ satisfies this

(2) $w=0$ at $y=0, b$ for any x " " "

(3) $M_x = 0$ at $x=0, a$ for any y ⑤

Eq(5.7-4a)

$$M_x = -D\left(\frac{\partial^2 w}{\partial x^2} + \nu\frac{\partial^2 w}{\partial y^2}\right) \overset{④}{=} \frac{P_o}{\pi^2}\left(\frac{a^2b^2}{a^2+b^2}\right)^2\left(\frac{1}{a^2}+\frac{\nu}{b^2}\right)\sin\frac{\pi x}{a}\sin\frac{\pi y}{b} \quad ⑥$$

Eq ⑥ satisfies ⑤

(4) $M_y = 0$ at $y=0, b$ for any x ⑦

$$M_y = -D\left(\frac{\partial^2 w}{\partial y^2} + \nu\frac{\partial^2 w}{\partial x^2}\right) = \frac{P_o}{\pi^2}\left(\frac{a^2b^2}{a^2+b^2}\right)^2\left(\frac{1}{b^2}+\frac{\nu}{a^2}\right)\sin\frac{\pi x}{a}\sin\frac{\pi y}{b} \quad ⑧$$

Eq ⑧ satisfies ⑦. Thus B.C.s are satisfied **ANS**

5.57 (a) Eq.(c) of Prob 5.56 is max when $x = \frac{a}{2}$, $y = \frac{b}{2}$

$$w(0.5a, 0.5b) = \frac{P_o}{\pi^4 D}\left(\frac{a^2b^2}{a^2+b^2}\right)^2 \quad \text{ANS}$$

(b) From the previous problem $(M_x)_{max}$, $(M_y)_{max}$ also occur at $x=a/2$, $y=b/2$. From Prob 5.56,

$$(M_x)_{max} = \frac{P_o}{\pi^2}\frac{a^2b^2}{(a^2+b^2)^2}(b^2+\nu a^2)$$

$$(M_y)_{max} = \frac{P_o}{\pi^2}\frac{a^2b^2}{(a^2+b^2)^2}(a^2+\nu b^2)$$

ANS

(c) If $a>b$, $(M_y)_{max} > (M_x)_{max}$ To see this,

$$\frac{(M_y)_{max}}{(M_x)_{max}} = \frac{a^2+\nu b^2}{b^2+\nu a^2} = \frac{r+\nu}{1+\nu r} \quad \text{where } r = \left(\frac{a}{b}\right)^2 > 1$$

Now, if $r+\nu > 1+\nu r$, $(M_y)_{max} > (M_x)_{max}$. Since $r>1$ and $\nu<1$ for real materials, $r = 1+\alpha$ and $\nu = 1-\beta$ where α, β are positive numbers.

$r+\nu = 2+\alpha-\beta$, $1+\nu r = 1+(1-\beta)(1+\alpha) = 2+\alpha-\beta-\alpha\beta$

$(r+\nu)-(1+\nu r) = \alpha\beta$ positive $\therefore r+\nu > 1+\nu r$ & $(M_y)_{max} > (M_x)_{max}$

$$M_{max} = \frac{P_o}{\pi^2}\frac{a^2b^2}{(a^2+b^2)^2}(a^2+\nu b^2) \quad \text{ANS}$$

5.58 $C_{mn} = \dfrac{4p_o}{ab}\left(-\dfrac{a}{m\pi}\cos\dfrac{m\pi x}{a}\Big|_0^a\right)\left(-\dfrac{b}{n\pi}\cos\dfrac{n\pi y}{b}\Big|_0^b\right)$

$$= \frac{4p_o}{ab}\left[\frac{ab}{\pi^2 mn}(\cos m\pi - 1)(\cos n\pi - 1)\right]$$

$$= \begin{cases} \dfrac{16p_o}{\pi^2 mn} & m,n = 1,3,5,\ldots \quad [a] \\ 0 & m,n = 2,4,6\ldots \end{cases}$$ **ANS**

In general, $\dfrac{\partial^4 w}{\partial x^4} = w_{mn}\left(\dfrac{m\pi}{a}\right)^4 \sin\dfrac{m\pi x}{a}\sin\dfrac{n\pi y}{b}$

$\dfrac{\partial^4 w}{\partial x^2 \partial y^2} = w_{mn}\left(\dfrac{m\pi}{a}\right)^2\left(\dfrac{n\pi}{b}\right)^2 \sin\dfrac{m\pi x}{a}\sin\dfrac{n\pi}{b}y$, $\dfrac{\partial^4 w}{\partial y^4} = w_{mn}\left(\dfrac{n\pi}{b}\right)^4 \sin\dfrac{m\pi x}{a}\sin\dfrac{n\pi y}{b}$

Eq (5.7-12) with [a] into eq(a) of Prob. 5.55 yields

$$\pi^4 w_{mn}\left[\left(\frac{m}{a}\right)^4 + 2\left(\frac{m}{a}\right)^2\left(\frac{n}{b}\right)^2 + \left(\frac{n}{b}\right)^4\right] = \frac{16p_o}{\pi^2 mnD} \quad m,n = 1,3,5\ldots$$

Simplifying yields

$$w_{mn} = \frac{16p_o}{\pi^6 mn}\left[\frac{a^2b^2}{(bm)^2+(an)^2}\right]^2 \quad m,n = 1,3,5\ldots$$

$$= 0 \quad m,n = 2,4\ldots$$

Thus,

$$w(x,y) = \frac{16p_o}{\pi^6 D}\sum_{m=1}^{\infty}\sum_{n=1}^{\infty}\frac{1}{mn}\left[\frac{a^2b^2}{(bm)^2+(an)^2}\right]^2 \sin\frac{m\pi x}{a}\cos\frac{n\pi y}{b}$$

ANS $m,n = 1,3,5\ldots$

5.59 w_{max} occurs at $x = a/2$, $y = b/2$

$$w_{max} = \frac{16p_o}{\pi^6 D}\sum_{m=1}^{\infty}\sum_{n=1}^{\infty}\frac{1}{mn}\left[\frac{a^2b^2}{(bm)^2+(an)^2}\right]^2 \sin\frac{m\pi}{2}\cos\frac{n\pi}{2}$$

Sub in $a = 1.6b$, $\nu = 0.3$ for D [Eq. (5.7-5)],

$$w_{max} = \frac{16p_o b^4(12)(1-0.3)^2(1.6)^4}{\pi^6(Et^3)}\sum_m\sum_n\frac{1}{mn}\left(\frac{1}{m^2+2.56n^2}\right)^2\sin\frac{m\pi}{2}\sin\frac{n\pi}{2}$$

$$= \frac{p_o b^4}{Et^3}\left[1.19103\sum_{m=1}^{\infty}\sum_{n=1}^{\infty}\frac{1}{mn}\left(\frac{1}{m^2+2.56n^2}\right)^2\sin\frac{m\pi}{2}\sin\frac{n\pi}{2}\right]$$

Using MathCad the above converges truncating to $m,n = 17$ to

$$w_{max} = 0.090725\frac{p_o b^4}{Et^3}$$ compare with Table 5.7-1 0.0906

(continued)

Stresses $\quad M_{max} = (M_y)_{max}$ (see Prob. 5.57 solution)

$$M_y = -D\left(\frac{\partial^2 w}{\partial y^2} + \frac{\partial^2 w}{\partial x^2}\right)$$

$$= \frac{16 p_o}{\pi^4} \sum_m \sum_n \frac{1}{mn}\left[\frac{a^2 b^2}{(bm)^2+(an)^2}\right]^2 \left[\left(\frac{n}{b}\right)^2 + \nu\left(\frac{m}{a}\right)^2\right] \sin\frac{m\pi x}{a} \sin\frac{n\pi y}{b}$$

For $(M_y)_{max}$ $x = a/2$, $y = b/2$. Sub $(M_y)_{max}$, $a = 1.6b$, $\nu = 1/3$ into $\sigma_{max} = 6(M_y)_{max}/t^2$ and simplify

$$\sigma_{max} = p_o\left(\frac{b}{t}\right)^2\left[2.52297 \sum_{m=1}^{\infty}\sum_{n=1}^{\infty} \frac{1}{mn}\frac{[2.56n^2+0.3m^2]}{(m^2+2.56n^2)^2} \sin\frac{m\pi}{2} \sin\frac{n\pi}{2}\right]$$

Takes $m, n = 55$ to converge to

$$\sigma_{max} = 0.51737\, p_o\left(\frac{b}{t}\right)^2 \quad \underline{\text{ANS}}$$

Table is 0.5172

5.60 (a) Table 5.7-1, Case 1, $k_1 = 0.6102$, $k_2 = 0.1110$

$$\sigma_{max} = 0.6102(100)(10^{-3})\left(\frac{200}{6}\right)^2 = \underline{67.8 \text{ MPa}}$$

$$w_{max} = \frac{(0.111)(100)(10^3)(200)^4}{210(10^9)(6)^3} = \underline{0.392 \text{ mm}}$$

(b) Case 2, $k_1 = 0.4974$, $k_2 = 0.0277$

$$\sigma_{max} = 0.4974(100)(10^{-3})\left(\frac{200}{6}\right)^2 = \underline{55.3 \text{ MPa}}$$

$$w_{max} = \frac{0.0277(100)(10^3)(200)^4}{210(10^9)(6)^3} = \underline{0.0977 \text{ mm}}$$

(c) Case 6, $k_1 = 0.7146$, $k_2 = 0.0922$

$$\sigma_{max} = 0.7146(100)(10^{-3})\left(\frac{200}{6}\right)^2 = \underline{79.4 \text{ MPa}}$$

$$w_{max} = \frac{0.922(100)(10^3)(200)^4}{210(10^9)(6)^3} = \underline{0.325 \text{ mm}}$$

(d) Consider a beam of length 400 mm, simply-supported

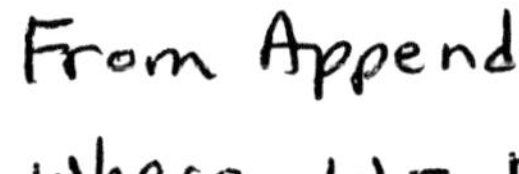

$$I = \frac{1}{12}(200)(6)^3 = 3600 \text{ mm}^4$$

$$R = \frac{1}{2}\, 100(10^3)(0.200)(0.400) = 4000 \text{ N}$$

$$M_{max} = 4000(0.200) - \frac{1}{2}(100)(10^3)(0.200)^3 = 400 \text{ N-m}$$

$$\sigma_{max} = \frac{Mc}{I} = \frac{(400)(10^3)(3)}{3600} = \underline{333.3 \text{ MPa}}$$

From Appendix C, beam C.6, $(v_c)_{max} = \frac{5}{384}\frac{wL^4}{EI}$,

where $w = p_o b$
b is width of plate

(Continued)

For wide plate, plane strain, replace E with $E/1-\nu^2$. See Section. 4.1.3.

$$w_{max} = \frac{5}{384}\frac{(1-0.3)^2(100)(10^3)(0.200)(0.400)^4}{210(10^9)(3600)(10^{-3})^4} = 8.025(10^{-3})m = \underline{8.025\ mm}$$

From Ref 5.10, tables yielded

$w_{max} = 8.796\ mm$, and $\sigma_{max} = \underline{335.7\ MPa}$

5.61 $w = -\dfrac{M_o r^2}{2(1+\nu)D}$ ①

(1) Since $p=0$, Eq(5.7-24) with Eq ① is

$$-\frac{1}{r^2}\left(-\frac{M_o}{(1+\nu)D}\right) + \frac{1}{r^3}\left(-\frac{M_o r}{(1+\nu)D}\right) \equiv 0 \quad \text{is satisfied.}$$

(2) $w=0$ at $r=0$ B.C Eq① satisfies this

(3) BC. $M_r = M_o$ at $r=r_o$

Eq(5.7-15a)

$$M(r) = -D\left[-\frac{M_o}{(1+\nu)D} + \nu\left(\frac{-M_o}{(1+\nu)D}\right)\right] = \frac{M_o(1+\nu)}{1+\nu} = M_o \quad \text{B.C... satisfied}$$

5.62 From Example 5.7-1, derivation up to Eq(a) holds where

$$w = \frac{p_o r^4}{64D} + C_1\frac{r^2}{4} + C_3 \quad \text{and} \quad \frac{dw}{dr} = \frac{p_o r^3}{16D} + C_1\frac{r}{2}$$

B.C.s (a) $w=0$, at $r=a$ (b) $\dfrac{dw}{dr}=0$, at $r=a$

From(b) $0 = \dfrac{p_o a^3}{16D} + C_1\dfrac{a}{2} \Rightarrow C_1 = -\dfrac{p_o a^2}{8D}$

From (a) with C_1 just obtained

$$0 = \frac{p_o a^4}{64D} - \frac{p_o a^4}{32D} + C_3 \Rightarrow C_3 = \frac{p_o a^4}{64D}$$

$$\therefore w = \frac{p_o r^4}{64D} - \frac{p_o a^2 r^2}{32D} + \frac{p_o a^4}{64D} = \frac{p_o}{64D}(a^2-r^2)^2 \quad \boxed{ANS}$$

At $r=0$, $w_{max} = \dfrac{p_o a^4}{64D}$ $\boxed{ANS}$

Eq(5.7-15)

$$M_r = -D\left[\frac{3}{16}\frac{p_o r^2}{D} - \frac{p_o a^2}{16D} + \nu\frac{p_o r^2}{16D} - \nu\frac{p_o a^2}{16D}\right] = \frac{p_o}{16}\left[(1+\nu)a^2-(3+\nu)r^2\right] \quad \boxed{ANS}$$

$$M_\theta = -D\left[\frac{p_o r^2}{16D} - \frac{p_o a^2}{16D} + \nu\frac{3}{16}\frac{p_o r^2}{D} - \nu\frac{p_o a^2}{16D}\right] = \frac{p_o}{16}\left[(1+\nu)a^2-(1+3\nu)r^2\right] \quad \boxed{ANS}$$

Maximas of M_r & M_θ occur at $r=0$. Must also check at $r=a$

$r=0$, $M_r = M_\theta = \dfrac{1+\nu}{16}p_o a^2$

$r=a$, $M_r = -\dfrac{p_o a^2}{8}$, $M_\theta = -\dfrac{\nu}{8}p_o a^2$

Since $\nu<1$, max mag. of M is $M_{max} = 0.125\, p_o a^2$ $\boxed{ans}$

(5.63)

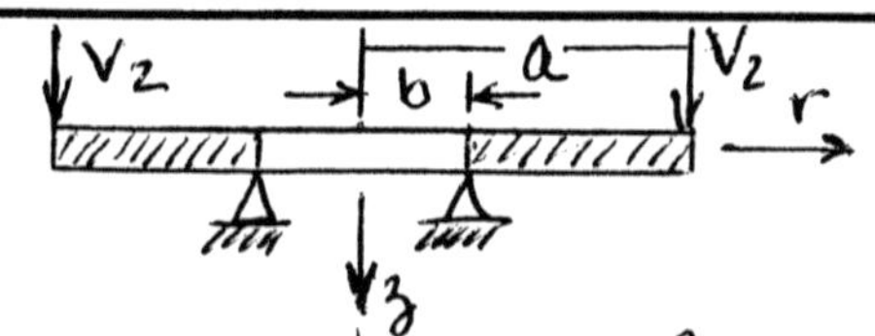

Table 5.7-2, case 7. Reverse loading 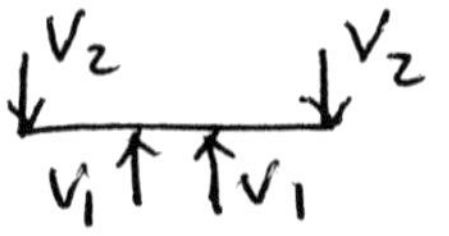

$\Sigma F_z = 0$ yields

$$V_1(2\pi b) = V_2(2\pi a) \Rightarrow V_1 b = V_2 a$$

Then, subtract the deflection at $r=b$ so that net deflection at $r=b$ is zero. From table, (reversed loading)

$$w\Big|_{r=b} = -\frac{V_2 a}{8D}\left[(a^2-b^2)\left(\frac{3+\nu}{1+\nu} - \frac{2b^2}{a^2-b^2}\ln\frac{b}{a}\right) + 2\left(b^2 + 2\frac{1+\nu}{1-\nu}\frac{a^2b^2}{a^2-b^2}\ln\frac{b}{a}\right)\ln\frac{b}{a}\right]$$

Subtract this from the negative of table eq (with $V_1 b = V_2 a$).

$$w = -\frac{V_2 a}{8D}\left[(b^2-r^2)\left(\frac{3+\nu}{1+\nu} - \frac{2b^2}{a^2-b^2}\ln\frac{b}{a}\right) + 2\left(r^2 + 2\frac{1+\nu}{1-\nu}\frac{a^2b^2}{a^2-b^2}\ln\frac{b}{a}\right)\ln\frac{r}{a}\right.$$
$$\left. - 2\left(b^2 + 2\frac{1+\nu}{1-\nu}\frac{a^2b^2}{a^2-b^2}\ln\frac{b}{a}\right)\ln\frac{b}{a}\right] \quad \boxed{\text{ANS}}$$

$$\frac{dw}{dr} = \frac{V_2 a}{8D}\left[2r\left(\frac{3+\nu}{1+\nu} - \frac{2b^2}{a^2-b^2}\ln\frac{b}{a}\right) - 4r\ln\frac{r}{a} - 2r - 4\left(\frac{1+\nu}{1-\nu}\frac{a^2b^2}{a^2-b^2}\ln\frac{b}{a}\right)\frac{1}{r}\right]$$

$$= \frac{V_2 a}{2D}\left[\left(\frac{1}{1+\nu} - \ln\frac{r}{a} - \frac{b^2}{a^2-b^2}\ln\frac{b}{a}\right)r - \left(\frac{1+\nu}{1-\nu}\frac{a^2b^2}{a^2-b^2}\ln\frac{b}{a}\right)\frac{1}{r}\right] \quad \boxed{\text{ANS}}$$

$$\frac{d^2w}{dr^2} = \frac{V_2 a}{2D}\left[\frac{1}{1+\nu} - \frac{b^2}{a^2-b^2}\ln\frac{b}{a} - \ln\frac{r}{a} - 1 + \left(\frac{1+\nu}{1-\nu}\frac{a^2b^2}{a^2-b^2}\ln\frac{b}{a}\right)\frac{1}{r^2}\right]$$

$$M_r = -D\left(\frac{d^2w}{dr^2} + \frac{\nu}{r}\frac{dw}{dr}\right)$$

$$= -\frac{V_2 a}{2}\left[\frac{1}{1+\nu} - \frac{b^2}{a^2-b^2}\ln\frac{b}{a} - \ln\frac{r}{a} - 1 + \left(\frac{1+\nu}{1-\nu}\frac{a^2b^2}{a^2-b^2}\ln\frac{b}{a}\right)\frac{1}{r^2}\right.$$
$$\left. + \frac{\nu}{1+\nu} - \nu\ln\frac{r}{a} - \nu\frac{b^2}{a^2-b^2}\ln\frac{b}{a} - \nu\left(\frac{1+\nu}{1-\nu}\frac{a^2b^2}{a^2-b^2}\ln\frac{b}{a}\right)\frac{1}{r^2}\right]$$

$$= \frac{(1+\nu)}{2}V_2 a\left[\frac{b^2}{a^2-b^2}\ln\frac{b}{a} + \ln\frac{r}{a} - \left(\frac{a^2b^2}{a^2-b^2}\ln\frac{b}{a}\right)\frac{1}{r^2}\right]$$

$$= \frac{(1+\nu)}{2}V_2 a\left[\frac{r^2-a^2}{a^2-b^2}\left(\frac{b}{r}\right)^2\ln\frac{b}{a} + \ln\frac{r}{a}\right] \quad \boxed{\text{ANS}}$$

$M_\theta = -D\left(\frac{1}{r}\frac{dw}{dr} + \nu\frac{d^2w}{dr^2}\right)$. Result is

$$M_\theta = \frac{V_2 a}{2}\left\{(1+\nu)\left[\ln\frac{r}{a} + \frac{a^2+r^2}{a^2-b^2}\left(\frac{b}{r}\right)^2\ln\frac{b}{a}\right] - 1 + \nu\right\} \quad \boxed{\text{ANS}}$$

$$M_{max} = (M_\theta)_{r=b} = \frac{V_2 a}{2}\left[(1+\nu)\frac{2a^2}{a^2-b^2}\ln\frac{b}{a} - 1 + \nu\right] \quad \boxed{\text{ANS}}$$

(5.64)

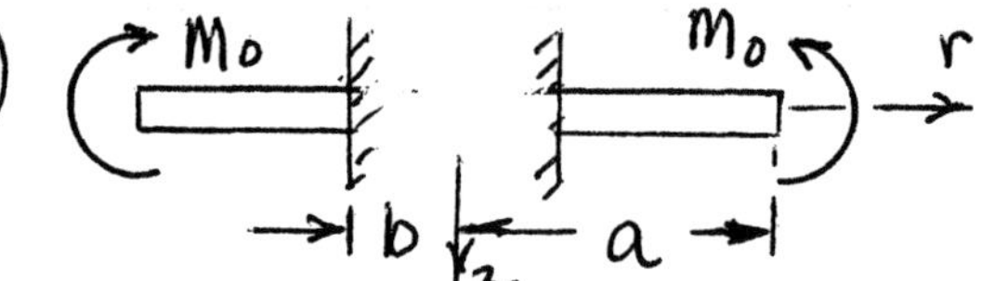

Can use Case 6 of Table 5.7-2 by subtracting deflection at $r=b$ and M_1 subject to constraint of $dw/dr=0$ at $r=b$. However, here we will use Eq. (5.7-21) with $V_r=0$.

$$\frac{1}{r}\frac{d}{dr}\left(r\frac{dw}{dr}\right)=C_1 \Rightarrow \frac{d}{dr}\left(r\frac{dw}{dr}\right)=C_1 r \Rightarrow r\frac{dw}{dr}=\frac{C_1 r^2}{2}+C_2$$

$$\frac{dw}{dr}=\frac{C_1}{2}r+\frac{C_2}{r} \Rightarrow \frac{d^2w}{dr^2}=\frac{C_1}{r}-\frac{C_2}{r^2}$$

$$\Downarrow$$

$$w=\frac{C_1}{4}r^2+C_2\ln r+C_3$$

$w=0$ at $r=b \Rightarrow \frac{C_1}{4}b^2+C_2\ln b+C_3=0$ ①

$\frac{dw}{dr}=0$ at $r=b \Rightarrow \frac{C_1}{2}b+\frac{C_2}{b}=0 \Rightarrow C_1 b^2+2C_2=0$ ②

$M_r=M_0$ at $r=a$, $M_r=-D\left(\frac{d^2w}{dr^2}+\frac{\nu}{r}\frac{dw}{dr}\right)$

$$M_0=-D\left(\frac{C_1}{2}-\frac{C_2}{a^2}+\frac{\nu}{2}C_1+\frac{\nu}{a^2}C_2\right)=-\frac{D}{2a^2}\left[(1+\nu)a^2C_1-2(1-\nu)C_2\right]$$

or, $(1+\nu)a^2C_1-2(1-\nu)C_2=-\frac{2a^2M_0}{D}$ ③

Solve ①, ②, and ③ simultaneously to obtain

$$C_1=\frac{2a^2M_0}{\left[(1+\nu)a^2+(1-\nu)b^2\right]D},\quad C_2=\frac{a^2b^2M_0}{\left[(1+\nu)a^2+(1-\nu)b^2\right]D}$$

$$C_3=\frac{a^2b^2(1-2\ln b)M_0}{\left[(1+\nu)a^2+(1-\nu)b^2\right]D}$$

sub C's into w eq and simplify

$$w=\frac{a^2M_0}{2D}\frac{\left[b^2\left(1+2\ln\frac{r}{b}\right)-r^2\right]}{\left[(1+\nu)a^2+(1-\nu)b^2\right]}$$ ANS

At $r=a$, maximum upward (negative) deflection is

$$w\big|_{r=a}=\frac{a^2M_0}{2D}\frac{\left[b^2\left(1+2\ln\frac{a}{b}\right)-a^2\right]}{\left[(1+\nu)a^2+(1-\nu)b^2\right]}$$ ANS

At $r=b$, $\frac{d^2w}{dr^2}=\frac{a^2M_0\left(-1-b^2/b^2\right)}{D\left[(1+\nu)a^2+(1-\nu)b^2\right]}$, $M_r\big|_{r=b}=-D\frac{d^2w}{dr^2}=\frac{2a^2M_0}{\left[(1+\nu)a^2+(1-\nu)b^2\right]}$ ANS

$\frac{dw}{dr}=0$

5.65

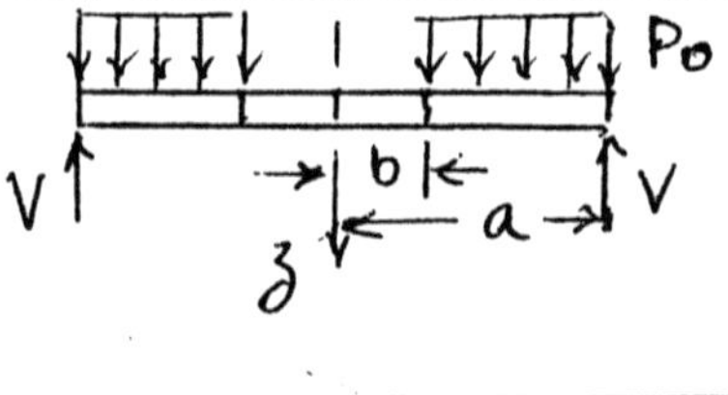

$\Sigma F_z = 0$

$p_o(\pi)(a^2 - b^2) - V(2\pi a) = 0$

$V = \dfrac{a^2 - b^2}{2a} p_o$ **ANS**

5.66

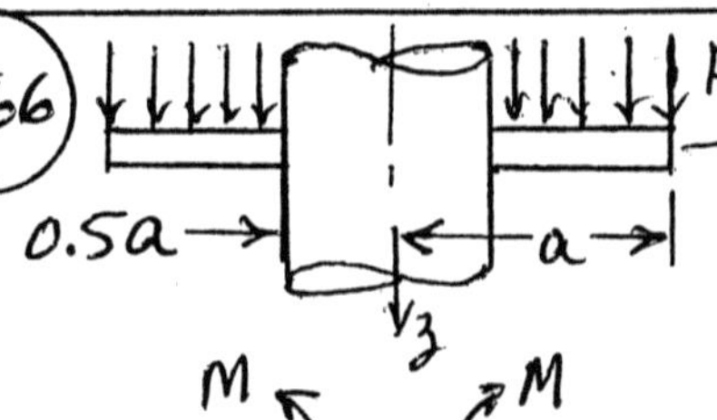

Strategy: Superpose Cases 6, 7, and 8 shown below, subject to V balancing p_o ①, $dw/dr = 0$ at $r = 0.5a$ ②. Then, add the negative of w at $r = 0.5a$ to obtain $w = 0$ ③.

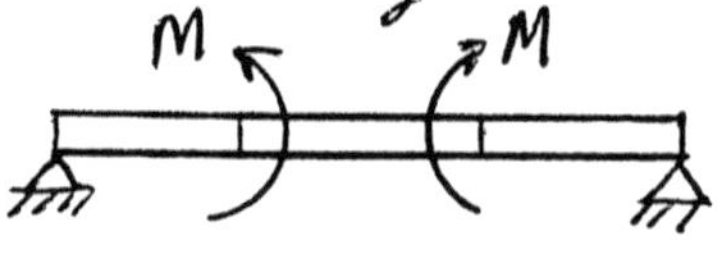

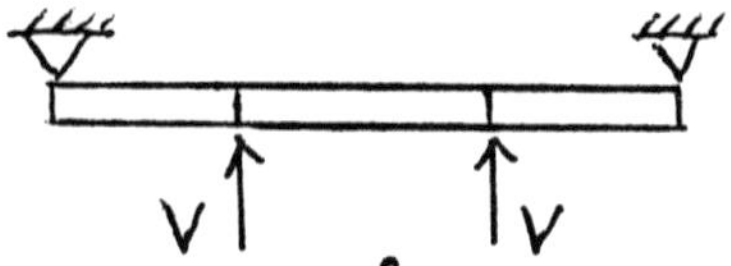

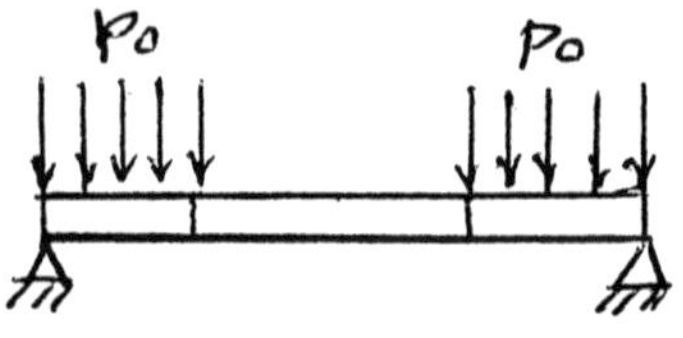

① $\Sigma F_z = 0 \Rightarrow V(2\pi)(0.5a) = p_o(\pi)\left[a^2 - (0.5a)^2\right] = \dfrac{3\pi}{4} p_o a^2$

$\therefore V = \dfrac{3}{4} p_o a$ ④

② $$\frac{dw}{dr} = \frac{\left[(1-\nu)b^2 r + (1+\nu)a^2 b^2/r\right]M}{(1-\nu^2)(a^2-b^2)D} - \frac{Vb}{2D}\left[\left(\ln\frac{r}{a} + \frac{b^2}{a^2-b^2}\ln\frac{b}{a} - \frac{1}{1+\nu}\right)r + \left(\frac{1+\nu}{1-\nu}\frac{a^2b^2}{a^2-b^2}\ln\frac{b}{a}\right)\frac{1}{r}\right] - \frac{p_o}{16D}\left[\left(\frac{3+\nu}{1+\nu}a^2 - \frac{1-\nu}{1+\nu}b^2 + 4\frac{b^4}{a^2-b^2}\ln\frac{b}{a} - r^2 + 4b^2\ln\frac{r}{a}\right)r + \left(\frac{3+\nu}{1-\nu}a^2b^2 + 4\frac{1+\nu}{1-\nu}\frac{a^2b^4}{a^2-b^2}\ln\frac{b}{a}\right)\frac{1}{r}\right] \quad ⑤$$

$\dfrac{dw}{dr} = 0$ at $r = b = 0.5a$. Also, $\nu = 0.3$ and $V = \dfrac{3}{4} p_o a$

After much algebra, $M = -0.1736\, p_o a^2$ max moment ⑥

Sub M back into ⑤ with $\nu = 0.3$ and $V = \frac{3}{4} p_o a$ at $r = a$ yields $(dw/dr)_{max}$. Again, after much algebra

$\left(\dfrac{dw}{dr}\right)_{max} = 0.02067 \dfrac{p_o a^3}{D}$ **ANS**

③ To determine the deflection equation, evaluate w at $r = b = 0.5a$ for the above 3 cases and subtract it from the superposed deflection equations of the 3 cases. At $r = b = 0.5a$ (with ④, ⑥, and $\nu = 0.3$)

$$w = \frac{0.7\left[-(0.5)^2a^2(-0.1736p_oa^2)(0.75a^2)\right] + 2(1.3)(0.5)^2a^4(-0.1736p_oa^2)(-0.6931)}{2(1-0.3^2)(0.75)a^2D}$$

$$-\frac{\left(\frac{3}{4}p_oa\right)(0.5a)}{8D}\left\{0.75a^2\left(\frac{3.3}{1.3} - \frac{2(0.5)^2a^2}{0.75a^2}(-0.6931)\right) + 2\left[(0.5)^2a^2 + 2\frac{1.3}{0.7}\frac{(0.5)^2a^4}{0.75a^2}(-0.6931)\right]\cdot(-0.693)\right\} + \frac{p_o}{64D}(0.75)a^2\left[\frac{5.3}{1.3}a^2 - 2\frac{3.3}{1.3}(0.5)^2a^2 - (0.5)^2a^2 + 8\frac{(0.5)^4a^4}{0.75a^2}(-0.6931)\right]$$

$$-\frac{p_o}{16D}\left[2(0.5)^4a^4 + \frac{3.3}{0.7}(0.5)^2a^4 + 4\frac{1.3}{0.7}\frac{(0.5)^4a^6}{0.75a^2}(-0.6931)\right](-0.6931)$$

(Continued)

$$w\Big|_{r=0.5a} = -0.00858\,\frac{p_0 a^4}{D}$$

Thus, $0.00858\,\frac{p_0 a^4}{D}$ must be added to the superposition of the deflections of the 3 cases. Since $w=0$ at $r=a$ for the three cases then at $r=a$

$$w_{max} = 0.00858\,\frac{p_0 a^4}{D} \quad \underline{\text{ANS}}$$

5.67 Superpose Cases 6 & 7 of Table 5.7-2 subject to the constraints that $\frac{dw}{dr}=0$ at $r=a,\ b$

For case 7, $V_1 = \frac{5000}{2\pi(10)} = \frac{250}{\pi}$ N/mm, $\quad \frac{b}{a}=0.2$

At $r=a$,

$$\frac{dw}{dr} = -\left[\frac{0.7(M_2-0.04M_1)\,a^3 + 1.3(0.04)\,a^4(M_2-M_1)\frac{1}{a}}{(0.91)(0.96)\,a^2 D}\right]$$

$$+\frac{(250/\pi)(0.2a)}{2D}\left[\left(\frac{0.04a^2}{0.96a^2}\ln 0.2-\frac{1}{1.3}\right)a+\left(\frac{1.3}{0.7}\,\frac{0.04a^4}{0.96a^2}\ln 0.2\right)\frac{1}{a}\right]=0$$

yields, $\quad M_1 - 9.4M_2 = 83.495a = 4.17475$ N ①

At $r=b$

$$\frac{dw}{dr} = -\left[\frac{0.7(M_2-0.04M_1)(0.2)a^3 + 1.3(0.04)a^4(M_2-M_1)\frac{5}{a}}{(0.91)(0.96)a^2 D}\right]$$

$$+\frac{(250/\pi)(0.2a)}{2D}\left[\left(\ln 0.2+\frac{0.04a^2}{0.96a^2}\ln 1.2-\frac{1}{1.3}\right)0.2a+\left(\frac{1.3}{0.7}\,\frac{(0.04a^4)}{0.96a^2}\ln 0.2\right)\frac{5}{a}\right]=0$$

yields $\quad M_1 - 1.506M_2 = 29.102a = 1455.1$ ②

Solving ① & ② simultaneously results in

$M_1 = 936$ N, $\quad M_2 = -344.55$ N

Deflection at $r=b$ with $E = 210(10^3)$ N/mm² is

$$w_{r=10\text{mm}} = \frac{(0.7)\left[-344.55-(0.04)(936)\right](50)^2(0.96)(50^2)-2(1.3)(0.04)(50)^4\cdot(-344.55-936)\ln 0.2}{2(1-0.3^2)(0.96)(50)^2 D}$$

$$+\frac{(250/\pi)(0.2)(50)}{8D}\left[0.96(50)^2\left(\frac{3.3}{1.3}-\frac{0.08}{0.96}\ln 0.2\right)+2\left(0.04(50)^2+2\,\frac{1.3}{0.7}\,\frac{(0.04)(50)^4}{0.96(50)^2}\ln 0.2\right)\cdot \ln 0.2\right] = \frac{131.38(10^3)}{D}$$

(continued)

$$w\Big|_{r=10} = \frac{131.38(10^3)(12)(1-0.3^2)}{210(10^3)(4)^3} = \underline{0.107\ mm}$$ ANS

Max bending stress. $M_1 = 936\ N$ so at $r = b$,

$$\sigma_\theta = \frac{6(936)}{4^2} = \underline{351\ MPa}$$

(5.68) Superpose Cases 3 & 5 from Table 5.7-2 subject to the constraint $w = 0$ at $r = 0$

$$w\Big|_{r=0} = \frac{p_0 a^2}{64D}\left(\frac{5+\nu}{1+\nu} a^2\right) - \frac{R}{16\pi D}\left(\frac{3+\nu}{1+\nu} a^2\right) = 0$$

$$R = \frac{\pi}{4}\left(\frac{5+\nu}{3+\nu}\right) p_0 a^2$$ ANS

(5.69) See Prob 5.68

$$\frac{p_0 a^2}{64D}\left(\frac{5+\nu}{1+\nu} a^2\right) - \frac{R}{16\pi D}\left(\frac{3+\nu}{1+\nu} a^2\right) = e$$ Solve for R

$$R = \frac{\pi}{4a^2(3+\nu)}\left[(5+\nu)p_0 a^4 - 64(1+\nu)De\right]$$ ANS

(5.70) For steel $\gamma = 0.283\ lb/in^3 = \frac{0.283(454)}{(0.0254)^3} = 7840\ kg/m^3$

$\omega = 100(2\pi) = 200\pi\ rad/s$

(a)

$$\sigma_r = \left(\frac{3.29}{8}\right)(7840)(200\pi)^2\left\{0.1^2 + 0.5^2 - \left[\frac{(0.1)(0.5)}{r}\right]^2 - r^2\right\}$$

$$= 1.2729(10^9)\left(0.26 - \frac{0.0025}{r^2} - r^2\right) \quad (1)$$ ANS

$$\sigma_\theta = \frac{3.29}{8}(7840)(200\pi)^2\left\{0.1^2 + 0.5^2 + \left[\frac{(0.1)(0.5)}{r}\right]^2 - \frac{1+3(0.29)}{3.29} r^2\right\}$$

$$= 1.2729(10^9)\left(0.26 + \frac{0.0025}{r^2} - 0.5844 r^2\right) \quad (2)$$ ANS

Plots on next page

At $r = \sqrt{(0.1)(0.5)} = 0.224\,m$, $(\sigma_r)_{max} = 203.7(10^6)\ Pa = 203.7\ MPa$

At $r = 0.100\,m$, $(\sigma_\theta)_{max} = 641.7(10^6)\ Pa = 641.7\ MPa$

r (mm)	σ_r (MPa)	σ_θ (MPa)
100	0.00	641.74
150	160.88	455.65
200	200.48	380.75
250	200.48	335.38
300	181.03	299.36
350	149.05	265.81
400	107.40	231.82
450	57.48	196.03
500	0.00	157.71

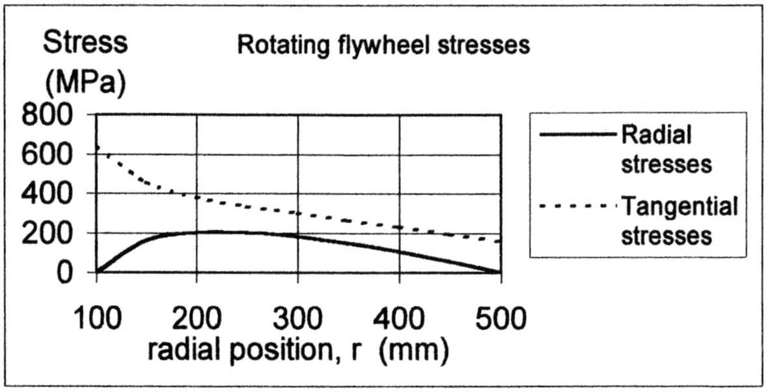

(b) At $r = 0.5\,m$, $\sigma_r = 0$, $\sigma_\theta = 157.7(10^6)\,Pa$

$$\epsilon_\theta = \frac{u_r}{r} = \frac{1}{E}(\sigma_\theta - \nu\sigma_r),\quad u_r = \frac{0.5}{210(10^9)}(157.7)(10^6) = 0.375(10^{-3})\,m$$

$\therefore$ $u_r = 0.375\,mm$

(5.71) $$\sigma_r = \frac{500(0.2)^2 + \left[\frac{(0.25)(0.2)}{r}\right]^2(-500)}{0.25^2 - 0.2^2} = 888.9 - \frac{55.56}{r^2}\ kPa\quad (r\text{ in }m)$$

$$\sigma_\theta = 888.9 + \frac{55.56}{r^2}\ kPa$$

r (mm)	σ_r (MPa)	σ_θ (MPa)
200	-0.500	2.28
205	-0.433	2.21
210	-0.371	2.15
215	-0.313	2.09
220	-0.259	2.04
225	-0.209	1.99
230	-0.161	1.94
235	-0.117	1.89
240	-0.076	1.85
245	-0.037	1.81
250	0.000	1.78

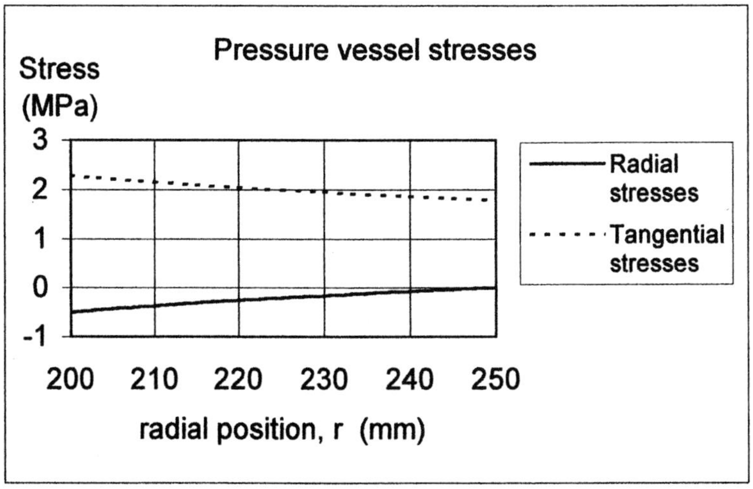

(5.72) (a) $$p = \frac{0.002}{4\left\{\frac{1}{30(10^6)}\left[\frac{(5^2+4^2)}{(5^2-4^2)} + 0.29\right] + \frac{1}{15(10^6)}\left[\frac{(4^2+2^2)}{(4^2-2^2)} - 0.32\right]\right\}}$$

$= 1990\ psi$ **ANS**

(Continued)

(b) Inner cylinder: $p_i = 0$, $p_o = 1990$ psi

$$\sigma_r = \frac{\left[-2^2 + \left(\frac{(1)(2)}{r}\right)^2\right]1990}{2^2 - 1^2} = -2653\left(1 - \frac{1}{r^2}\right)$$

$$\sigma_\theta = -2653\left(1 + \frac{1}{r^2}\right)$$

ANS

Outer cylinder: $p_i = 1990$ psi, $p_o = 0$

$$\sigma_r = \frac{1990\left[2^2 - \left(\frac{(2)(2.5)}{r}\right)^2\right]}{2.5^2 - 2^2} = 3537 - \frac{22{,}108}{r^2}$$

$$\sigma_\theta = 3537 + \frac{22{,}108}{r^2}$$

ANS

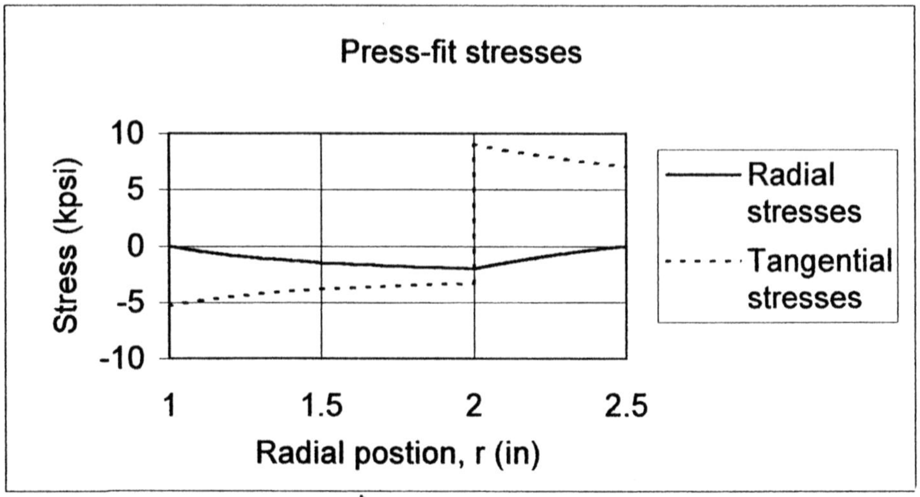

(c) $u_r = \frac{r}{E}(\sigma_\theta - \nu \sigma_r)$

Inner cylinder: $u_r\big|_{r=1} = \frac{1}{15(10^6)}\left[-5.306(10^3)\right] = -3.54(10^{-4})$ in

$$u_{r=2} = \frac{2}{15(10^6)}\left[-(3.316)(10^3) - (0.32)(-1.99)(10^3)\right] = -3.57(10^{-4}) \text{ in}$$

Outer cylinder: $u_r\big|_{r=2.5} = \frac{2.5}{30(10^6)}(7.074)(10^3) = 5.90(10^{-4})$ in

$$u_r\big|_{r=2} = \frac{2}{30(10^6)}\left[9.064(10^3) - (0.29)(-1.99)(10^3)\right] = 6.43(10^{-4}) \text{ in}$$

Note: At $r = 2$, u_r for the outer $- u_r$ for inner $= 0.001$ in

5.73

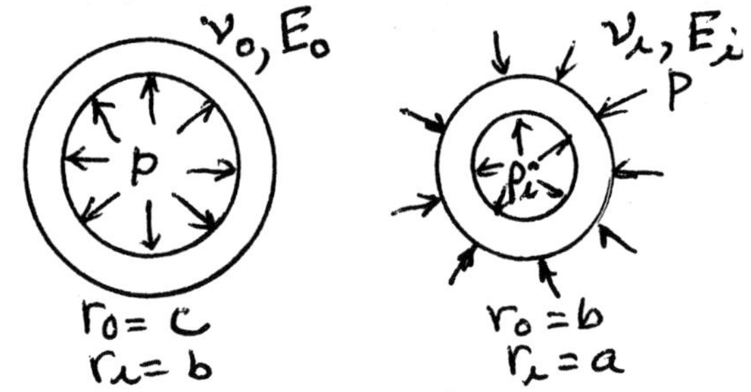

At the interface, $r = b$

$$(u_r)_{outer} = (u_r)_{inner} \quad (1)$$

Outer: $$(\sigma_\theta)_{r=b} = \frac{pb^2 + \left(\frac{bc}{b}\right)^2 p}{c^2 - b^2} = \frac{c^2+b^2}{c^2-b^2}\,p \quad (2)$$

Inner: $$(\sigma_\theta)_{r=b} = \frac{p_i a^2 - pb^2 - \left(\frac{ab}{b}\right)^2 (p - p_i)}{b^2 - a^2} = \frac{2a^2 p_i - (b^2 - a^2)p}{b^2 - a^2} \quad (3)$$

$u_r = \frac{r}{E}(\sigma_\theta - \nu\sigma_r)$. At $r = b$ $\sigma_r = -p$ for both members

Eq (1)
$$\frac{b}{E_o}\left[\left(\frac{c^2+b^2}{c^2-b^2}\right)p + \nu_o p\right] = \frac{b}{E_i}\left[\frac{2a^2 p_i - (b^2+a^2)p}{b^2-a^2} + \nu_i p\right] \quad (4)$$

Reduces to
$$p = \frac{2a^2 p_i}{(b^2-a^2)\left\{\frac{E_i}{E_o}\left[\frac{(c^2+b^2)}{(c^2-b^2)} + \nu_o\right] + \left[\frac{(b^2+a^2)}{(b^2-a^2)} - \nu_i\right]\right\}}$$ **ANS**

5.74

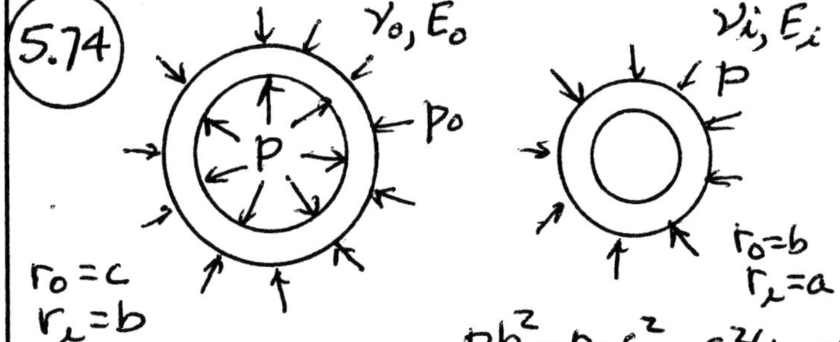

At the interface, $r = b$

$$(u_r)_{outer} = (u_r)_{inner} \quad (1)$$

Outer: $$(\sigma_\theta)_{r=b} = \frac{pb^2 - p_o c^2 - c^2(p_o - p)}{c^2 - b^2} = \frac{(b^2+c^2)p - 2c^2 p_o}{c^2-b^2} \quad (2)$$

Inner: $$(\sigma_\theta)_{r=b} = -\frac{pb^2 - a^2 p}{b^2 - a^2} = -\frac{(a^2+b^2)}{b^2-a^2}\,p \quad (3)$$

$u_r = \frac{r}{E}(\sigma_\theta - \nu\sigma_r)$, At $r = b$ $\sigma_r = -p$ for both members

$\therefore$ (1) $$\frac{b}{E_o}\left[\frac{(b^2+c^2)p - 2c^2 p_o}{c^2-b^2} + \nu_o p\right] = \frac{b}{E_i}\left[-\frac{(a^2+b^2)}{(b^2-a^2)}p + \nu_i p\right]$$

Reduces to:
$$p = \frac{2c^2 p_o}{(c^2-b^2)\left\{\left[\frac{(c^2+b^2)}{(c^2-b^2)} + \nu_o\right] + \frac{E_o}{E_i}\left[\frac{(b^2+a^2)}{(b^2-a^2)} - \nu_i\right]\right\}}$$ **ANS**

5.75 $\sigma_{mat} = 0.6(80) = 48$ kpsi occurs at inner radius of inner member where

$$\sigma_\theta|_{r=} \frac{-p(0.75)^2 - (0.75)^2 p}{0.75^2 - 0.5^2} = -3.6p = -48 \Rightarrow p = 13.33 \text{ kpsi}$$

(a) For two cylinders of the same material Eq(5.8-26) reduces to $p = \dfrac{E\delta}{b\left[\dfrac{c^2+b^2}{c^2-b^2} + \dfrac{b^2+a^2}{b^2-a^2}\right]}$ Solve for δ

$$\delta = \frac{13.33(10^3)(0.75)}{30(10^6)}\left[\frac{(1^2+0.75^2)}{(1^2-0.75^2)} + \frac{(0.75^2+0.5^2)}{(0.75^2-0.5^2)}\right] = \underline{2.057(10^{-3})\text{ in}}$$

(b) Outer member:

$$\sigma_r = \frac{13.33\left[0.75^2 - \left(\frac{(1)(0.75)}{r}\right)^2\right]}{1^2 - 0.75^2} = 17.14\left(1 - \frac{1}{r^2}\right)\text{ kpsi}$$

$$\sigma_\theta = 17.14\left(1 + \frac{1}{r^2}\right)\text{ kpsi}$$

Inner member:

$$\sigma_r = \frac{13.33\left[-0.75^2 + \left(\frac{(0.5)(0.75)}{r}\right)^2\right]}{0.75^2 - 0.5^2} = \frac{6}{r^2} - 24 \text{ kpsi}$$

$$\sigma_\theta = -24 - \frac{6}{r^2} \text{ kpsi}$$

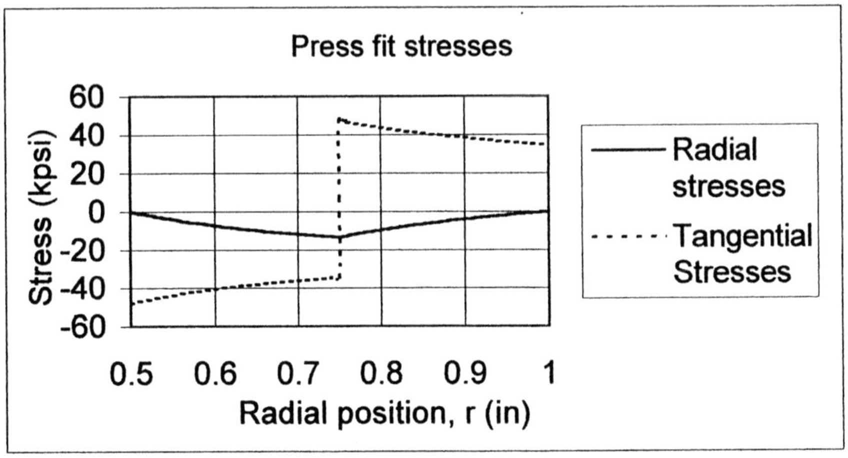

(c) For a solid cylinder of $r_i = 0.5$, $r_o = 1$ with $p_i = 5$ kpsi.

$$\sigma_r = \frac{5\left[0.5^2 - \left(\frac{(0.5)(1)}{r^2}\right)^2\right]}{1 - 0.5^2} = 1.667\left(1 - \frac{1}{r^2}\right)$$

$$\sigma_\theta = 1.667\left(1 + \frac{1}{r^2}\right)$$

(continued)

Add these to the stresses of part b

Outer member:

$$\sigma_r = 18.61\left(1-\frac{1}{r^2}\right), \quad \sigma_\theta = 18.61\left(1+\frac{1}{r^2}\right)$$

Inner member:

$$\sigma_r = \frac{4.333}{r^2} - 22.333, \quad \sigma_\theta = -22.333 - \frac{4.333}{r^2}$$

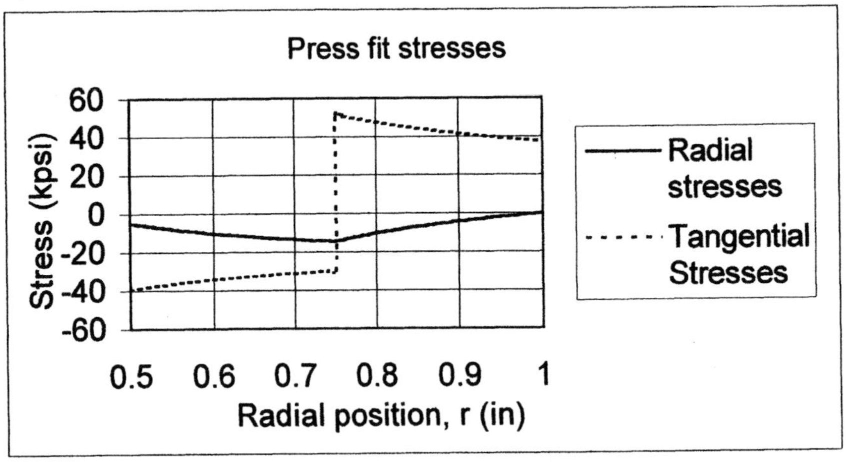

5.76 (a) $dF = f\,dN = f p\,dA \Rightarrow F = \int f p\,dA = f p A$

$A = 2\pi b t \quad \therefore F = 2\pi b t f p$ **ANS**

(b) $dT = b\,dF \Rightarrow T = 2\pi b^2 t f p$ **ANS**

5.77

thickness $= t \qquad dF = \tau\,dA = \tau t c\,d\theta$

$$T = \int c\,dF = \int_0^{2\pi} \tau t c^2\,d\theta = 2\pi \tau t c^2$$

$$\tau = \frac{T}{2\pi t c^2} \quad \textbf{ANS}$$

Derivation is independent of r, $\therefore \tau = \dfrac{T}{2\pi t r^2}$ **ANS**

5.78

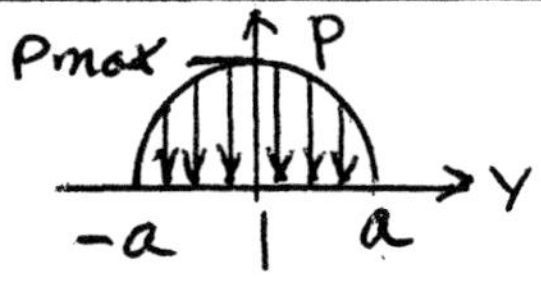

(a) Eq. for an ellipse

$$\left(\frac{p}{p_{max}}\right)^2 + \left(\frac{y}{a}\right)^2 = 1$$

$$\therefore p = p_{max}\sqrt{1-\left(\frac{y}{a}\right)^2} \quad \textbf{ANS}$$

(continued)

(b) $P = 2\int_0^a p_{max}\sqrt{1-\left(\frac{y}{a}\right)^2}(L)\,dy$ let $y = a\sin\theta$, $dy = a\cos\theta\, d\theta$

$$P = 2p_{max}aL\int_0^{\pi/2}\cos^2\theta\, d\theta = p_{max}aL\left(\theta+\frac{1}{2}\sin 2\theta\right)\Big|_0^{\pi/2} = p_{max}aL\frac{\pi}{2}$$

$$p_{max} = \frac{2P}{\pi a L} \quad \boxed{\text{ANS}}$$

(5.79) Eq (5.9-4a) with $y = 0$

$$d\sigma_x = -\frac{2}{\pi}\frac{p_{max}}{a}\frac{x^3\sqrt{a^2-y^2}}{(x^2+y^2)^2}\,dy$$ Distribution is

Symmetric. Thus,

$$\sigma_x = -\frac{4}{\pi}\frac{p_{max}}{a}x^3\int_0^a\frac{\sqrt{a^2-y^2}}{x^2+y^2}\,dy$$

Using MathCad the integral $\int_0^a \frac{\sqrt{a^2-y^2}}{x^2+y^2}\,dy$ was

evaluated symbolically and a very complicated result emerged. The software "simplified" the expression to $\frac{\pi}{4}a^2\frac{\sqrt{a^2+x^2}}{\left[(ix+a)(-ix+a)x^3\right]}$. The complex terms multiplied out resulted in $\frac{\pi a^2}{4x^3\sqrt{a^2+x^2}}$. Thus,

$$\sigma_x = -\frac{4}{\pi}\frac{p_{max}}{a}x^3\frac{\pi a^2}{4x^3\sqrt{a^2+x^2}} = -\frac{p_{max}a}{\sqrt{a^2+x^2}} \quad \text{or,}$$

$$\sigma_x = -\frac{p_{max}}{\sqrt{1+\left(\frac{x}{a}\right)^2}} \quad \boxed{\text{ANS}}$$ Same as Eq. (5.9-18)

(5.80) (a) $a = 1.522\sqrt{\frac{500(10)(50)}{5(210)(10^3)(10+50)}} = 0.0959\text{ mm}$ $\boxed{\text{ANS}}$

$$p_{max} = \frac{2(500)}{\pi(0.0959)(5)} = 664\text{ MPa} \quad \boxed{\text{ANS}}$$

(b) $\sigma_x = \frac{-664}{\sqrt{1+\zeta^2}}$, $\sigma_y = -664\left(\frac{1+2\zeta^2}{\sqrt{1+\zeta^2}} - 2\zeta\right)$ ANS

$$\sigma_z = -0.6(664)\left(\sqrt{1-\zeta^2}-\zeta\right) = -398\left(\sqrt{1-\zeta^2}-\zeta\right)$$ (continued)

(c) $\sigma_{VM} = 0.558(664) = \underline{371\ MPa}$

(d) Eq. (5.9-8)

$$\delta \approx \frac{(0.0959)^2}{2}\left(\frac{1}{10}+\frac{1}{50}\right) = 5.52(10^{-4})\,\text{mm} \qquad \underline{\text{ANS}}$$

5.81 Divide eq. (5.9-16) by R_2 and let $R_2 \to \infty$. This yields

$$a = 1.522\sqrt{\frac{PR_1}{LE}}$$

(b) $a = 1.522\sqrt{\dfrac{2000(4)}{2.5(30)(10^6)}} = \underline{0.01572\ in}$

(a) $p_{max} = \dfrac{2(2)}{\pi(0.01572)(2.5)} = \underline{32.4\ kpsi}$

(c) $\tau_{max} = 0.3(32.4) = \underline{9.72\ kpsi}$

5.82 (a) $a = 1.109\sqrt[3]{\dfrac{50}{210(10^3)}\dfrac{20(22)}{(20+22)}} = \underline{0.150\ mm}$

(b) $p_{max} = \dfrac{3(50)}{2\pi(0.150)^2} = \underline{1055\ MPa}$

(c) $\tau_{max} = 0.310(1055) = \underline{327\ MPa}$

5.83 See Figure 74 of Ref. 5.8

$D = 4.5$, $r = 0.5$, $2t = 2$, $d = 2.5$, $\frac{t}{r} = \frac{1}{0.5} = 2$, $\frac{r}{d} = \frac{0.5}{2.5} = 0.2$

$K_t = 2.55$, $\quad \sigma_{nom} = \dfrac{P}{dt} = \dfrac{200(4.5)}{(2.5)(0.25)} = 1440\ psi$

$\sigma_{max} = 2.55(1440) = 3670\ psi = \underline{3.67\ kpsi}$

5.84

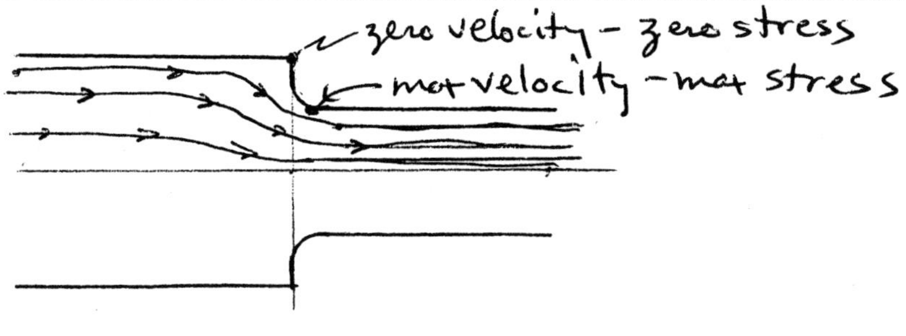

(5.85)

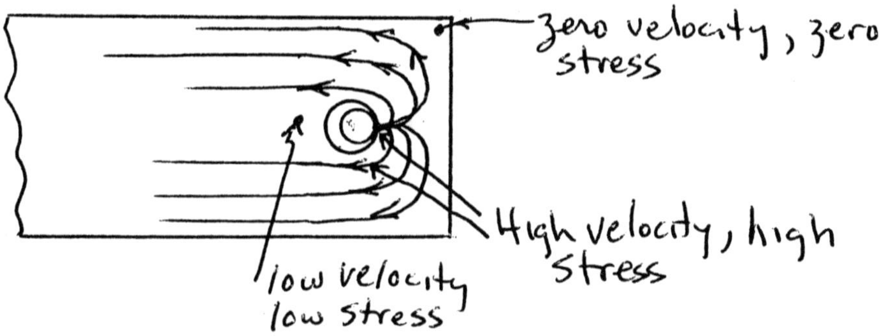

Improves flow will not reduce contact stress

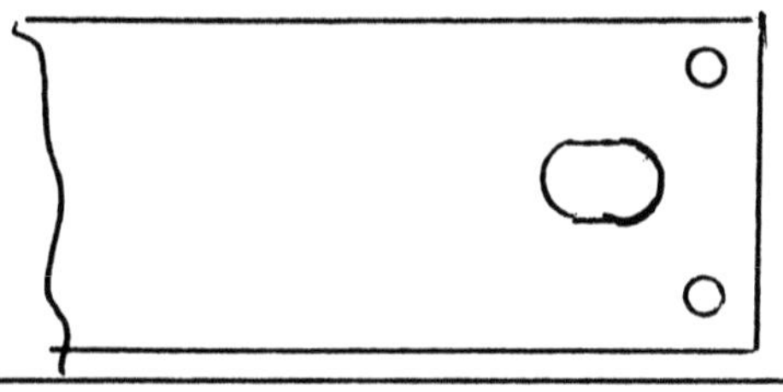

Slot and/or relief holes will improve flow but will not reduce contact stress

(5.86) $\frac{D}{d} = \frac{30}{20} = 1.5, \quad r/d = 1/20 = 0.05$

Bending: Chart F.10, $K_t \approx 2.55$, $\sigma_{nom} = \frac{32M}{\pi d^3} = \frac{32(50)(10^3)}{\pi(20)^3} = 63.7\ \text{MPa}$

$\sigma_{max_M} \approx 2.55(63.7) = 162\ \text{MPa}$

Torsion: Chart F.11, $K_{ts} \approx 1.8$, $\tau_{nom} = \frac{16T}{\pi d^3} = \frac{16(75)(10^3)}{\pi(20)^3} = 47.8\ \text{MPa}$

$\tau_{max_T} \approx 1.8(47.8) = 85.9\ \text{MPa}$

$\tau_{max} = \sqrt{\left(\frac{\sigma}{2}\right)^2 + \tau^2} = \sqrt{\left(\frac{162}{2}\right)^2 + 85.9^2} = \underline{118\ \text{MPa}}$

Maximum stress occurs at the root of the fillet at the top and bottom of bar

(5.87) $\sigma_{nom} = \frac{12(10^3)}{10(100-d)}\ \text{MPa} \quad (d \text{ in mm})$

From Chart F.1, approximating K_t

d/w	0	0.1	0.2	0.3	0.4	0.5	0.6	0.7	0.8
K_t	3	2.68	2.5	2.37	2.27	2.19	2.12	2.1	2.08

(continued)

For the cases tabulated on the previous page

σ_{nom}	12	13.3	15	17.1	20	24	30	40	60
σ_{max}	36	35.7	37.5	40.6	45.4	52.6	63.6	84	125

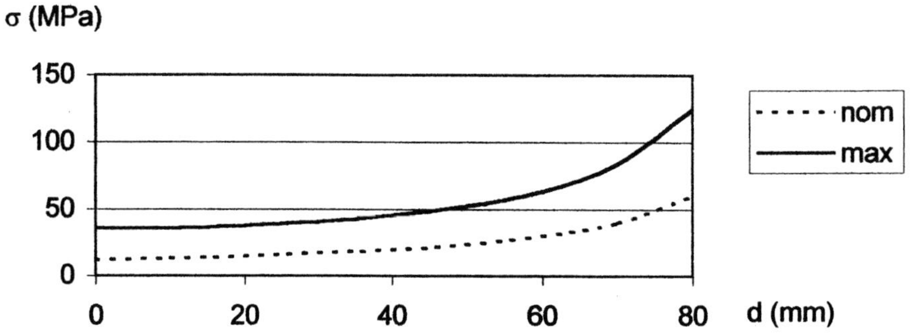

(5.88)

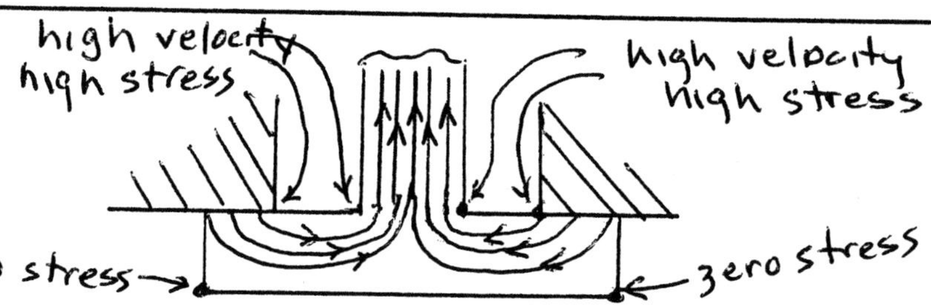

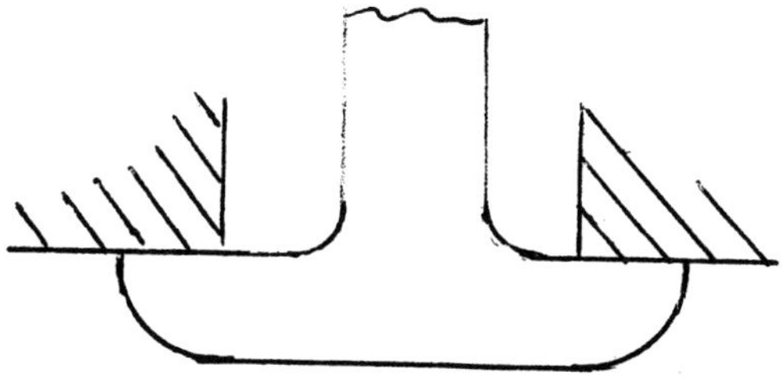

will not reduce contact stress

(5.89)

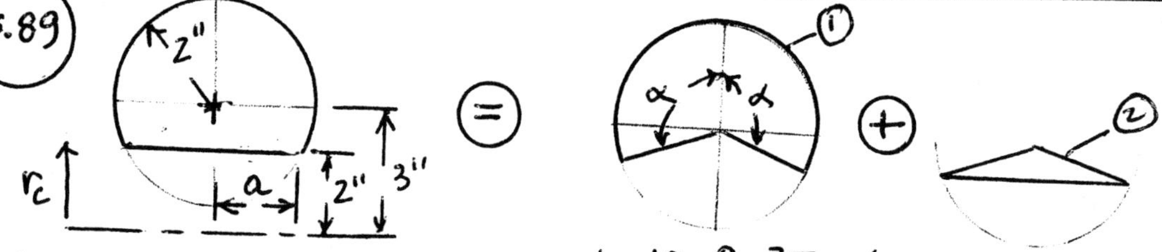

$a = 2\sin 60 = \sqrt{3}$ $\qquad \alpha = 120^\circ = \frac{2\pi}{3}$ rad

$(r_c)_1 = 3 + \frac{2(2)}{3}\frac{\sin 120}{2\pi/3}$, $A_1 = \frac{2\pi}{3}(2)^2$

$(r_c)_2 = 2 + \frac{1}{3}(1)$, $A_2 = \frac{1}{2}(2\sqrt{3})(1)$

$$r_c = \frac{(r_c)_1 A_1 + (r_c)_2 A_2}{A_1 + A_2} = \underline{3.342653608 \text{ in}}$$

(continued)

From Table 5.6-1 for $C \geq R$

$$\int \frac{dA}{r} = 2(3)\left(\frac{2\pi}{3}\right) - 2(2)\sin 120 - \sqrt{9-4}\left[\pi - 2\sin^{-1}\left(\frac{2+3\cos 120}{3+2\cos 120}\right)\right]$$

$$r_n = \frac{A_{①} + A_{②}}{\int \frac{dA}{r}} = \underline{3.151897394 \text{ in}}$$

$$e = r_c - r_n = \underline{0.190756214 \text{ in}}$$

(b) Notation for program for centroid.

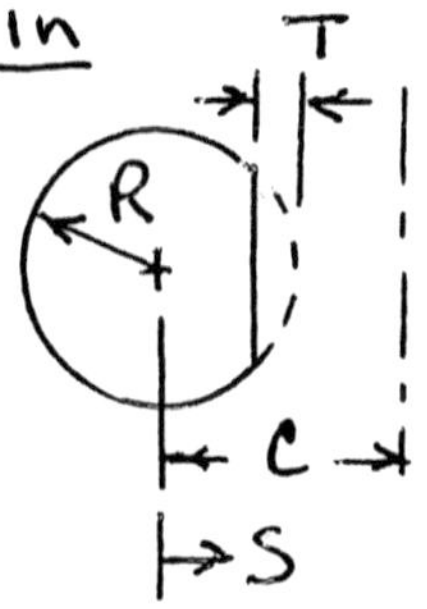


```
Function cen(C, R, T)
DS = (2 * R - T) / 1000
S = -R
SUM1 = 0
For I = 1 To 999 Step 1
   S = S + DS
   SUM1 = SUM1 + DS * (C - S) * Sqr (R ^ 2 - S ^ 2)
   Next I
S = -R
SUM2 = 0
For I = 1 To 999 Step 1
   S = S + DS
   SUM2 = SUM2 + DS * Sqr (R ^ 2 - S ^ 2)
   Next I
cen = SUM1 / SUM2
End Function
```

For e, S must be taken from centroid

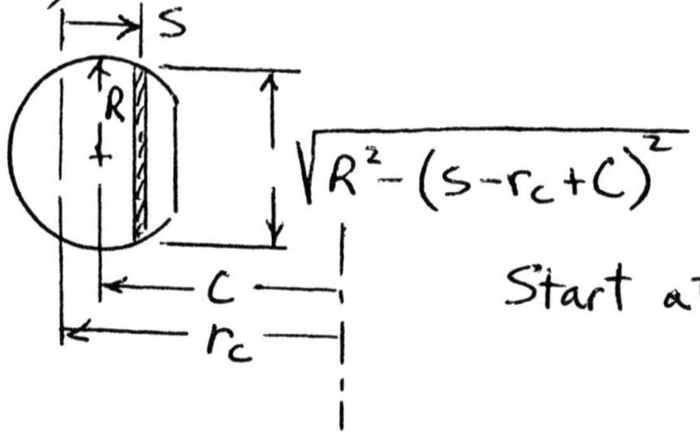

Start at $S = -R + r_c - C$

(continued)

```
Function ecc(C,  rc, R, T)
DS = (2 * R - T) / 1000
S = -R + rc - C
SUM1 = 0
For I = 1 To 999 Step 1
   S = S + DS
   SUM1 = SUM1 + DS * (S * Sqr (R ^ 2 -(S - rc + C) ^ 2)) / (rc - S)
   Next I
S = -R + rc - C
SUM2 = 0
For I = 1 To 999 Step 1
   S = S + DS
   SUM2 = SUM2 + DS * (Sqr (R ^ 2 - (S - rc + C) ^ 2)) / (rc - S)
   Next I
cen = SUM1 / SUM2
End Function
```

r_n is calculated from $r_n = r_c - e$

Input $C = 3$, $R = 2$, and $T = 1$ to above program for centroid r_c which is also input to e program

Yields,

$r_c = \underline{3.343321522 \text{ in}}$ (0.2% high)

$e = \underline{0.190506559 \text{ in}}$ (0.13% low)

$r_n = \underline{3.152814963 \text{ in}}$

(c) Stresses obtained from parts (a) & (b) are basically the same which are

$$\sigma_\theta\Big|_{r_i} = \frac{-(2000)(3.1519-2)}{(10.11)(0.1908)(2)} = \underline{-597 \text{ psi}}$$

$$\sigma_\theta\Big|_{r_o} = -\frac{(2000)(3.1519-5)}{(10.11)(0.1908)(5)} = \underline{383 \text{ psi}}$$

For the approximate method need $I_{\bar{3}}$

Section ①

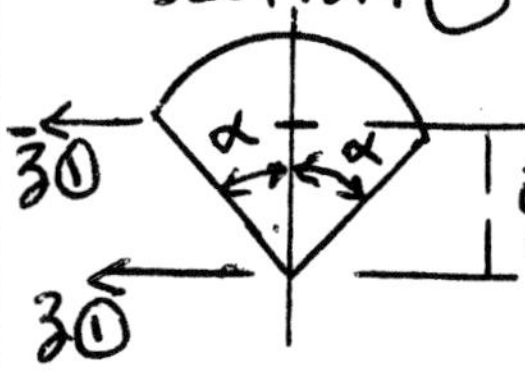

$A = \alpha r^2$

$\bar{y}_{①} = \frac{2}{3}\frac{r\sin\alpha}{\alpha}$

From Statics book such as "Meriam"

$$I_{3①} = \frac{r^4}{4}\left(\alpha + \frac{1}{2}\sin\alpha\right)$$

$$I_{\bar{3}①} = I_{3①} - \alpha r^2\left(\frac{2}{3} r \frac{\sin\alpha}{\alpha}\right)^2$$

(continued)

$$I_{\bar{z}①} = \frac{(2)^4}{4}\left(\frac{2\pi}{3} + \frac{1}{2}\sin 240\right) - \frac{2\pi}{3}(2)^2\left[\frac{2}{3}(2)\frac{\sin 120}{2\pi/3}\right]^2 = 4.099051 \text{ in}^4$$

Section ②

$$I_{\bar{z}②} = \frac{(2\sqrt{3})(1)^3}{36} = 0.096225 \text{ in}^4 \qquad A_② = \sqrt{3} \text{ in}^2$$

$$I_{\bar{z}} = 4.099051 + \frac{2\pi}{3}(2)^2(3.55132 - 3.34265)^2$$

$$+ 0.096225 + \sqrt{3}(3.34265 - 2.33333)^2$$

$$= 6.3245 \text{ in}^4$$

$$e \approx \frac{I_{\bar{z}}}{r_c A} = \frac{6.3245}{3.34265(10.11)} = 0.18715 \text{ in} \qquad (1.9\% \text{ low})$$

$$r_n \quad r_c - e \approx 3.34265 - 0.18715 = 3.1555 \text{ in}$$

$$(\sigma_\theta)_{r_i} \approx \frac{-2000(3.1555 - 2)}{(10.11)(0.18715)(2)} = \underline{-611 \text{ psi}}$$

$$(\sigma_\theta)_{r_o} \approx \frac{-2000(3.1555 - 5)}{(10.11)(0.18715)(5)} = \underline{390 \text{ psi}}$$

6.1

$$u = \frac{1}{2(70)(10^9)}\left\{30^2 + (-25)^2 + 40^2 - 2(0.33)[30(-25) + (-25)(40) + 40(30)] + 2(1.33)[(-15)^2 + 10^2 + 20^2]\right\}(10^6)^2 = 38.7(10^3) \text{ N/m}^2$$

$$= \underline{38.7 \text{ kJ/m}^3}$$

6.2 From the solution of Prob. 1.17

$$\sigma = \begin{bmatrix} 56.7 & -2.63 & 5.26 \\ -2.63 & 40.9 & -10.53 \\ 5.26 & -10.53 & 67.2 \end{bmatrix} \text{MPa}$$

$$u = \frac{1}{2(70)(10^9)}\left\{56.7^2 + 40.9^2 + 67.2^2 - 2(0.33)[56.7(40.9) + (40.9)(67.2) + 67.2(56.7)] + 2(1.33)[(-2.63)^2 + (-10.53)^2 + 5.26^2]\right\}(10^6)^2 = 28.1(10^3) \text{ N/m}^2$$

$$= \underline{28.1 \text{ kJ/m}^3}$$

6.3 $u = \dfrac{1}{2(210)(10^9)}\left[3(20)^2 - 2(0.33)(3)(20^2) + 2(1.33)(3)(10^2)\right](10^6)^2$

$= 2.87(10^3)\,N/m^2 = \underline{2.87\ kJ/m^3}$

For principal stresses see solution to Prob. 2.26 where

$(\sigma_1, \sigma_2, \sigma_3) = (40, 10, 10)$ MPa

$u = \dfrac{1}{2(210)(10^9)}\left\{40^2 + 10^2 + 10^2 - 2(0.33)\left[40(10) + 10(10) + 10(40)\right]\right\}(10^6)^2$

$= 2.87(10^3)\,N/m^2 = \underline{2.87\ kJ/m^3}$ Same as above

6.4 $\sigma_p = \dfrac{\sigma_x + \sigma_y}{2} \pm \sqrt{\left(\dfrac{\sigma_x - \sigma_y}{2}\right)^2 + \tau_{xy}^2}$

Eq. (6.2-14)

$$u = \frac{1}{2E}\left\{\left(\frac{\sigma_x+\sigma_y}{2}\right)^2 + 2\left(\frac{\sigma_x+\sigma_y}{2}\right)\sqrt{\left(\frac{\sigma_x-\sigma_y}{2}\right)^2 + \tau_{xy}^2} + \left(\frac{\sigma_x-\sigma_y}{2}\right)^2 + \tau_{xy}^2 + \left(\frac{\sigma_x+\sigma_y}{2}\right)^2 - 2\left(\frac{\sigma_x+\sigma_y}{2}\right)\sqrt{\left(\frac{\sigma_x-\sigma_y}{2}\right)^2 + \tau_{xy}^2} + \left(\frac{\sigma_x-\sigma_y}{2}\right)^2 + \tau_{xy}^2 - 2\nu\left[\left(\frac{\sigma_x+\sigma_y}{2}\right)^2 - \left(\frac{\sigma_x-\sigma_y}{2}\right)^2 - \tau_{xy}^2\right]\right\}$$

$$= \frac{1}{2E}\left[\left(\frac{\sigma_x^2}{4} + \frac{\sigma_x\sigma_y}{2} + \frac{\sigma_y^2}{4} + \frac{\sigma_x^2}{4} - \frac{\sigma_x\sigma_y}{2} + \frac{\sigma_y^2}{4} + \tau_{xy}^2\right)(2) - 2\nu\left(\frac{\sigma_x^2}{4} + \frac{\sigma_x\sigma_y}{2} + \frac{\sigma_y^2}{4} - \frac{\sigma_x^2}{4} + \frac{\sigma_x\sigma_y}{2} - \frac{\sigma_y^2}{4} - \tau_{xy}^2\right)\right]$$

$$= \frac{1}{2E}\left[\sigma_x^2 + \sigma_y^2 + 2\tau_{xy}^2 - 2\nu(\sigma_x\sigma_y - \tau_{xy}^2)\right]$$

$$u = \frac{1}{2E}\left[\sigma_x^2 + \sigma_y^2 - 2\nu\sigma_x\sigma_y + 2(1+\nu)\tau_{xy}^2\right] \quad \boxed{ANS}$$

6.5 Using variable from free end 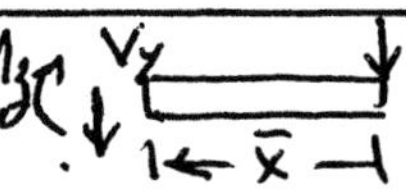 *

$V_y = -1.5$ kN, $M_z = -1.5\bar{x}$ kN·m

$$U_b = \frac{1}{2(70)(50)}\left[\int_0^{0.4}(2.5x - 5x^2)(10^3)^2 dx + \int_0^{0.2}(-1.5\bar{x})(10^3)^2 d\bar{x}\right]$$

$$= \frac{10^6}{7000}\left[\frac{2.5^2}{3}(0.4)^3 - \frac{2}{4}(2.5)(5)(0.4)^4 + \frac{25}{5}(0.4)^5 + \frac{1.5^2}{3}(0.2)^3\right] = \underline{15.33\ J}$$

$$U_s = \frac{(1.33)(1.0)}{70(10^9)(150)(10^{-3})^2}\left[\int_0^{0.4}(10x - 2.5)^2(10^3)^2 dx + \int_0^{0.2}(-1.5)^2(10^3)^2 d\bar{x}\right]$$

$$= 0.1267\left[\frac{100}{3}(0.4)^3 - \frac{2}{2}(10)(2.5)(0.4)^2 + (2.5)^2(0.4) + (1.5)^2(0.2)\right] = \underline{0.68\ J}$$

Bending energy is approx 23 times shear energy

* See Example 3.4-1 for V_y & M_z for left end

$V_y = 10x - 2.5$ kN $\quad M_z = 2.5x - 5x^2$ kN·m

6.6

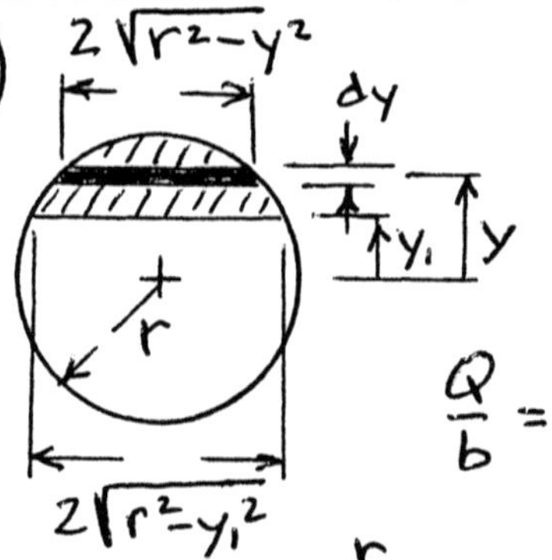

$$Q = \int_{y_1}^{y_{max}} y\,dA = \int_{y_1}^{r} y\left(2\sqrt{r^2-y^2}\right)dy$$

$$= \frac{2}{3}\left(r^2-y_1^2\right)^{3/2}$$

$$\frac{Q}{b} = \frac{\frac{2}{3}(r^2-y_1^2)^{3/2}}{2\sqrt{r^2-y_1^2}} = \frac{1}{3}\left(r^2-y^2\right)$$ can drop subscript on y_1

$$\int_A \left(\frac{Q}{b}\right)^2 dA = \frac{1}{9}\int_{-r}^{r}(r^2-y^2)^2\left(2\sqrt{r^2-y^2}\right)dy = \frac{2}{9}\int_{-r}^{r}(r^2-y^2)^{5/2}dy$$

Evaluate symbolically with software or change variable using $y = r\sin\theta$, $dy = r\cos\theta\,d\theta$ with limits $\pm\pi/2$. Result is

$$\int_A \left(\frac{Q}{b}\right)^2 dA = \frac{5\pi}{72}r^6$$

$$U_s = \int_A\int_0^L \frac{1+\nu}{E}\left(\frac{V_y}{I_z}\right)^2\left(\frac{Q}{b}\right)^2 dx\,dA = \int_0^L \frac{1+\nu}{E}\,\frac{V_y}{\left(\frac{\pi}{4}r^4\right)^2}\,\frac{5\pi}{72}r^6\,dx$$

Factor and multiply by $(\pi r^2/A)$. This yields

$$U_s = \frac{5\pi^2 r^8}{(\pi^2/16)r^8(72)}\,\frac{1+\nu}{EA}\int_0^L V_y^2\,dx = \frac{10}{9}\,\frac{(1+\nu)}{EA}\int_0^L V_y^2\,dx$$

Thus, $k = 10/9 = 1.11$ **ANS**

6.7 Force in cable is given by

$$F = C\left[\delta/L - 500(\delta/L)^2\right] \quad (1) \quad \text{where } C = 200(10^9)A$$

$$U = C\int_0^{\delta_{AB}}\left[\left(\frac{\delta}{L}\right)_{AB} - 500\left(\frac{\delta}{L}\right)^2_{AB}\right]d\delta_{AB} + C\int_0^{\delta_{BC}}\left[\left(\frac{\delta}{L}\right)_{BC} - 500\left(\frac{\delta}{L}\right)^2_{BC}\right]d\delta_{BC}$$

$$= C\left(\frac{\delta_{AB}^2}{2L_{AB}} - 500\frac{\delta_{AB}^3}{3L_{AB}^2} + \frac{\delta_{BC}^2}{2L_{BC}} - 500\frac{\delta_{BC}^3}{3L_{BC}^2}\right)$$

$$P = \frac{\partial U}{\partial(\delta_B)_V} = C\left\{\left[\frac{\delta_{AB}}{L_{AB}} - 500\left(\frac{\delta_{AB}}{L_{AB}}\right)^2\right]\frac{\partial\delta_{AB}}{\partial(\delta_B)_V} + \left[\frac{\delta_{BC}}{L_{BC}} - 500\left(\frac{\delta_{BC}}{L_{BC}}\right)^2\right]\frac{\partial\delta_{BC}}{\partial(\delta_B)_V}\right\} \quad (2)$$

Apply $\Delta(\delta_B)_V$ to system

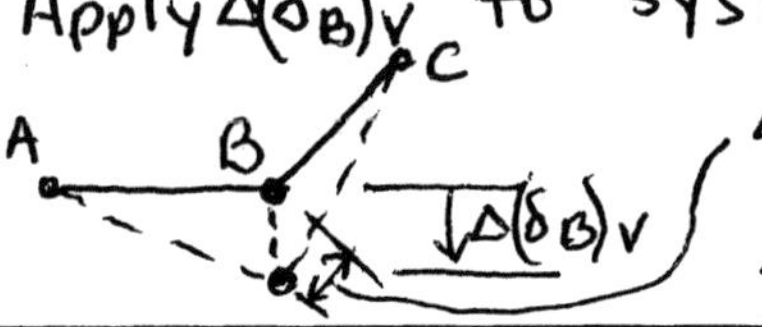

$$\Delta\delta_{BC} = \Delta(\delta_B)_V\cos 45 = \frac{\sqrt{2}}{2}\Delta(\delta_B)_V$$

$$\Delta\delta_{AB} = 0$$

(continued)

Thus, $\frac{\partial \delta_{AB}}{\partial(\delta_B)_V} = 0$, $\frac{\partial \delta_{BC}}{\partial(\delta_B)_V} = \frac{\sqrt{2}}{2}$ sub into (2)

$$8000 = 200(10^9)(125)(10^{-3})^2\left[\frac{\delta_{BC}}{0.750} - 500\left(\frac{\delta_{BC}}{0.750}\right)^2\right]\frac{\sqrt{2}}{2}$$

or, $\delta_{BC}^2 - 1.5(10^{-3})\delta_{BC} + 5.0912(10^{-7}) = 0$

with roots, $\delta_{BC} = 5.19(10^{-4})\,m,\ 9.81(10^{-4})\,m$ Lower reached first

Repeating analysis in horizontal direction with applied force = 0

$$0 = C\left\{\left[\frac{\delta_{AB}}{L_{AB}} - 500\left(\frac{\delta_{AB}}{L_{AB}}\right)^2\right]\frac{\partial \delta_{AB}}{\partial(\delta_B)_H} + \left[\frac{\delta_{BC}}{L_{BC}} - 500\left(\frac{\delta_{BC}}{L_{BC}}\right)^2\right]\frac{\partial \delta_{BC}}{\partial(\delta_B)_H}\right\} \quad (3)$$

Apply $\Delta(\delta_B)_H \rightarrow$ to system. Will get $\Delta\delta_{AB} = \Delta(\delta_B)_H$

$\Delta\delta_{BC} = -\frac{\sqrt{2}}{2}(\Delta\delta_B)_H$ $\therefore$ $\frac{\partial \delta_{AB}}{\partial(\delta_B)_H} = 1$ & $\frac{\partial \delta_{BC}}{\partial(\delta_B)_H} = -\frac{\sqrt{2}}{2}$ sub into (3)

yields

$$\frac{\delta_{AB}}{L_{AB}} - 500\left(\frac{\delta_{AB}}{L_{AB}}\right)^2 = \frac{\sqrt{2}}{2}\left[\frac{\delta_{BC}}{L_{BC}} - 500\left(\frac{\delta_{BC}}{L_{BC}}\right)^2\right] \quad (4)$$

Sub $\delta_{BC} = 5.19(10^{-4})\,m$ into (4). Yields,

$$\delta_{AB}^2 - 1(10^{-3})\delta_{AB} + 1.6(10^{-7}) = 0$$

roots are: $\delta_{AB} = 2(10^{-4})\,m,\ 8(10^{-4})\,m$

Lower roots are $\delta_{AB} = 0.2\,mm$, $\delta_{BC} = 0.519\,mm$

From geometry $(\delta_B)_V = \delta_{AB} + \frac{\delta_{BC}}{\cos 45} = 0.2 + \sqrt{2}(0.519)$

$\therefore$ $(\delta_B)_V = \underline{0.934\ mm}$

(6.8) Using Excel, plotting ϵ vs σ with σ in MPa, $C_1\,(MPa)^{-1}$, $C_2\,(MPa)^{-2}$, $\epsilon = 1.985(10^{-6})\sigma + 6.866(10^{-8})\sigma^2$

Force analysis is part of Prob 3.4 solution

$F_{BC} = 11.313\ kN = \sqrt{2}\,P$, $F_{AB} = 8\,kN = P$

Similar to Example 6.5-2,

$$\Phi = A\left\{L_{AB}\left[\frac{1}{2}C_1\left(\frac{F_{AB}}{A}\right)^2 + \frac{1}{3}C_2\left(\frac{F_{AB}}{A}\right)^3\right] + L_{BC}\left[\frac{1}{2}C_1\left(\frac{F_{BC}}{A}\right)^2 + \frac{1}{3}C_2\left(\frac{F_{BC}}{A}\right)^3\right]\right\}$$

$$= A\left\{L_{AB}\left[\frac{1}{2}C_1\left(\frac{P}{A}\right)^2 + \frac{1}{3}C_2\left(\frac{P}{A}\right)^3\right] + L_{BC}\left[C_1\left(\frac{P}{A}\right)^2 + \frac{2\sqrt{2}}{3}C_2\left(\frac{P}{A}\right)^3\right]\right.$$

(continued)

$$(\delta_B)_V = \frac{\partial \Phi}{\partial P} = \frac{L_{AB}+2L_{BC}}{A} C_1 P + \frac{L_{AB}+2\sqrt{2}\,L_{BC}}{A^2} C_2 P^2$$

$$= \frac{(500+1500)}{125}(1.985)(10^{-6})(8000) + \frac{[500+2\sqrt{2}(750)]}{(125)^2}(6.866)(10^{-9})(8000)^2$$

$$= \underline{0.991 \text{ mm}}$$

This answer is 6% larger than the solution to Prob. 6.7 as the curve-fit overestimates strain of members BC at 90.5 MPa

---

(6.9) From statics $F_{CB} = F_{CD} = -\frac{P}{\sqrt{2}}$, $F_{BD} = \frac{P}{2}$

$$W_P = \frac{1}{2}P(\delta_C)_V = \frac{PL}{AE} + \frac{\sqrt{2}}{4}\frac{PL}{AE} = 1.354\frac{PL}{AE} \downarrow \quad \boxed{\text{ANS}}$$

Prior to loading, place an infinitesimal load H at D

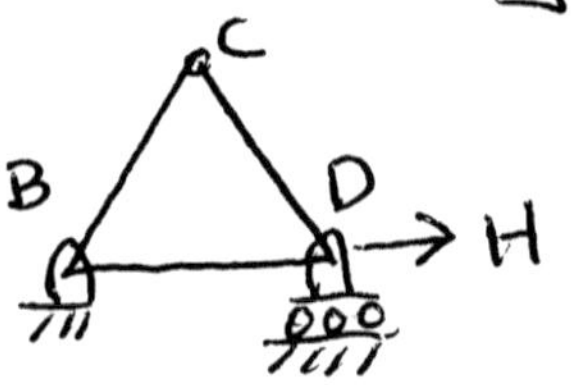

$F_{BC} = F_{CD} = 0$, $F_{BD} = H$

$$H(\delta_D)_H = \frac{H\left(\frac{P}{2}\right)\sqrt{2}\,L}{AE}$$

$$(\delta_D)_H = \frac{\sqrt{2}}{2}\frac{PL}{AE} \rightarrow \quad \boxed{\text{ANS}}$$

Geometric Solution: Disconnect joints and allow each member to deflect according to FL/(AE)

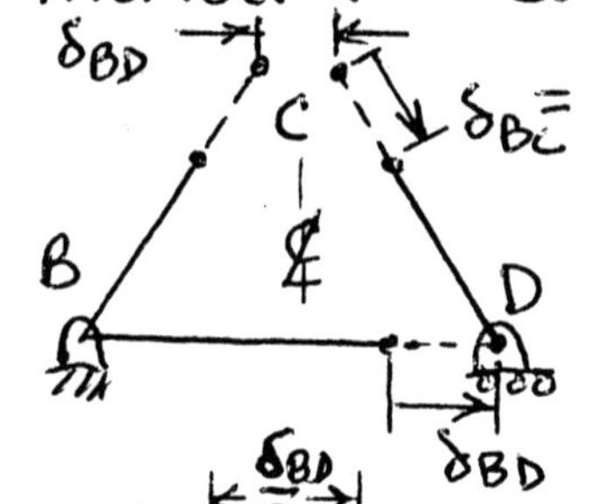

$\delta_{BC} = \frac{PL}{\sqrt{2}AE}$

$$\delta_{BD} = \frac{\frac{P}{2}\sqrt{2}\,L}{AE} = \frac{\sqrt{2}}{2}\frac{PL}{AE} = (\delta_D)_H \quad \boxed{\text{ANS}}$$

Next, rotate members to rejoin at C

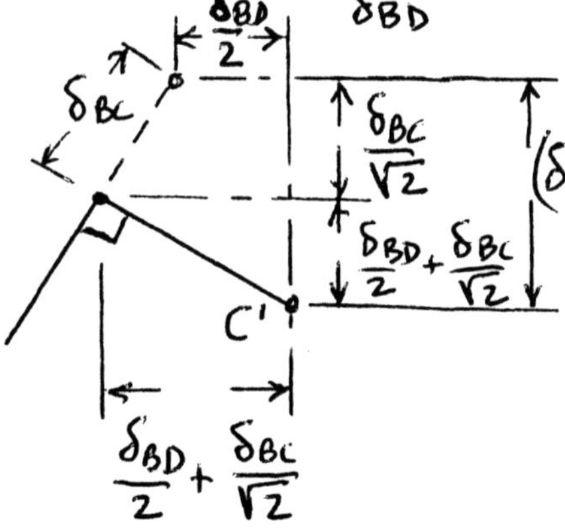

$$(\delta_C)_V = \frac{\delta_{BC}}{\sqrt{2}} + \frac{\delta_{BD}}{2} + \frac{\delta_{BC}}{\sqrt{2}} = \sqrt{2}\,\delta_{BC} + \frac{\delta_{BD}}{2}$$

$$= \sqrt{2}\frac{PL}{\sqrt{2}AE} + \frac{1}{2}\frac{\sqrt{2}}{2}\frac{PL}{AE} = 1.354\frac{PL}{AE}$$

$\boxed{\text{ANS}}$

6.10 $\Phi = \Sigma \int \delta_i dF_i$, $F_{CD} = F_{BC}$

$$\Phi = 2\left[\frac{1}{2}C_{11}F_{BC}^2 + \frac{1}{3}C_{12}F_{BC}^2\right] + \frac{1}{2}C_{21}F_{BD}^2 + \frac{1}{3}C_{22}F_{BD}^3$$

From statics $F_{BC} = -P/\sqrt{2}$, $F_{BD} = P/2$ Thus,

$$\Phi = 2\left(\frac{1}{4}C_{11}P^2 - \frac{1}{6\sqrt{2}}C_{12}P^3\right) + \frac{1}{8}C_{21}P^2 + \frac{1}{24}C_{22}P^3$$

$$(\delta_c)_V = \frac{\partial \Phi}{\partial P} = 2\left[\frac{1}{2}C_{11}P - \frac{1}{2\sqrt{2}}C_{12}P^2\right] + \frac{1}{4}C_{21}P + \frac{1}{8}C_{22}P^2$$

$$= \frac{\sqrt{2}}{16}\left[2\sqrt{2}\,(4C_{11} + C_{21}) + (\sqrt{2}C_{22} - 8C_{12})P\right]P \quad \boxed{\text{ANS}}$$

6.11 $R = R_1 + \dfrac{R_2 - R_1}{2\pi N}\theta$ Torsion of a section $T = \dfrac{PR}{2\pi N}$ $dL = Rd\theta$

$$\delta_P = \frac{\partial U}{\partial P} = \frac{2(1+\nu)}{EJ}\int T\frac{\partial T}{\partial P}dL = \frac{2(1+\nu)}{EJ}\int_0 PR^3 d\theta$$

$$= \frac{2(1+\nu)P}{EJ}\int_0^{2\pi N}\left(R_1 + \frac{R_2 - R_1}{2\pi N}\theta\right)^3 d\theta = \frac{2(1+\nu)}{EJ}\frac{1}{4}\left(\frac{2\pi N}{R_2 - R_1}\right)\left(R_1 + \frac{R_2 - R_1}{2\pi N}\theta\right)^4\Bigg|_0^{2\pi N}$$

$$= \frac{(1+\nu)\pi PN}{EJ(R_2 - R_1)}(R_2^4 - R_1^4) = \frac{(1+\nu)\pi PN}{EJ}(R_1 + R_2)(R_1^2 + R_2^2)$$

$J = \pi/32\, d^4$ $\therefore$ $\delta_P = \dfrac{32(1+\nu)PN(R_1 + R_2)(R_1^2 + R_2^2)}{Ed^4}$

$$k = \frac{P}{\delta_P} = \frac{Ed^4}{32(1+\nu)N(R_1 + R_2)(R_1^2 + R_2^2)} \quad \boxed{\text{ANS}}$$

6.12 $M_z = -Px$, $\dfrac{\partial M_z}{\partial P} = -x$

$\underline{a < x < L}$ $h(x) = \dfrac{h}{L}x$, $I = \dfrac{1}{12}b\left(\dfrac{h}{L}\right)^3 x^3$

$0 < x < a$ $h = \dfrac{h}{L}a$, $I = \dfrac{1}{12}b\left(\dfrac{ha}{L}\right)^3$

$$\delta = \frac{\partial U}{\partial P} = \frac{1}{E}\left[\int_0^a \frac{Px^2\,dx}{b(ha/L)^3/12} + \int_a^L \frac{Px^2}{b(h/L)^3 x^3/12}\right]$$

$$= \frac{12PL^3}{bh^3E}\left[\frac{1}{3} + \ln\left(\frac{L}{a}\right)\right] = \frac{4PL^3}{bh^3E}\left[1 + 3\ln\left(\frac{L}{a}\right)\right] \quad \boxed{\text{ANS}}$$

(6.13)

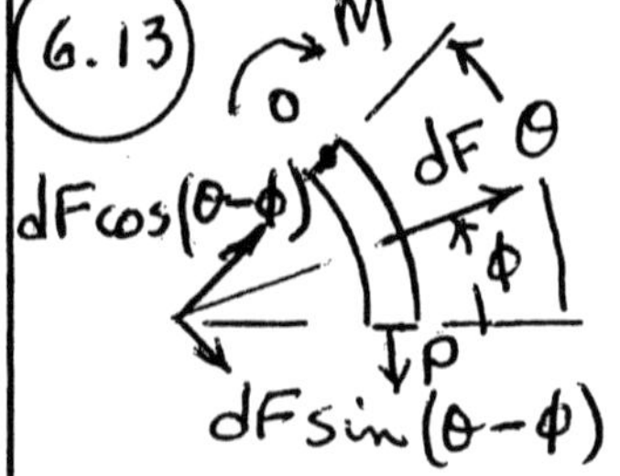

$$dF = qR\,d\theta$$

$$M = \int R\,dF\sin(\theta-\phi) - PR(1-\cos\theta)$$

$$= \int_0^\theta qR^2\sin(\theta-\phi)\,d\phi - PR(1-\cos\theta)$$

$$= qR^2(1-\cos\theta) - PR(1-\cos\theta)$$

$$\frac{\partial M}{\partial P} = -R(1-\cos\theta)$$

$$\delta_P = \frac{\partial U}{\partial Q}\bigg|_{P=0} = \frac{1}{EI}\int_0^\pi qR^2(1-\cos\theta)\left[-R(1-\cos\theta)\right]R\,d\theta$$

$$= -\frac{qR^4}{EI}\int_0^\pi (1-\cos\theta)^2 d\theta = -\frac{3\pi}{2}\frac{qR^4}{EI} = \boxed{\frac{3\pi}{2}\frac{qR^4}{EI}\uparrow}$$

(6.14) Consider half and double results

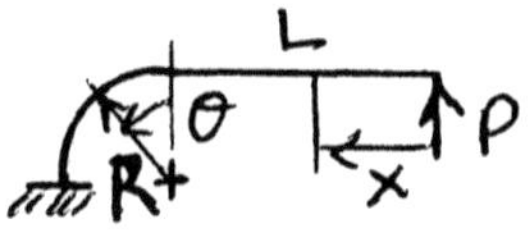

Straight section: $M = Px$, $\frac{\partial M}{\partial P} = x$

Curved section: $M = PL + PR\sin\theta$

$$\frac{\partial M}{\partial P} = L + R\sin\theta$$

$$\delta = \frac{2}{EI}\left[\int_0^L Px^2dx + \int_0^{\pi/2} P(L+R\sin\theta)^2 R\,d\theta\right]$$

$$= \frac{2}{EI}\left[\frac{PL^3}{3} + PR\int_0^{\pi/2}(L^2 + 2LR\sin\theta + R^2\sin^2\theta)\,d\theta\right]$$

$$= \frac{2P}{EI}\left\{\frac{L^3}{3} + R\left[L^2\frac{\pi}{2} - 2LR\left(\cos\frac{\pi}{2} - \cos 0\right) + \frac{R^2}{2}\left(\frac{\pi}{2} - \frac{1}{2}\sin\pi + \frac{1}{2}\sin 0\right)\right]\right\}$$

$$= \frac{P}{6EI}\left(4L^3 + 6\pi L^2R + 24LR^2 + 3\pi R^3\right) \quad \boxed{\text{ANS}}$$

(6.15) Consider half and double results

Straight section: 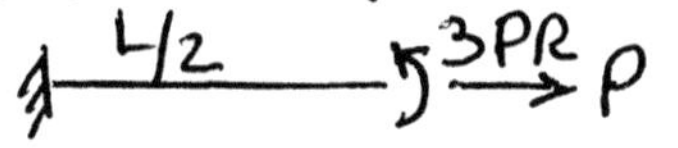

$M = 3PR$, $\frac{\partial M}{\partial P} = 3R$

Upper 180° section:

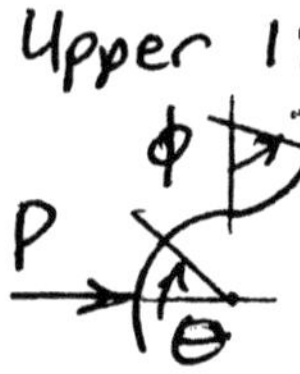

$$M = P[R + R(1-\cos\phi)] = PR(2-\cos\phi), \quad \frac{\partial M}{\partial P} = R(2-\cos\phi)$$

Lower section: $M = PR\sin\theta$

$$\frac{\partial M}{\partial P} = R\sin\theta$$

(continued)

Considering bending only

$$\delta = \frac{2}{EI}\left[\int_0^{L/2} 9PR^2 dx + \int_0^{\pi} PR^2(2-\cos\phi)^2 R d\phi + \int_0^{\pi/2} P(R\sin\theta)^2 R d\theta\right]$$

$$= \frac{2P}{EI}\left[\frac{9}{2}R^2 L + R^3\left(4\pi - 4\sin\phi\Big|_0^{\pi} + \frac{\pi}{2}\right) + R^3\left(\frac{\pi}{4}\right)\right]$$

$$= \frac{2PR^2}{EI}\left[\frac{19\pi}{4}R + \frac{9}{2}L\right] = \frac{PR^2}{2EI}(19\pi R + 18L) \quad \boxed{ANS}$$

(6.16) $I_{AB} = \frac{\pi}{64}(40^4 - 34^4) = 60.07(10^3)\text{ mm}^4$

$A_{AB} = \frac{\pi}{4}(40^2 - 34^2) = 348.7\text{ mm}^2$

From the solution of Prob. 1.5, $F_{BC} = \frac{1.423}{3}P = 0.4743P$, $F_{BD} = \frac{2.174}{3}P = 0.7247P$, transverse force at A is $P/2$,

Axial force $= \sqrt{2.4^2 + 1.8^2} = -3\text{ kN} = -P$

$$\delta_P = \frac{\partial U}{\partial P} = \frac{2}{210(10^3)(60.07)(10^3)}\int_0^{500}\frac{P}{2}x\left(\frac{x}{2}\right)dx + \frac{(0.4743)^2 P(948.7)}{50(210)(10^3)}$$

$$+ \frac{(0.7247)^2 P(1449)}{50(210)(10^3)} + \frac{(-1)^2 P(1000)}{348.7(210)(10^3)} \quad *$$

$$= 1.758(10^{-3})P = 1.758(10^{-3})(3000) = \underline{5.27\text{ mm}}$$

* where $L_{BC} = \sqrt{0.8^2 + 0.5^2 + 0.1^2} = 0.9487\text{ m}$

and $L_{BD} = \sqrt{0.8^2 + 0.5^2 + 1.1^2} = 1.449\text{ m}$

(6.17) Place a dummy load Q at point B in the negative y direction and determine forces in members. No bending of AB occurs due to Q. Let axial force in AB be F_{AB} where $\bar{F}_{AB} = -F_{AB}\left[\frac{0.8\bar{i} + 0.6\bar{k}}{1}\right]$. For other force definitions see the solution of Prob 1.5. $\Sigma\bar{F} = \bar{0}$ yields

$$-Q\bar{j} - F_{AB}(0.8\bar{i} + 0.6\bar{k}) + (-0.8433\bar{i} + 0.527\bar{j} - 0.1054)F_{BC}$$

$$+ (-0.5521\bar{i} + 0.345\bar{j} - 0.7591)F_{BD} = \bar{0} \qquad \bar{i}, \bar{j}, \bar{k} \text{ eqns are}$$

$$-0.8F_{AB} - 0.8433F_{BC} - 0.5521F_{BD} = 0$$

$$0.527F_{BC} + 0.345F_{BD} = Q$$

$$-0.6F_{AB} - 0.1054F_{BC} - 0.7591F_{BD} = 0$$

(continued)

Solving simultaneously yields

$F_{AB} = -2Q$, $F_{BC} = 0.9487\,Q$, $F_{BD} = 1.449\,Q$

$$\frac{\partial F_{AB}}{\partial Q} = -2, \quad \frac{\partial F_{BC}}{\partial Q} = 0.9487, \quad \frac{\partial F_{BD}}{\partial Q} = 1.449$$

Using forces determined in Prob 1.5

$$(\delta_B)_V = \sum \left.\frac{F\frac{\partial F}{\partial Q}L}{AE}\right|_{Q=0} = \frac{(-3000)(-2)(1000)}{348.7(210)(10^3)} + \frac{1}{50(210)(10^3)}\Big[1423(0.9487)(948.7)$$
$$+ (2174)(1.449)(1449)\Big] \quad *$$

$= \underline{0.639 \text{ mm}}$

* For A_{AB}, L_{BC}, and L_{BD} see solution to Prob. 6.16

(6.18) Add dummy moment M_A

$$M_z = -PR\sin\theta - M_A \qquad \frac{\partial M_z}{\partial M_A} = -1$$

$$\theta_A = \left.\frac{\partial U}{\partial M_A}\right|_{M_A=0} = \frac{1}{EI}\int_0^{\pi}(-PR\sin\theta)(-1)R\,d\theta$$

$$\theta_A = -\frac{PR^2}{EI}(\cos\theta)\Big|_0^{\pi} = \frac{2PR^2}{EI} \quad \boxed{\text{ANS}}$$

(6.19)

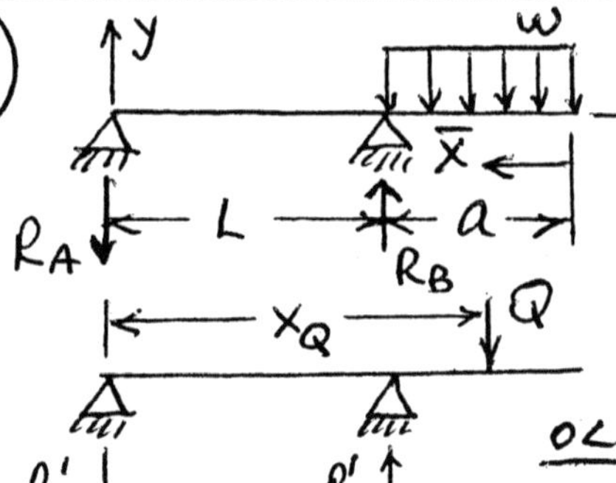

$$R_A = \frac{wa^2}{2L}$$

$\underline{0<x<L}$ $\quad M_z = -\dfrac{wa^2}{2L}x$

$\underline{0<\bar{x}<a}$ $\quad M_z = -\dfrac{w\bar{x}^2}{2}$

$$R_A' = \frac{Q(x_Q - L)}{L}$$

$\underline{0<x<L}$

$$M_z = -\frac{Q(x_Q-L)}{L}x, \quad \frac{\partial M_z}{\partial Q} = -\frac{(x_Q-L)}{L}x$$

$\underline{0<\bar{x}<(L-x_Q)}$ $\quad M_z = 0$

$\underline{L+a-x_Q<\bar{x}<a}$ $\quad M_z = -Q[\bar{x}-(L+a-x_Q)]$, $\quad \dfrac{\partial M_z}{\partial Q} = -[\bar{x}-(L+a-x_Q)]$

$$\delta_Q = \left.\frac{\partial U}{\partial Q}\right|_{Q=0} = \frac{1}{EI}\left\{\int_0^L\left(-\frac{wa^2}{2L}x\right)(-)\left(\frac{x_Q-L}{L}\right)x\,dx + \int_{L+a-x_Q}^{a}\left(-\frac{w\bar{x}^2}{2}\right)(-1)\left[\bar{x}-(L+a-x_Q)\right]d\bar{x}\right\}$$

$$= \frac{w}{EI}\left\{\frac{a^2L^3}{6L}\left(\frac{x_Q-L}{L}\right) + \frac{1}{2}\left[\frac{\bar{x}^4}{4} - \frac{1}{3}(L+a-x_Q)\bar{x}^3\right]\Big|_{L+a-x_Q}^{a}\right\}$$

Substitute limits and simplify. At this point we can let $x_Q = x$

(continued)

Reduces to $L < x < L+a$

$$\delta_Q = \frac{w}{24EI}\left[4a^2L(x-L) + 3a^4 + (L+a-x)^4 + 4(x-L)a^3 - 4a^4\right]$$ ANS

To check Example 3.7-2 expand the above

$$\delta_Q = \frac{w}{24EI}\left(L^4 + 4aL^3 + 2a^2L^2 - 4L^3x - 12aL^2x - 8a^2Lx + 6L^2x^2 + 12aLx^2 + 6a^2x^2 - 4Lx^3 - 4ax^3 + x^4\right) \quad ①$$

Expand Eq (b) of Example 3.7-2

$$(v_c)_{BC} = \frac{w}{24EI}\left(-L^4 - 4aL^3 - 2a^2L^2 + 4L^3x + 12aL^2x + 8a^2Lx - 6L^2x^2 - 12aLx^2 - 6a^2x^2 + 4Lx^3 + 4ax^3 - x^4\right) \quad ②$$

We see that ① = −② Thus the two approaches agree since δ_Q is defined positive downward (negative y direction)

6.20 From the solution of part (a) of Prob. 3.3 the internal force is $F = \frac{1}{2}\rho\omega^2 A(L^2 - x^2)$. Place a dummy force Q at $x = x_Q$

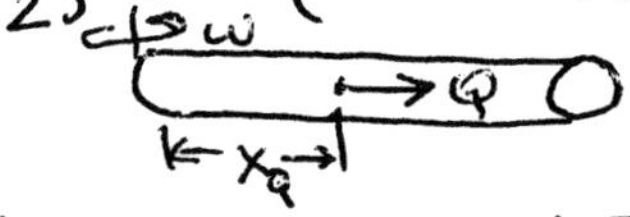

The internal force is

$0 < x < x_Q$ $\quad F = \frac{\rho\omega^2 A}{2}(L^2 - x^2) + Q, \quad \frac{\partial F}{\partial Q} = 1$

$x_Q < x < L$ $\quad F = \frac{\rho\omega^2 A}{2}(L^2 - x^2), \quad \frac{\partial F}{\partial Q} = 0$

$$\delta_Q = \left.\frac{\partial U}{\partial Q}\right|_{Q=0} = \frac{1}{AE}\int_0^{x_Q}\left[\frac{1}{2}\rho\omega^2 A(L^2 - x^2)\right](1)dx = \frac{\rho\omega^2}{2E}\left(L^2 x_Q - \frac{x_Q^3}{3}\right)$$

Let $x_Q = x$ and simplify

$$\delta_Q = \frac{\rho\omega^2 x}{6E}(3L^2 - x^2)$$ ANS

6.21 $F_{AB} = P, \quad \frac{\partial F_{AB}}{\partial P} = 1, \quad F_{BC} = \sqrt{2}P, \quad \frac{\partial F_{BC}}{\partial P} = \sqrt{2}$

Vertical deflection

$$(\delta_B)_V = \frac{\partial U}{\partial P} = \sum \frac{F\,\partial F/\partial P\,L}{AE} = \frac{1}{AE}\left[P(1)L_{AB} + (\sqrt{2}P)(\sqrt{2})L_{BC}\right]$$

$$= \frac{8000}{125(200)(10^3)}\left[500 + 2(750)\right] = 0.640\text{ mm}\downarrow$$ ANS

For horizontal displacement place horizontal dummy load Q on B acting to the right. $(F_{AB})_Q = Q \quad \partial F_{AB}/\partial Q = 1$

$(F_{BC})_Q = 0$

$$(\delta_B)_H = \left.\frac{\partial U}{\partial Q}\right|_{Q=0} = \frac{1}{AE}\left[P(1)L_{AB}\right] = \frac{8000(500)}{125(200)(10^3)} = 0.160\text{ mm}\rightarrow$$ ANS

6.22 Let $\bar{F} = 32\bar{i} - P\bar{j} + 16\bar{k}$ where $P = 22$ lb.

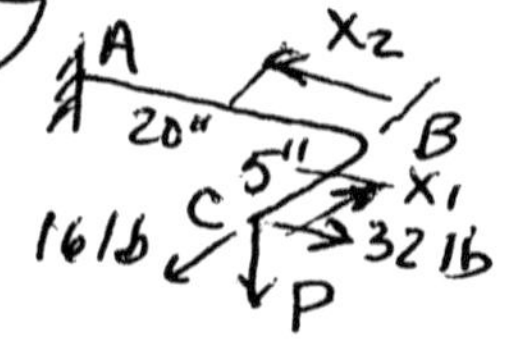

<u>CB:</u> Only loading from P is bending & shear (ignore)

$$M = Px_1 \quad \frac{\partial M}{\partial P} = x_1$$

<u>AB:</u> Torque and bending

$$M = Px_2, \quad \frac{\partial M}{\partial P} = x_2, \quad T = 5P, \quad \frac{\partial T}{\partial P} = 5$$

$$\delta_P = \frac{\partial U}{\partial P} = \frac{1}{EI}\int_0^5 Px_1(x_1)dx_1 + \frac{1}{EI}\int_0^{20} Px_2(x_2)dx_2 + \frac{2(1+\nu)(5P)(5)(20)}{EJ}$$

From Prob. 3.32, $I = 0.02022 \text{ in}^4$, $J = 2I$

$$\delta_P = \frac{P}{10(10^6)(0.02022)}\left[\frac{5^3}{3} + \frac{20^3}{3} + (1.33)(25)(20)\right] \quad \text{Sub in } P = 22 \text{ lb}$$

$\delta_P = 0.367$ in **ANS**

6.23

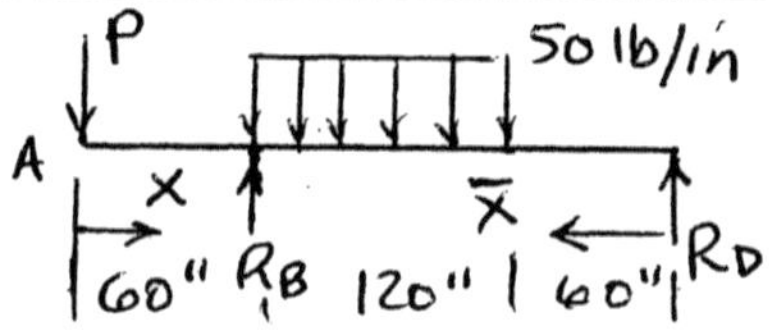

$P = 3000$ lb

$R_A = \frac{4}{3}P + 4000$ lb, $R_D = 2000 - \frac{P}{3}$ lb

From solution of Prob 3.31, $I = 136 \text{ in}^4$

<u>$0 < x < 60$</u> $\quad M = -Px, \quad \partial M/\partial P = -x$

<u>$0 < \bar{x} < 60$</u> $\quad M = \left(2000 - \frac{P}{3}\right)\bar{x}, \quad \partial M/\partial P = -\bar{x}/3$

<u>$60 < \bar{x} < 180$</u> $\quad M = \left(2000 - \frac{P}{3}\right) - \frac{50}{2}(\bar{x} - 60)^2, \quad \frac{\partial M}{\partial P} = -\frac{\bar{x}}{3}$ $\quad$ Can now sub $P = 3000$ lb

$$\delta_A = \frac{\partial U}{\partial P} = \frac{1}{EI}\left\{\int_0^{60} -3000x(-x)dx + \int_0^{60}(2000-1000)\bar{x}\left(-\frac{\bar{x}}{3}\right)d\bar{x} + \int_{60}^{180}\left[(2000-1000)\bar{x} - 25(\bar{x}-60)^2\right]\left(-\frac{\bar{x}}{3}\right)d\bar{x}\right\}$$

$$= \frac{1}{30(10^6)(136)}\left[\frac{3000}{3}(60)^3 - \frac{1000}{9}(60)^3 - \frac{1000}{9}(180^3 - 60^3) + \frac{25}{12}(180^4 - 60^4) - \frac{25}{9}(120)(180^3 - 60^3) + \frac{25(60)^2}{6}(180^2 - 60^2)\right]$$

$= 0.0706$ in **ANS**

6.24 $\tau = T/(2\bar{A}t)$

$$\theta = \frac{\partial U}{\partial T} = \int_0^{\bar{s}} \frac{2(1+\nu)}{E}\tau\frac{\partial \tau}{\partial T}tL\,ds = \int_0^{\bar{s}} \frac{2(1+\nu)}{E}\frac{T}{4\bar{A}^2t^2}tL\,ds$$

Assuming $\nu, E, T, L, \bar{A}$ to be constants

$$\theta = \frac{(1+\nu)TL}{2E\bar{A}^2}\int_0^{\bar{s}}\frac{ds}{t}$$ **ANS**

Agrees with Eq (5.2-6) where a geometric assumption was necessary making the derivation more abstract than energy method.

6.25 AI, DI, BC, CD, EF, and FG are zero force members and can be ignored.

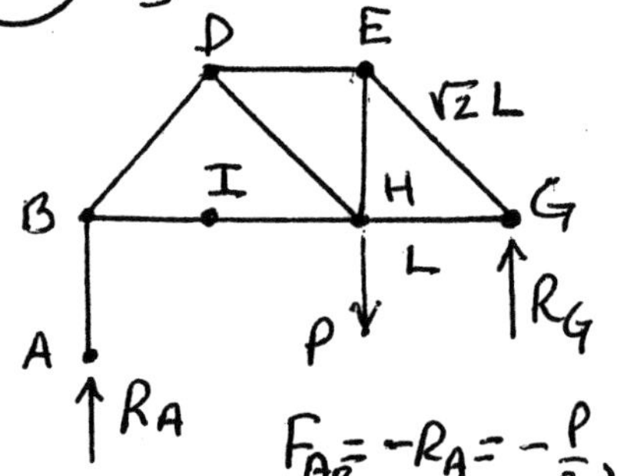

$R_A = \frac{P}{3}$, $R_G = \frac{2}{3}P$

$F_{EG} = -\sqrt{2}R_G = -\frac{2\sqrt{2}}{3}P$

$F_{GH} = R_G = \frac{2}{3}P$

$F_{DE} = \frac{F_{EG}}{\sqrt{2}} = -\frac{2}{3}P$, $F_{EH} = -F_{DE} = \frac{2}{3}P$

$F_{AB} = -R_A = -\frac{P}{3}$, $F_{BD} = \sqrt{2}F_{AB} = -\frac{\sqrt{2}}{3}P$, $F_{BI} = F_{HI} = -F_{AB} = \frac{P}{3}$

Joint D:

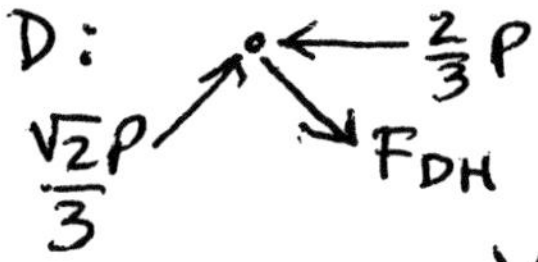

$F_{DH} = \frac{\sqrt{2}}{3}P$

$$(\delta_H)_V = \frac{\partial U}{\partial P} = \frac{1}{AE}\sum F\frac{\partial F}{\partial P}L = \frac{1}{AE}\Big[\underbrace{\left(-\tfrac{P}{3}\right)\left(-\tfrac{1}{3}\right)L}_{AB} + \underbrace{\left(-\tfrac{\sqrt{2}}{3}P\right)\left(-\tfrac{\sqrt{2}}{3}\right)\sqrt{2}L}_{BD} + \underbrace{2\left(\tfrac{P}{3}\right)\left(\tfrac{1}{3}\right)L}_{BI \,\&\, HI}$$

$$+ \underbrace{\left(\tfrac{\sqrt{2}}{3}P\right)\left(\tfrac{\sqrt{2}}{3}\right)\sqrt{2}L}_{DH} + \underbrace{\left(-\tfrac{2}{3}P\right)\left(-\tfrac{2}{3}\right)L}_{DE} + \underbrace{2\left(\tfrac{2}{3}P\right)\left(\tfrac{2}{3}\right)L}_{EH \,\&\, GH} + \underbrace{\left(-\tfrac{2\sqrt{2}}{3}P\right)\left(-\tfrac{2\sqrt{2}}{3}\right)\sqrt{2}L}_{EG}\Big]$$

$$= \frac{PL}{9AE}(15 + 12\sqrt{2}) = \frac{5+4\sqrt{2}}{3}\frac{PL}{AE} = 3.552\frac{PL}{AE}$$

$$= 3.552\frac{(8000)(3)}{200(10^{-6})(210)(10^9)} = 2.03(10^{-3})\text{m} = \underline{2.03\text{ mm}}$$

6.26 Place a horizontal dummy force Q at point G. Members BC, CD, EF, & FG are zero force members and can be ignored. Determine forces in remaining members due to Q and superpose forces from solution of Prob. 6.25 due to P.

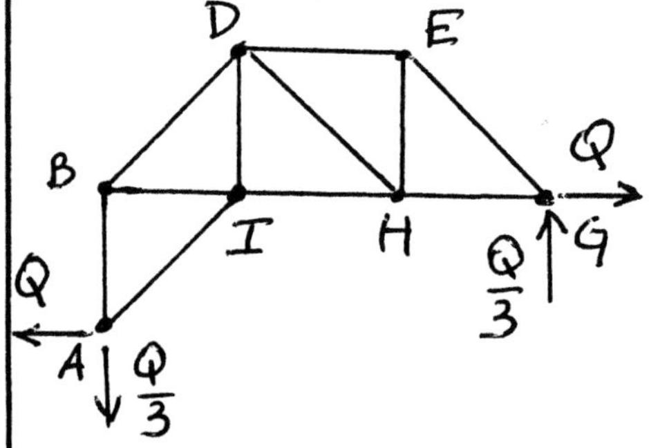

$F_{AI} = \sqrt{2}Q$, $F_{EG} = -\frac{\sqrt{2}}{3}Q$

$F_{AB} = -\frac{2}{3}Q$, $F_{GH} = \frac{4}{3}Q$

$F_{BI} = \frac{2}{3}Q$, $F_{ED} = -\frac{1}{3}Q$

$F_{BD} = -\frac{2\sqrt{2}}{3}Q$, $F_{EH} = \frac{1}{3}Q$

$F_{DH} = -\frac{\sqrt{2}}{3}Q$, $F_{DI} = Q$, $F_{HI} = \frac{5}{3}Q$

$\frac{\partial F_{AI}}{\partial Q} = \sqrt{2}$, $\frac{\partial F_{AB}}{\partial Q} = -\frac{2}{3}$, $\frac{\partial F_{BI}}{\partial Q} = \frac{2}{3}$, $\frac{\partial F_{BD}}{\partial Q} = \frac{-2\sqrt{2}}{3}$, $\frac{\partial F_{EG}}{\partial Q} = \frac{-\sqrt{2}}{3}$

$\frac{\partial F_{GH}}{\partial Q} = \frac{4}{3}$, $\frac{\partial F_{ED}}{\partial Q} = -\frac{1}{3}$, $\frac{\partial F_{EH}}{\partial Q} = \frac{1}{3}$, $\frac{\partial F_{DH}}{\partial Q} = \frac{-\sqrt{2}}{3}$, $\frac{\partial F_{DI}}{\partial Q} = 1$, $\frac{\partial F_{HI}}{\partial Q} = \frac{5}{3}$

(Continued)

$$(\delta_G)_H = \frac{\partial U}{\partial Q}\bigg|_{Q=0} = \frac{1}{AE}\left(\Sigma F \frac{\partial F}{\partial Q} L\right)\bigg|_{Q=0}$$

For F use forces determined in Prob. 6.25

$$(\delta_G)_H = \frac{1}{AE}\Big[\underset{AI}{(0)(\sqrt{2})\sqrt{2}L} + \underset{AB}{\left(-\frac{P}{3}\right)\left(-\frac{2}{3}\right)L} + \underset{BI}{\frac{P}{3}\left(\frac{2}{3}\right)L} + \underset{BD}{\left(-\frac{\sqrt{2}}{3}P\right)\left(-\frac{2\sqrt{2}}{3}\right)\sqrt{2}L}$$

$$+ \underset{EG}{\left(-\frac{2\sqrt{2}}{3}P\right)\left(-\frac{\sqrt{2}}{3}\right)\sqrt{2}L} + \underset{GH}{\left(\frac{2}{3}P\right)\left(\frac{4}{3}\right)L} + \underset{ED}{\left(-\frac{2}{3}P\right)\left(-\frac{1}{3}\right)L} + \underset{EH}{\left(\frac{2}{3}P\right)\left(\frac{1}{3}\right)L}$$

$$+ \underset{DH}{\frac{\sqrt{2}}{3}P\left(-\frac{\sqrt{2}}{3}\right)\sqrt{2}L} + \underset{DI}{(0)(1)L} + \underset{HI}{\frac{P}{3}\left(\frac{5}{3}\right)L}\Big] = \frac{PL}{9AE}(21+6\sqrt{2})$$

$$= \frac{7+2\sqrt{2}}{3}\frac{PL}{AE} = 3.276\frac{PL}{AE}$$

$$= 3.276\frac{(8000)(3)}{200(10^{-6})(210)(10^9)} = 1.872(10^{-3})\text{m} = \underline{1.872\text{ mm} \rightarrow}$$

(6.27) Place a dummy moment M on member DH. members AI, DI, BC, CD, EF, and FG are zero-force members and can be ignored. Determine forces in members due to M only and superpose forces determined in Prob. 6.25

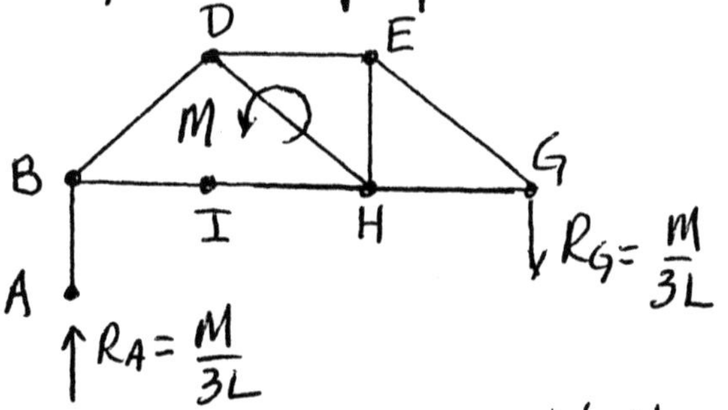

$F_{AB} = -\frac{M}{3L}$ $\quad F_{GH} = -\frac{M}{3L}$

$F_{BD} = -\frac{\sqrt{2}}{3}\frac{M}{L}$ $\quad F_{EG} = \frac{\sqrt{2}}{3}\frac{M}{L}$

$F_{BI} = F_{HI} = \frac{M}{3L}$ $\quad F_{DE} = \frac{M}{3L}$

$F_{EH} = -\frac{M}{3L}$

Member DH at Joint H

Ignore bending of DH as the actual loading yields no moment and dummy moment will be set to zero at the conclusion of solution.

$$\frac{\partial F_{AB}}{\partial M} = \frac{\partial F_{GH}}{\partial M} = \frac{\partial F_{EH}}{\partial M} = -\frac{1}{3L}, \quad \frac{\partial F_{BD}}{\partial M} = -\frac{\sqrt{2}}{3L}, \quad \frac{\partial F_{EG}}{\partial M} = \frac{\sqrt{2}}{3L}$$

$$\frac{\partial F_{BI}}{\partial M} = \frac{\partial F_{HI}}{\partial M} = \frac{\partial F_{DE}}{\partial M} = \frac{1}{3L}, \quad \frac{\partial F_{DH}}{\partial M} = \frac{\sqrt{2}}{6L}$$

(continued)

$$\theta_{DH} = \left.\frac{\partial U}{\partial M}\right|_{M=0} = \frac{1}{AE}\sum F\frac{\partial F}{\partial M}L\Big|_{M=0}$$

For F substitute forces obtained in Prob. 6.25

$$\theta_{DH} = \frac{1}{AE}\Big[\underset{AB}{-\frac{P}{3}\left(-\frac{1}{3L}\right)L} + \underset{BD}{\left(-\frac{\sqrt{2}}{3}P\right)\left(-\frac{\sqrt{2}}{3L}\right)\sqrt{2}L} + \underset{DE}{\left(-\frac{2}{3}P\right)\left(\frac{1}{3L}\right)L} + \underset{EG}{\left(-\frac{2\sqrt{2}}{3}P\right)\left(\frac{\sqrt{2}}{3L}\right)\sqrt{2}L}$$

$$+ \underset{GH\,\&\,EH}{2\left(\frac{2}{3}P\right)\left(-\frac{1}{3L}\right)L} + \underset{DH}{\left(\frac{\sqrt{2}}{3}P\right)\left(-\frac{\sqrt{2}}{6L}\right)\sqrt{2}L} + \underset{BI\,\&\,HI}{2\left(\frac{P}{3}\right)\left(\frac{1}{3L}\right)L}\Big]$$

$$= \frac{P}{9AE}\left(-3-3\sqrt{2}\right) = -\frac{1+\sqrt{2}}{3}\frac{P}{AE} = -\frac{(1+\sqrt{2})(8000)}{3(200)(10^{-6})(210)(10^{9})} = \underline{-1.533(10^{-4})\text{ rad}}$$

$$= \underline{-8.78(10^{-3})\text{ deg}}$$

(6.28) Apply a dummy moment to member 2 and evaluate forces due to moment.

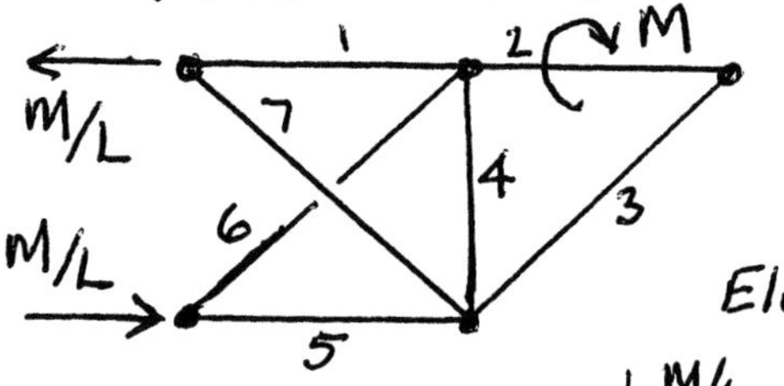

$F_1 = F_2 = F_4 = M/L$, $F_5 = -M/L$

Element 2:

Element 3:

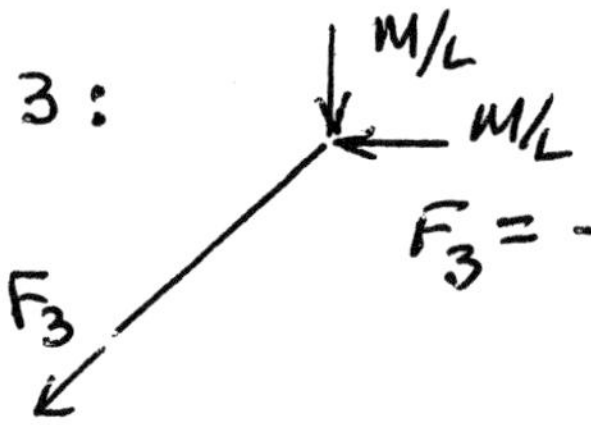

$F_3 = -\sqrt{2}\dfrac{M}{L}$

Ignore bending energy in member 2 as it will disappear when M is set to zero.

$$\frac{\partial F_1}{\partial M} = \frac{\partial F_2}{\partial M} = \frac{\partial F_4}{\partial M} = \frac{1}{L},\quad \frac{\partial F_5}{\partial M} = -\frac{1}{L},\quad \frac{\partial F_3}{\partial M} = -\frac{\sqrt{2}}{L}$$

$$\theta_2 = \left.\frac{\partial U}{\partial M}\right|_{M=0} = \frac{1}{AE}\sum F\frac{\partial F}{\partial M}L\Big|_{M=0}$$

For the F terms use forces determined in Example 6.6-1

$$\theta_2 = \frac{1}{AE}\Big[\underset{①\&②}{2P\left(\frac{1}{L}\right)L} + \underset{③}{(-\sqrt{2}P)\left(-\frac{\sqrt{2}}{L}\right)\sqrt{2}L} + \underset{④}{0} + \underset{⑤}{(-2P)\left(-\frac{1}{L}\right)L} + \underset{⑥}{0} + \underset{⑦}{0}\Big]$$

$$= 2(2+\sqrt{2})\frac{P}{AE} = 6.828\frac{P}{AE}\quad (+ \therefore \text{clockwise}) \quad \boxed{\text{ANS}}$$

6.29 For slope add dummy moment M

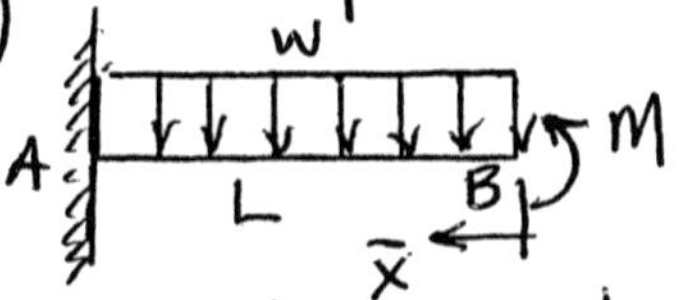

$$M_z = M - \frac{w\bar{x}^2}{2}, \quad \frac{\partial M_z}{\partial M} = 1$$

$$\theta_B = \left.\frac{\partial U}{\partial M}\right|_{M=0} = \frac{1}{EI}\int_0^L \left(-\frac{w\bar{x}^2}{2}\right)(1)\,d\bar{x} = -\frac{wL^3}{6EI} \quad \boxed{\text{ANS}}$$

For end deflection add dummy force Q

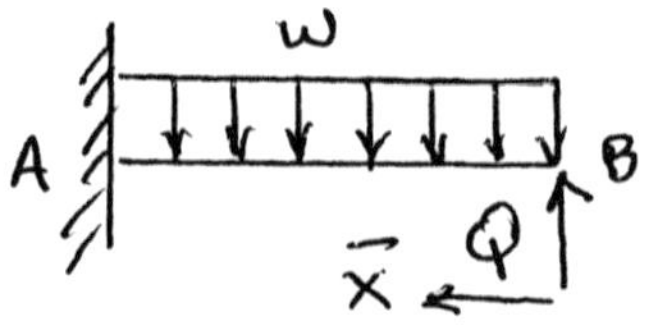

$$M_z = Q\bar{x} - \frac{w\bar{x}^2}{2}, \quad \frac{\partial M_z}{\partial Q} = \bar{x}$$

$$(v_c)_B = \left.\frac{\partial U}{\partial Q}\right|_{Q=0} = \frac{1}{EI}\int_0^L \left(-\frac{w\bar{x}^2}{2}\right)(\bar{x})\,d\bar{x} = -\frac{wL^4}{8EI} \quad \boxed{\text{ANS}}$$

6.30

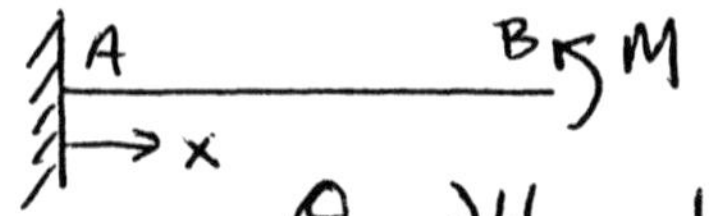

$$M_z = M, \quad \frac{\partial M_z}{\partial M} = 1$$

$$\theta_B = \frac{\partial U}{\partial M} = \frac{1}{EI}\int M(1)(dx) = \frac{ML}{EI} \quad \boxed{\text{ANS}}$$

For end deflection add a dummy force Q

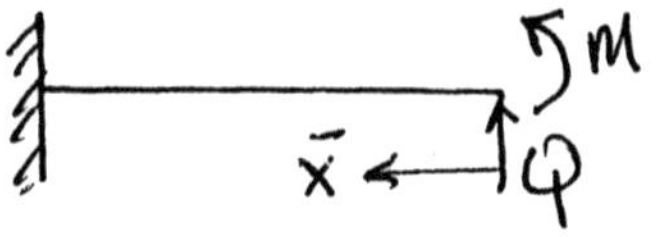

$$M_z = M + Q\bar{x}, \quad \frac{\partial M_z}{\partial Q} = \bar{x}$$

$$(v_c)_B = \left.\frac{\partial U}{\partial Q}\right|_{Q=0} = \frac{1}{EI}\int_0^L M(\bar{x})\,d\bar{x} = \frac{ML^2}{2EI} \quad \boxed{\text{ANS}}$$

6.31

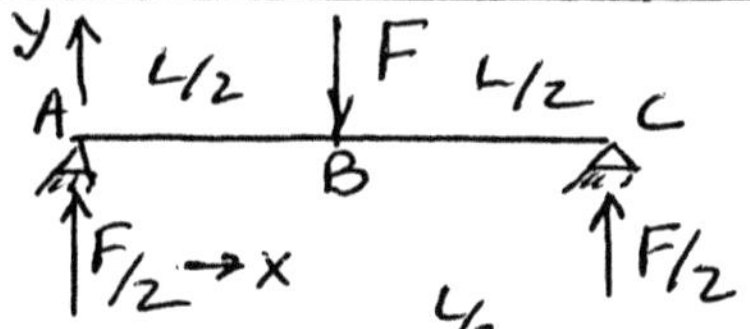

Consider half of beam and double results

$0 < x < L/2 \quad M_z = \frac{F}{2}x, \quad \frac{\partial M_z}{\partial F} = \frac{x}{2}$

$$(v_c)_B = -\frac{\partial U}{\partial F} = -\frac{2}{EI}\int_0^{L/2}\left(\frac{F}{2}x\right)\left(\frac{x}{2}\right)dx = -\frac{2F}{12EI}\left(\frac{L}{2}\right)^3 = -\frac{FL^3}{48EI} \quad \boxed{\text{ANS}}\ *$$

For slope at A add dummy moment M

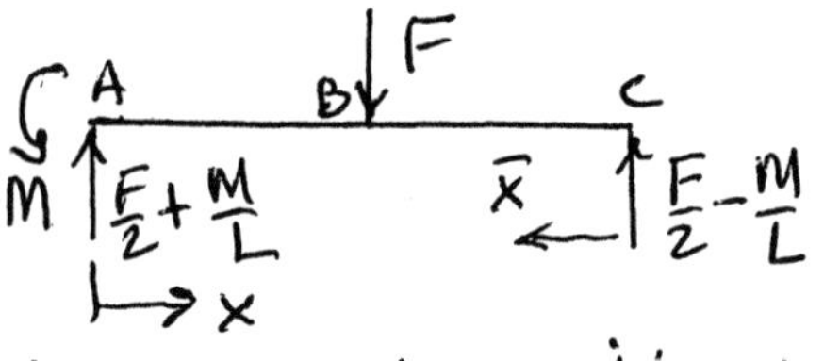

$0 < x < L/2, \quad M_z = \left(\frac{F}{2} + \frac{M}{L}\right)x - M, \quad \frac{\partial M_z}{\partial M} = \frac{x}{L} - 1$

$0 < \bar{x} < L/2, \quad M_z = \left(\frac{F}{2} - \frac{M}{L}\right)\bar{x}, \quad \frac{\partial M_z}{\partial M} = -\frac{\bar{x}}{L}$

* Note v_c is positive in y direction opposite direction of F

(continued)

$$\theta_A = \left.\frac{\partial U}{\partial M}\right|_{M=0} = \frac{1}{EI}\left[\int_0^{L/2}\left(\frac{F}{2}x\right)\left(\frac{x-L}{L}\right)dx + \int_0^{L/2}\left(\frac{F}{2}\bar{x}\right)\left(-\frac{\bar{x}}{L}\right)d\bar{x}\right]$$

$$= \frac{F}{EI}\left\{\frac{1}{2L}\left[\frac{1}{3}\left(\frac{L}{2}\right)^3 - \frac{1}{2}L\left(\frac{L}{2}\right)^2\right] - \frac{1}{6L}\left(\frac{L}{2}\right)^3\right\} = -\frac{FL^2}{16EI} \quad \boxed{ANS}$$

Checking with Appendix C, beam C.5, with $a = b = L/2$

$$(v_c)_{x=\frac{L}{2}} = \frac{F\left(\frac{L}{2}\right)\left(\frac{L}{2}\right)}{6EIL}\left[\left(\frac{L}{2}\right)^2 + \left(\frac{L}{2}\right)^2 - L^2\right] = -\frac{FL^3}{48EI} \quad \underline{\text{checks}}$$

$$\theta|_{x=0} = \frac{F(L/2)}{6EIL}\left[\left(\frac{L}{2}\right)^2 - L^2\right] = -\frac{FL^2}{16EI} \quad \underline{\text{checks}}$$

6.32

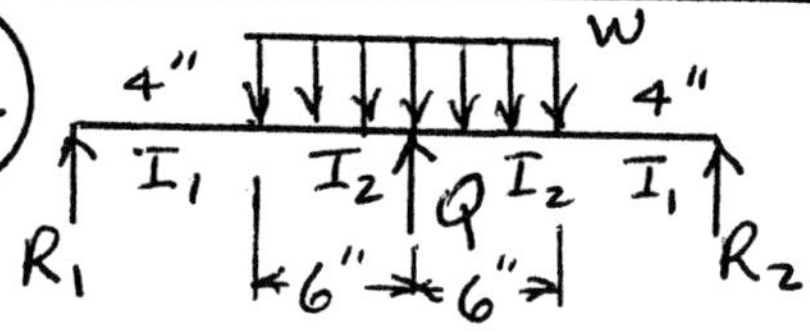

$R_1 = R_2 = 6w - \frac{Q}{2}$

Since symmetric, consider half beam and double results

$\underline{0<x<4}$ $\quad M_z = \left(6w - \frac{Q}{2}\right)x, \quad \frac{\partial M}{\partial Q} = -\frac{x}{2}$

$\underline{4<x<10}$ $\quad M_z = \left(6w - \frac{Q}{2}\right)x - \frac{w}{2}(x-4)^2, \quad \frac{\partial M}{\partial Q} = -\frac{x}{2}$

$$(v_c)_{x=10} = \left.\frac{\partial U}{\partial Q}\right|_{Q=0} = \frac{2}{E}\left\{\frac{1}{I_1}\int_0^4 6wx\left(-\frac{x}{2}\right)dx\right.$$

$$= \frac{2w}{E}\left[-\frac{64}{I_1} - \frac{783}{I_2}\right] \qquad I_1 = \frac{\pi}{64}(1.5)^4,\ I_2 = \frac{\pi}{64}(2)^4,\ w = 200$$

$E = 30(10^6)$

$(v_c)_{x=10} = -0.0167$ in $\boxed{ANS}$

6.33 $\quad F_{BG} = F_{DH} = F_{GH} = -\frac{P}{\sqrt{3}},\ F_{CG} = F_{CH} = \frac{P}{\sqrt{3}},\ F_{BC} = F_{CD} = \frac{P}{2\sqrt{3}}$

$$(\delta_c)_v = \frac{\partial U}{\partial P} = \frac{L}{AE}\sum F\frac{\partial F}{\partial P} = \frac{L}{AE}\left[3\left(-\frac{P}{\sqrt{3}}\right)\left(-\frac{1}{\sqrt{3}}\right) + 2\left(\frac{P}{\sqrt{3}}\right)\left(\frac{1}{\sqrt{3}}\right) + 2\left(\frac{P}{2\sqrt{3}}\right)\left(\frac{1}{2\sqrt{3}}\right)\right]$$

$$= \frac{22}{12}\frac{PL}{AE} = \frac{11}{6}\frac{PL}{AE} \quad \boxed{ANS}$$

6.34 Place a dummy force Q at D and determine associated forces

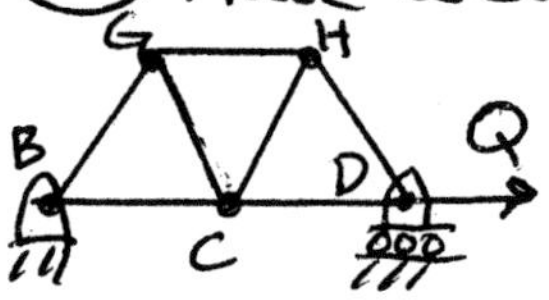

$F_{BG} = F_{DH} = F_{GH} = F_{CG} = F_{CH} = 0,\ F_{BC} = F_{CD} = Q$

$\frac{\partial F_{BC}}{\partial Q} = \frac{\partial F_{CD}}{\partial Q} = 1$, For forces due to P see Prob. 6.33 solution

$$(\delta_D)_H = \left.\frac{\partial U}{\partial Q}\right|_{Q=0} = 2\frac{P}{2\sqrt{3}}(1)\frac{L}{AE} = \frac{\sqrt{3}}{3}\frac{PL}{AE} \quad \boxed{ANS}$$

6.35 Apply a dummy moment M to BG and evaluate forces due to M alone.

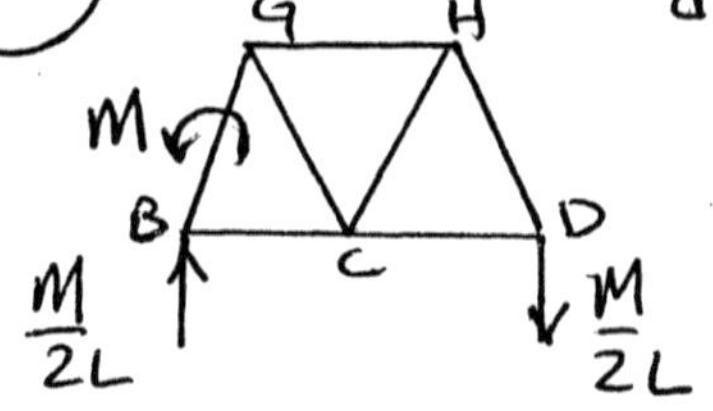

member BG:

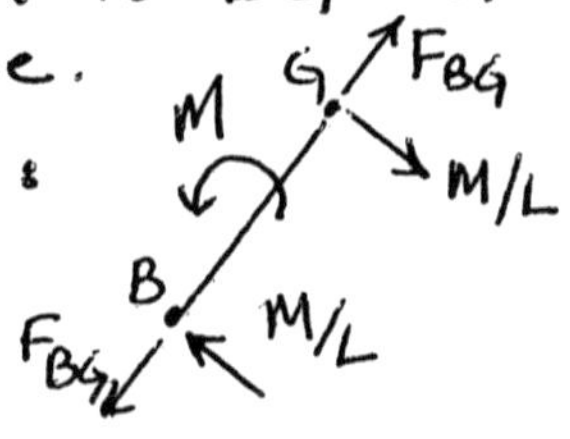

Joint B:

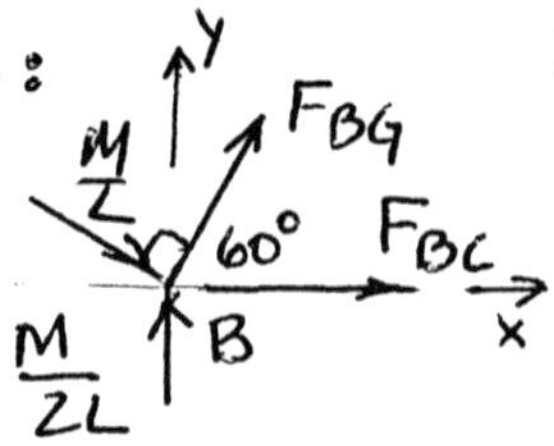

$\Sigma F_y = 0$

$-\frac{M}{L}\sin 30 + \frac{M}{2L} + F_{BG}\sin 60 = 0 \Rightarrow F_{BG} = 0$

$\Sigma F_x = 0$

$F_{BC} + \frac{M}{L}\cos 30 = 0 \Rightarrow F_{BC} = -\frac{\sqrt{3}}{2}\frac{M}{L}$

Remaining joints yield $F_{DH} = F_{GH} = F_{CG} = \frac{\sqrt{3}}{3}\frac{M}{L}$, $F_{CD} = -\frac{\sqrt{3}}{6}\frac{M}{L}$, $F_{CH} = -\frac{\sqrt{3}}{3}\frac{M}{L}$. Thus, $\frac{\partial F_{BC}}{\partial M} = -\frac{\sqrt{3}}{2L}$, $\frac{\partial F_{DH}}{\partial M} = \frac{\partial F_{GH}}{\partial M} = \frac{\partial F_{CG}}{\partial M} = \frac{\sqrt{3}}{3L}$

$\frac{\partial F_{CD}}{\partial M} = -\frac{\sqrt{3}}{6L}$, $\frac{\partial F_{CH}}{\partial M} = -\frac{\sqrt{3}}{3L}$.

$$\theta_{BG} = \left.\frac{\partial U}{\partial M}\right|_{M=0} = \frac{L}{AE}\sum F \left.\frac{\partial F}{\partial M}\right|_{M=0}$$

F are obtained from the solution of Prob 6.33

$$\theta_{BC} = \frac{L}{AE}\Big[\underbrace{\frac{P}{2\sqrt{3}}\left(-\frac{\sqrt{3}}{2L}\right)}_{BC} + \underbrace{\left(-\frac{P}{\sqrt{3}}\right)\left(\frac{\sqrt{3}}{3L}\right)}_{GH} + \underbrace{\left(\frac{P}{\sqrt{3}}\right)\left(\frac{\sqrt{3}}{3L}\right)}_{CG} + \underbrace{\frac{P}{\sqrt{3}}\left(-\frac{\sqrt{3}}{3L}\right)}_{CH} + \underbrace{\frac{P}{2\sqrt{3}}\left(-\frac{\sqrt{3}}{6L}\right)}_{CD} + \underbrace{\left(-\frac{P}{\sqrt{3}}\right)\left(\frac{\sqrt{3}}{3L}\right)}_{DH} + \underbrace{0}_{BG}\Big]$$

$$= -\frac{P}{AE} \quad (\therefore \text{clockwise}) \quad \underline{\text{ANS}}$$

6.36 Place a dummy force Q at point A

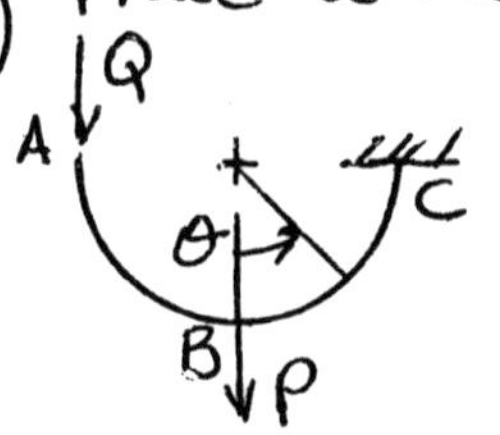

Only section BC bends due to applied load P.

$\underline{0 < \theta < \frac{\pi}{2}}$ $\quad M = PR\sin\theta + QR(1+\sin\theta)$

$\frac{\partial M}{\partial Q} = R(1+\sin\theta)$

$$(\delta_A)_v = \left.\frac{\partial U}{\partial Q}\right|_{Q=0} = \frac{1}{EI}\int_0^{\pi/2} PR\sin\theta(R)(1+\sin\theta)R\,d\theta$$

$$= \frac{PR^3}{EI}\left[-\cos\theta + \frac{\theta}{2} + \frac{\sin 2\theta}{4}\right]\Big|_0^{\pi/2} = \frac{PR^3}{EI}\left(2+\frac{\pi}{4}\right)$$

$$= \frac{\pi+8}{4}\frac{PR^3}{EI} \quad \boxed{\text{ANS}}$$

6.37

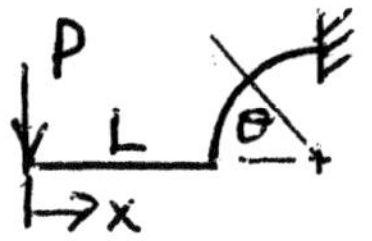

$0<x<L$ $\quad M_z = -Px, \quad \frac{\partial M}{\partial P} = -x$

$0<\theta<\frac{\pi}{2}$ $\quad M_z = P[L+R(1-\cos\theta)], \quad \frac{\partial M_z}{\partial P} = L+R(1-\cos\theta)$

$$(\delta_A)_v = \frac{\partial U}{\partial P} = \frac{1}{EI}\left\{\int_0^L -Px(-x)dx + \int_0^{\pi/2} P[L+R(1-\cos\theta)]^2 R\,d\theta\right\}$$

$$= \left\{4L^3 + 3R\left[2\pi L^2 + 4(\pi-2)LR + (3\pi-8)R^2\right]\right\}\frac{P}{12EI} \qquad \boxed{\text{ANS}}$$

6.38

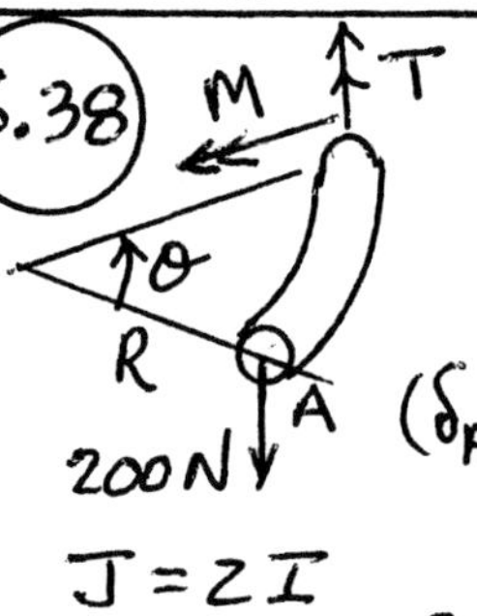

$0<\theta<\frac{\pi}{2}: \quad M = PR\sin\theta, \quad \frac{\partial M}{\partial P} = R\sin\theta$

$T = PR(1-\cos\theta), \quad \frac{\partial T}{\partial P} = R(1-\cos\theta)$

$$(\delta_A)_y = -\frac{\partial U}{\partial P} = -\left\{\frac{1}{EI}\int_0^{\pi/2} P(R\sin\theta)^2 R\,d\theta + \frac{2(1+\nu)}{EJ}\int_0^{\pi/2} P[R(1-\cos\theta)]^2 R\,d\theta\right\}$$

$J = 2I$

$$(\delta_A)_y = -\frac{PR^3}{EI}\left[\frac{\pi}{4} + (1+\nu)\left(\frac{3\pi}{4}-2\right)\right] = -\left[4\pi-8+(3\pi-8)\nu\right]\frac{PR^3}{4EI}$$

$$= -\left[4\pi-8+(3\pi-8)(0.29)\right]\frac{(200)(100)^3}{4(200)(10^3)\frac{\pi}{64}(5)^4} = \underline{-40.6\text{ mm}}$$

6.39

$$U_{DH} = \frac{1}{2}\frac{F_{DH}^2 L_{DH}}{AE} + F_{DH}\alpha L_{DH}\Delta T, \quad \text{From Prob. 6.25 } F_{DH} = \frac{\sqrt{2}}{3}P, \quad \frac{\partial F_{DH}}{\partial P} = \frac{\sqrt{2}}{3}$$

$$\frac{\partial U_{DH}}{\partial P} = \left(\frac{F_{DH}}{AE} + \alpha\Delta T\right)L_{DH}\frac{\partial F_{DH}}{\partial P} = \left(\frac{\frac{\sqrt{2}}{3}P}{AE} + \alpha\Delta T\right)(\sqrt{2}L)\left(\frac{\sqrt{2}}{3}\right)$$

$$= \frac{2}{9}\left(\sqrt{2}\frac{PL}{AE} + 6\alpha L\Delta T\right)$$

{Add this to remaining member terms in the solution to Prob 6.25}

$$(\delta_H)_v = \frac{\partial U}{\partial P} = \frac{5+4\sqrt{2}}{3}\frac{PL}{AE} + \frac{2}{3}\alpha L\Delta T$$

$$= \left(\frac{5+4\sqrt{2}}{3}\right)\left(\frac{(8000)(3000)}{200(210)(10^3)}\right) + \frac{2}{3}11.7(10^{-6})(3000)(100) = \underline{4.37\text{ mm}}$$

6.40

Similar to Prob. 6.39 add $\alpha L_{GC}\Delta T\frac{\partial F_{GC}}{\partial P}$ to $\frac{\partial U}{\partial P}$

From the solution to Prob 6.33

$F_{GC} = P/\sqrt{3} \quad \therefore \partial F_{GC}/\partial P = P/\sqrt{3}$ and $\frac{\partial U}{\partial P} = \frac{11}{6}\frac{PL}{AE}$

Thus $(\delta_C)_v = \frac{11}{6}\frac{PL}{AE} + \frac{\sqrt{3}}{3}\alpha L\Delta T \qquad \boxed{\text{ANS}}$

6.41 With symmetry consider half the structure and double results. From the solution to Prob 5.49 $r_c = 2.961 \text{ in} = R$, $e = 0.1447$ $A = 2.0354 \text{ in}^2$, $P = 200 \text{ lb}$, $E = 30(10^6) \text{ psi}$, $\nu = 0.3$

$k = 1.0$

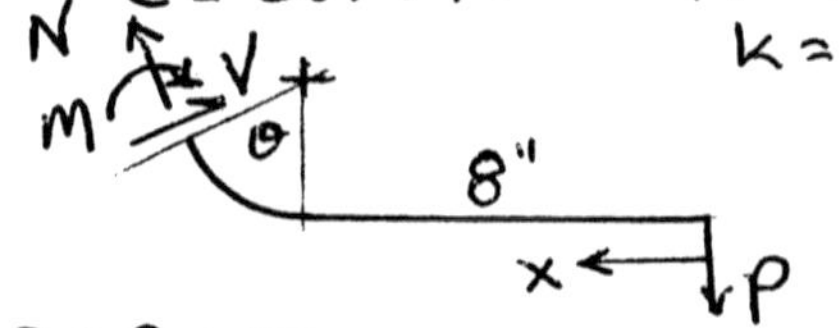

$\underline{0 < x < 8}$ $V = P$, $\frac{\partial V}{\partial P} = 1$

$M = -Px$, $\frac{\partial M}{\partial P} = -x$

$\underline{0 < \theta < \frac{\pi}{2}}$ $V = P\cos\theta$, $\frac{\partial V}{\partial P} = \cos\theta$, $N = P\sin\theta$, $\frac{\partial N}{\partial P} = \sin\theta$

$M = -P(8 + R\sin\theta)$, $\partial M/\partial P = -(8 + R\sin\theta)$

$$\delta_P = \frac{\partial U}{\partial P} = \frac{2}{E}\left\{\frac{1}{I}\int_0^8 Px^2 dx + \frac{2(1+\nu)k}{A}\int_0^8 P dx\right.$$

$$+ \frac{1}{A}\left[\frac{1}{e}\int_0^{\pi/2} P(8+R\sin\theta)^2 d\theta - \int_0^{\pi/2} P(8+R\sin\theta)\sin\theta d\theta\right.$$

$$- \int_0^{\frac{\pi}{2}} P\sin\theta(8+R\sin\theta)d\theta + R\int_0^{\pi/2} P\sin^2\theta d\theta$$

$$\left.\left. + 2(1+\nu)kR\int_0^{\pi/2} P\cos^2\theta d\theta\right]\right\}$$ Sub in $R = 2.961$ and other terms*

$$\delta_P = \frac{2(200)}{30(10^6)}\left\{\frac{8^3}{3(0.9714)} + \frac{2(1.3)(1)}{2.0354}8 + \frac{1}{2.0354}\left[\frac{1}{0.1447}\left(64\cdot\frac{\pi}{2} + 47.38 + 8.768\frac{\pi}{4}\right)\right.\right.$$

$$\left.\left. - 2\left(8 + 2.961\cdot\frac{\pi}{4}\right) + 2.961\cdot\frac{\pi}{4} + 2(1.3)(1)(2.961)\frac{\pi}{4}\right]\right\}$$

$= 9.41(10^{-3})$ in **ANS**

* where $I = \frac{1}{12}(0.5)(2.5)^3 + 0.5(2.5)(3.25 - 2.9606)^2 + \frac{\pi}{4}(0.5)^4 + \pi(0.5)^2(2.9606 - 2.5)^2 = 0.9714 \text{ in}^4$

6.42 With symmetry consider half the structure and double the results. From the solution to Prob. 5.50, $R = r_c = 55$ mm $e = 9.035$ mm, $A = 2400 \text{ mm}^2$, $P = 30$ kN

$$I = \frac{1}{12}60(20)^3 + 60(20)(55-35)^2 + 2\left[\frac{1}{12}10(60)^3 + 10(60)(75-55)^2\right] = 1.36(10^6)\text{mm}^4$$

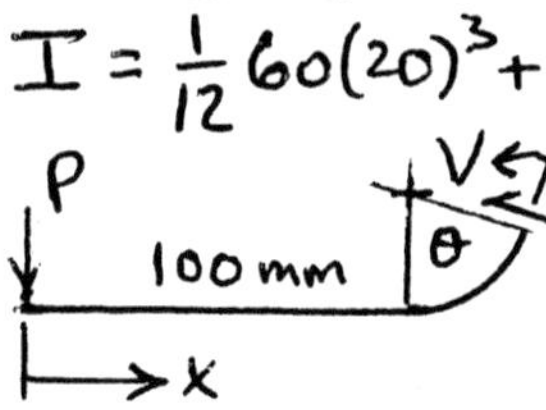

$\underline{0 < x < 100}$ $M = -Px$, $\frac{\partial M}{\partial P} = -x$, $V = P$, $\frac{\partial V}{\partial P} = 1$

$\underline{0 < \theta < \pi/2}$ $V = P\cos\theta$, $\frac{\partial V}{\partial P} = \cos\theta$, $N = P\sin\theta$, $\frac{\partial N}{\partial P} = \sin\theta$

$M = -P(100 + 55\sin\theta)$, $\frac{\partial M}{\partial P} = -(100 + 55\sin\theta)$

(Continued)

$$\delta_P = \frac{\partial U}{\partial P} = \frac{2}{E}\left\{\frac{1}{I}\int_0^{100} P x^2 dx + \frac{2(1+\nu)k}{A}\int_0^{100} P dx + \frac{1}{A}\left[\frac{1}{e}\int_0^{\pi/2} P(100+55\sin\theta)^2 d\theta \right.\right.$$

$$- \int_0^{\pi/2} P(100+55\sin\theta)\sin\theta\, d\theta - \int_0^{\pi/2} P\sin\theta(100+55\sin\theta)d\theta$$

$$\left.\left. + 55\int_0^{\pi/2} P\sin^2\theta\, d\theta + 2(1+\nu)k(55)\int_0^{\pi/2} P\cos^2\theta\, d\theta\right]\right\}$$ sub in values and integrate

$$\delta_P = \frac{2(30)(10^3)}{210(10^3)}\left\{\frac{100^3}{3(1.36)(10^6)} + \frac{2(1.3)(1)(100)}{2400} + \frac{1}{2400}\left[\frac{1}{9.035}\left(100^2\frac{\pi}{2} + 2(100)(55)\right.\right.\right.$$

$$\left.\left.\left. + 55^2\frac{\pi}{4}\right) - 2\left(100 + 55\cdot\frac{\pi}{4}\right) + 2(1.3)(1)(55)\left(\frac{\pi}{4}\right)\right]\right\} = \underline{0.469\text{ mm}}$$

6.43 With symmetry consider half the structure and double the results. From the solution to Prob. 5.52

$A = 0.375\text{ in}^2$, $R = r_c = 3.444\text{ in}$, $e = 0.02299\text{ in}$, $P = 100\text{ lb}$.

From Appendix B, Example B.2-1, eq. (e)

$$I = \frac{(1)^3}{36}\left[\frac{0.5^2 + 4(0.5)(0.25) + (0.25)^2}{0.5 + 0.25}\right] = 0.03009\text{ in}^4$$

(a)

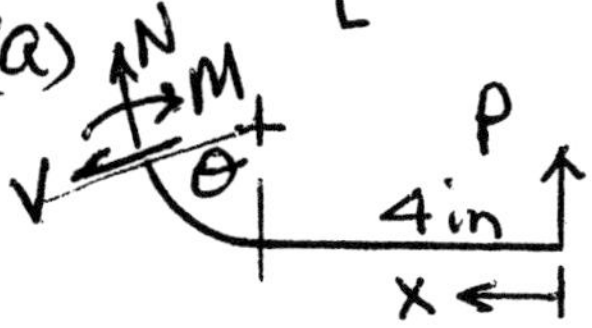

considering only bending

$0 < x < 4$: $M = Px$, $\frac{\partial M}{\partial P} = x$

$0 < \theta < \frac{\pi}{2}$: $M = P(4 + R\sin\theta)$, $\frac{\partial M}{\partial P} = 4 + R\sin\theta$

$$\delta_P = \frac{\partial U}{\partial P} = \frac{2}{EI}\left[\int_0^4 P x^2 dx + \int_0^{\pi/2} P(4+R\sin\theta)^2 R\, d\theta\right]$$

$$= \frac{2(100)}{(30)(10^6)(0.03009)}\left\{\frac{4^3}{3} + 3.444\left[16\cdot\frac{\pi}{2} + 2(4)(3.444) + 3.444^2\left(\frac{\pi}{4}\right)\right]\right\}$$

$$= \underline{0.0520\text{ in}}$$

(b)

$$\delta_P = \frac{2(100)}{30(10^6)}\left\{\frac{1}{0.03009}\frac{4^3}{3} + \frac{1}{0.375(0.02299)}\left[16\cdot\frac{\pi}{2} + 2(4)(3.444) + 3.444^2\left(\frac{\pi}{4}\right)\right]\right\}$$

$$= \underline{0.0527\text{ in}}$$

(c) In straight section: $V = P$, $\frac{\partial V}{\partial P} = 1$

curved: $V = P\cos\theta$, $\frac{\partial V}{\partial P} = \cos\theta$, $N = -P\sin\theta$, $\frac{\partial N}{\partial P} = -\sin\theta$

(continued)

$$\delta_P = \frac{2(100)}{30(10^6)}\left\{\frac{1}{0.03009}\frac{4^3}{3} + \frac{2(1.3)(1)(4)}{0.375} + \frac{1}{0.375}\left[\frac{1}{0.02299}\left(16\cdot\frac{\pi}{2} + 2\cdot 4\cdot 3.444 + 3.444^2\cdot\frac{\pi}{4}\right)\right.\right.$$

$$\left.\left. - 2\left(4 + 3.444\cdot\frac{\pi}{4}\right) + 3.444\left(\frac{\pi}{4}\right) + 2(1.3)(1)(3.444)\left(\frac{\pi}{4}\right)\right]\right\} = \underline{0.0528 \text{ in}}$$

(d) We see that the results of parts a–c are quite close indicating that r_c is large enough to consider only straight beam bending formulation

6.44 With symmetry consider half the structure and double the results. From the solution to prob. 5.51,

$A = 3 \text{ in}^2$, $r_c = 3 \text{ in}$, $e = 0.1146 \text{ in}$, $P = 750 \text{ lb}$

$I = \frac{1}{12}(1.5)(2)^3 = 1 \text{ in}^4$

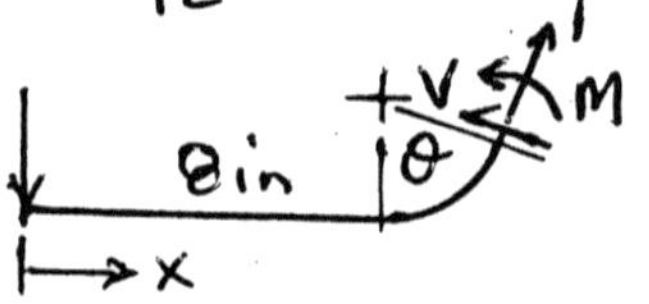

$\underline{0 < x < 8}$ $M = -Px$, $\frac{\partial M}{\partial P} = -x$, $V = P$, $\frac{\partial V}{\partial P} = 1$

$\underline{0 < \theta < \frac{\pi}{2}}$ $V = P\cos\theta$, $\frac{\partial V}{\partial P} = \cos\theta$, $N = P\sin\theta$, $\frac{\partial N}{\partial P} = \sin\theta$

$M = -P(8 + 3\sin\theta)$, $\frac{\partial M}{\partial P} = -(8 + 3\sin\theta)$

$$\delta_P = \frac{\partial U}{\partial P} = \frac{2}{E}\left\{\frac{1}{I}\int_0^8 Px^2\,dx + \frac{2(1+\nu)k}{A}\int_0^8 P\,dx + \frac{1}{A}\left[\frac{1}{e}\int_0^{\pi/2} P(8+3\sin\theta)^2\,d\theta\right.\right.$$

$$\left.\left. - 2\int_0^{\pi/2} P(8+3\sin\theta)\sin\theta\,d\theta + R\int_0^{\pi/2} P\sin^2\theta\,d\theta + 2(1+\nu)kR\int_0^{\pi/2} P\cos^2\theta\,d\theta\right]\right\}$$

$$= \frac{2(750)}{30(10^6)}\left\{\frac{1}{1}\frac{8^3}{3} + \frac{2(1.3)(1.2)8}{3} + \frac{1}{3}\left[\frac{1}{0.1146}\left(64\cdot\frac{\pi}{2} + 2(8)(3) + 3^2\cdot\frac{\pi}{4}\right)\right.\right.$$

$$\left.\left. - 2\left(8 + 3\cdot\frac{\pi}{4}\right) + 3\left(\frac{\pi}{4}\right) + 2(1.3)(1.2)(3)\left(\frac{\pi}{4}\right)\right]\right\} = \underline{0.0314 \text{ in}}$$

6.45 As in Example 6.8-3

$M_z = -M_A + \frac{PR}{2}(1 - \cos\theta)$, $\frac{\partial M_z}{\partial M_A} = -1$

(a) $N = \frac{P}{2}\cos\theta$, $V = \frac{P}{2}\sin\theta$, $\partial N/\partial M_A = \partial V/\partial M_A = 0$

$$\frac{\partial U}{\partial M_A} = \frac{1}{EA}\left\{\frac{1}{e}\int_0^{\pi/2}\left[-M_A + \frac{PR}{2}(1-\cos\theta)\right](-1)\,d\theta + \int_0^{\pi/2}\left(\frac{P}{2}\cos\theta\right)(-1)\,d\theta\right\} = 0$$

$$\frac{1}{e}\left(\frac{\pi}{2}M_A - \frac{\pi}{4}PR + \frac{PR}{2}\right) - \frac{P}{2} = 0 \quad \text{solve for } M_A$$

$$M_A = \frac{2P}{\pi}\left(\frac{\pi}{4}R - \frac{R}{2} + \frac{e}{2}\right) = \frac{P}{2\pi}\left[(\pi - 2)R + 2e\right] \quad \boxed{\text{ANS}}$$ Sub back into M_z

$$M_z = -\frac{P}{2\pi}\left[(\pi-2)R + 2e\right] + \frac{PR}{2}(1 - \cos\theta) = \frac{P}{2\pi}(2R - 2e - \pi R\cos\theta)$$

(continued)

At point C, $\theta = \pi/2$ and

$$M_C = \frac{P}{\pi}(R-e) \quad \boxed{\text{ANS}}$$

For deflection consider ¼ structure

$$(\delta_B)_v = \frac{4}{E}\left[\frac{1}{e}\int_0^{\pi/2}\frac{P}{4\pi^2}(2R-2e-\pi R\cos\theta)^2 d\theta + \int_0^{\pi/2}\frac{P}{2\pi}(2R-2e-\pi R\cos\theta)\left(\frac{\cos\theta}{2}\right)d\theta\right.$$

$$+\int_0^{\pi/2}\frac{P\cos\theta}{2}\left(\frac{1}{2\pi}\right)(2R-2e-\pi R\cos\theta)d\theta + R\int_0^{\pi/2}\frac{P}{4}\cos^2\theta\, d\theta$$

$$\left. + 2(1+\nu)kR\int_0^{\pi/2}\frac{P}{4}\sin^2\theta\, d\theta\right]$$

$$(\delta_B)_v = \frac{4P}{EA}\left[\frac{1}{4\pi^2 e}\left(-2\pi R^2+\frac{\pi^3}{4}R^2+2\pi e^2\right)+\frac{2}{4\pi}\left(2R-2e-\frac{\pi^2 R}{2}\right)+\frac{\pi}{16}R+\frac{\pi}{8}(1+\nu)kR\right]$$

Simplifying yields

$$(\delta_B)_v = \frac{P}{4\pi EAe}\left\{(\pi^2-8)R^2+\left[16-\pi^2+2\pi^2(1+\nu)k\right]Re-8e^2\right\} \quad \boxed{\text{ANS}}$$

(b) $R = 37.5\text{ mm}$, $r_n = \dfrac{25(12)}{12\ln\frac{50}{25}} = 36.06738\text{ mm}$, $e = R - r_n = 1.4326\text{ mm}$

$$M_A = \frac{200}{2\pi}\left[(\pi-2)37.5+2(1.433)\right] = 1454\text{ N-m} = \underline{1.45\text{ kN-m}}$$

$$M_C = \frac{200}{\pi}(37.5-1.433) = 2296\text{ N-m} = \underline{2.30\text{ kN-m}}$$

$$(\delta_B)_v = \frac{200(10^3)\{(\pi^2-8)(37.5^2)+[16-\pi^2+2\pi^2(1.3)(1.2)](37.5)(1.433)-8\pi(1.433^2)\}}{4\pi(210)(10^3)(25)(12)(1.4326)}$$

$$= \underline{0.8106\text{ mm}}$$

(c) From Example 6.8-3

$$M_A = \frac{200(37.5)}{2}\left(1-\frac{2}{\pi}\right) = 1363\text{ N-m} = \underline{1.36\text{ kN-m} \quad 6.3\%\text{ low}}$$

$$M_C = \frac{200(37.5)}{\pi} = 2387\text{ N-m} = \underline{2.39\text{ kN-m} \quad 4\%\text{ high}}$$

$$(\delta_B)_v = 0.1488\frac{200(10^3)(37.5)^3}{210(10^3)\frac{1}{12}(12)(25)^3} = \underline{0.4783\text{ mm} \quad 41\%\text{ low}}$$

6.46

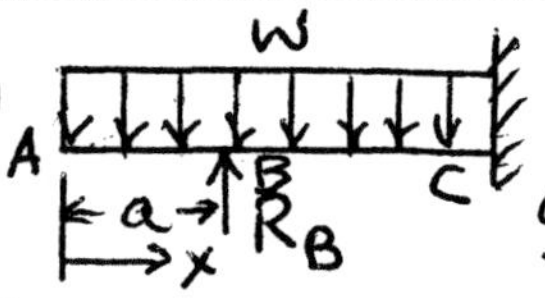

$0<x<a$: $M = -\dfrac{wx^2}{2}$, $\dfrac{\partial M}{\partial R_B} = 0$

$a<x<L$: $M = -\dfrac{wx^2}{2} + R_B(x-a)$, $\dfrac{\partial M}{\partial R_B} = x - a$

Note:
Since $\dfrac{\partial M}{\partial R_B} = 0$ for $0<x<a$, can start variable at R_B (see Prob. 6.48)

(Continued)

$$\delta_B = \frac{\partial U}{\partial R_B} = \frac{1}{EI}\int_a^L \left[-\frac{wx^2}{2} + R_B(x-a)\right](x-a)\,dx = 0$$

$$-\frac{w}{2}\left[\frac{1}{4}(L^4-a^4) - \frac{a}{3}(L^3-a^3)\right] + \frac{R_B}{3}\left[(L-a)^3 - (a-a)^3\right] = 0$$

$$R_B = \frac{w}{8(L-a)^3}\left[3(L^4-a^4) - 4a(L^3-a^3)\right] = \frac{w}{8(L-a)}(3L^2+2aL+a^2) \quad \boxed{\text{ANS}}$$

$$R_C = wL - R_B = \frac{w}{8(L-a)}(5L^2-10aL-a^2) \quad \boxed{\text{ANS}}$$

$$M_C = \frac{wL^2}{2} - R_B(L-a) = \frac{w}{8}(L^2-2aL-a^2) \quad \boxed{\text{ANS}}$$

6.47

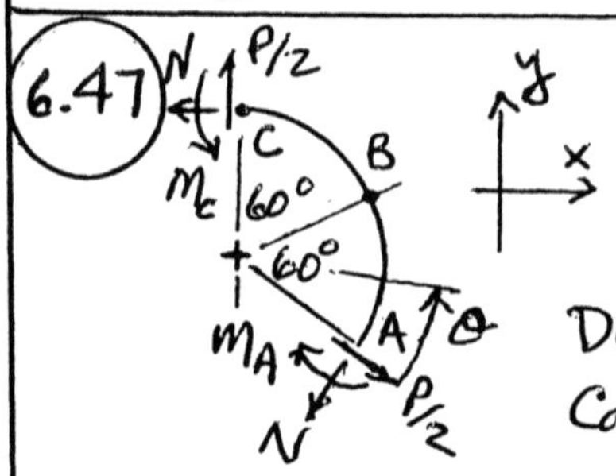

$$\Sigma F_y = 0 \Rightarrow \frac{P}{2} - \frac{P}{2}\sin 30 - N\cos 30 = 0$$

$$N = \frac{P}{4\cos 30} = \frac{\sqrt{3}}{6}P$$

Due to symmetry $M_C = M_A$ and $\theta_A = \theta_B = \theta_C = 0$

Can deal with segment AB

$\underline{0 < \theta < \pi/3}$ $\quad M = M_A - \frac{P}{2}R\sin\theta + \frac{\sqrt{3}}{6}PR(1-\cos\theta), \quad \frac{\partial M}{\partial M_A} = 1$

$$\theta_B = \frac{1}{EI}\int_0^{\pi/3}\left[M_A - \frac{PR}{2}\sin\theta + \frac{\sqrt{3}}{6}PR(1-\cos\theta)\right]R\,d\theta = 0 \qquad \text{Factor } \frac{R}{EI}$$

$$M_A\frac{\pi}{3} + \frac{PR}{2}\cos\theta\Big|_0^{\pi/3} + \frac{\sqrt{3}}{6}PR\frac{\pi}{3} - \frac{\sqrt{3}}{6}PR\sin\theta\Big|_0^{\pi/3} = 0$$

$$\frac{\pi}{3}M_A - \frac{PR}{4} + \frac{\sqrt{3}\pi}{18}PR - \frac{PR}{4} = 0 \qquad \text{Solve for } M_A$$

$$M_A = \frac{9-\sqrt{3}\pi}{6\pi}PR \quad \boxed{\text{ANS}} \qquad \text{sub back into } M$$

$\underline{0 < \theta < \pi/3}$ $\quad M = \frac{9-\sqrt{3}\pi}{6\pi}PR - \frac{PR}{2}\sin\theta + \frac{\sqrt{3}}{6}PR(1-\cos\theta)$

$$M = \frac{9-3\pi\sin\theta-\sqrt{3}\pi\cos\theta}{6\pi}PR \quad \boxed{\text{ANS}}$$

Due to symmetry other segments are identical

(b) Due to symmetry each force P deflects the same and the deflection is related to the energy of each 120° segment.

$$\delta_P = \frac{\partial U}{\partial P} = \frac{2}{EI}\int_0^{\pi/3} M\frac{\partial M}{\partial P}R\,d\theta = \frac{2}{EI}\int_0^{\pi/3}\left(\frac{9-3\pi\sin\theta-\sqrt{3}\pi\cos\theta}{6\pi}\right)^2 PR^3\,d\theta$$

$$= \frac{2PR^3}{36\pi^2 EI}\int_0^{\pi/3}\left(81 + 9\pi^2\sin^2\theta + 3\pi^2\cos^2\theta - 54\pi\sin\theta - 18\sqrt{3}\pi\cos\theta + 6\sqrt{3}\pi^2\sin\theta\cos\theta\right)d\theta$$

(continued)

$$\delta_P = \frac{PR^3}{18\pi^2 EI}\left[27\pi + 9\pi^2\left(\frac{\pi}{6} - \frac{\sqrt{3}}{8}\right) + 3\pi^2\left(\frac{\pi}{6} + \frac{\sqrt{3}}{8}\right) + 54\pi\left(-\frac{1}{2}\right) - 18\sqrt{3}\,\pi\frac{\sqrt{3}}{2} + 3\sqrt{3}\,\pi^2\left(\frac{3}{4}\right)\right]$$

$$= 0.01594\,\frac{PR^3}{EI} \quad \boxed{\text{ANS}}$$

(6.48)

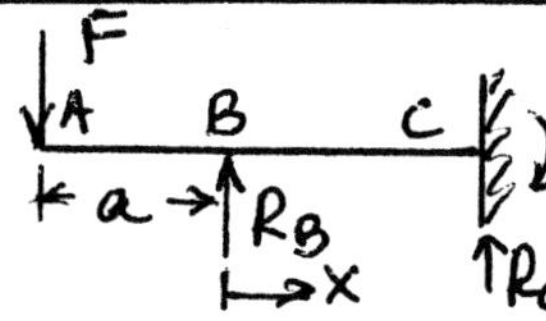

For $\delta_B = \frac{\partial U}{\partial R_B} = 0$, M_z is not a function of R_B in region AB. So region BC only needs to be considered.

$\underline{0 < x < L-a}$ $\quad M_z = -P(a+x) + R_B x, \quad \frac{\partial M_z}{\partial R_B} = x$

$$\delta_B = \frac{\partial U}{\partial R_B} = \frac{1}{EI}\int_0^{L-a}\left[-P(a+x) + R_B x\right]x\,dx = 0$$

$$-P\left[\frac{a}{2}(L-a)^2 + \frac{1}{3}(L-a)^3\right] + \frac{R_B}{3}(L-a)^3 = 0$$

$$R_B = \frac{3P}{L-a}\left[\frac{a}{2} + \frac{L-a}{3}\right] = \frac{P(2L+a)}{2(L-a)} \quad \boxed{\text{ANS}}$$

$$R_C = P - \frac{P(2L+a)}{2(L-a)} = -\frac{3}{2}\frac{Pa}{L-a} \quad \boxed{\text{ANS}}$$

$$M_C = PL - \frac{P(2L+a)}{2(L-a)}(L-a) = -\frac{Pa}{2} \quad \boxed{\text{ANS}}$$

(6.49) See solution to Prob. 6.48. The only difference is $\delta_B = \frac{\partial U}{\partial R_B} = -\epsilon$. Thus

$$\frac{1}{EI}\left\{-P\left[\frac{a}{2}(L-a)^2 + \frac{1}{3}(L-a)^3\right] + \frac{R_B}{3}(L-a)^3\right\} = -\epsilon$$

$$R_B = \frac{3}{(L-a)^3}\left\{\frac{P}{6}\left[3a(L-a)^2 + 2(L-a)^3\right] - \epsilon EI\right\} \quad \text{or,}$$

$$R_B = \frac{2L+a}{2(L-a)}P - \frac{3\epsilon EI}{(L-a)^3} \quad \boxed{\text{ANS}}$$

$$R_C = P - \frac{2L+a}{2(L-a)}P + \frac{3\epsilon EI}{(L-a)^3} = \frac{3\epsilon EI}{(L-a)^3} - \frac{3}{2}\frac{Pa}{L-a} \quad \boxed{\text{ANS}}$$

$$M_C = PL - \left[\frac{P(2L+a)}{2(L-a)} - \frac{3\epsilon EI}{(L-a)^3}\right](L-a) = \frac{3\epsilon EI}{(L-a)^2} - \frac{Pa}{2} \quad \boxed{\text{ANS}}$$

6.50 (a) Imagine an infinitesimal gap between the two beams. Applying the load P to the upper beam creates curvature where the upper beam contacts the lower beam at point B. Increasing P, the curvature of the upper beam remains greater than the lower thereby maintaining contact at B.

(b) Upper beam: We will be equating the deflection of B for both beams using $\frac{\partial U}{\partial R_B}$. Since M_z is not a function of R_B in region AB the variable x can be as shown where the moment in region BC need only be considered.

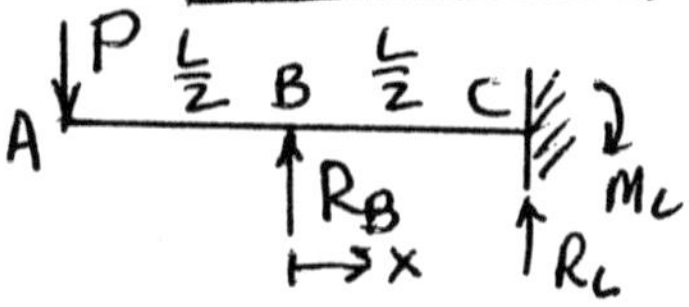

$0<x<L/2$ $\quad M_z = -P\left(x+\frac{L}{2}\right) + R_B x, \quad \frac{\partial M}{\partial R_B} = x$

$$(\delta_B)_u = \frac{\partial U}{\partial R_B} = \frac{1}{EI}\int_0^{L/2}\left[-P\left(x+\frac{L}{2}\right)+R_B x\right]x\,dx = \frac{1}{EI}\left(-\frac{5}{48}PL^3 + \frac{R_B L^3}{24}\right) \quad (1)$$

Lower beam:

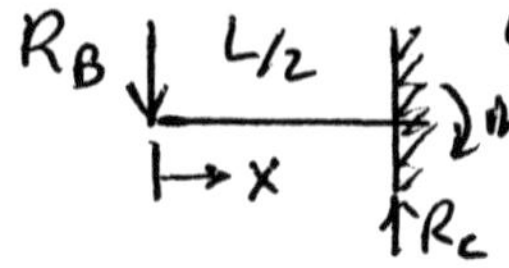

$0<x<\frac{L}{2}$: $\quad M_z = -R_B x, \quad \frac{\partial M_z}{\partial R_B} = -x$

$$(\delta_B)_L = \frac{\partial U}{\partial R_B} = \frac{1}{EI}\int_0^{L/2} R_B x^2\,dx = \frac{R_B L^3}{24EI} \quad (2)$$

Because R_B is defined in opposite directions on the upper and lower beams $(\delta_B)_L = -(\delta_B)_u$ From (1) & (2)

$$\frac{R_B L^3}{24EI} = -\frac{1}{EI}\left(-\frac{5}{48}PL^3 + R_B L^3\right) \Rightarrow R_B = \frac{5}{4}P \quad \boxed{ANS}$$

Upper beam: $R_C = P - \frac{5}{4}P = -\frac{P}{4}, \quad M_C = PL - \frac{5}{4}P\left(\frac{L}{2}\right) = \frac{3}{8}PL$ $\boxed{ANS}$

Lower beam: $R_C = \frac{5}{4}P, \quad M_C = \frac{5}{4}P\left(\frac{L}{2}\right) = \frac{5}{8}PL$ $\boxed{ANS}$

6.51 For a spring, $U = \frac{1}{2}F^2/k, \quad \frac{\partial U}{\partial P} = \frac{1}{k}F\,\partial F/\partial P$

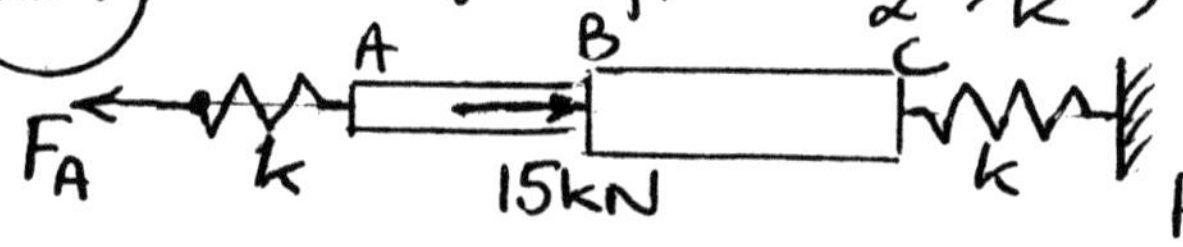

$F = F_A$ in left spring and AB

$F = F_A - 15$ in right spring and BC

$$\delta_A = \frac{\partial U}{\partial F_A} = 0 = \frac{F_A(10^3)(1)}{k} + \frac{F_A(10^3)(1)L_{AB}}{EA_{AB}} + \frac{(F_A-15)(10^3)(1)L_{BC}}{EA_{BC}} + \frac{(F_A-15)(10^3)(1)}{k}$$
$$+ 11.7(10^6)(250+400)(5)(1)$$

$$\frac{F_A(10^3)}{1.5(10^6)} + \frac{F_A(10^3)(250)}{207(10^3)\frac{\pi}{4}(12)^2} + \frac{(F_A-15)(10^3)(400)}{207(10^3)\frac{\pi}{4}(25)^2} + \frac{(F_A-15)(10^3)}{1.5(10^6)} + 3.8025(10^{-2}) = 0$$

(Continued)

Multiply by 10^4

$$6.667F_A + 106.79F_A + 39.366F_A - 590.49 + 6.667F_A - 100 + 380.25 = 0$$

$F_A = 1.945$ kN **ANS**

$F_C = 15 - 1.95 = 13.05$ kN **ANS**

6.52 Release bar AD at D. Turn nut to create gap at D. Apply F_D at D to return to ground. Thus

(a) $\delta_D = \dfrac{\partial U}{\partial F_D} = 1.5$ mm, $F_{AD} = F_D$, $\dfrac{\partial F_{AD}}{\partial F_D} = 1$, $F_{CE} = \dfrac{200}{150}F_D = \dfrac{4}{3}F_D$

$= \dfrac{1}{AE}\sum F\dfrac{\partial F}{\partial F_D}L$ $\qquad \partial F_{CE}/\partial F_D = 4/3$

$$\frac{1}{\frac{\pi}{4}(10)^2(200)(10^3)}\left[F_D(1)(600) + \frac{4}{3}F_D\left(\frac{4}{3}\right)(800)\right] = 1.2874(10^{-4})F_D = 1.5$$

$F_D = 11.65(10^3)N = \underline{11.65\text{ kN}}$, $F_C = \frac{4}{3}F_D = \underline{15.54\text{ kN}}$

(b) Add bending of ABC. Let x be directed from A to B and $\bar{x}$ be directed from C to B.

$\underline{0 < x < 200\text{mm}}$ $\quad M_z = -F_D x$, $\dfrac{\partial M_z}{\partial F_D} = -x$

$\underline{0 < \bar{x} < 150\text{mm}}$ $\quad M_z = -\frac{4}{3}F_D\bar{x}$, $\dfrac{\partial M_z}{\partial F_D} = -\frac{4}{3}\bar{x}$ $\qquad$ Add bending to part(a)

$$1.2874(10^{-4})F_D + \frac{1}{EI}\left[\int_0^{200} F_D x^2 dx + \int_0^{150}\frac{16}{9}F_D\bar{x}^2 d\bar{x}\right] = 1.5$$

$$1.2874(10^{-4})F_D + \frac{1}{200(10^3)(62.5)(10^3)}\left[\frac{F_D}{3}(200^3) + \frac{16}{27}F_D(150^3)\right] = 1.5$$

$F_D = 2988$ N $= \underline{2.99\text{ kN}}$, $F_C = \frac{4}{3}F_D = \underline{3.98\text{ kN}}$

6.53 Consider the reaction at E, T_E, as applied subject to the constraint that $\theta_E = 0$. The torque in shaft AB is T_A and $\partial T_A/\partial T_E = 0$. The torque in shaft CD is $T_D = \frac{100}{25}T_A - \frac{100}{50}T_E = 4T_A - 2T_E$ and $\partial T_D/\partial T_E = -2$. The torque in EF is T_E and $\partial T_E/\partial T_E = 1$

$$\theta_E = \frac{\partial U}{\partial T_E} = 0 = \frac{2(1+\nu)}{JE}\sum T\frac{\partial T}{\partial T_E}L = \frac{2(1+\nu)}{JE}\left[T_E(1)(500) + (4T_A - 2T_E)(-2)(250)\right]$$

(Continued)

$$500T_E + 1000T_E - 2000T_A = 0 \qquad T_A = 10 \text{ N-m}$$

$$T_E = \frac{4}{3}T_A = 13.33 \text{ N-m} \qquad \partial T_E/\partial T_A = 4/3$$

$$T_D = 4T_A - 2T_E = 4T_A - \frac{8}{3}T_A = \frac{4}{3}T_A, \quad \frac{\partial T_D}{\partial T_A} = \frac{4}{3}$$

$$T_A = T_A, \quad \frac{\partial T_A}{\partial T_A} = 1$$

$$\theta_A = \frac{\partial U}{\partial T_A} = \frac{2(1+\nu)}{JE}\left[T_A(1)(500) + \left(\frac{4}{3}\right)^2 T_A(250) + \left(\frac{4}{3}\right)^2 T_A(500)\right]$$

$$= \frac{2(1+0.29)(1833.3)(10)(10^3)}{\frac{\pi}{32}(15)^4(200)(10^3)} = \underline{4.758(10^{-2})\text{rad} = 2.73^\circ}$$

(6.54)

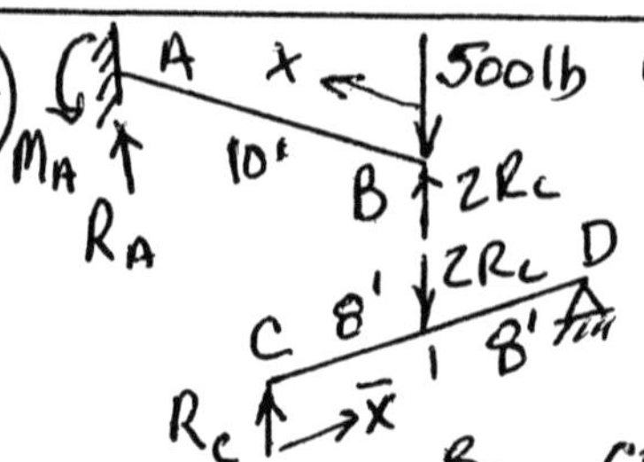

Consider R_C as applied subject to the constraint $\delta_C = \frac{\partial U}{\partial R_C} = 0$

Beam AB

$$M = (2R_C - 500)x, \quad \frac{\partial M}{\partial R_C} = 2x$$

Beam CD $0 < \bar{x} < 8'$

$$M = R_C\bar{x}, \quad \frac{\partial M}{\partial R_C} = \bar{x}$$

$$\delta_C = \frac{\partial U}{\partial R_C} = 0 = \frac{2}{EI}\int_0^8 R_C\bar{x}^2 d\bar{x} + \frac{1}{EI}\int_0^{10}(2R_C - 500)x(2x)dx$$

$$\frac{2}{3}R_C(8)^3 + \frac{2}{3}(2R_C - 500)(10^3) = 0 \Rightarrow \underline{R_C = 199 \text{ lb}}$$

Statics: $R_A = 500 - 2R_C = \underline{102 \text{ lb}}$, $M_A = 102(10) = \underline{1020 \text{ ft-lb}}$

(6.55)

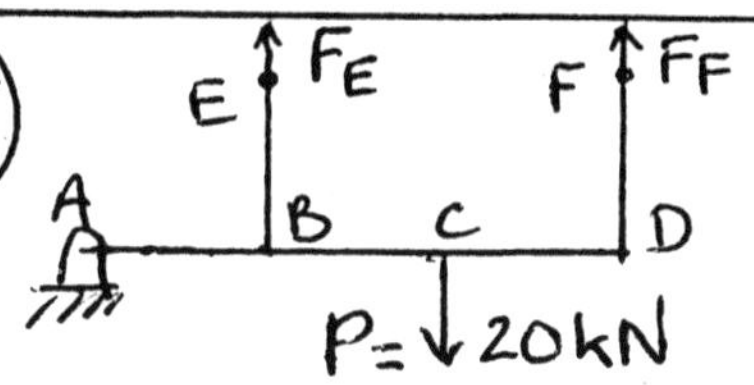

Consider F_F applied subject to the constraint $\delta_F = \frac{\partial U}{\partial F_F} = 0$

$\Sigma M_A = 0$ yields

$$F_E = 40 - 3F_F \text{ kN}, \quad \frac{\partial F_E}{\partial F_F} = -3$$

(a) Cables only

$$\delta_F = \frac{\partial U}{\partial F_F} = 0 = \frac{L}{AE}\Sigma F\frac{\partial F}{\partial F_F} = \frac{L}{AE}\left[F_F(1) + (40 - 3F_F)(-3)\right]$$

$$10F_F - 120 = 0 \Rightarrow \underline{F_F = 12 \text{ kN}}, \quad F_E = 40 - 3F_F = \underline{4 \text{ kN}}$$

$$\delta_B = \left(\frac{FL}{AE}\right)_{BE} = \frac{4000(1000)}{\frac{\pi}{4}(12)^2(209)(10^3)} = \underline{0.169 \text{ mm}\downarrow}$$

(continued)

$$\delta_D = \left(\frac{FL}{AE}\right)_{DF} = \frac{12(10^3)(1000)}{\frac{\pi}{4}(12^2)(209)(10^3)} = \underline{0.508\text{ mm}\downarrow}$$

In terms of P, $F_{BE} = \frac{4}{20}P = \frac{P}{5}$, $\frac{\partial F_{BE}}{\partial P} = \frac{1}{5}$, $F_{DF} = \frac{12}{20}P = \frac{3}{5}P$, $\frac{\partial F_{DF}}{\partial P} = \frac{3}{5}$

$$\delta_C = \frac{\partial U}{\partial P} = \frac{1000}{\frac{\pi}{4}(12^2)(209)(10^3)}\left[\left(\frac{1}{5}\right)^2 P + \left(\frac{3}{5}\right)^2 P\right] = 1.692(10^{-5})P = \underline{0.338\text{ mm}\downarrow}$$

(b) including bending of ABCD

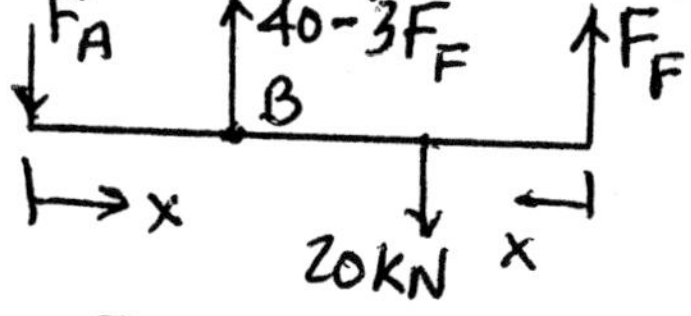

$\Sigma M_B = 0 \Rightarrow F_A = 20 - 2F_F$

$\underline{0 < x < 500\text{mm}}$ $\quad M = -(20-2F_F)x$, $\frac{\partial M}{\partial F_F} = 2x$

$\underline{0 < \bar{x} < 500\text{ mm}}$ $\quad M = F_F\bar{x}$, $\frac{\partial M}{\partial F_F} = \bar{x}$

$\underline{500 < \bar{x} < 1000\text{mm}}$ $\quad M = F_F\bar{x} - 20(\bar{x}-500)$, $\frac{\partial M}{\partial F_F} = \bar{x}$

$$\delta_F = \frac{\partial U}{\partial F_F} = 0 = \frac{1000}{\frac{\pi}{4}(12)^2(209)(10^3)}\left[F_F + (40-3F)(-3)\right]$$

$$+ \frac{1}{209(10^3)\,8(10^5)}\left\{\int_0^{500}(2F_F-20)x(2x)dx + \int_0^{500}F_F\bar{x}^2d\bar{x} + \int_{500}^{1000}\left[F_F\bar{x}^2 - 20(\bar{x}-500)\right]\bar{x}d\bar{x}\right\}$$

Multiply by $209(10^3)$ (combine)

$$8.842F_F - 1061 + 79.58F_F + \frac{1}{8(10^5)}\left[\frac{4}{3}(500)^3F_F - \frac{40}{3}(500)^3 + \frac{F_F}{3}(1000)^3 - \frac{20}{3}\left[1000^3 - 500^3\right] + \frac{10000}{2}(1000^2 - 500^2)\right] = 0$$

$$88.42F_F - 1061 + 208.3F_F - 2083 + 416.7F_F - 2604 = 0$$

$\underline{F_F = 8.06\text{ kN}}$, $F_E = 40 - 3F_F = \underline{15.83\text{ kN}}$

$$\delta_B = \left(\frac{FL}{AE}\right)_{BE} = \frac{15.83(10^3)(1000)}{\pi/4(12^2)(209)(10^3)} = \underline{0.670\text{ mm}}$$

$$\delta_D = \left(\frac{FL}{AE}\right)_{DF} = \frac{8.06(10^3)(1000)}{\pi/4(12^2)(209)(10^3)} = \underline{0.341\text{ mm}}$$

$F_{DF} = \frac{8.06}{20}P = 0.403P$, $\frac{\partial F_{DF}}{\partial P} = 0.403P$, $F_{BE} = \frac{15.83}{20}P = 0.792P$,

$\frac{\partial F_{BE}}{\partial P} = 0.792$, $F_A = P - 2F_{DF} = 0.194P$ (continued)

$0 < x < 500\text{ mm}$ $\quad M = -0.194Px, \quad \dfrac{\partial M}{\partial P} = -0.194x$

$0 < \bar{x} < 500\text{ mm}$ $\quad M = 0.403P\bar{x}, \quad \dfrac{\partial M}{\partial P} = 0.403\bar{x}$

$500 < \bar{x} < 1000\text{ mm}$ $\quad M = 0.403P\bar{x} - P(x-500) = P(500 - 0.597\bar{x})$

$$\partial M/\partial P = 500 - 0.597\bar{x}$$

$$\delta_C = \frac{\partial U}{\partial P} = \frac{P}{209(10^3)}\left\{\frac{\left[(0.403)^2 + (0.792)^2\right]1000}{\pi/4(12^2)} + \frac{1}{8(10^5)}\left[\int_0^{500}(0.194)^2x^2dx + \int_0^{500}(0.403)^2\bar{x}^2d\bar{x} + \int_{500}^{1000}(500-0.597\bar{x})^2d\bar{x}\right]\right\}$$

$$= \frac{20(10^3)}{209(10^3)}\left\{6.982 + \frac{1}{8(10^5)}\left[(0.194)^2\frac{500^3}{3} + (0.403)^2\frac{500^3}{3} - \frac{(500-0.597\bar{x})^3}{3(0.597)}\Bigg|_{500}^{1000}\right]\right\}$$

$$= 2.27\text{ mm}$$

6.56 Due to symmetry consider 1/4 segment

Considering bending only

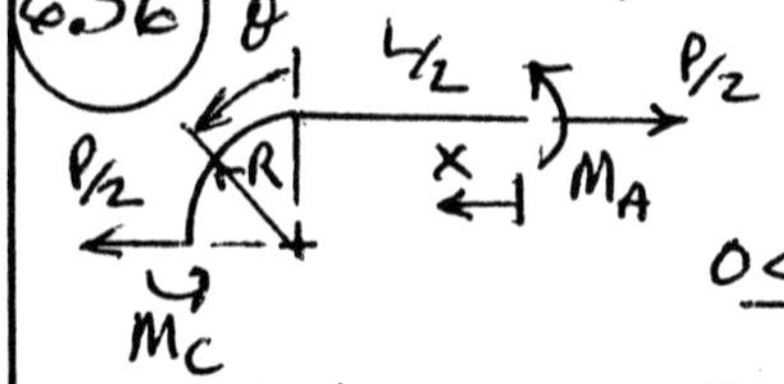

$0 < x < L/2 \quad M = M_A, \quad \dfrac{\partial M}{\partial M_A} = 1$

$0 < \theta < \pi/2 \quad M = -M_A + \dfrac{PR}{2}(1-\cos\theta), \quad \dfrac{\partial M}{\partial M_A} = -1$

$$\theta_A = \frac{\partial U}{\partial M_A} = 0 = \frac{1}{EI}\left\{\int_0^{L/2} M_A(1)dx + \int_0^{\pi/2}\left[-M_A + \frac{PR}{2}(1-\cos\theta)\right](-1)R\,d\theta\right\}$$

$$M_A\frac{L}{2} + M_A R\frac{\pi}{2} - \frac{PR^2}{2}\cdot\frac{\pi}{2} + PR^2 = 0 \qquad \text{solve for } M_A$$

$$M_A = \frac{\pi - 2}{2(L+\pi R)}PR^2 \quad \boxed{\text{ANS}} \text{ for } M \text{ for } 0 < x < \frac{L}{2}$$

For $0 < \theta < \pi/2$, $M = \dfrac{PR}{2}\left[\dfrac{L+2R}{L+\pi R} - \cos\theta\right]$ $\boxed{\text{ANS}}$

For $0 < x < L/2 \quad \partial M/\partial P = \dfrac{\pi - 2}{2(L+\pi R)}R^2$

For $0 < \theta < \pi/2 \quad \partial M/\partial P = R/2\left[\dfrac{L+2R}{L+\pi R} - \cos\theta\right]$

$$\delta_P = \frac{\partial U}{\partial P} = \frac{4}{EI}\left\{\int_0^{L/2}\left[\frac{\pi-2}{2(L+\pi R)}\right]^2 PR^4dx + \int_0^{\pi/2}\left[\frac{L+2R}{L+\pi R} - \cos\theta\right]^2 P\left(\frac{R}{2}\right)^2 R\,d\theta\right\}$$

(Continued)

$$\delta_P = \frac{4PR^3}{EI}\left\{\left[\frac{\pi-2}{2(L+\pi R)}\right]^2 R\frac{L}{2}+\frac{1}{4}\left[\left(\frac{L+2R}{L+\pi R}\right)^2\frac{\pi}{2}-2\left(\frac{L+2R}{L+\pi R}\right)+\frac{\pi}{4}\right]\right\}$$

This simplifies to

$$\delta_P = \frac{(3\pi-8)L^2+4(\pi^2-2\pi-2)RL+\pi(\pi^2-8)R^2}{4(L+\pi R)^2}\,\frac{PR^3}{EI} \quad \boxed{\text{ANS}}$$

(6.57) As in Prob. 6.56

$0<x<\frac{L}{2}$ $\quad M=M_A,\ \partial M/\partial M_A=1,\ N=\frac{P}{2},\ V=0,\ \frac{\partial N}{\partial M_A}=\frac{\partial V}{\partial M_A}=0$

$0<\theta<\pi/2$ $\quad M=-M_A+\frac{PR}{2}(1-\cos\theta),\ \frac{\partial M}{\partial M_A}=-1,\ N=\frac{P}{2}\cos\theta,\ V=\frac{P}{2}\sin\theta$

$\frac{\partial N}{\partial M_A}=\frac{\partial V}{\partial M_A}=0$

$$\theta_A=\frac{\partial U}{\partial M_A}=0=\frac{1}{EI}\int_0^{L/2}M_A(1)\,dx+\frac{1}{EA}\left[\frac{1}{e}\int_0^{\pi/2}\left[M_A-\frac{PR}{2}(1-\cos\theta)\right]d\theta+\int_0^{\pi/2}\left(\frac{P}{2}\cos\theta\right)(-1)\,d\theta\right]$$

$$\frac{M_A}{I}\left(\frac{L}{2}\right)+\frac{1}{A}\left[\frac{1}{e}\left(\frac{\pi}{2}M_A-\frac{\pi}{4}PR+\frac{PR}{2}\right)-\frac{P}{2}\right]=0$$

$$\left(\frac{L}{I}+\frac{\pi}{Ae}\right)\frac{M_A}{2}=\frac{P}{2A}\left(\frac{\pi R}{2e}-\frac{R}{e}+1\right)$$

$$M_A=\frac{[(\pi-2)R+2e]IP}{2(eAL+\pi I)}$$

$$M=\left\{R(1-\cos\theta)-\frac{[(\pi-2)R+2e]I}{eAL+\pi I}\right\}\frac{P}{2} \quad \boxed{\text{ANS}}$$

(6.58) Consider the horizontal reaction at B to be applied subject to the constraint that $(\delta_B)_H=0$.

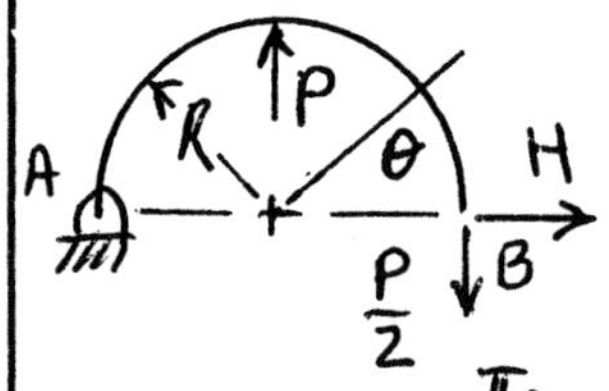

$(\delta_B)_H=\frac{\partial U}{\partial H}=0$ Due to symmetry consider half structure. P does not deflect horizontally.

$0<\theta<\frac{\pi}{2}$

$$M=\frac{PR}{2}(1-\cos\theta)-HR\sin\theta,\quad \frac{\partial M}{\partial H}=-R\sin\theta$$

$$\frac{\partial U}{\partial H}=\frac{1}{EI}\int_0^{\pi/2}\left[\frac{PR}{2}(1-\cos\theta)-HR\sin\theta\right](-R\sin\theta)R\,d\theta=0$$

$$-\frac{P}{2}+\frac{P}{4}+H\frac{\pi}{4}=0 \Rightarrow H=\frac{P}{\pi} \quad \boxed{\text{ANS}}$$

Reaction at A the same where H goes to the left

(b) For $0<\theta<\pi/2$, $M=\frac{PR}{2}(1-\cos\theta)-\frac{PR}{\pi}\sin\theta$

$$M=\frac{PR}{2\pi}\left[\pi(1-\cos\theta)-2\sin\theta\right] \quad \boxed{\text{ANS}}$$

Due to symmetry the left side is identical

(Continued)

(c) $\dfrac{\partial M}{\partial P} = \dfrac{R}{2\pi}\left[\pi(1-\cos\theta) - 2\sin\theta\right]$

$$\delta_P = \frac{\partial U}{\partial P} = \frac{2}{EI}\int_0^{\pi/2} \frac{PR^2}{4\pi^2}\left[\pi(1-\cos\theta) - 2\sin\theta\right]^2 R\,d\theta$$

$$= \frac{PR^3}{2\pi^2 EI}\int_0^{\pi/2}\left(\pi^2 + \pi^2\cos^2\theta + 4\sin^2\theta - 2\pi^2\cos\theta - 4\pi\sin\theta + 4\pi\sin\theta\cos\theta\right)d\theta$$

$$= \frac{PR^3}{2\pi^2 EI}\left(\pi^2\frac{\pi}{2} + \pi^2\frac{\pi}{4} + 4\frac{\pi}{4} - 2\pi^2 - 4\pi + 2\pi\right)$$

$$= \frac{(3\pi^2 - 8\pi - 4)}{8\pi}\,\frac{PR^3}{EI}$$ **ANS**

6.59

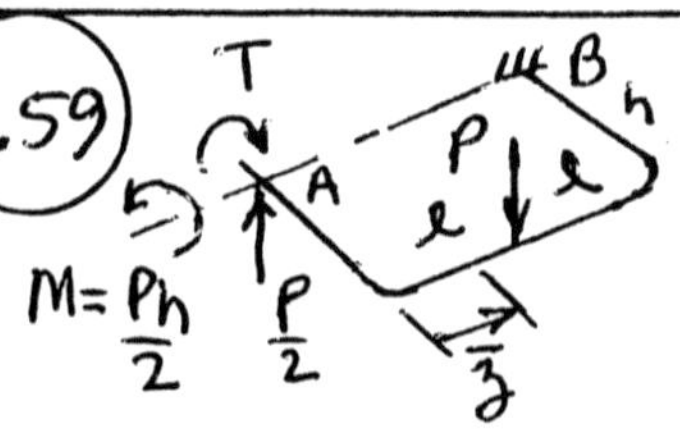

$(\theta_A)_x = \dfrac{\partial U}{\partial T} = 0$

In members of length h only torque is a function of T $\quad \dfrac{\partial U_h}{\partial T} = 2(1+\nu)\dfrac{Th}{JE}$ each member

In members of length ℓ only bending is a function of T

$M = \frac{P}{2}\bar{z} + T, \quad \frac{\partial M}{\partial T} = 1 \quad 0 < \bar{z} < \ell$

$$(\theta_A)_x = \frac{\partial U}{\partial T} = \frac{2}{EI}\int_0^{\ell}\left(\frac{P}{2}\bar{z} + T\right)(1)\,d\bar{z} + 4(1+\nu)\frac{Th}{JE} = 0$$

Let $J = 2I$. Thus $\dfrac{P\ell^2}{4} + T\ell + (1+\nu)Th = 0$

$$T = \frac{-P\ell^2}{4\left[\ell + (1+\nu)h\right]}$$ **ANS**

Torque is actually in the opposite direction shown. Reactions at B are the same with T being in the opposite direction of that at A.

6.60 Apply a horizontal force H at point G subject to the constraint that $\partial U/\partial H = 0$. Determine forces in members due to H and using superposition add these to the forces determined in Prob. 6.25. Ignore zero force members.

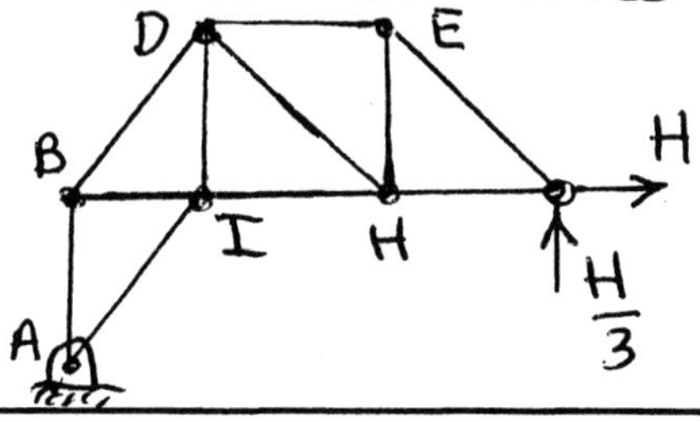

$F_{AI} = \sqrt{2}H$, $F_{AB} = -\frac{2}{3}H$, $F_{BD} = -\frac{2\sqrt{2}}{3}H$

$F_{BI} = \frac{2}{3}H$, $F_{DI} = H$, $F_{HI} = \frac{5}{3}H$, $F_{DE} = -\frac{1}{3}H$

$F_{DH} = -\frac{\sqrt{2}}{3}H$, $F_{EG} = -\frac{\sqrt{2}}{3}H$, $F_{GH} = \frac{4}{3}H$, $F_{DH} = -\frac{\sqrt{2}}{3}H$

(continued)

Add the forces obtained in Prob 6.25 due to $P = 8\,kN$ at H.

$F_{AI} = \sqrt{2}H$, $F_{AB} = -\frac{P}{3} - \frac{2}{3}H$, $F_{BD} = -\frac{\sqrt{2}}{3}P - \frac{2\sqrt{2}}{3}H$, $F_{BI} = \frac{P}{3} + \frac{2}{3}H$

$F_{DI} = H$, $F_{HI} = \frac{P}{3} + \frac{5}{3}H$, $F_{DE} = -\frac{2}{3}P - \frac{H}{3}$, $F_{DH} = \frac{\sqrt{2}}{3}P - \frac{\sqrt{2}}{3}H$

$F_{EG} = -\frac{\sqrt{2}}{3}H$, $F_{GH} = \frac{4}{3}H$, $F_{DH} = -\frac{\sqrt{2}}{3}H$

$$(\delta_G)_H = \frac{\partial U}{\partial H} = \frac{1}{AE}\sum F\frac{\partial F}{\partial H}L = 0, \quad \text{AE cancels}$$

$$(\sqrt{2}H)(\sqrt{2})\sqrt{2}L + \left(-\frac{P}{3} - \frac{2}{3}H\right)\left(-\frac{2}{3}\right)L + \left(-\frac{\sqrt{2}}{3}P - \frac{2\sqrt{2}}{3}H\right)\left(-\frac{2\sqrt{2}}{3}\right)\sqrt{2}L$$
$$+\left(\frac{P}{3} + \frac{2}{3}H\right)\left(\frac{2}{3}\right)L + (H)(1)L + \left(\frac{P}{3} + \frac{5}{3}H\right)\left(\frac{5}{3}\right)L + \left(-\frac{2}{3}P - \frac{H}{3}\right)\left(-\frac{1}{3}\right)L$$
$$+\left(\frac{\sqrt{2}}{3}P - \frac{\sqrt{2}}{3}H\right)\left(-\frac{\sqrt{2}}{3}\right)\sqrt{2}L + \left(-\frac{2\sqrt{2}}{3}P - \frac{\sqrt{2}}{3}H\right)\left(-\frac{\sqrt{2}}{3}\right)\sqrt{2}L + \left(\frac{2}{3}P + \frac{4}{3}H\right)\left(\frac{4}{3}\right)L$$
$$+\left(\frac{2}{3}P + \frac{H}{3}\right)\left(\frac{1}{3}\right)L = 0$$

Factor L out and collect P & H terms

$$\frac{10}{3}(2+\sqrt{2})H + \frac{1}{3}(7+2\sqrt{2})P = 0$$

$$H = -\frac{1}{10}\,\frac{7+2\sqrt{2}}{2+\sqrt{2}}P = -0.2879P$$

sub back into above force equations

$F_{AI} = -0.4071P$, $F_{AB} = -0.1414P$, $F_{BD} = -0.2P$, $F_{BI} = 0.1414P$

$F_{DI} = -0.2879P$, $F_{HI} = -0.1464P$, $F_{DE} = -0.5707P$, $F_{DH} = 0.6071P$

$F_{EG} = -0.8071P$, $F_{GH} = 0.2828P$, $F_{EH} = 0.5707P$

$$\delta_P = \frac{\partial U}{\partial P} = \frac{1}{AE}\sum F\frac{\partial F}{\partial P}L$$

$$\delta_P = \frac{PL}{AE}\Big[(-0.4071)^2\sqrt{2} + (-0.1414)^2 + (-0.2)^2\sqrt{2} + (0.1414)^2 + (-0.2879)^2$$
$$+ (-0.1464)^2 + (-0.5707)^2 + (0.6071)^2\sqrt{2} + (-0.8071)^2\sqrt{2}$$
$$+ (0.2828)^2 + (0.5707)^2\Big]$$

$$\delta_P = 2.609\frac{PL}{AE} = 2.609\frac{8(10^3)(3)(10^3)}{200(210)(10^3)} = \underline{1.491\,mm}$$

6.61

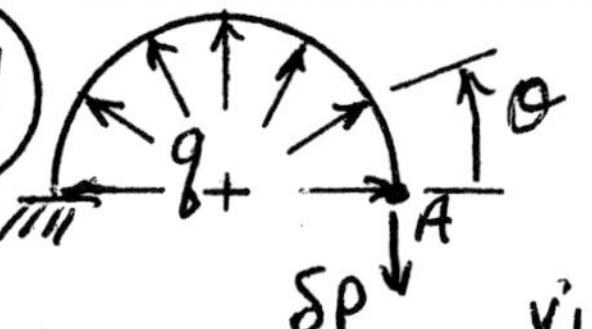

The bending moment due to q is determined in Prob 6.13 and is $M_q = qR^2(1-\cos\theta)$. The moment due to the virtual force is $\delta M = -(\delta P)R(1-\cos\theta)$

$$\delta\Phi = \delta_A\,\delta P = \frac{1}{EI}\int_0^{\pi} qR^2(1-\cos\theta)(-\delta P)R(1-\cos\theta)\,R\,d\theta$$

$$\delta_A = -\frac{qR^4}{EI}\int_0^{\pi}(1-\cos\theta)^2 d\theta = -\frac{3\pi}{2}\frac{qR^4}{EI} \quad \boxed{\text{ANS}}$$

(6.62) From Prob 6.16 $A_C = 50\,mm^2$, $I_{AB} = 60.07(10^3)\,mm^4$

$E = 210(10^3)\,N/mm^2$, $P = 3\,kN$, $A_{AB} = 348.7\,mm^2$, $L_{BC} = 948.7\,mm$

$F_{BC} = 0.4743P = 1.423\,kN$, $F_{BD} = 0.7247P = 2.174\,kN$ $L_{BD} = 1449\,mm$

Add δP to P. $\delta F_{BC} = 0.4743\,\delta P$, $\delta F_{BD} = 0.7247\,\delta P$

In the beam, $0 < x < 500\,mm$, $M = 1500x$ $\delta M = \frac{\delta P}{2}x$

↑ symmetric

$$\delta\Phi = \frac{2}{210(10^3)(60.07)(10^3)}\int_0^{500}(1500x)\left(\frac{\delta P x}{2}\right)dx$$

$$+\frac{1}{50(210)(10^3)}\left[(1423)(0.4743\,\delta P)(948.7) + 2174(0.7247\,\delta P)(1449)\right]$$

$$= \underline{5.27\ mm}$$

(6.63)

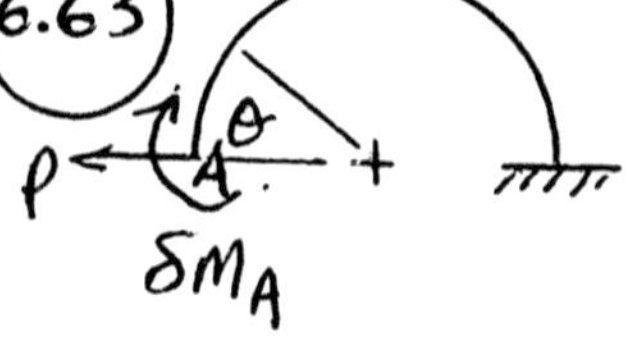

$M = -PR\sin\theta$ $\delta M = -\delta M_A$

$$\delta\Phi = \theta_A\,\delta M_A = \frac{1}{EI}\int_0^{\pi}(-PR\sin\theta)(-\delta M_A)R\,d\theta$$

$$\theta_A = \frac{PR^2}{EI}\int_0^{\pi}\sin\theta\,d\theta = \frac{2PR^2}{EI} \quad \boxed{ANS}$$

(6.64) $F_{AB} = 8\,kN$, $F_{BC} = 11.314\,kN$

Add δP in same direction of the applied load.

$\delta F_{AB} = \delta P$, $\delta F_{BC} = \sqrt{2}\,\delta P$

$$\delta\Phi = (\delta_B)_V\,\delta P = \frac{1}{AE}\left[8000\,\delta P L_{AB} + (11{,}314)\sqrt{2}\,\delta P L_{BC}\right]$$

$$(\delta_B)_V = \frac{1}{125(200)(10^3)}\left[8000(500) + (11{,}314)(\sqrt{2})(750)\right] = \underline{0.640\,mm}\downarrow$$

(6.65) From the solution to Prob 6.25

$F_{AB} = -2667\,N$, $F_{BD} = -3771\,N$, $F_{BI} = F_{HI} = 2667\,N$, $F_{EG} = -7542\,N$

$F_{GH} = 5333\,N$, $F_{DE} = -5333\,N$, $F_{DH} = 3771\,N$

Add δP to point H in downward vertical direction. The forces due to this are:

$\delta F_{AB} = -\frac{\delta P}{3}$, $\delta F_{BD} = -\frac{\sqrt{2}}{3}\delta P$, $\delta F_{BI} = \delta F_{HI} = \frac{\delta P}{3}$, $\delta F_{EG} = -\frac{2\sqrt{2}}{3}\delta P$, $\delta F_{GH} = \frac{2}{3}\delta P$,

$\delta F_{DE} = -\frac{2}{3}\delta P$, $\delta F_{DH} = \frac{\sqrt{2}}{3}\delta P$. For member DH add $\alpha\Delta T$ to F/AE.

(Continued)

$$\delta\Phi = (\delta_H)_V\,\delta P = \frac{1}{AE}\Big(F_{AB}\,\delta F_{AB}L + F_{BD}\,\delta F_{BD}\sqrt{2}L + F_{BI}\,\delta F_{BI}L + F_{HI}\,\delta F_{HI}L$$
$$+ F_{EG}\,\delta F_{EG}\sqrt{2}L + F_{GH}\,\delta F_{GH}L + F_{DE}\,\delta F_{DE}L\Big) + \left(\frac{F_{DH}}{AE} + \alpha\Delta T\right)\sqrt{2}L\,\delta F_{DH}$$

Sub in values for forces, $\alpha = 11.7(10^{-6})$, $\Delta T = 100$, $A = 200\ mm^2$, and $E = 210(10^3)\ N/mm^2$ results in

$(\delta_H)_V = 4.37\ mm$

(6.66) From the solution to Prob 6.33

$F_{BG} = F_{DH} = F_{GH} = -\frac{P}{\sqrt{3}}$, $F_{CG} = F_{CH} = \frac{P}{\sqrt{3}}$, $F_{BC} = F_{CD} = \frac{P}{2\sqrt{3}}$

Placing δP at the same point & direction as P yields

$\delta F_{BG} = \delta F_{DH} = \delta F_{GH} = -\frac{\delta P}{\sqrt{3}}$, $\delta F_{CG} = \delta_{CH} = \frac{\delta P}{\sqrt{3}}$, $\delta F_{BC} = \delta F_{CD} = \frac{\delta P}{2\sqrt{3}}$

$$\delta\Phi = (\delta_C)_V\,\delta P = \frac{L}{AE}\Big(F_{BG}\,\delta F_{BG} + F_{DH}\,\delta F_{DH} + F_{GH}\,\delta F_{GH} + F_{CH}\,\delta F_{CH} + F_{BC}\,\delta F_{BC}$$
$$+ F_{CD}\,\delta F_{CD}\Big) + \left(\frac{F_{CG}}{AE} + \alpha\Delta T\right)L\,\delta F_{CG}$$

$$(\delta_C)_V = \frac{L}{AE}\left[3\left(-\frac{P}{\sqrt{3}}\right)\left(-\frac{1}{\sqrt{3}}\right) + \frac{P}{\sqrt{3}}\left(\frac{1}{\sqrt{3}}\right) + 2\left(\frac{P}{2\sqrt{3}}\right)\left(\frac{1}{2\sqrt{3}}\right)\right] + \left(\frac{P/\sqrt{3}}{AE} + \alpha\Delta T\right)L\left(\frac{1}{\sqrt{3}}\right)$$
$$= \frac{11}{6}\frac{PL}{AE} + \frac{\sqrt{3}}{3}\alpha L\Delta T \quad \boxed{ANS}$$

(6.67)

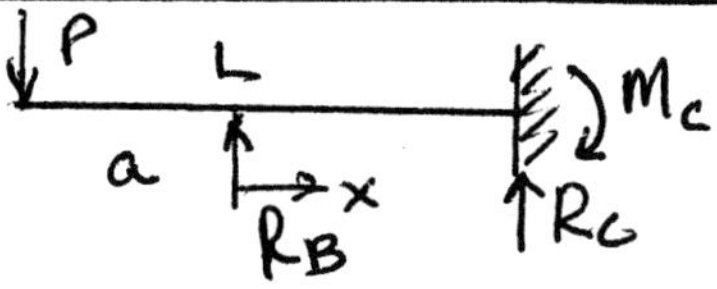

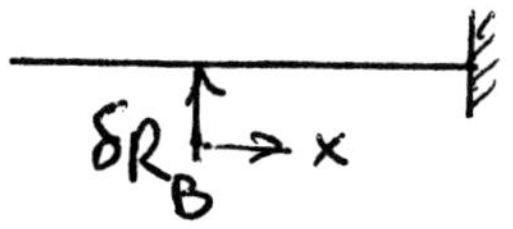

$\underline{-a < x < 0}$

$M_z = -P(a+x)$

$\underline{0 < x < (L-a)}$

$M_z = -P(a+x) + R_B x$

$\underline{-a < x < 0}$

$\delta M_z = 0$

$\underline{0 < x < (L-a)}$

$\delta M_z = \delta R_B x$

$$\delta\Phi = (\delta_B)\,\delta R_B = \frac{1}{EI}\int_0^{L-a}\left[-P(a+x) + R_B x\right]\delta R_B x\,dx = 0$$

$$\therefore \int_0^{L-a}\left[-P(a-x) + R_B x\right]x\,dx = 0 \Rightarrow -\frac{Pa}{2}(L-a)^2 - \frac{P}{3}(L-a)^3 + \frac{R_B}{3}(L-a)^3 = 0$$

$$R_B = \frac{2L+a}{2(L-a)}P \quad \boxed{ANS}$$

$$R_C = P - R_B = -\frac{3Pa}{2(L-a)} \quad \boxed{ANS}$$

$$M_C = PL - \frac{P(2L+a)}{2(L-a)}(L-a) = -\frac{Pa}{2} \quad \boxed{ANS}$$

6.68 See prob. 6.67 solution. Only change is,

$$\delta\Phi = -\epsilon\,\delta R_B = \frac{1}{EI}\int_0^{L-a}\left[-P(a+x)+R_B x\right]\delta R_B\, x\,dx \quad \text{results in}$$

$$R_B = \frac{2L+a}{2(L-a)}P - \frac{3\epsilon EI}{(L-a)^3} \quad \boxed{\text{ANS}}$$

$$R_C = P - R_B = \frac{3\epsilon EI}{(L-a)^3} - \frac{3}{2}\frac{Pa}{L-a} \quad \boxed{\text{ANS}}$$

$$M_C = PL - R_B(L-a) = \frac{3\epsilon EI}{(L-a)^2} - \frac{Pa}{2} \quad \boxed{\text{ANS}}$$

6.69 (a) Release bar AD at D and turn nut. Then apply force F_D to return D such that $\delta_D = 0$. For this also apply δF_D. $F_{AD} = F_D$, $\delta F_{AD} = \delta F_D$, $F_{CE} = \frac{4}{3}F_D$, $\delta F_{CE} = \frac{4}{3}\delta F_D$
$L_{AD} = 600$ mm, $L_{CE} = 800$ mm, $A = \frac{\pi}{4}(10)^2\ \text{mm}^2$, $E = 200(10^3)\ \text{N/mm}^2$

$$\delta\Phi = 1.5\,\delta F_D = \frac{1}{\frac{\pi}{4}(10)^2(200)(10^3)}\left[F_D\,\delta F_D(600) + \left(\tfrac{4}{3}F_D\right)\left(\tfrac{4}{3}\delta F_D\right)(800)\right]$$

$$1.5 = 1.2874(10^{-4})F_D \Rightarrow F_D = 11.65(10^3)\,\text{N} = \underline{11.65\ \text{kN}}$$

$$F_E = \tfrac{4}{3}F_D = \underline{15.54\ \text{kN}}$$

(b) Add bending of ABC, $I = 62.5(10^3)\ \text{mm}^4$

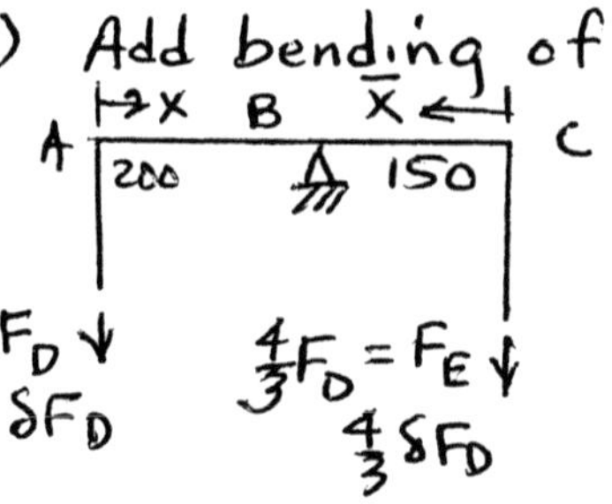

$0 < x < 200$ mm

$$M_z = -F_D x, \quad \delta M_z = -\delta F_D x$$

$0 < \bar{x} < 150$ mm

$$M_z = -\tfrac{4}{3}F_D\bar{x}, \quad \delta M_z = -\tfrac{4}{3}\delta F_D\bar{x}$$

$$\delta\Phi = 1.5\,\delta F_D = 1.2874(10^{-4})F_D\,\delta F_D + \frac{1}{EI}\left[\int_0^{200}F_D\,\delta F_D\,x^2dx + \int_0^{150}\left(\tfrac{4}{3}\right)^2F_D\,\delta F_D\,\bar{x}^2d\bar{x}\right]$$

$$1.5 = F_D\left[1.2874(10^{-4}) + \frac{1}{200(10^3)(62.5)(10^3)}\left(\frac{200^3}{3} + \frac{16}{9}\cdot\frac{150^3}{3}\right)\right]$$

$$F_D = 2988\,\text{N} = \underline{2.99\ \text{kN}}, \quad F_E = \tfrac{4}{3}F_D = \underline{3.98\ \text{kN}}$$

6.70 See diagram in solution to Prob. 6.53. Apply virtual load δT_E at point E

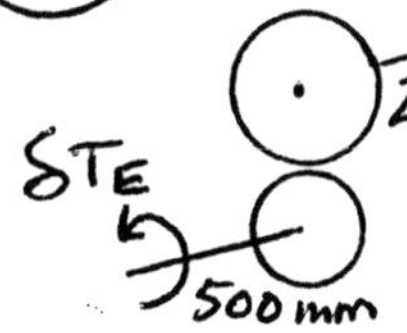

$$\delta\Phi = \theta_E\,\delta T_E = \frac{2(1+\nu)}{JE}\left[T_E\,\delta T_E(500) + (4T_A - 2T_E)(-2\delta T_E)(200)\right] = 0$$

(continued)

$500T_E + 1000T_E - 2000T_A = 0 \Rightarrow T_E = \frac{4}{3}T_A = \frac{4}{3}(10) = \underline{13.33\text{ N-m}}$

Shaft AB: $T_{AB} = T_A$, $\delta T_{AB} = \delta T_A$

CD: $T_{CD} = 4T_A - 2T_E = \frac{4}{3}T_A$, $\delta T_{CD} = \frac{4}{3}\delta T_A$

EF: $T_{EF} = \frac{4}{3}T_A$, $\delta T_{EF} = \frac{4}{3}\delta T_A$

$$\delta\Phi = \theta_A \delta T_A = \frac{2(1+\nu)}{JE}\left[T_A\delta T_A(500) + \frac{4}{3}T_A\left(\frac{4}{3}\delta T_A\right)(250) + \frac{4}{3}T_A\left(\frac{4}{3}\delta T_A\right)(500)\right]$$

Sub in $\nu = 0.29$, $E = 200(10^3)\text{ N/mm}^3$, $J = \frac{\pi}{32}(15^4)\text{ mm}^4$, $T_A = 10(10^3)\text{ N-mm}$ and solve for θ_A:

$\theta_A = 0.0476\text{ rad} = \underline{2.73^\circ}$

(6.71)

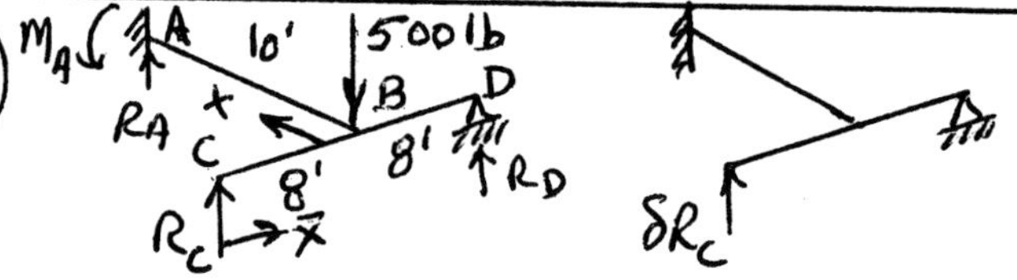

Beam AB: $M = (2R_C - 500)x$ $\delta M = 2\delta R_C x$

Beam CD: $\underline{0 < \bar{x} < 8'}$ $M = R_C\bar{x}$, $\delta M = \delta R_C \bar{x}$

$$\delta\Phi = \delta_C \delta R_C = \frac{2}{EI}\int_0^8 (R_C\bar{x})(\delta R_C\bar{x})d\bar{x} + \frac{1}{EI}\int_0^{10}(2R_C - 500)x(2\delta R_C x)dx = 0$$

$$2\frac{R_C}{3}8^3 + \frac{2}{3}(2R_C - 500)(10^3) = 0 \Rightarrow R_C = \frac{500(10^3)}{8^3 + 2(10^3)} = \underline{199\text{ lb} = R_D}$$

$R_A = 500 - 2R_C = \underline{102\text{ lb}}$, $M_A = 102(10) = \underline{1020\text{ lb-ft}}$

(6.72) Consider F_F applied at point F such that $\delta_F = 0$

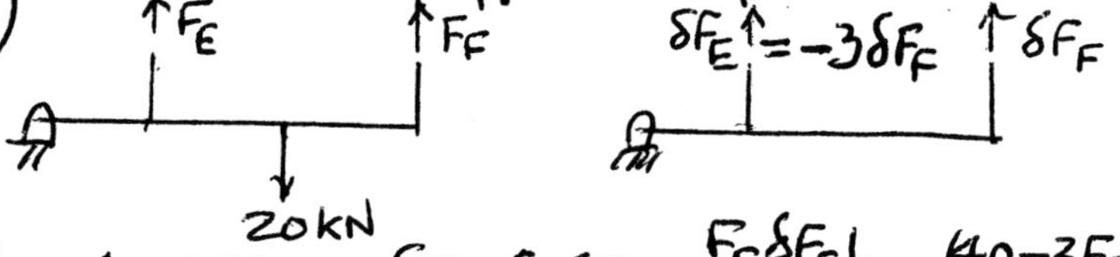

$F_E = 40 - 3F_F$, $\delta\Phi = \delta_F\delta F_F = \frac{F_F\delta F_F L}{AE} + \frac{(40 - 3F_F)(-3\delta F_F)L}{AE} = 0$

A, E, & L cancel and

$F_F + 9F_F - 120 = 0 \Rightarrow \underline{F_F = 12\text{ kN}}$, $F_E = 40 - 3F_F = \underline{4\text{ kN}}$

$$\delta_B = \left(\frac{FL}{AE}\right)_{BE} = \frac{4000(1000)}{\pi/4(12)^2(209)(10^3)} = \underline{0.169\text{ mm}\downarrow}$$

$$\delta_D = \left(\frac{FL}{AE}\right)_{DF} = \frac{12{,}000(1000)}{\pi/4(12)^2(209)(10^3)} = \underline{0.508\text{ mm}\downarrow}$$

(Continued)

For point C place a virtual force $\delta F_C \downarrow$ at point C.

$\delta F_{FD} = \frac{12}{20}\delta F_C = \frac{3}{5}\delta F_C$, $\delta F_{BE} = \frac{4}{20}\delta F_C = \frac{1}{5}\delta F_C$

$$\delta\Phi = \delta_C \, \delta F_C = \frac{1000}{\frac{\pi}{4}(12)^2(209)(10^3)}\left[(4000)\left(\frac{1}{5}\delta F_C\right) + 12000\left(\frac{3}{5}\delta F_C\right)\right]$$

$\delta_C = \underline{0.338 \text{ mm} \downarrow}$

(6.73) (a) $v_C'' = -\delta\left(\frac{\pi}{2L}\right)^2 \cos\frac{\pi x}{2L}$

$$W_w = -\frac{1}{2}w\int_0^L -\delta\left(1-\cos\frac{\pi x}{2L}\right)dx = \frac{1}{2}w\delta L\left(1-\frac{2}{\pi}\right)$$

$$U = \frac{1}{2}EI\delta^2\left(\frac{\pi}{2L}\right)^4\int_0^L \cos^2\frac{\pi x}{2L}\,dx = \frac{\pi^4}{64}\delta^2\frac{EI}{L^3}, \quad U = W$$

$$\frac{\pi^4}{64}\delta^2\frac{EI}{L^3} = \frac{1}{2}w\delta L\left(1-\frac{2}{\pi}\right) \Rightarrow \delta = 0.1194\frac{wL^4}{EI} \quad \boxed{ANS}$$

Appendix C gives $(v_C)_{x=L} = \frac{wL^4}{8EI}$. Approximation is 4.5% low

(b) $v_C'' = 4\frac{\delta}{L^4}(2Lx - x^2 - L^2)$

$$W_w = -\frac{1}{2}w\int_0^L \frac{\delta}{3}\frac{x^2}{L^4}(4Lx - x^2 - 6L^2)dx = \frac{w}{5}L\delta$$

$$U = \frac{1}{2}EI\left(4\frac{\delta}{L^4}\right)^2\int_0^L (2Lx - x^2 - L^2)dx = \frac{8}{5}\delta^2\frac{EI}{L^3}, \quad U = W$$

$$\frac{8}{5}\delta^2\frac{EI}{L^3} = \frac{w}{5}L\delta \Rightarrow \delta = \frac{wL^4}{8EI} \quad \boxed{ANS}$$ Exact answer as shape is exact.

(6.74) $W_M = \frac{1}{2}M(v_C')|_{x=L}$, $v_C' = \delta\frac{\pi}{2L}\sin\frac{\pi x}{2L}$, $(v_C')_{x=L} = \delta\frac{\pi}{2L}$

$U = \frac{\pi^4 EI}{64L^3}\delta^2$ (see Example 6.11-1), $U = W$

$$\frac{\pi^4 EI}{64L^3}\delta^2 = \frac{1}{2}M\left(\delta\frac{\pi}{2L}\right) \Rightarrow \delta = 0.516\frac{ML^2}{EI} \quad \boxed{ANS}$$

From Appendix C, $(v_C)_{x=L} = \frac{ML^2}{2EI}$, Approximation is 3% high

---

(6.75) $W_w = -\frac{1}{2}w\int_0^L -\delta\sin\frac{\pi x}{L}dx = \frac{wL}{\pi}\delta$, $v_C'' = \left(\frac{\pi}{L}\right)^2\delta\sin\frac{\pi x}{L}$

$$U = \frac{1}{2}EI\int_0^L\left(\frac{\pi}{L}\right)^4\delta^2\sin^2\frac{\pi x}{L}dx = \frac{\pi^4}{4}\frac{EI}{L^3}\delta^2, \quad U = W$$

$$\frac{\pi^4}{4}\frac{EI}{L^3}\delta^2 = \frac{wL}{\pi}\delta \Rightarrow \delta = \frac{4}{\pi^5}\frac{wL^4}{EI} \quad \boxed{ANS}$$

From Appendix C, $(v_C)_{x=L/2} = -\frac{5}{384}wL^4/EI$. Approximation is 0.4% high

6.76 $W_M = \frac{1}{2} M v_c'\big|_{x=\frac{L}{2}} = \frac{1}{2}M\left(-\frac{2\pi}{L}\delta\cos\pi\right) = \frac{\pi}{L}M\delta$, $v_c'' = \frac{4\pi^2}{L^2}\delta\sin\frac{2\pi x}{L}$

$\mathcal{U} = \frac{1}{2}EI\int_0^L \frac{16\pi^4}{L^4}\delta^2\sin^2\frac{2\pi x}{L}dx = \frac{4\pi^4}{L^3}EI\delta^2$, $\mathcal{U} = W$

$\frac{4\pi^4}{L^3}EI\delta^2 = \frac{\pi}{L}M\delta \Rightarrow \delta = \frac{1}{4\pi^3}\frac{ML^2}{EI}$ **ANS**

From Appendix C, $(v_c)_{x=L/4} = \frac{M(L/4)}{6EIL}\left[\left(\frac{L}{4}\right)^2 + 3\left(\frac{L}{2}\right)^2 - 6\left(\frac{L}{2}\right)L + 2L^2\right] = -\frac{1}{128}\frac{ML^2}{EI}$

Approximation is 3% larger in magnitude

6.77 at $x=0$, $v_c = v_c' = 0$ yields $a_0 = a_1 = 0$

$v_c'' = 2(a_2 + 3a_3x + 6a_4x^2)$ sub into $\mathcal{U} = \frac{1}{2}EI\int_0^L (v_c'')^2 dx$

$\mathcal{U} = 2EI\left(\frac{36}{5}L^5a_4^2 + 3L^3a_3^2 + La_2^2 + 9L^4a_3a_4 + 4L^3a_2a_4 + 3L^2a_2a_3\right)$

$d\mathcal{U} = 2EI\left[\left(\frac{72}{5}L^5a_4 + 9L^4a_3 + 4L^3a_2\right)da_4 + \left(9L^4a_4 + 6L^3a_3 + 3L^2a_2\right)da_3 + \left(4L^3a_4 + 3L^2a_3 + 2La_2\right)da_2\right]$ ①

$dW = d\left(-w\int_0^L v_c dx\right) = -w\left(\frac{L^5}{5}da_4 + \frac{L^4}{4}da_3 + \frac{L^3}{3}da_2\right)$ ②, $d\mathcal{U} = dW$

Equating coefficients of da_2, da_3, da_4 yields

$2EI\left(\frac{72}{5}L^5a_4 + 9L^4a_3 + 4L^3a_2\right) = -wL^5/5$

$2EI\left(9L^4a_4 + 6L^3a_3 + 3L^2a_2\right) = -wL^4/4$

$2EI\left(4L^3a_4 + 3L^2a_3 + 2La_2\right) = -wL^3/3$

Solving three equations simultaneously yields

$a_4 = -\frac{w}{24EI}$, $a_3 = \frac{wL}{6EI}$, $a_2 = -\frac{wL^2}{4EI}$. This gives

$v_c = -\frac{wx^2}{24EI}\left(x^2 - 4Lx + 6L^2\right)$ **ANS** Which agrees exactly with beam C.3 of Appendix C

6.78 $v_c = v_c' = 0$ at $x=0$ yields $a_0 = a_1 = 0$

Thus $v_c = x^2(a_2 + a_3x)$, $v_c' = 2a_2x + 3a_3x^2$

$v_c'' = 2a_2 + 6a_3x$, sub into $\mathcal{U} = \frac{1}{2}EI\int_0^L (v_c'')^2dx$

$\mathcal{U} = 2EI(3L^3a_3^2 + La_2^2 + 3L^2a_2a_3)$

$d\mathcal{U} = 2EI\left[(6L^3a_3 + 3L^2a_2)da_3 + (3L^2a_3 + 2La_2)da_2\right]$ ①

$dW = d\left(Mv_c'\big|_{x=L}\right) = M(2L\,da_2 + 3L^2da_3)$ ②. $d\mathcal{U} = dW$

(continued)

Equating coefficients of da_2 and da_3 of Eqs ① and ②

$$6L^3a_3 + 3L^2a_2 = \frac{3}{2}\frac{ML^2}{EI}, \quad 3L^2a_3 + 2La_2 = \frac{ML}{EI}$$ Solve simultaneously

$a_3 = 0$, $a_2 = \frac{1}{2}\frac{M}{EI}$ Sub into equation for v_c

$v_c = \frac{1}{2}\frac{Mx^2}{EI}$ **ANS** Agrees exactly with beam C.4 of Appendix C

6.79

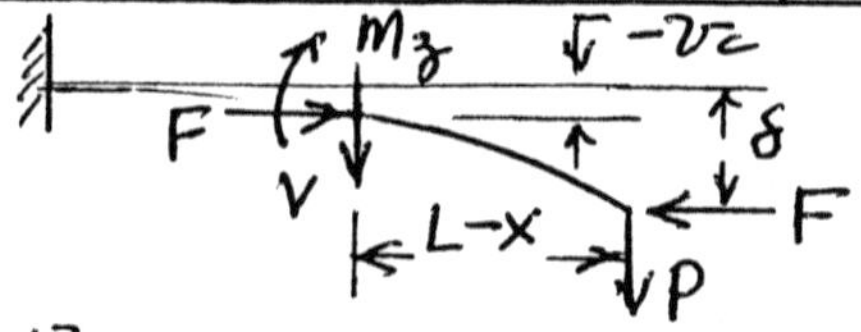

$$M_z = -P(L-x) - F[\delta - (-v_c)] = -P(L-x) - F(\delta + v_c)$$

$$EI\frac{d^2v_c}{dx^2} = -P(L-x) - F(\delta + v_c)$$

$$\frac{d^2v_c}{dx^2} + \frac{F}{EI}v_c = -\frac{F}{EI}\delta - \frac{P}{EI}(L-x)$$

Solution: $v_c = A\sin kx + B\cos kx - \delta - \frac{P}{F}(L-x)$ where $k = \sqrt{\frac{F}{EI}}$

B.C. $v_c = 0$ at $x = 0$ yields

$B = \delta + \frac{P}{F}L$ ①

B.C. $v_c' = 0$ at $x = 0$ yields $A = -\frac{P}{Fk}$ ②

B.C. $v_c = -\delta$ at $x = L$

$-\delta = -\frac{P}{kL}\sin kL + B\cos kL - \delta \Rightarrow B = \frac{P}{Fk}\tan kL$ ③

Equating ① & ③ yields $\delta = \frac{P}{Fk}(\tan kL - kL)$. Thus

$$v_c = -\frac{P}{Fk}\left[\sin kx + \tan kL(1 - \cos kx) - kx\right]$$

$F = 3$ kN, $P = 2$ kN, $L = 1$ m, $EI = 30$ kN-m^2

$$k = \sqrt{\frac{3}{30}} = 0.3162 \text{ m}^{-1}$$

$$\delta = \frac{2}{(3)(0.3162)(10^{-3})}\left[\tan(0.3162)(1) - (0.3162)(1)\right] = \underline{23.15 \text{ mm}}$$

From Example 6.13-3, eq (d)

$$\delta = \frac{5(2000)(1)^3}{3[5(30)(10^3) - 2(3000)(1)^2]} = 23.15(10^{-3}) m = \underline{23.15 \text{ mm}}$$

Approximation is actually $2(10^{-3})\%$ low

6.80 (a) $v_c = -\delta \sin\frac{\pi x}{L}$, $v_c' = -\delta\frac{\pi}{L}\cos\frac{\pi x}{L}$, $v_c'' = \delta\left(\frac{\pi}{L}\right)^2 \sin\frac{\pi x}{L}$

$$W_N = \frac{1}{2}\frac{N^2L}{AE} - \frac{AE}{L}\left[\int_0^L \delta^2\left(\frac{\pi}{L}\right)^2\cos^2\frac{\pi x}{L}dx\right]^2 = \frac{1}{2}\frac{N^2L}{AE} - \frac{\pi^4}{16}\frac{AE}{L^3}\delta^4$$

$$W_P = \frac{1}{2}P\delta$$

$$U = \frac{1}{2}\frac{N^2L}{AE} + \frac{1}{2}EI\int_0^L \delta^2\left(\frac{\pi}{L}\right)^4\sin^2\frac{\pi x}{L}dx = \frac{1}{2}\frac{N^2L}{AE} + \frac{\pi^4}{4}\frac{EI}{L^3}\delta^2$$

$U = W$

$$\frac{1}{2}\frac{N^2L}{AE} + \frac{\pi^4}{4}\frac{EI}{L^3}\delta^2 = \frac{1}{2}\frac{N^2L}{AE} - \frac{\pi^4}{16}\frac{AE}{L^3}\delta^4 + \frac{1}{2}P\delta$$

Simplifying yields, $\delta^3 + 4\frac{I}{A}\delta - \frac{8}{\pi^4}\frac{PL^3}{AE} = 0$

$I = \frac{1}{12}(1)(2)^3 = 0.667\text{ in}^4$, $A = 2\text{ in}^2$, $P = 1000$ lb, $L = 40$ in, $E = 30$ Mpsi

$\delta^3 + 1.333\delta - 0.08760 = 0$ yields only one real root

$\delta = 0.06579$ in

$$N \approx \frac{1}{2}\frac{AE}{L}\int_0^L (v_c')^2dx = \frac{\pi^2}{4}\frac{AE}{L^2}\delta^2 = \frac{\pi^2}{4}\frac{(2)(30)(10^6)}{40^2}(0.0658)^2 = 400\text{ lb}$$

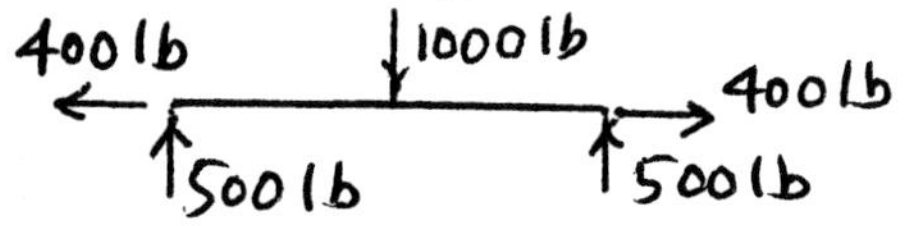

(b) $0 < x < 20$ in $\quad v_c = Kx\left(x^2 - \frac{3}{4}L^2\right) = \frac{K}{4}x(4x^2 - 3L^2)$

For $v_c = -\delta$ at $x = \frac{L}{2}$: $-\delta = \frac{K}{4}\left(\frac{L}{2}\right)\left[4\left(\frac{L}{2}\right)^2 - 3L^2\right] \Rightarrow K = \frac{4\delta}{L^3}$

Thus $v_c = \delta\frac{x}{L^3}(4x^2 - 3L^2)$, $v_c' = \frac{3\delta}{L^3}(4x^2 - L^2)$, $v_c'' = 24\frac{\delta x}{L^3}$

$$W_N = \frac{1}{2}\frac{N^2L}{AE} - \frac{AE}{4L}\left[2\int_0^{L/2}\frac{9\delta^2}{L^6}(4x^2 - L^2)^2dx\right]^2 = \frac{1}{2}\frac{N^2L}{AE} - 5.76\frac{AE}{L^3}\delta^4$$

$$W_P = \frac{1}{2}P\delta$$

$$U = \frac{1}{2}\frac{N^2L}{AE} + EI\int_0^{L/2}\left(\frac{24}{L^3}\right)^2\delta^2x^2dx = \frac{1}{2}\frac{N^2L}{AE} + 24\frac{EI}{L^3}\delta^2, \quad U = W$$

$$\frac{1}{2}\frac{N^2L}{AE} + 24\frac{EI}{L^3}\delta^2 = \frac{1}{2}\frac{N^2L}{AE} - 5.76\frac{AE}{L^3}\delta^4 + \frac{1}{2}P\delta$$

Simplifying yields: $\delta^3 + 4.1667\frac{I}{A} - 0.08681\frac{PL^3}{AE} = 0$

sub in values: $\delta^3 + 1.3889\delta - 0.09260 = 0$

yields one real root $\delta = 0.06646$ in

$$N \approx \frac{AE}{L}\int_0^{L/2}\frac{9\delta^2}{L^6}(4x^2 - L^2)^2dx = 2.4\frac{AE}{L^2}\delta^2 = 397.5\text{ lb}$$

6.81

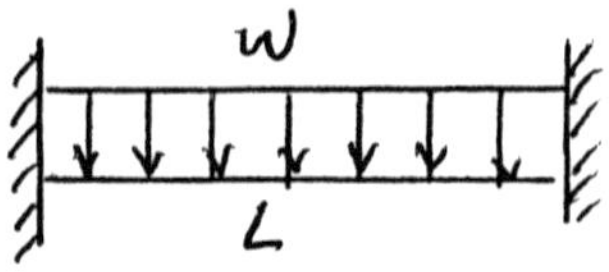

$v_c = Cx^2(L-x)^2$, Let $v_c = \delta$ at $x = \frac{L}{2}$

$\delta = C\left(\frac{L}{2}\right)^2\left(\frac{L}{2}\right)^2 \Rightarrow C = \frac{16\delta}{L^4}$ Thus,

$$v_c = \frac{16\delta}{L^4}x^2(L-x)^2, \quad v_c' = \frac{32\delta}{L^4}(L^2x - 3Lx^2 + 2x^3)$$

$$v_c'' = \frac{32\delta}{L^4}(L^2 - 6Lx + 6x^2)$$

$$W_N = \frac{1}{2}\frac{N^2L}{AE} - \frac{AE}{4L}\left[\int_0^L \left(\frac{32\delta}{L^4}\right)^2 (L^2x - 3Lx^2 + 2x^3)^2 dx\right]^2 = \frac{1}{2}\frac{N^2L}{AE} - 5.9443\frac{AE}{L^3}\delta^4$$

$$W_w = \frac{1}{2}\int_0^L w\left[\frac{16}{L^4}\delta x^2(L-x)^2\right]dx = \frac{8w\delta}{L^4}\int_0^L (x^2L^2 - 2Lx^3 + x^4)dx = \frac{4}{15}w\delta L$$

$$U = \frac{1}{2}\frac{N^2L}{AE} + \frac{1}{2}EI\delta^2\left(\frac{32}{L^4}\right)^2\int_0^L (L^2 - 6Lx + 6x^2)^2 dx = \frac{1}{2}\frac{N^2L}{AE} + \frac{(32)^2}{10}\frac{EI}{L^3}\delta^2$$

$U = W$

$$\frac{1}{2}\frac{N^2L}{AE} + 102.4\frac{EI}{L^3}\delta^2 = \frac{1}{2}\frac{N^2L}{AE} - 5.9443\frac{AE}{L^3}\delta^4 + \frac{4}{15}wL\delta \quad \text{or,}$$

$$5.9443\frac{AE}{L^3}\delta^3 + 102.4\frac{EI}{L^3}\delta - \frac{4}{15}wL = 0$$

$A = 250\text{ mm}^2$, $I = 30(10^3)\text{ mm}^4$, $L = 1\text{ m}$, $E = 210(10^3)\text{ N/mm}^2$, $w = 20(10^3)\text{ N/m}$ yields,

$$0.31208\delta^3 + 645.12\delta - 5.3333(10^3) = 0$$

$$\delta^3 + 2.0672(10^3)\delta - 1.7089(10^4) = 0$$

This yields one real root, $\underline{\delta = 8.017\text{ mm}}$ **ANS**

$$N \approx \frac{1}{2}\frac{AE}{L}\int_0^L \left(\frac{32\delta}{L^4}\right)^2 (L^2x - 3Lx^2 + 2x^3)^2 dx = \frac{1}{2}\frac{AE}{L}\left(\frac{32\delta}{L^4}\right)^2\frac{L^7}{120}$$

$$\approx 4.2667\frac{AE}{L^2}\delta^2 = 4.2667\frac{(250)(210)(10^3)}{(1000)^2}(8.017)^2 = 14.4(10^3)\text{ N}$$

$\approx \underline{14.4\text{ kN}}$ **ANS**

Approximate the end moments as follows. From Appendix C, beam C.14, $M_A = M_B = \frac{20(10^3)(1)^2}{12} = 1667\text{ N}\cdot\text{m}$

$$v_c\big|_{x=\frac{L}{2}} = -\frac{20(10^3)(0.5)^2(0.5)^2}{24(210)(10^9)(30)(10^{-9})} = -8.267(10^{-3})\text{m} = -8.267\text{mm}.$$

With constraint v_c is less. Thus, $M_A = M_B \approx \frac{8.017}{8.267}1667 = 1617\text{ N}\cdot\text{m}$

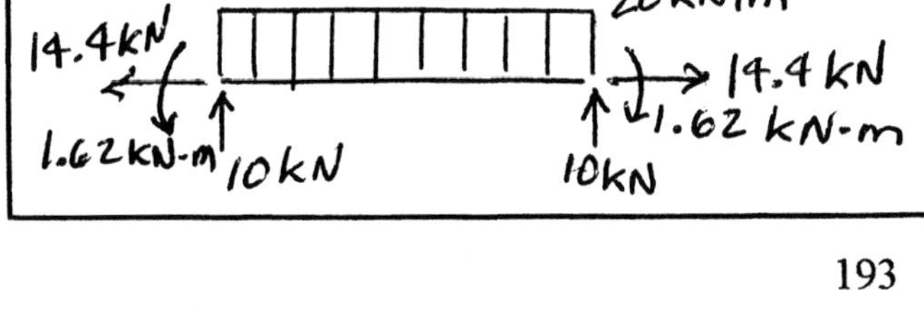

ANS Can also calculate M using $M = EIv_c''$ and obtain same result.

6.82

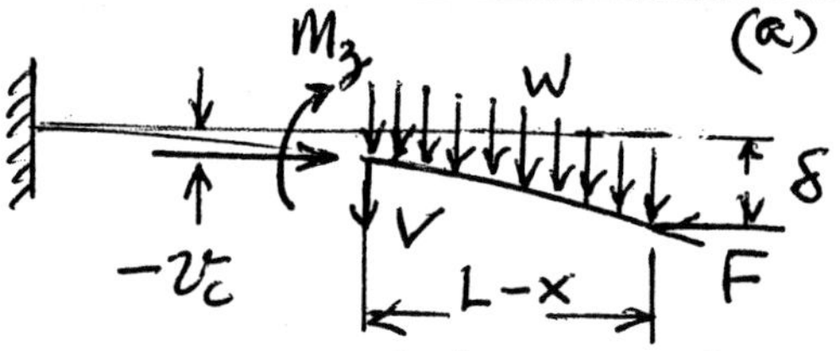

(a) $M_z = -\frac{w}{2}(L-x)^2 - F(\delta + v_c) = EI\frac{d^2 v_c}{dx^2}$

$$\frac{d^2 v_c}{dx^2} + \frac{F}{EI}v_c = -\frac{F}{EI}\delta - \frac{w}{2EI}(L-x)^2 \quad (1)$$

Homogeneous solution $(v_c)_h = A_1 \sin kx + A_2 \cos kx$ where $k = \sqrt{\frac{F}{EI}}$

Particular solution let $(v_c)_p = B_1 + B_2 x + B_3 x^2$ sub into (1) and solve for B's. Final form of v_c is

$$v_c = A_1 \sin kx + A_2 \cos kx + \frac{w}{F}\left(\frac{1}{k^2} - \frac{L^2}{2}\right) - \delta + \frac{wL}{F}x - \frac{w}{2F}x^2$$

$$v_c' = A_1 k \cos kx - A_2 k \sin kx + wL/F - \frac{w}{F}x$$

$v_c' = 0$ at $x = 0 \Rightarrow A_1 = -wL/kF$

$v_c = 0$ at $x = 0 \Rightarrow A_2 = -\frac{w}{F}\left(\frac{1}{k^2} - \frac{L^2}{2}\right) + \delta$

Thus, $v_c = \frac{wL}{F}\left(x - \frac{\sin kx}{k}\right) + \left[\frac{w}{F}\left(\frac{1}{k^2} - \frac{L^2}{2}\right) - \delta\right](1 - \cos kx) - \frac{w}{2F}x^2$ (2) **ANS**

For δ, $v_c = -\delta$ at $x = L$. Sub into (2) and solve for δ.

Result is, $\delta = \frac{w}{F\cos kL}\left[L^2 - \frac{1}{k^2} + \left(\frac{1}{k^2} - \frac{L^2}{2}\right)\cos kL - L\left(L - \frac{\sin kL}{k}\right)\right]$ or,

$$\delta = \frac{w}{Fk^2 \cos kL}\left[kL \sin kL + \left(1 - \frac{k^2L^2}{2}\right)\cos kL - 1\right]$$

(b) $v_c = -\delta\left[1 - \cos\left(\frac{\pi x}{2L}\right)\right]$, $v_c' = -\frac{\pi}{2L}\delta \sin\left(\frac{\pi x}{2L}\right)$, $v_c'' = -\left(\frac{\pi}{2L}\right)^2 \delta \cos\left(\frac{\pi x}{2L}\right)$

$$W_w = \frac{1}{2}\int_0^L w\delta\left(1 - \cos\frac{\pi x}{2L}\right)dx = \frac{\pi - 2}{2\pi} wL\delta$$

$$W_N = \frac{1}{2}\frac{N^2L}{AE} - \frac{1}{2}N\int_0^L \left(\frac{\pi}{2L}\right)^2 \delta^2 \sin^2\frac{\pi x}{2L}dx = \frac{1}{2}\frac{N^2L}{AE} - \frac{\pi^2 N}{16L}\delta^2$$

$$U = \frac{1}{2}\frac{N^2L}{AE} + \frac{1}{2}EI\int_0^L \left(\frac{\pi}{2L}\right)^4 \delta^2 \cos^2\frac{\pi x}{2L}dx = \frac{1}{2}\frac{N^2L}{AE} + \frac{\pi^4}{64}\frac{EI}{L^3}\delta^2$$

$U = W$

$$\frac{1}{2}\frac{N^2L}{AE} + \frac{\pi^4}{64}\frac{EI}{L^3}\delta^2 = \frac{\pi - 2}{2\pi}wL\delta + \frac{1}{2}\frac{N^2L}{AE} - \frac{\pi^2 N}{16L}\delta^2 \quad \text{Let } N = -F$$

$$\delta = \frac{\frac{\pi - 2}{2\pi}wL}{\frac{\pi^4}{64L^3}EI - \frac{\pi^2 F}{16L}} = \frac{32(\pi - 2)}{\pi^3}\frac{wL^4}{(\pi^2 EI - 4FL^2)}$$ **ANS**

(c) Part (a) $k = \sqrt{\frac{750}{10(10^5)}} = 2.7386(10^{-2})$, $kL = 0.5477$

$$\delta = \frac{20\left[0.5477 \sin 0.5477 + \left(1 - \frac{0.5477^2}{2}\right)\cos 0.5477 - 1\right]}{(750)(2.7386)^2(10^{-4})\cos 0.5477} = \underline{0.453 \text{ in}}$$

(continued)

From part (b)

$$\delta = \frac{32}{\pi^3}\frac{(\pi-2)(20)(20^4)}{[\pi^2(10)(10^5)-4(750)(20^2)]} = \underline{0.435 \text{ in}}$$ 4% lower than exact

7.1 Stress uncertainties

- Stress calculation at fixed end
- Stress concentrations due to rapid geometric changes
- Force position and location
- Assembly stresses at fixed end and load support
- Weight of beam
- Flaws in material (macroscopic or microscopic)
- Dynamic application of load

Strength Uncertainties

- Properties of material
- Flaws in material
- Temperature and corrosion effects

7.2 (1) $f_{\sigma_1} = \dfrac{\sigma_{nom}}{\sigma_{max}} = \dfrac{4F/\pi d^2}{4F/\pi(0.998)^2} = 0.996$

(2) $\sigma_{nom} = \dfrac{4F}{\pi d^2}$, $\sigma_{max} = \dfrac{4F}{\pi d^2} + \dfrac{F(0.004d)\frac{d}{2}}{\frac{\pi}{64}d^4} = 1.032\dfrac{4F}{\pi d^2}$

$f_{\sigma_2} = \dfrac{\sigma_{nom}}{\sigma_{max}} = \dfrac{1}{1.032} = 0.969$

(3) $\sigma_{nom} = 4F/\pi d^2$, $\sigma_{max} = \dfrac{4(1.1F)}{\pi d^2}$, $f_{\sigma_3} = \dfrac{1}{1.1} = 0.909$

(4) $f_s = \dfrac{S_{nom}}{S_m} = \dfrac{1}{0.99} = 1.01$

$n = \dfrac{\Pi f_s}{\Pi f_\sigma} = \dfrac{1.01}{(0.996)(0.969)(0.909)} = \underline{1.15}$

7.3 For curved rectangular beam the maximum stress occurs at the inner radius as given by Eq. (5.6-3)

$\sigma_{max} = \dfrac{M(r_n - r_i)}{Aer_i}$, $r_n = \dfrac{A}{\int dA/r} = \dfrac{bh}{b\ln\frac{r_i+h}{r_i}} = \dfrac{h}{\ln(1+\frac{h}{r_i})}$

$e = r_c - r_n = r_c - \dfrac{h}{\ln(1+h/r_i)}$ $\therefore \sigma_{max} =$

$$\therefore \sigma_{max} = \frac{M\left[\frac{h}{\ln(1+h/r_i)} - r_i\right]}{bhr_i\left[r_c - \frac{h}{\ln(1+\frac{h}{r_i})}\right]} = \frac{M\,[h - r_i\ln(1+h/r_i)]}{bhr_i\,[r_c\ln(1+h/r_i) - h]}$$

(continued)

$$f_\sigma = \frac{\sigma_{nom}}{\sigma_{max}} = \frac{6M/bh^2}{\frac{M}{bhr_i}\frac{[h - r_i \ln(1+h/r_i)]}{[r_c \ln(1+h/r_i) - h]}} = \frac{6r_i\left[r_c \ln\left(1+\frac{h}{r_i}\right) - h\right]}{h\left[h - r_c \ln\left(1+h/r_i\right)\right]}$$

ANS

7.4 $\Delta U = \Delta U_{weight} + \Delta U_{elastic} = -W\delta_{max} + \frac{1}{2}k\delta_{max}^2 = 0$

where $k = AE/L$ $\therefore$ $\delta_{max} = 2\frac{WL}{AE}$ Since the deflection is double the nominal, the max stress is $2\sigma_{nom}$

$$f_\sigma = \frac{\sigma_{nom}}{\sigma_{max}} = \frac{\sigma_{nom}}{2\sigma_{nom}} = \underline{0.5}$$

7.5 (a) $\sigma_1 = 10$ kpsi, $\sigma_2 = \sigma_3 = 0$

(b) $\tau_{max} = \frac{10}{2} = \underline{5\text{ kpsi}} = \tau_{Tresca}$

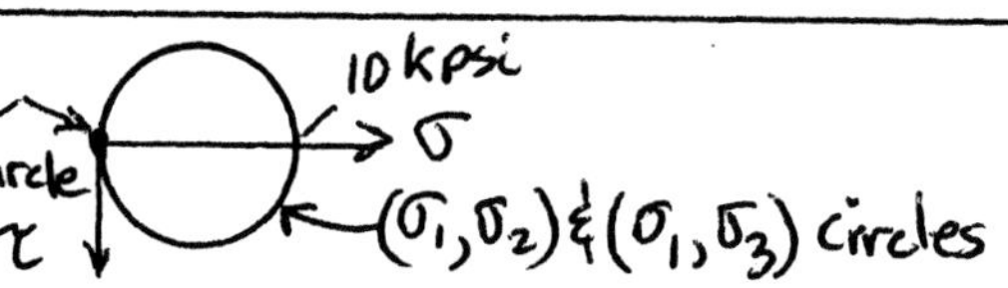

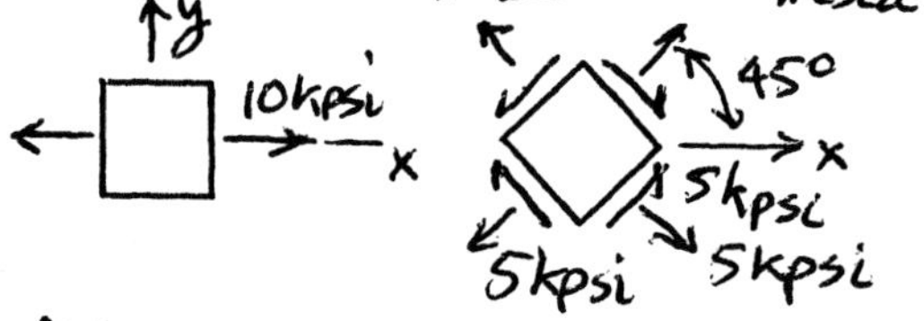

View down $\underline{\text{any}}$ axis ⊥ to x

(c) $\sigma_{VM} = 10\text{ kpsi} = 2\tau_{Tresca}$

7.6 (a) Plane stress, $\sigma_p = \pm\sqrt{\tau_{xy}^2} = \pm 20$ MPa.

$(\sigma_1, \sigma_2, \sigma_3) = (20, 0, -20)$ MPa

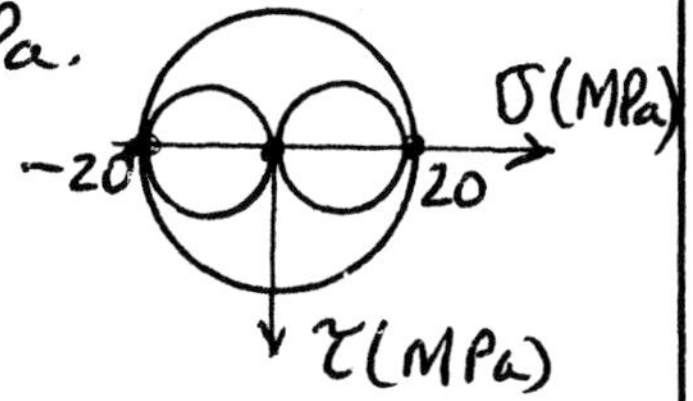

(b) Given state is τ_{max}, $\therefore$ $\tau_{Tresca} = \underline{20\text{ MPa}}$

(c) $\sigma_{VM} = \sqrt{20^2 + (-20)^2 - 20(-20)} = 20\sqrt{3} = \underline{34.6\text{ MPa}}$

$$\frac{\sigma_{VM}}{2\tau_{Tresca}} = \frac{20\sqrt{3}}{40} = \underline{0.866}$$

7.7

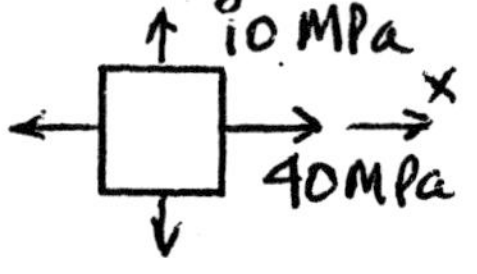

$(\sigma_1, \sigma_2, \sigma_3) = (40, 10, 0)$ MPa

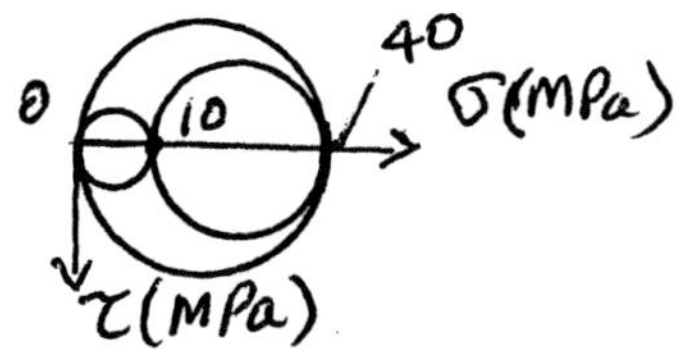

(b) $\tau_{max} = \frac{40}{2} = \underline{20\text{ MPa}} = \tau_{Tresca}$

View down y axis

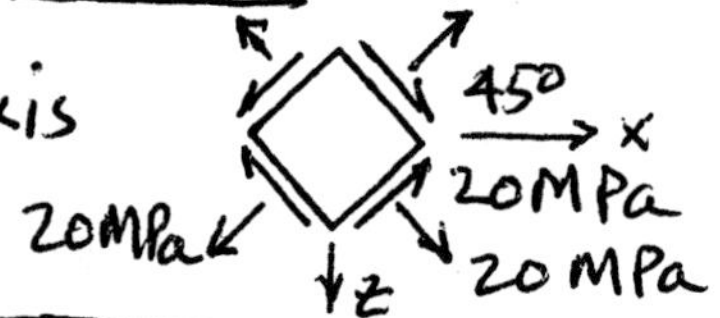

(c) $\sigma_{vm} = \sqrt{40^2 + 10^2 - 40(10)} = \underline{36.1\text{ MPa}}$ $\quad \frac{\sigma_{VM}}{2\tau_{Tresca}} = \frac{36.1}{40} = \underline{0.901}$

7.8 (a)

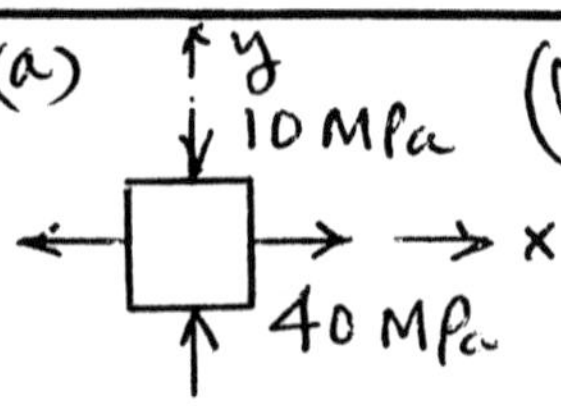

$(\sigma_1, \sigma_2, \sigma_3) = (40, 0, -10)\ \text{MPa}$

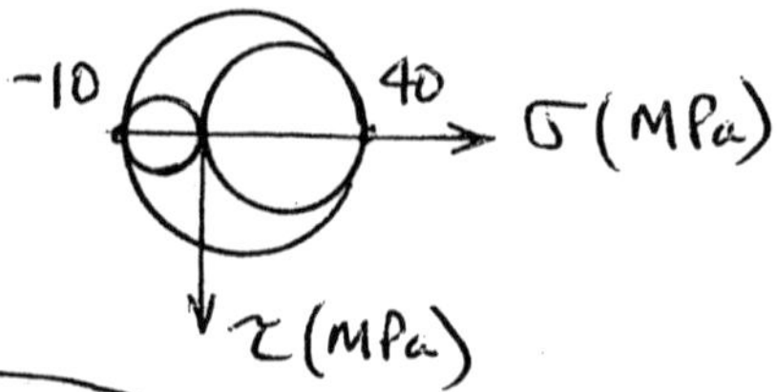

(b) $\tau_{max} = \dfrac{40+10}{2} = \underline{25\ \text{MPa} = \tau_{Tresca}}$

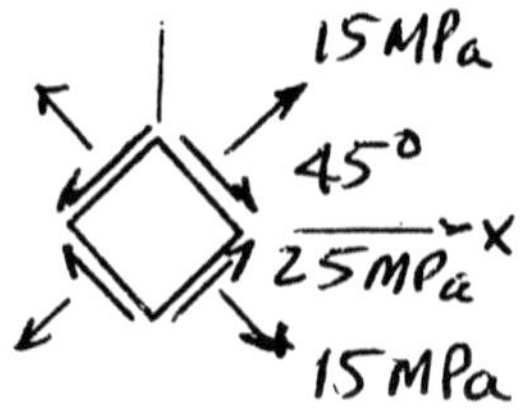

(c) $\sigma_{VM} = \sqrt{40^2 + (-10^2) - 40(-10)} = \underline{45.8\ \text{MPa}}$

$\dfrac{\sigma_{VM}}{2\tau_{Tresca}} = \dfrac{45.8}{50} = \underline{0.917}$

7.9 (a)

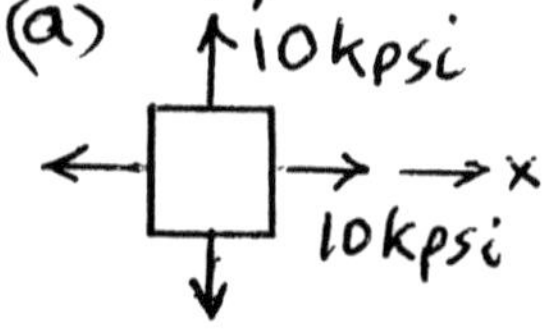

$(\sigma_1, \sigma_2, \sigma_3) = (10, 10, 0)\ \text{kpsi}$

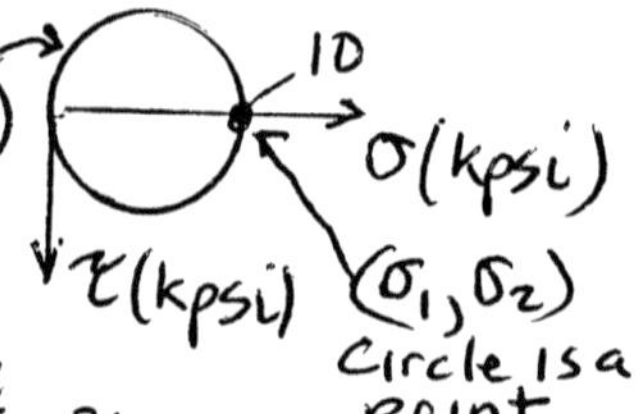

(b) $\tau_{max} = \dfrac{10}{2} = \underline{5\ \text{kpsi} = \tau_{Tresca}}$

View along any axis ⊥ to z axis

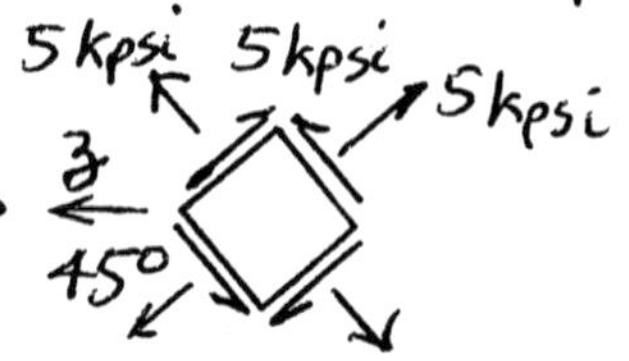

(c) $\sigma_{VM} = \sqrt{10^2 + 10^2 - (10)(10)} = \underline{10\ \text{kpsi}}$

$\dfrac{\sigma_{VM}}{2\tau_{Tresca}} = \underline{1}$

7.10 (a)

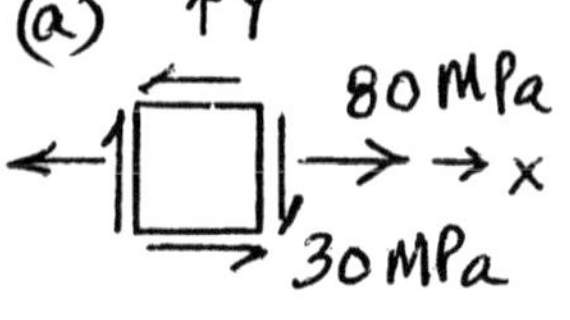

$\sigma_p = \dfrac{80}{2} \pm \sqrt{\left(\dfrac{80}{2}\right)^2 + 30^2} = 90, -10\ \text{MPa}$

$(\sigma_1, \sigma_2, \sigma_3) = (90, 0, -10)\ \text{MPa}$

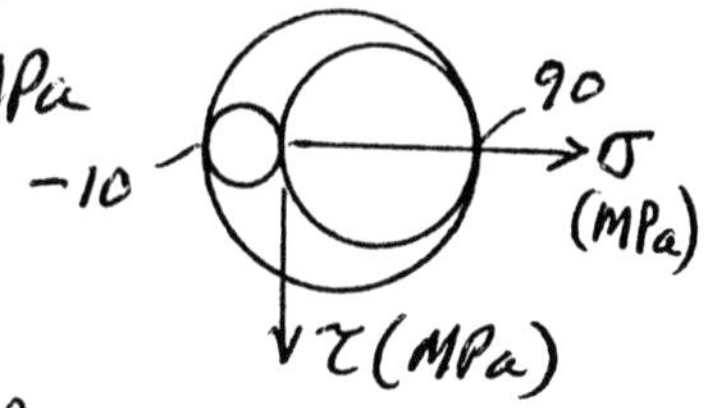

(b) $\tau_{max} = \dfrac{90+10}{2} = \underline{50\ \text{MPa} = \tau_{Tresca}}$

$\theta_s = \dfrac{1}{2}\tan^{-1}\left(\dfrac{-80}{2(-30)}\right) = 26.6^\circ, -63.4^\circ \qquad \sigma_{av} = \dfrac{80}{2} = 40\ \text{MPa}$

At $\theta = 26.6^\circ$

$\tau = -80 \sin 26.6 \cos 26.6 + (-30)(\cos^2 26.6 - \sin^2 26.6) = -50\ \text{MPa}$

(c) $\sigma_{VM} = \sqrt{90^2 + (-10)^2 - 90(-10)} = \underline{95.4\ \text{MPa}}$

$\dfrac{\sigma_{VM}}{2\tau_{Tresca}} = \dfrac{95.4}{100} = \underline{0.954}$

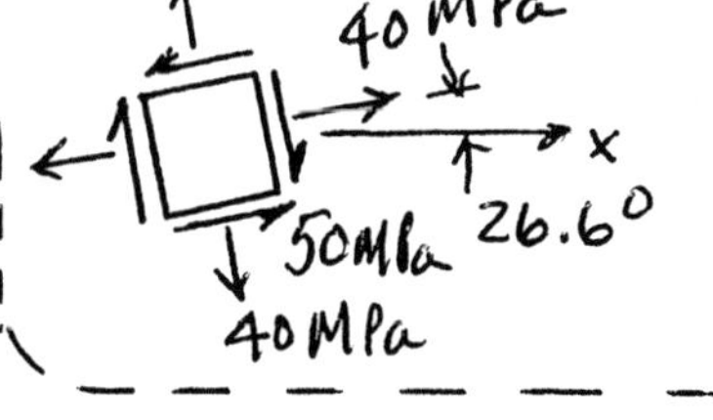

7.11 (a)

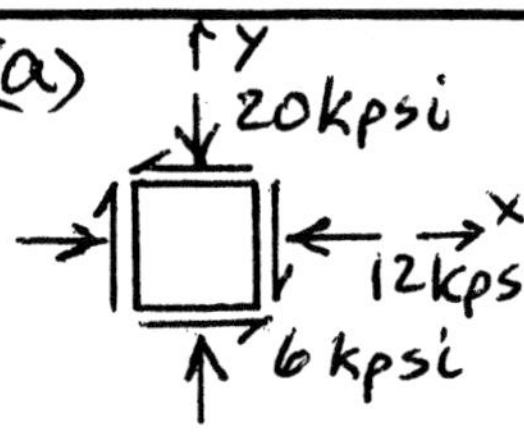

$$\sigma_p = \frac{-12-20}{2} \pm \sqrt{\left(\frac{-12+20}{2}\right)^2 + (-6)^2} = -8.79, -23.21 \text{ kpsi}$$

$$(\sigma_1, \sigma_2, \sigma_3) = (0, -8.79, -23.21) \text{ kpsi}$$

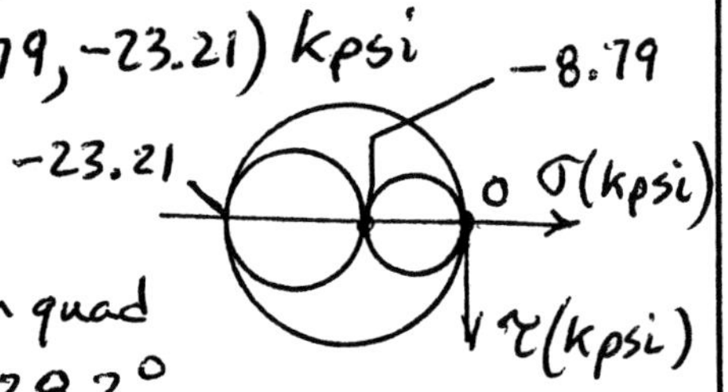

(b) $\tau_{max} = \frac{23.21}{2} = \underline{11.6 \text{ kpsi} = \tau_{Tresca}}$

$$\theta_p = \frac{1}{2}\tan^{-1}\left(\frac{2(-6)}{-12+20}\right) = \frac{1}{2}\tan^{-1}\frac{-12}{8} = -28.2^\circ$$

← 4th quad

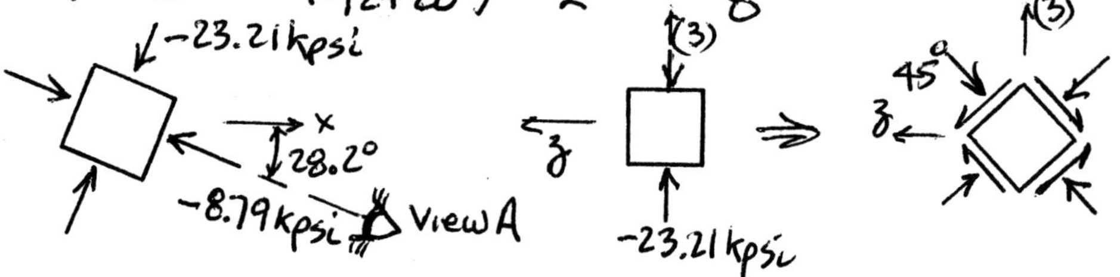

All stresses are 11.6 kpsi

(c) $\sigma_{VM} = \sqrt{(-8.79)^2 + (-23.21)^2 - (-8.79)(-23.21)} = \underline{20.3 \text{ kpsi}}$

$$\frac{\sigma_{VM}}{2\tau_{Tresca}} = \frac{20.3}{23.21} = \underline{0.875}$$

7.12 From the solution of Prob. 3.37,

cylinder: $\tau_{max} = \frac{146}{2} = \underline{73 \text{ MPa}}$ Sphere: $\tau_{max} = \frac{73}{2} = \underline{36.5 \text{ MPa}}$

Cylinder: $\sigma_{VM} = \sqrt{146^2 + 73^2 - 146(73)} = \underline{126.4 \text{ MPa}}$

Spherical ends: $\sigma_{VM} = \sqrt{73^2 + 73^2 - 73(73)} = \underline{73 \text{ MPa}}$

7.13 From the solution of Prob. 5.51

Curved beam, $\sigma_{max} = \sigma_\theta|_{r=2} = 10.87$ kpsi

$\tau_{max} = \frac{10.87}{2} = \underline{5.44 \text{ kpsi}}$, $\underline{\sigma_{VM} = 10.87 \text{ kpsi}}$

7.14

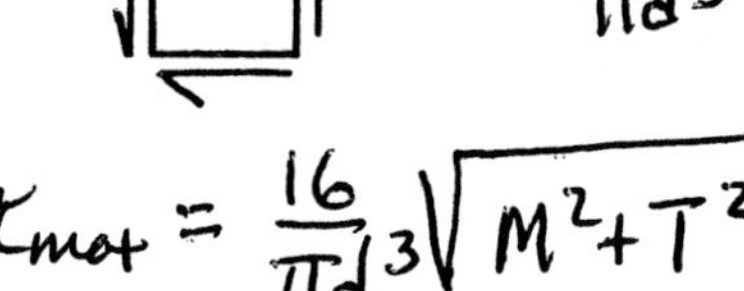

$$\sigma_p = \frac{1}{2}\left(\frac{32M}{\pi d^3}\right) \pm \sqrt{\left[\frac{1}{2}\left(\frac{32M}{\pi d^3}\right)\right]^2 + \left(\frac{16T}{\pi d^3}\right)^2}$$

$$= \frac{16}{\pi d^3}\left(M \pm \sqrt{M^2 + T^2}\right)$$

(a) $\tau_{max} = \frac{16}{\pi d^3}\sqrt{M^2 + T^2}$ **ANS**

(b) $\sigma_{VM} = \frac{16}{\pi d^3}\sqrt{\left(M + \sqrt{M^2+T^2}\right)^2 + \left(M - \sqrt{M^2+T^2}\right)^2 - \left(M + \sqrt{M^2+T^2}\right)\left(M - \sqrt{M^2+T^2}\right)}$

$= \frac{16}{\pi d^3}\sqrt{4M^2 + 3T^2}$ **ANS** $\frac{\sigma_{VM}}{2\tau_{Tresca}} = \sqrt{\frac{M^2 + 0.75T^2}{M^2 + T^2}}$ **ANS**

7.15 From the solution of Prob. 3.32 $M_{max} = \sqrt{160^2 + 440^2} = 468$ lb-in
Due to bending and axial loading the maximum normal stress is

$$\sigma = \frac{Mc}{I} + \frac{P}{A} = \frac{(468)(0.5)}{0.02022} + \frac{32}{0.1841} = 11.75(10^3)\text{ psi}$$

From torsion

$$\tau = \frac{Tc}{J} = \frac{110(0.5)}{0.04045} = 1360\text{ psi}$$

$$\sigma_p = \frac{11.75}{2} \pm \sqrt{\left(\frac{11.75}{2}\right)^2 + 1.36^2} = 11.90, -0.155\text{ kpsi}$$

$$\sigma_{VM} = \sqrt{11.9^2 + (-0.155)^2 - (11.9)(-0.155)} = \underline{11.98\text{ kpsi}}$$

7.16 From the solution of Prob. 3.33
$\sigma = 10.5$ kpsi, $\tau = 1.02$ kpsi

$$\sigma_p = \frac{10.5}{2} \pm \sqrt{\left(\frac{10.5}{2}\right)^2 + 1.02^2} = 10.6, -0.1\text{ kpsi}$$

$$\sigma_{VM} = \sqrt{10.6^2 + (-0.1)^2 - 10.6(-0.1)} = \underline{10.7\text{ kpsi}}$$

7.17 From the solution of Prob. 3.34
$\sigma = 37.6$ MPa, $\tau = 1.89$ MPa

$$\sigma_p = \frac{37.6}{2} \pm \sqrt{\left(\frac{37.6}{2}\right)^2 + 1.89^2} = 37.7, -0.1\text{ MPa}$$

$$\sigma_{VM} = \sqrt{37.7^2 + (-0.1)^2 - 37.7(-0.1)} = \underline{37.7\text{ MPa}}$$

7.18 From the solution of Prob. 3.35
$\sigma_x = 240.6$ MPa, $\sigma_y = 28.5$ MPa, $\tau_{xy} = 34.8$ MPa

$$\sigma_p = \frac{240.6 + 28.5}{2} \pm \sqrt{\left(\frac{240.6 - 28.5}{2}\right)^2 + 34.8^2} = 246.2, 22.9\text{ MPa}$$

$$\sigma_{VM} = \sqrt{246.2^2 + 22.9^2 - 246.2(22.9)} = \underline{236\text{ MPa}}$$

7.19 Max σ_θ and min σ_r occur at the inner radius.
From the solution of Prob 5.71, $\sigma_r = -0.5$ MPa and

$$\sigma_\theta = 888.9 + \frac{55.56}{(0.2)^2} = 2278\text{ kPa} = 2.278\text{ MPa}$$

$$\tau_{Tresca} = \frac{2.278 + 0.5}{2} = \underline{1.39\text{ MPa}}$$

$$\sigma_{VM} = \sqrt{2.278^2 + (-0.5)^2 - 2.278(-0.5)} = \underline{2.56\text{ MPa}}$$

7.20 From the solution of Prob 5.72, a plot of radial and tangential stresses would reveal that the worst case in each cylinder occurs at their inner radii. In this case, the worst case occurs in the outer cylinder. At $r = 2$

$$\sigma_r = -1990 \text{ psi}, \quad \sigma_\theta = 3537.2 + \frac{22108}{(2)^2} = 9064 \text{ psi}$$

$$\tau_{Tresca} = \frac{9.064 + 1.99}{2} = \underline{5.53 \text{ kpsi}}$$

$$\sigma_{VM} = \sqrt{9.064^2 + (-1.99)^2 - 9.064(-1.99)} = \underline{10.2 \text{ kpsi}}$$

7.21 Bending: $\sigma = \dfrac{32M}{\pi d^3} = \dfrac{32(150)(10^3)}{\pi(20)^3} = 191 \text{ MPa}$

Torsion: $\tau = \dfrac{16T}{\pi d^3} = \dfrac{16(100)(10^3)}{\pi(20)^3} = 63.7 \text{ MPa}$

Axial: $\sigma = \dfrac{4P}{\pi d^2} = \dfrac{4P}{\pi(20)^2} = 3.183(10^{-3})P \text{ MPa}$ (P in N)

(a) $\sigma_{Total} = 191 + 3.183(10^{-3})P$

$$\tau_{max} = \sqrt{\left(\frac{\sigma}{2}\right)^2 + \tau^2} = \sqrt{\left[\frac{191 + 3.183(10^{-3})P}{2}\right]^2 + 63.7^2} = S_{ys} = \frac{350}{2}$$

square both sides and simplify

$$P^2 + 1.20(10^5)P - 6.888(10^9) = 0 \quad \text{Solution is}$$

$42.41(10^3), -162.4(10^3)$ N   Negative root erroneous

Thus, $\underline{P = 42.4 \text{ kN}}$

(b) $\sigma_p = \dfrac{\sigma}{2} \pm \tau_{max}$, $\sigma_{VM} = \sqrt{\left(\frac{\sigma}{2} + \tau_{max}\right)^2 + \left(\frac{\sigma}{2} - \tau_{max}\right)^2 - \left(\frac{\sigma}{2} + \tau_{max}\right)\left(\frac{\sigma}{2} - \tau_{max}\right)}$

$$\sigma_{VM} = \sqrt{\left(\frac{\sigma}{2}\right)^2 + 3\tau_{max}^2} = S_y \quad \text{Again, square both sides}$$

$$\left[\frac{191 + 3.183(10^{-3})P}{2}\right]^2 + 3\left\{\left[\frac{191 + 3.183(10^{-3})P}{2}\right]^2 + 63.7^2\right\} = 350^2$$

$$\left[191 + 3.183(10^{-3})P\right]^2 + 3(63.7)^2 = 350^2 \quad \text{simplification yields}$$

$$P^2 + 1.20(10^5)P - 7.289(10^9) = 0$$

The positive root is $\underline{P = 44.35 \text{ kN}}$

7.22 $\sigma_r = -p$, $\sigma_\theta = \frac{pr}{t} = \frac{p}{0.09375}\frac{(3-0.09375)}{2} = 15.5p$, $\sigma_z = \frac{pr}{2t} = 7.75p$

Using von Mises theory

$$53 = p\sqrt{\frac{1}{2}\left[(15.5-7.75)^2 + (7.75+1)^2 + (-1-15.5)^2\right]}$$

$p = \underline{3.71 \text{ kpsi}}$

7.23 Reviewing Eqs. (5.8-29) and (5.8-30) σ_r and σ_θ are always positive with $\sigma_\theta > \sigma_r$ for a given r. Since σ_θ is maximum at inner surface, $r = r_i$, where $\sigma_r = 0$ $(\sigma_{vm})_{max} = (\sigma_\theta)_{r_i}$ and $\tau_{max} = 0.5(\sigma_\theta)_{r_i}$

7.24 $(\sigma_1, \sigma_2, \sigma_3) = (140, 140, 0)$ MPa

(a) $\tau_{max} = \frac{140}{2} = 70$ MPa, $n = \frac{280}{2(70)} = \underline{2}$

(b) $\sigma_{vm} = \sqrt{140^2 + 140^2 - 140(140)} = 140$ MPa, $n = \frac{280}{140} = \underline{2}$

7.25 $\sigma_p = \frac{105+105}{2} \pm \sqrt{\left(\frac{105-105}{2}\right)^2 + 105^2} = 210, 0$ MPa

$(\sigma_1, \sigma_2, \sigma_3) = (210, 0, 0)$ MPa

(a) $\tau_{max} = \frac{210}{2} = 105$ MPa, $n = \frac{280}{2(105)} = \underline{1.33}$

(b) $\sigma_{vm} = 210$ MPa, $n = \frac{280}{210} = \underline{1.33}$

7.26 $\sigma_p = \pm\sqrt{140^2} = 140, -140$ MPa

$(\sigma_1, \sigma_2, \sigma_3) = (140, 0, -140)$ MPa

(a) $\tau_{max} = 140$ MPa, $n = \frac{280}{2(140)} = \underline{1}$

(b) $\sigma_{VM} = \sqrt{140^2 + (-140)^2 - 140(-140)} = 140\sqrt{3} = 242.5$ MPa

$n = \frac{280}{242.5} = \underline{1.15}$

7.27 $\sigma_p = \frac{-14+70}{2} \pm \sqrt{\left(\frac{-14-70}{2}\right)^2 + (-56)^2} = 98, -42 \text{ MPa}$

(a) $\tau_{max} = \frac{98+42}{2} = 70 \text{ MPa}, \quad n = \frac{280}{2(70)} = \underline{2}$

(b) $\sigma_{VM} = \sqrt{98^2 + (-42)^2 - 98(-42)} = 124.4 \text{ MPa}$

$n = \frac{280}{124.4} = \underline{2.25}$

7.28 $\sigma_p = \frac{12}{2} \pm \sqrt{\left(\frac{12}{2}\right)^2 + (-8)^2} = 16, -4 \text{ kpsi}$

(a) $n = \frac{20}{16} = \underline{1.25}$

(b) Case 2, Eq. (7.3-20b)

$\frac{16}{20} - \frac{-4}{80} = \frac{1}{n} \Rightarrow \underline{n = 1.18}$

7.29 $\sigma_p = \frac{-39-31}{2} \pm \sqrt{\left(\frac{-39+31}{2}\right)^2 + 3^2} = -30, -40 \text{ kpsi}$

(a) $n = \frac{80}{40} = \underline{2}$

(b) Case 3, Eq (7.3-20c)

$40 = \frac{80}{n} \Rightarrow \underline{n = 2}$

7.30 (a) $n = \frac{20}{18} = \underline{1.11}$ (b) Case 1, Eq (7.3-20a), $18 = \frac{20}{n} \Rightarrow \underline{n = 1.11}$

7.31 $\sigma_p = \frac{12-40}{2} \pm \sqrt{\left(\frac{12+40}{2}\right)^2 + (-15)^2} = 16.0, -44.0 \text{ kpsi}$

(a) $n_T = \frac{20}{16} = 1.25, \quad n_C = \frac{80}{44} = 1.82 \quad \therefore \quad n = \underline{1.25}$

(b) Case 2, Eq (7.3-20b)

$\frac{16}{20} - \frac{(-44)}{80} = \frac{1}{n} \Rightarrow n = \underline{0.74}$ Predicting failure

7.32 The stress state of a tensile specimen at yield is $(\sigma_1, \sigma_2, \sigma_3) = (S_Y, 0, 0)$ Thus, $\tau_{oct} = \frac{1}{3}\sqrt{S_Y^2 + S_Y^2} = \frac{\sqrt{2}}{3} S_Y$

Thus, yield condition for a general state of stress is,

$$\frac{1}{3}\sqrt{(\sigma_1-\sigma_2)^2+(\sigma_2-\sigma_3)^2+(\sigma_3-\sigma_1)^2} = \frac{\sqrt{2}}{3} S_Y \quad \text{or,}$$

$$\sqrt{0.5\left[(\sigma_1-\sigma_2)^2+(\sigma_2-\sigma_3)^2+(\sigma_3-\sigma_1)^2\right]} = S_Y \quad \underline{\text{ANS}}$$

Thus, max-oct-shear theory is identical to the von Mises theory [see Eq. (7.3-11)]

7.33 From the solution to Prob. 2.26 $(\sigma_1, \sigma_2, \sigma_3) = (40, 10, 10)$ MPa

$$\tau_{Tresca} = \frac{40-10}{2} = \underline{15 \text{ MPa}}$$

$$\sigma_{VM} = \sqrt{0.5\left[(40-10)^2+(10-10)^2+(10-40)^2\right]} = \underline{30 \text{ MPa}}$$

$$\frac{\sigma_{VM}}{2\tau_{Tresca}} = \underline{1}$$

7.34 From the solution of Prob. 2.21, $(\sigma_1, \sigma_2, \sigma_3) = (7.012, 1.89, -8.90)$ kpsi

$$\tau_{Tresca} = \frac{7.012+8.90}{2} = \underline{7.96 \text{ kpsi}}$$

$$\sigma_{VM} = \sqrt{0.5\left[(7.012-1.89)^2+(1.89+8.90)^2+(-8.90-7.012)^2\right]} = \underline{14.07 \text{ kpsi}}$$

$$\frac{\sigma_{VM}}{2\tau_{Tresca}} = \frac{14.07}{2(7.96)} = \underline{0.884}$$

7.35 From the solution of Prob. 2.23, $(\sigma_1, \sigma_2, \sigma_3) = (30, 10, -20)$ MPa

$$\tau_{Tresca} = \frac{30+20}{2} = \underline{25 \text{ MPa}}$$

$$\sigma_{VM} = \sqrt{0.5\left[(30-10)^2+(10+20)^2+(-20-30)^2\right]} = \underline{43.6 \text{ MPa}}$$

$$\sigma_{VM}/2\tau_{Tresca} = 43.6/50 = \underline{0.872}$$

7.36 From the solution of Prob. 2.25, $(\sigma_1, \sigma_2, \sigma_3) = (9, 0, 0)$ kpsi

$$\tau_{Tresca} = \frac{9}{2} = \underline{4.5 \text{ kpsi}}$$

$$\sigma_{VM} = \underline{9 \text{ kpsi}}$$

$$\frac{\sigma_{VM}}{2\tau_{Tresca}} = \underline{1}$$

7.37 From the solution of Prob. 7.21

$$\sigma_p = \frac{191+0.003183P}{2} \pm \sqrt{\left(\frac{191+0.003183P}{2}\right)^2 + 63.7^2}$$

Coulomb-Mohr theory, Case 2, $\frac{\sigma_{p1}}{S_{uT}} - \frac{\sigma_{p2}}{S_{uc}} = 1$ results in

$$\left(\frac{1}{S_{uT}} - \frac{1}{S_{uc}}\right)(95.5+0.001591P) + \left(\frac{1}{S_{uT}} + \frac{1}{S_{uc}}\right)\sqrt{(95.5+0.001591P)^2 + 4053} = 1$$

$S_{uT} = 340$ MPa, $S_{uc} = 600$ MPa

$$1.2745(10^{-3})(95.5+0.001591P) + 4.6078(10^{-3})\sqrt{(95.5+0.001591P)^2 + 4053} = 1$$

Divide by $4.6078(10^{-3})$, place 1st product on RHS of eq. & square. Then simplify. The resulting equation is

$$P^2 + 2.018(10^5)P - 9.916(10^9) = 0$$ yields one positive root of $P = 40.86(10^3)\text{N} = \underline{40.9 \text{ kN}}$

7.38

$$\sigma_p = \frac{1}{2}\left(2\frac{K_I}{\sqrt{2\pi r}}\cos\frac{\theta}{2}\right) \pm \left[\left(\frac{K_I}{\sqrt{2\pi r}}\sin\frac{\theta}{2}\cos\frac{\theta}{2}\sin\frac{3\theta}{2}\right)^2 + \left(\frac{K_I}{\sqrt{2\pi r}}\sin\frac{\theta}{2}\cos\frac{\theta}{2}\cos\frac{3\theta}{2}\right)^2\right]^{1/2}$$

$$= \frac{K_I}{\sqrt{2\pi r}}\left[\cos\frac{\theta}{2} \pm \left(\sin^2\frac{\theta}{2}\cos^2\frac{\theta}{2}\sin^2\frac{3\theta}{2} + \sin^2\frac{\theta}{2}\cos^2\frac{\theta}{2}\cos^2\frac{3\theta}{2}\right)^{1/2}\right]$$

$$= \frac{K_I}{\sqrt{2\pi r}}\left(\cos\frac{\theta}{2} \pm \cos\frac{\theta}{2}\sin\frac{\theta}{2}\right) = \frac{K_I}{\sqrt{2\pi r}}\cos\frac{\theta}{2}\left(1 \pm \sin\frac{\theta}{2}\right)$$

Plane stress: Third principal stress is zero and

$$\sigma_1 = \frac{K_I}{\sqrt{2\pi r}}\cos\frac{\theta}{2}\left(1+\sin\frac{\theta}{2}\right),\quad \sigma_2 = \frac{K_I}{\sqrt{2\pi r}}\cos\frac{\theta}{2}\left(1-\sin\frac{\theta}{2}\right),\quad \sigma_3 = 0$$

which are eqs (7.4-15) **ANS**

Plane strain: σ_1 and σ_2 equations still valid however,

$$\sigma_z = \sigma_3 = \nu(\sigma_x + \sigma_y) = 2\nu\frac{K_I}{\sqrt{2\pi r}}\cos\frac{\theta}{2}$$ which is Eq (7.4-15c) for plane strain **ANS**

7.39 For plane stress $\sqrt{\sigma_1^2 + \sigma_2^2 - \sigma_1\sigma_2} = S_y$

$$S_Y = \frac{K_I}{\sqrt{2\pi r}}\sqrt{\cos^2\frac{\theta}{2}\left(2+2\sin^2\frac{\theta}{2} - 1 + \sin^2\frac{\theta}{2}\right)} = \frac{K_I}{\sqrt{2\pi r}}\sqrt{\cos^2\frac{\theta}{2}\left(1+3\sin^2\frac{\theta}{2}\right)}$$

But $\cos^2\frac{\theta}{2} = \frac{1+\cos\theta}{2}$, $\sin^2\frac{\theta}{2} = \frac{1-\cos\theta}{2}$. Thus,

$$S_Y = \frac{K_I}{\sqrt{2\pi r}}\sqrt{\left(\frac{1+\cos\theta}{2}\right)\left(1+\frac{3}{2}-\frac{3}{2}\cos\theta\right)}$$ multiply & simplify (continued)

$$S_Y = \frac{K_I}{\sqrt{2\pi r}}\sqrt{0.5\left(\frac{5}{2}+\frac{5}{2}\cos\theta-\frac{3}{2}\cos\theta-\frac{3}{2}\cos^2\theta\right)} \quad \text{Let } \cos^2\theta = 1-\sin^2\theta$$

$$= \frac{K_I}{\sqrt{2\pi r}}\sqrt{0.5\left(1+\frac{3}{2}\sin^2\theta+\cos\theta\right)} \quad \text{Solve for } r = r_p^*$$

$$r_p^* = \frac{1}{4\pi}\left(\frac{K_I}{S_Y}\right)^2\left(1+\frac{3}{2}\sin^2\theta+\cos\theta\right) \quad \boxed{\text{ANS}} \quad \text{Eq 7.4-16}$$

(7.40) At $\theta = 0$ and plane strain $\sigma_1 = \sigma_2 = \frac{K_I}{\sqrt{2\pi r}}$, $\sigma_3 = 2\nu\frac{K_I}{\sqrt{2\pi r}}$

Substitute σ_1, $\sigma_2 = \sigma_1$, and $\sigma_3 = 2\nu\sigma_1$ into Eq. (7.3-11)

$$\sqrt{0.5\left[(\sigma_1-\sigma_1)^2+(\sigma_1-2\nu\sigma_1)^2+(2\nu\sigma_1-\sigma_1)^2\right]} = S_Y, \text{ or,}$$

$$\sigma_1 - 2\nu\sigma_1 = (1-2\nu)\sigma_1 = S_Y. \quad \text{With } \nu = 1/3 \Rightarrow \sigma_1 = 3S_Y \quad \boxed{\text{ANS}}$$

(7.41) Based on yield, $P = S_Y A = 160(4-0.625)(0.75) = 405$ kips

$r_p^* = \frac{1}{6\pi}\left(\frac{70}{160}\right)^2 = 0.0319$ in, $a_{eff} = 0.625+0.032 = 0.657$ in

Table 7.4-1, case 3, $\eta\frac{\pi}{2} = \frac{0.657}{4}\frac{\pi}{2} = 0.258$

$$C_3 = \left[\frac{0.752+2.02\left(\frac{0.657}{4}\right)+0.37(1-\sin 0.258)^3}{\cos 0.258}\right]\sqrt{\frac{\tan 0.258}{0.258}} = 1.294$$

$K_I = 1.294\sigma\sqrt{\pi(0.657)} = 1.858\sigma = 70 \quad \therefore \sigma = 37.7$ kpsi

$P = 37.7(4)(0.75) = \underline{113 \text{ kips}}$

Checking Eqs. (7.4-12) & (7.4-14), $2.5\left(\frac{70}{160}\right)^2 = 0.479$ in

Since t and a are greater solution is acceptable

(7.42) $r = \frac{350-15}{2} = 162.5$ mm, $t = 25$ mm, $a = 12.5$ mm

$\sigma_\theta = \frac{pr}{t} = \frac{162.5}{25}p = 6.5p$, $r_p^* = \frac{1}{6\pi}\left(\frac{80}{1200}\right)^2 = 2.36(10^{-4})$ m

$a_{eff} = 12.5+0.24 = 12.74$ mm

No correction can be made as width is not given.

$K_I = \sigma\sqrt{\pi(12.74)(10^{-3})} = 0.200\sigma = 80 \Rightarrow \sigma = 400$ MPa

$6.5p = 400 \Rightarrow p = \underline{61.5 \text{ MPa}}$

7.43 $\eta = \frac{15}{100} = 0.15$, $\frac{\pi}{2}\eta = 0.2356$. Use Case 4 of Table 7.4-1

$$C_4 = \left[\frac{0.923 + 0.199(1-\sin 0.2356)^4}{\cos 0.2356}\right]\sqrt{\frac{\tan 0.2356}{0.2356}} = 1.030$$

$$K_I = 1.03\sigma\sqrt{\pi(15)(10^{-3})} = 0.223\sigma = 60 \Rightarrow \sigma = 269 \text{ MPa}$$

$$\sigma = \frac{6M}{bh^2} = \frac{6M}{(0.025)(0.1)^2} = 269(10^6) \Rightarrow M = 11.2(10^3)\ \text{N}\cdot\text{m}$$

$M = 11.2$ kN-m

If moment is reversed crack will not propagate.

$$M = \frac{S_u bh^2}{6} = \frac{1}{6}(1200)(10^6)(0.025)(0.100-0.015) = 36.1(10^3)\ \text{N-m}$$

$M = 36.1$ kN-m

7.44 To establish the minimum radius of curvature consider the ellipse shown where $a > b$. Min. radius of curvature occurs at $x=0$, $y=\pm a$

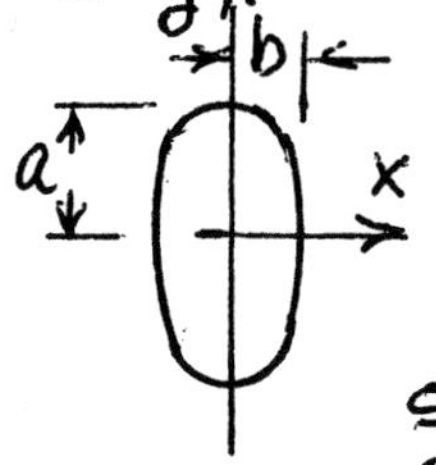

Eq. for an ellipse,

$$\frac{x^2}{a^2} + \frac{y^2}{b^2} = 1 \Rightarrow y = \frac{a}{b}\sqrt{b^2-x^2},\ \frac{dy}{dx} = -\frac{a}{b}\frac{x}{\sqrt{b^2-x^2}}$$

$$\frac{d^2y}{dx^2} = -\frac{a}{b}\left(\frac{x^2}{\sqrt{(b^2-x^2)^3}} + \frac{1}{\sqrt{b^2-x^2}}\right)$$

At $x=0$, $\frac{d^2y}{dx^2} = -\frac{a}{b^2}$, $\frac{dy}{dx} = 0$. The radius of curvature is given by

$$r = |\rho| = \left|\frac{[1+(dy/dx)^2]^{3/2}}{d^2y/dx^2}\right| = \left|\frac{1}{-a/b^2}\right| = \frac{b^2}{a}$$

or $b = \sqrt{ar}$ ①

From Eq. (7.4-1) $K_t = 1 + 2\frac{a}{b} = 1 + 2\frac{a}{\sqrt{ar}} = 1 + 2\sqrt{\frac{a}{r}}$

Before stop hole $r \approx 0$: $K_t \to \infty$ **ANS**

After stop hole $r = d/2$: $K_t = 1 + 2\sqrt{\frac{2a}{d}}$

For $a = 25$ mm and $K_t = 5$

$$d = \frac{8a}{(K_t-1)^2} = \frac{8(25)}{(5-1)^2} = 12.5\text{mm}$$ **ANS**

7.45 $\sigma = \dfrac{250}{20(0.75)} = 16.67 \text{ kpsi}$

Case 1, Table 7.4-1, $\eta = \frac{1}{20} = 0.05$

$$C_1 = \left[1 - 0.1(0.05)^2 + 0.96(0.05)^4\right]\sqrt{\sec(0.05\pi)} = 1.012$$

$$K_I = 1.012(16.67)\sqrt{\pi(1)} = \underline{29.9 \text{ ksi}\cdot\sqrt{\text{in}}} \quad \boxed{\text{ANS}}$$

For yield,

$$\sigma = \frac{250}{(20-2)(0.75)} = 18.5 \text{ kpsi} < 75 \text{ kpsi} \quad \underline{\text{no yielding}} \quad \boxed{\text{ANS}}$$

For plane strain $t > 2.5\left(\dfrac{29.9}{75}\right)^2 = 0.397 \text{ in}$

Since $t = 0.75$, plane strain exists. $\boxed{\text{ANS}}$

$$r_p^* = \frac{1}{6\pi}\left(\frac{29.9}{75}\right)^2 = 0.00843 \text{ in} \quad \boxed{\text{ANS}}$$

7.46 $\sigma_{nom} = \dfrac{F}{(40)(6)} = 4.167(10^{-3})F \text{ MPa}$

From Appendix F, $K_t = 2.35$ at hole, $K_t = 2.1$ at fillet ∴ hole governs

$S_u = \dfrac{560(10^6)}{6.895(10^3)} = 81.22(10^3)$ psi. From fig. 7.5-7 the surface factor is 0.77. Thus,

$$S_E \simeq (0.77)\left(\frac{1}{2.35}\right)\left(\frac{560}{2}\right) = 91.7 \text{ MPa}$$

$$S_F = \frac{[0.9(560)]^2}{91.7}(10^5)^{-\frac{1}{3}\log_{10}\left[\frac{0.9(560)}{91.7}\right]} = 162 \text{ MPa}$$

$$4.167(10^{-3})F = 162 \Rightarrow F = 38.8(10^3)\text{N} = \underline{38.8 \text{ kN}}$$

7.47 From the solution of Prob. 7.46 $S_E \simeq 91.7$ MPa

$$\sigma_{nom} = \frac{50(10^3)}{(60-20)(6)} = 208 \text{ MPa}$$

$$N \simeq \left\{\frac{(208)(91.7)}{[0.9(560)]^2}\right\}^{-3/\{\log_{10}[0.9(560)/91.7]\}}$$

$$\simeq \underline{36(10^3) \text{ cycles}}$$

(7.48) From the solution of Prob. 7.46 $S_f = 162$ MPa

$$\sigma_a = \sigma_m = \frac{1}{2}\frac{F_{max}}{40(6)} = 2.083(10^{-3})F_{max}$$

Eq(7.5-16)
$$(\sigma_a)_e = \frac{2.083(10^{-3})F_{max}}{1-\left[\frac{2.083(10^{-3})F_{max}}{560}\right]} = 162 \quad \text{Solve for } F_{max}$$

$F_{max} = 60.3(10^3)\text{N} = \underline{60.3 \text{ kN}}$

(7.49) From Prob. 3.33, $\sigma = 10.5$ kpsi, $\tau = 1.02$ kpsi
We will apply the stress concentrations to the stress calculations. The surface factor for $S_u = 80$ kpsi is 0.755

$S_E \approx 0.755\left(\frac{80}{2}\right) = 30.2$ kpsi.

$\sigma = K_t\sigma_{nom} = 1.8(10.5) = 18.9$ kpsi
$\tau = K_{ts}\,{}_{nom} = 1.8(1.02) = 1.83$ kpsi

(a) von Mises theory. $\sigma_p = \frac{18.9}{2} \pm \sqrt{\left(\frac{18.9}{2}\right)^2 + 1.83^2} = 19.1, -0.2$ kpsi

$\sigma_{VM} = \sqrt{19.1^2 + (-0.2)^2 - 19.1(-0.2)} = 19.2$ kpsi

$n = \frac{60}{19.2} = 3.1$ ANS

(b) $\sigma_a = 18.9$ kpsi, $\tau_m = 1.83$ kpsi (convert to equiv. von Mises)

$\sigma_{1m} = 1.83$ kpsi, $\sigma_{2m} = -1.83$ kpsi, $(\sigma_m)_{vm} = \sqrt{3}(1.83) = 3.17$ kpsi

$\frac{\sigma_a}{\sigma_m} = \frac{18.9}{3.17} = 5.96$, $\frac{S_E(S_u - S_Y)}{S_u(S_Y - S_E)} = \frac{30.2(80-60)}{80(60-30.2)} = 0.25$

Since $\frac{\sigma_a}{\sigma_m} > \frac{S_E(S_u - S_Y)}{S_u(S_Y - S_E)}$ use $\frac{\sigma_m}{S_u} + \frac{\sigma_a}{S_E} = \frac{1}{n}$

$\frac{3.17}{80} + \frac{18.9}{30.2} = \frac{1}{n} \Rightarrow \underline{n = 1.5}$ ANS

(7.50) $\sigma_m = \frac{200+100}{2} = 150$ MPa, $\sigma_a = \frac{200-100}{2} = 50$ MPa

$(\sigma_a)_e = \frac{50}{1 - \frac{150}{560}} = 68.3$ MPa

$$S_f = \frac{[0.9(560)]^2}{210}\left[5(10^5)\right]^{-\frac{1}{3}\log_{10}\left[\frac{0.9(560)}{210}\right]} = 229 \text{ MPa} \Rightarrow n = \frac{229}{68.3} = \underline{3.35}$$

(7.51) S-N diagram is linear on a log-log scale

$\log S_F = m \log N + b$ Substitute points $(10^3, 0.9S_u)$ & $(10^6, S_E)$

$\log(0.9S_u) = m \log 10^3 + b = 3m + b$ ①

$\log S_E = m \log 10^6 + b = 6m + b$ ②

②-① yields $3m = \log S_E - \log(0.9S_u) = \log \frac{S_E}{0.9S_u}$

$\therefore m = \frac{1}{3} \log \frac{S_E}{0.9S_u}$

2①-② yields $b = 2\log(0.9S_u) - \log S_E = \log\left[\frac{(0.9S_u)^2}{S_E}\right]$

Thus

$$\log S_F = -\frac{1}{3}\log\left(\frac{0.9S_u}{S_E}\right)\log N + \log\left[\frac{(0.9S_u)^2}{S_E}\right]$$

$$= \log\left\{\left[\frac{(0.9S_u)^2}{S_E}\right] N^{-\frac{1}{3}\log\left(\frac{0.9S_u}{S_E}\right)}\right\}$$

$$\therefore \quad S_F = \frac{(0.9S_u)^2}{S_E} N^{-\frac{1}{3}\log\left(\frac{0.9S_u}{S_E}\right)} \quad \text{Eq. (7.5-2)} \quad \boxed{\text{ANS}}$$

Rearranging $N^{-\frac{1}{3}\log\left(\frac{0.9S_u}{S_E}\right)} = \frac{S_F S_E}{(0.9S_u)^2}$ or,

$$N = \left[\frac{S_F S_E}{(0.9S_u)^2}\right]^{-3/\log(0.9S_u/S_E)} \quad \text{Eq. (7.5-3)} \quad \boxed{\text{ANS}}$$

(7.52) $S_E \approx \frac{560}{2} = 280$ MPa (nothing stated about surface)

$D/d = \frac{60}{40} = 1.5$, $\frac{r}{d} = \frac{4}{40} = 0.1$ From Appendix F,

$K_t \approx 1.95$, $K_{ts} \approx 1.5$

$S_u = \frac{560}{6.895} = 81.2$ kpsi. From table 7.51, $\sqrt{\rho'} \approx 0.080$

For torsion, $\sqrt{\rho'} \approx 0.062$

$$q_s = \frac{1}{1 + \frac{0.062}{\sqrt{4/25.4}}} = 0.865, \quad q = \frac{1}{1 + \frac{0.080}{\sqrt{4/25.4}}} = 0.832$$

(Continued)

$$K_{fs} = 1 + 0.865(1.5-1) = 1.43, \quad K_f = 1 + 0.832(1.95-1) = 1.79$$

$$\sigma_{nom} = 32M/\pi d^3, \quad \tau_{nom} = 16T/\pi d^3$$

$$\sigma_a = 1.79\,\frac{32(500)(10^3)}{\pi(40)^3} = 142.4 \text{ MPa}$$

$$\tau_m = 1.43\,\frac{16(750)(10^3)}{\pi(40)^3} = 85.3 \text{ MPa}$$

$$(\sigma_m)_{VM} = \sqrt{3}\,\tau_m = 147.8 \text{ MPa}$$

(a) Soderberg: $\dfrac{147.8}{420} + \dfrac{142.4}{280} = \dfrac{1}{n} \Rightarrow \underline{n = 1.16}$

(b) Modified Goodman:

$$\sigma_a/\sigma_m = \frac{142.4}{147.8} = 0.96, \quad \frac{S_E}{S_u}\frac{(S_u - S_Y)}{(S_Y - S_E)} = \frac{280}{560}\frac{(560-420)}{(420-280)} = 0.5$$

$$\therefore \frac{147.8}{560} + \frac{142.4}{280} = \frac{1}{n} \Rightarrow \underline{n = 1.29}$$

(7.53)

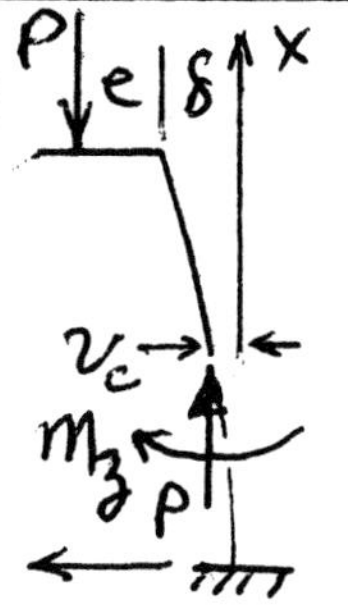

$$M_z = P(\delta + e - v_c) = EI\frac{d^2 v_c}{dx^2}$$

$$\frac{d^2 v_c}{dx^2} + \frac{P}{EI}v_c = \frac{P}{EI}(\delta + e)$$

$$v_c = C_1 \sin kx + C_2 \cos kx + \delta + e \quad \text{where } k = \sqrt{\frac{P}{EI}}$$

$$v_c' = C_1 k \cos kx - C_2 k \sin kx$$

$v_c = 0$ at $x = 0 \Rightarrow C_2 = -(\delta + e)$, $v_c' = 0$ at $x = 0 \Rightarrow C_1 = 0$

$\therefore v_c = (\delta + e)(1 - \cos kx)$ ① But $v_c = \delta$ at $x = L$

$$\delta = (\delta + e)(1 - \cos kL) \Rightarrow \delta = \frac{1 - \cos kL}{\cos kL}\,e \quad \text{sub into ①}$$

$$v_c = \left[\left(\frac{1 - \cos kL}{\cos kL}\right)e + e\right](1 - \cos kL) = \frac{e}{\sec kL}(1 - \cos kx) \quad \boxed{\text{ANS}}$$

where $k = \sqrt{P/EI}$

For stress, $M_{max} = P(e + \delta) = P\left(e + \dfrac{1 - \cos kL}{\cos kL}e\right) = Pe \sec kL$

Max compressive stress

$$\sigma_{max} = \frac{P}{A} + \frac{Mc}{I_z} = \frac{P}{A} + \frac{Pec}{I_z}\sec kL \qquad \text{(continued)}$$

Since $k = \sqrt{\frac{P}{EI}} = \frac{1}{r_g}\sqrt{\frac{P}{EA}}$ where $I_z = A r_g^2$

$$\sigma_{max} = \frac{P}{A}\left[1 + \frac{ec}{r_g^2}\sec\left(\frac{L}{r_g}\sqrt{\frac{P}{EA}}\right)\right] \quad \boxed{\text{ANS}}$$

Note: Can use Eq (7.6-9) with $L_{eff} = 2L$ (See Table 3.10-1, K=2)

(7.54)

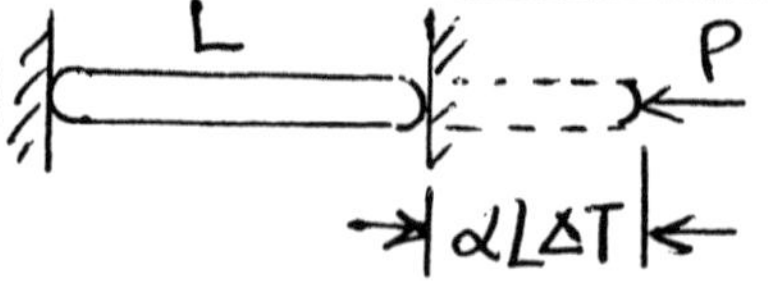

$$\delta = \frac{PL}{AE} = \alpha L \Delta T$$

$$P = AE\alpha\Delta T = \frac{\pi^2 E I_z}{L^2}$$

Solve for $\Delta T = \frac{\pi^2 I_z}{\alpha A L^2}$ $\boxed{\text{ANS}}$

(7.55) Check KL/r_g for each axis.

Pinned-pinned: $I = \frac{1}{12}(30)(50^3) = 312.5(10^3)\ mm^4$

$A = 50(30) = 1500\ mm^2$

$$KL/r_g = \frac{1(500)}{\sqrt{312.5(10^3)/1500}} = 34.6$$

Fixed-fixed: $I = \frac{1}{12}50(30)^3 = 112.5(10^3)\ mm^4$

$$KL/r_g = \frac{(0.5)(500)}{\sqrt{112.5(10^3)/1500}} = 28.9$$

Since $34.6 > 28.9$ will buckle in pinned-pinned mode

$$\sigma_{cr} = \frac{\pi^2 E_t}{(L/r_g)^2} = \frac{\pi^2 E_t}{(500)^2(1500/312.5(10^3))} = 8.2115(10^{-3})E_t \quad (1)$$

From a curve fit of the given data for $294 \le \sigma \le 510$ MPa

$$\sigma = -151.171 + 4.3983(10^5)\epsilon - 9.6946(10^7)\epsilon^2 + 7.07783(10^9)\epsilon^3 \quad (2)$$

$$E_t = \frac{d\sigma}{d\epsilon} = 4.3983(10^5) - 193.892(10^6)\epsilon + 2.12335(10^{10})\epsilon^2 \quad (3)$$

Sub (2) & (3) into (1) and factoring yields

$$1 - 540.01\epsilon + 72.101(10^3)\epsilon^2 - 7.07783(10^9)\epsilon^3 = 0$$

The only solution in range is $\epsilon = 2.868(10^{-3})$ Sub this into 2 yields $\sigma_{cr} = 479.8$ MPa. Thus

$$P_{cr} = 479.8(50)(30) = 719.7(10^3)N = \underline{720\ kN}$$

(7.56) For $\sigma > 294$ MPa, $\frac{d\sigma}{d\epsilon}$ is obtained from Eq ③ of the solution to Prob 7.55. Then KL/r_g is obtained from $\pi^2 E_t/\sigma$.

ε	σ	dσ/dε	KL/r_g	σ	σ
0	0				0
0.0002	42	210000	222.1441		42
0.0004	84	210000	157.0796		84
0.0006	126	210000	128.255		126
0.0008	168	210000	111.0721		168
0.001	210	210000	99.34588		210
0.0011	231	210000	94.72258		231
0.0012	252	210000	90.68997		252
0.0013	273	210000	87.1321		273
0.0014	294	210000	83.9626		294
0.0014	294	209998.9	83.96237	294	
0.0015	314	196767.4	78.64328	314	
0.0017	351	171578.4	69.45886	351	
0.002	397	136980	58.35569	397	
0.0023	434	106203.6	49.14447	434	
0.0027	469	71113.82	38.68481	469	
0.0032	495.5	36806.64	27.07641	495.5	
0.0036	506	17004.96	18.2122	506	
0.004	510	3998	8.796019	510	

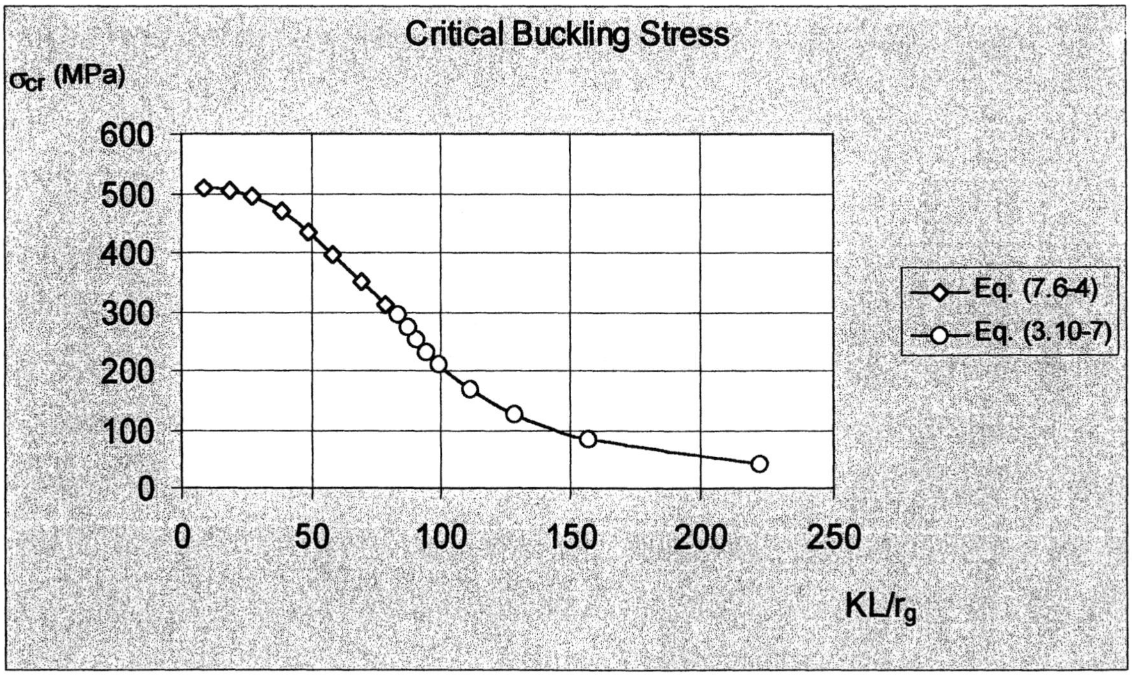

7.57 $\left(\dfrac{KL}{r_g}\right)_c = \sqrt{\dfrac{2\pi^2(29)(10^6)}{60(10^3)}} = 97.7$ [Eq. (7.6-5)]

Pinned-pinned: $K=1$, $\dfrac{KL}{r_g} = \dfrac{(1)(50)}{0.65} = 76.9$ use parabolic eq. for critical stress.

$$\sigma_{cr} = \left[1 - \frac{1}{2}\left(\frac{76.9}{97.7}\right)^2\right]60 = 41.4 \text{ kpsi}$$

$$P_{cr} = 41.4(0.85) = \underline{35.2 \text{ kips}}$$

For plot, Euler eq.: $\sigma_{cr} = \dfrac{\pi^2(29)(10^6)}{(L/r_g)^2}$ $\quad L/r_g > 97.7$

Parabolic eq: $\sigma_{cr} = 60\left[1 - \dfrac{1}{2}\left(\dfrac{L/r_g}{97.7}\right)^2\right]$ $\quad L/r_g < 97.7$

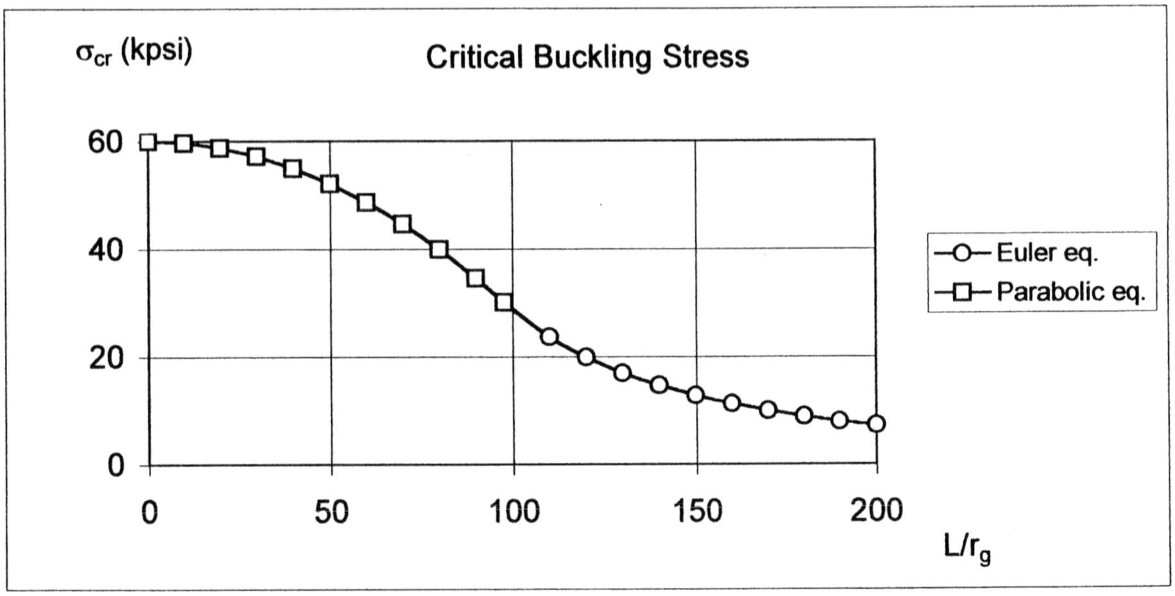

7.58 $\sigma_{max} = 60 = \dfrac{P}{0.85}\left\{1 + \dfrac{0.5(1.2)}{(0.65)^2}\sec\left[\dfrac{50}{2(0.65)}\sqrt{\dfrac{P}{29(10^3)(0.85)}}\right]\right\}$

Root searching method is easy. Using a spreadsheet program increment P and solve for σ_{max}. When σ_{max} goes over 60 drop back to previous value of P and increment at a finer value. Continue process until desired accuracy is obtained. The value of P obtained should be $\underline{P = 14.9 \text{ kips}}$

7.59 From the solution of Prob. 7.57, $C_c = 97.7$, and $KL/r_g = 76.9$. Then, from Eq. (7.6-10a) with $A = 0.85$

$$\frac{P}{0.85} = \frac{\left[1-\frac{1}{2}\left(\frac{76.9}{97.7}\right)^2\right]60}{\frac{5}{3}+\frac{3(76.9)}{8(97.7)}-\frac{(76.9)^3}{3(97.7)^3}} \Rightarrow \underline{P = 18.5 \text{ kips}}$$

7.60 Substituting $N_x = N_y = -P$ into Eq. 7.6-15, $V_{xy} = 0$

$$\frac{\partial^4 w}{\partial x^4} + 2\frac{\partial}{\partial x^2 \partial y^2} + \frac{\partial^4 w}{\partial y^4} + \frac{P}{D}\left(\frac{\partial^2 w}{\partial x^2} + \frac{\partial^2 w}{\partial y^2}\right) = 0$$

Boundary conditions of Eq (7.6-19) are valid as well as Eq. (7.6-20). With $b = a$

$$\sum\sum C_{mn}\left\{\left(\frac{m\pi}{a}\right)^4 + 2\left(\frac{mn\pi^2}{a^2}\right)^2 + \left(\frac{n\pi}{a}\right)^4 - \frac{P}{D}\left[\left(\frac{m\pi}{a}\right)^2 + \left(\frac{n\pi}{a}\right)^2\right]\right\}\sin\frac{m\pi x}{a}\sin\frac{n\pi y}{a} = 0$$

$$\frac{\pi^2}{a^2}\sum\sum C_{mn}\left[\frac{\pi^2}{a^2}(m^2+n^2)^2 - \frac{P}{D}(m^2+n^2)\right]\sin\frac{m\pi x}{a}\sin\frac{n\pi y}{a} = 0$$

$$\frac{\pi^2}{a^2}(m^2+n^2)^2 - \frac{P}{D}(m^2+n^2) = 0 \Rightarrow P = \frac{\pi^2}{a^2}D(m^2+n^2)$$

P_{cr} occurs for lowest values of m, n. For $m = n = 1$

$$P_{cr} = \frac{2\pi^2}{a^2}D \quad \boxed{\text{ANS}}$$

7.61 $d/w = 20/100 = 0.2$

(a) $K_t \approx 2.5$, $\sigma_{nom} = \frac{P}{(100-20)(10)} = 1.25(10^{-3})P$ MPa

$\sigma_{max} = 2.5\sigma_{nom} = 3.125(10^{-3})P$

$P_Y = \frac{300}{3.125(10^{-}} = 96(10^3)N = \underline{96 \text{ kN}}$

(b) $P_P = S_Y A = 300(100-20)(10) = 240(10^3)N = \underline{240 \text{ kN}}$

7.62 (a) $T = \int_0^{r_0}\left[\int_0^{2\pi} r\,\tau\, r\,d\theta\right]dr = 2\pi\int_0^{r_0} r^2\tau\,dr$ where $\tau = \tau(r)$

$$T = 2\pi\left(\int_0^{r_0-h_p}\frac{S_{YS}}{r_0-h_p}r^3dr + \int_{r_0-h_p}^{r_0}S_{YS}r^2dr\right)$$

$$= 2\pi\left\{\frac{1}{4}S_{YS}(r_0-h_p)^3 + \frac{1}{3}S_{YS}\left[r_0^3-(r_0-h_p)^3\right]\right\} = \frac{\pi}{6}S_{YS}\left[4r_0^3-(r_0-h_p)^3\right]$$

(continued)

$$(r_o - h_p)^3 = 4r_o^3 - \frac{6T}{\pi S_{YS}} \Rightarrow r_o - h_p = \sqrt[3]{4r_o^3 - 6T/\pi S_{YS}}$$

$$h_p = r_o - \sqrt[3]{4r_o^3 - 6T/\pi S_{YS}}, \quad T > \frac{S_{YS} J}{r_o} \qquad \boxed{\text{ANS}}$$

(b) Linear unloading, $\tau = \frac{Tr}{J} = \frac{2}{\pi}\frac{Tr}{r_o^4}$ where $J = \frac{\pi}{2} r_o^4$

$$\tau_{res} = \frac{S_{YS}\, r}{r_o - h_p} - \frac{2}{\pi}\frac{Tr}{r_o^4} \qquad 0 < r < r_o - h_p$$

$$\tau_{res} = S_{YS} - \frac{2Tr}{\pi r_o^4} \qquad r_o - h_p < r < r_o$$

$$h_p = 0.5 - \sqrt[3]{4(0.5)^3 - \frac{6(6500)}{\pi(28.85)(10^3)}} = 0.0885 \text{ in}$$

$$\tau_{res} = \left[\frac{28.85(10^3)}{0.5 - 0.0885} - \frac{2(6500)}{\pi(0.5)^4}\right] r = 3893 r \qquad 0 < r < 0.4115 \text{ in}$$

$$= 28.85(10^3) - \frac{2(6500)}{\pi(0.5)^4} r = (28.85 - 66.21 r)(10^3) \qquad 0.4115 \text{ in} < r < 0.5$$

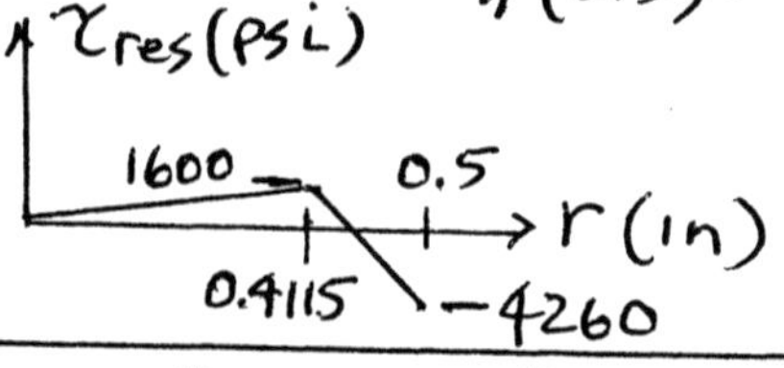

$\boxed{\text{ANS}}$

(7.63) $A = b^2 - (0.8b)^2 = 0.36 b^2$, $I = \frac{1}{12}\left[b^4 - (0.8b)^4\right] = 0.0492 b^4$

$$M_Y = \frac{S_Y I}{c} = \frac{S_Y (0.0492 b^4)}{0.5b} = 0.0984 b^3 S_Y$$

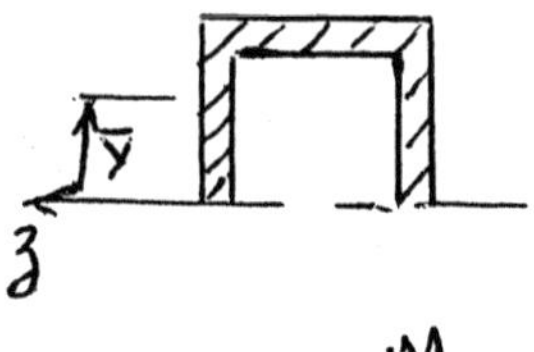

$$\bar{y} = \frac{(0.25b)(0.5b)(b) - (0.2b)(0.4b)(0.8b)}{0.18b^2} = 0.3389 b$$

$$M_P = 2(0.3389 b) S_Y (0.18 b^2) = 0.122 b^3 S_Y$$

$$SF = \frac{M_P}{M_Y} = \frac{0.122 b^3 S_Y}{0.0984 b^3 S_Y} = \underline{1.24}$$

SF is a "good" value as more material is placed away from the z axis. The section is equivalent to an "I" section

7.64 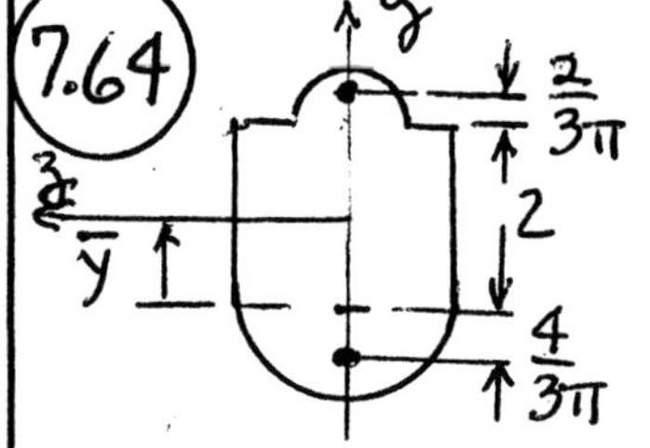

$A = 2(2) + \frac{\pi}{2}(1^2 - 0.5^2) = 5.9635\ in^2$

$\bar{y} = \frac{1(2)(2) + (2 + \frac{2}{3\pi})(\frac{\pi}{2})(0.5^2) - \frac{4}{3\pi}(\frac{\pi}{2})(1^2)}{5.9635} = 0.70463\ in$

$I_z = \frac{1}{12}(2)^4 + 2^2(1 - 0.70463)^2 + (\frac{\pi}{8} - \frac{8}{9\pi})(0.5)^4 + \frac{\pi}{2}(0.5)^2(2 + \frac{2}{3\pi} - 0.70463)^2$

$+ (\frac{\pi}{8} - \frac{8}{9\pi})(1)^4 + \frac{\pi}{2}(1)^2(\frac{4}{3\pi} + 0.70463)^2 = 4.694\ in^4$

$y_{max} = 2.5 - 0.70463 = 1.795\ in$

$M_Y = \frac{S_Y I}{y_{max}} = \frac{40(4.694)}{1.795} = \underline{104.6\ kip\text{-}in}$

(b) Determine where areas balance. $\frac{1}{2}A = 2.9817\ in^2$

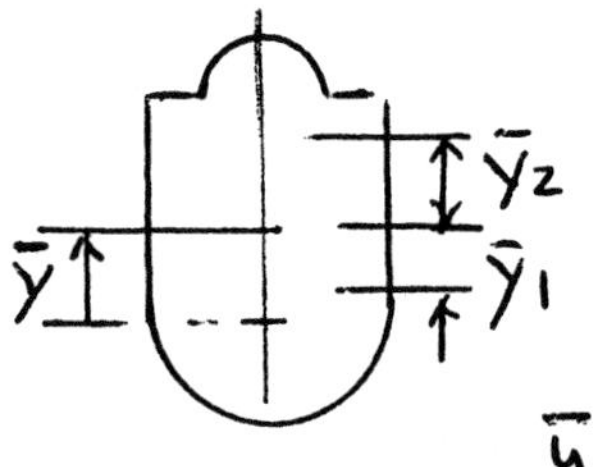

$2\bar{y} + \frac{\pi}{2}(1)^2 = 2.9817 \Rightarrow \bar{y} = 0.70548\ in$

$\bar{y}_1 = \frac{\frac{(0.70548)^2}{2}(2) + (0.70548 + \frac{4}{3\pi})\frac{\pi}{2}(1^2)}{2.9817} = 0.7622\ in$

$\bar{y}_2 = \frac{\frac{(2 - 0.70548)^2}{2}(2) + (2 - 0.70548 + \frac{2}{3\pi})(\frac{\pi}{2})(\frac{1}{2})^2}{2.9817} = 0.7605\ in$

$M_P = (\bar{y}_1 + \bar{y}_2)S_Y(0.5A) = (0.7622 + 0.7605)(40)(2.9817) = 181.6\ kip\text{-}in$

$SF = \frac{M_P}{M_Y} = \frac{181.6}{104.6} = \underline{1.74}$

7.65 Without a steel handbook, a Mechanics of materials text book can be used. For example a W920×446 WF beam section was selected from "Mechanics of Materials," Beer, F.P., and E.R. Johnston Jr., 2nd ed., McGraw-Hill, 1992. Neglecting fillets the section is

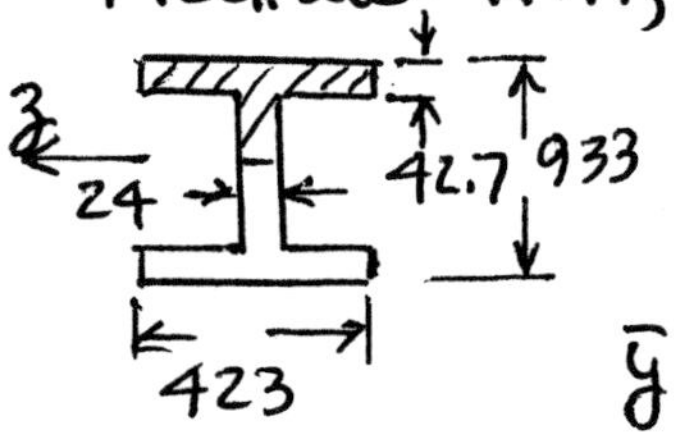

Section modulus, $S_z = 18.11(10^6)\ mm^3$

$M_Y = S_z S_Y = 18.11(10^6)\ S_Y$

For top section,

$\bar{y}A = \frac{(933 - 42.7)}{2}(423)(42.7) + \frac{(\frac{933}{2} - 42.7)^2}{2}(24) = 10.2(10^6)\ mm^3$

$M_P = 2\bar{y}AS_Y = 2(10.2)(10^6)S_Y = 20.4(10^6)S_Y$

$SF = \frac{M_P}{M_Y} = \frac{20.4(10^6)S_Y}{18.11(10^6)S_Y} = 1.126 \simeq 1.13$ **ANS**

(7.66) $I \approx \frac{\pi}{8} t \bar{d}^3$, $M_Y = \frac{S_Y I}{\bar{d}/2} = \frac{\pi}{4} t \bar{d}^2 S_Y$

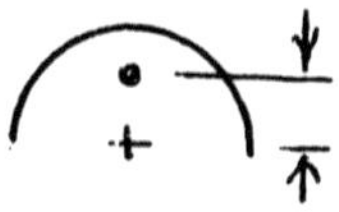

$\frac{2r}{\pi} = \frac{\bar{d}}{\pi}$ $\frac{A}{2} = \frac{\pi}{2}\bar{d}t$, $M_P = 2\left(\frac{\bar{d}}{\pi}\right) S_Y \left(\frac{\pi}{2}\bar{d}t\right) = t\bar{d}^2 S_Y$

$$SF = \frac{M_P}{M_Y} = \frac{t\bar{d}^2 S_Y}{\pi/4\, t\bar{d}^2 S_Y} = \frac{4}{\pi} = \underline{1.27}$$

(7.67) $S_{YS} = \frac{16 T_Y}{\pi d_o^3} \Rightarrow T_Y = \frac{\pi}{16} d_o^3 S_{YS}$

Inelastic:

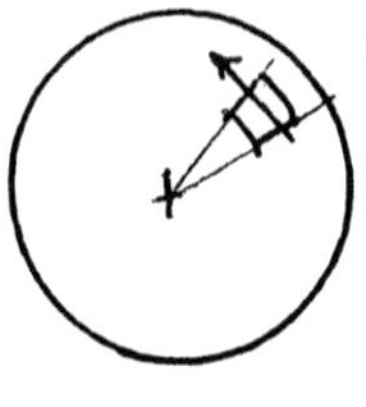

$dA = r\,d\theta\,dr$ $dT_P = r S_{YS} dA = r^2 S_{YS}\, dr\, d\theta$

$$T_P = \int_0^{r_o}\int_0^{2\pi} r^2 S_{YS}\, dr\, d\theta = 2\pi S_{YS} \frac{r_o^3}{3} = \frac{\pi}{12} S_{YS} d_o^3$$

$$SF = \frac{T_P}{T_Y} = \frac{\frac{\pi}{12} S_{YS} d_o^3}{\frac{\pi}{16} d_o^3 S_{YS}} = \frac{4}{3} = 1.333$$

(7.68) $J \approx \frac{\pi}{4} t \bar{d}^3$, $S_{YS} = \frac{T_Y \bar{d}/2}{\pi/4\, t\bar{d}^3} \Rightarrow T_Y = \frac{\pi}{2} t \bar{d}^2 S_{YS}$

See the solution of Prob. 7.67. $T_P = r_o S_{YS} A$

$$T_P = \frac{\bar{d}}{2} S_{YS} (\pi \bar{d} t) = \frac{\pi}{2} t \bar{d}^2 S_{YS}$$

$$SF = \frac{T_P}{T_Y} = \frac{\pi/2\, t\bar{d}^2 S_{YS}}{\pi/2\, t\bar{d}^2 S_{YS}} = 1 \quad \boxed{ANS}$$

(7.69) From the solution of Prob. 5.51, due to bending only, $\sigma_{max} = \sigma_\theta|_{r=2} = S_Y = \frac{M_Y(2.8854-2)}{3(0.1146)(2)}$

$\therefore M_Y = 0.7766\, S_Y$

Plastic: For rect. section $M_P = S_Y bc^2 = S_Y(1.5)(1)^2 = 1.5 S_Y$

$$SF = \frac{M_P}{M_Y} = \frac{1.5 S_Y}{0.7766 S_Y} = \underline{1.93}$$

For yield of a curved beam, M_Y is lower. For example, for a straight rect. beam $M_Y = \frac{2}{3} bc^2 S_Y = \frac{2}{3}(1.5)(1)^2 S_Y = S_Y$ which is greater than $0.7766 S_Y$

(7.70) From the solution of Prob. 3.10, $I = 274.2(10^3)\,mm^4$, $y_{max} = 60 - 23.026 = 36.974\,mm$

$$M_Y = \frac{S_Y I}{y_{max}} = \frac{274.2(10^3)}{36.974} S_Y = 7.416(10^3) S_Y$$

Inelastic:

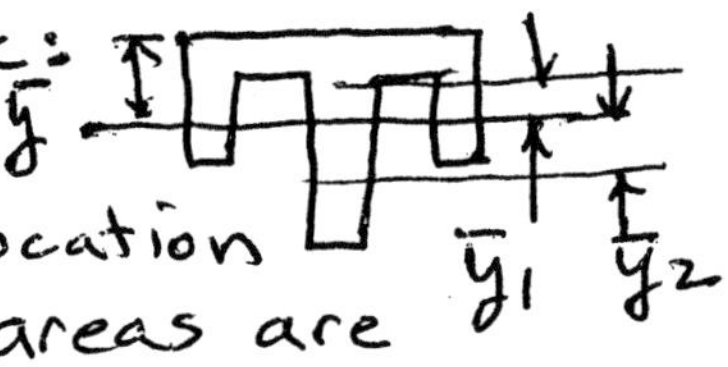

$$A = 45(5) + 55(5) + 2(45)(5) = 950\,mm^2$$

$\bar{y}$ is location where areas are equal

$$\frac{950}{2} = 45(5) + 3(\bar{y} - 5)(5) \Rightarrow \bar{y} = 21.667\,mm$$

$$\bar{y}_1 = \frac{(21.667 - 2.5)(45)(5) + 3\frac{(21.667-5)^2}{2}(5)}{475} = 13.465\,mm$$

$$\bar{y}_2 = \frac{\frac{2(50-21.667)^2(5)}{2} + \frac{(60-21.667)^2(5)}{2}}{475} = 16.184\,mm$$

$$M_p = (13.465 + 16.184)(475) S_Y = 14.08(10^3) S_Y$$

$$SF = \frac{M_p}{M_Y} = \frac{14.08(10^3) S_Y}{7.416(10^3) S_Y} = \underline{1.90}$$

(7.71) Elastic: $S_Y = \dfrac{32 M_Y}{\pi d^3} \Rightarrow M_Y = \dfrac{\pi d^3}{32} S_Y$

Inelastic:

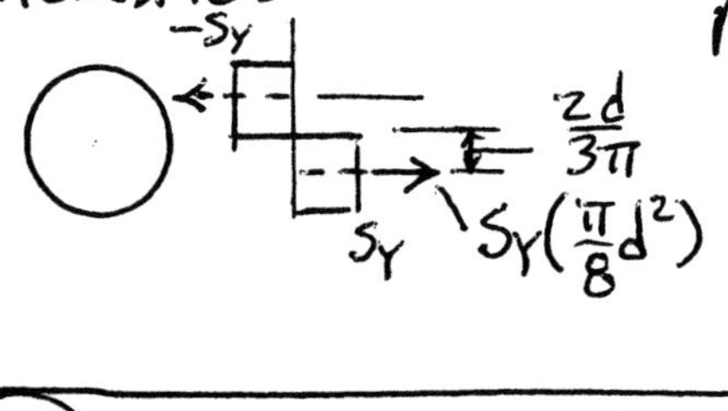

$$M_p = 2\left(\frac{2}{3}\frac{d}{\pi}\right) S_Y \left(\frac{\pi}{8} d^2\right) = \frac{d^3}{6} S_Y$$

$$SF = \frac{M_p}{M_Y} = \frac{\frac{d^3}{6} S_Y}{\frac{\pi d^3}{32} S_Y} = \frac{16}{3\pi} = 1.698 = \underline{1.7}$$

(7.72) Triangle about its base, $I = \frac{1}{12} bh^3$. For diamond, double this. Thus, $I = \frac{1}{6} bh^3$.

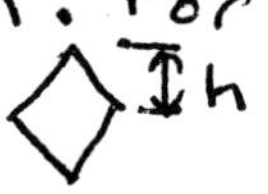

$$S_Y = \frac{M_Y h}{1/6\, bh^3} \Rightarrow M_Y = \frac{bh^2}{6} S_Y$$

Inelastic:

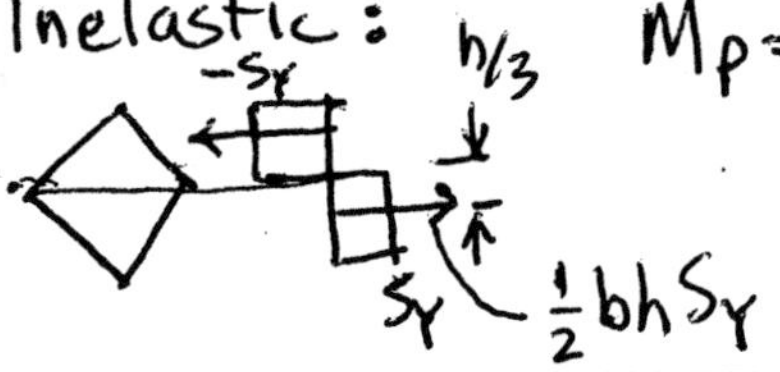

$$M_p = 2\frac{h}{3}\left(\frac{1}{2} bh S_Y\right) = \frac{bh^2}{3} S_Y$$

$$SF = \frac{M_p}{M_Y} = \frac{bh^2/3\, S_Y}{bh^2/6\, S_Y} = \underline{2}$$

(7.73) Elastic: $A = 20(50+80+100) = 4600\ \text{mm}^2$

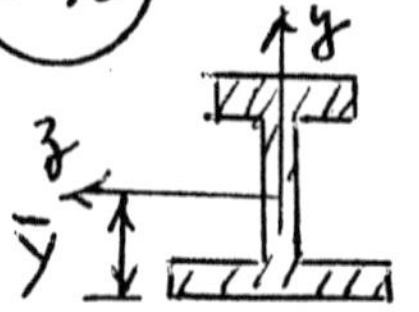

$$\bar{y} = \frac{10(20)(100) + 60(20)(80) + 110(20)(50)}{4600} = 49.13\ \text{mm}$$

$$I_z = \frac{1}{12}100(20^3) + (100)(20)(49.13-10)^2 + \frac{1}{12}20(80)^3 + 20(80)(60-49.13)^2$$
$$+ \frac{1}{12}50(20^3) + 50(20)(110-49.13)^2 = 7.9099(10^6)\ \text{mm}^4$$

$$S_Y = \frac{M_Y(120-49.13)}{7.9099(10^6)} \Rightarrow M_Y = 3.906(10^7)\ \text{N-mm}$$

(S_Y = 350 MPa)

Inelastic: $\bar{y}$ for equal area above and below axis

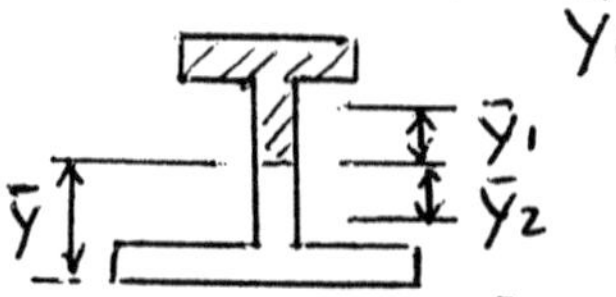

$\bar{y}_1, \bar{y}_2$ centroids for areas above and below axis, respectively

$$20(100) + (\bar{y}-20)(20) = 2300 \Rightarrow \bar{y} = 35\ \text{mm}$$

$$\bar{y}_1 = \frac{\frac{65^2}{2}(20) + 75(50)(20)}{2300} = 50.978\ \text{mm}$$

$$\bar{y}_2 = \frac{25(20)(100) + \frac{15^2}{2}(20)}{2300} = 22.717\ \text{mm}$$

$$M_P = (\bar{y}_1 + \bar{y}_2)(0.5A)S_Y = (50.978 + 22.717)(2300)(350) = 59.32(10^6)\ \text{N-mm}$$

$$SF = \frac{M_P}{M_Y} = \frac{59.32(10^6)}{3.906(10^7)} = \underline{1.52}$$

(7.74) (a) $$\frac{h_P}{c} = 1 - \sqrt{3 - \frac{2M_z}{M_Y}} = 1 - \sqrt{3 - \frac{2.4M_Y}{M_Y}} = 0.225 \quad \therefore h_P = \underline{0.225c}$$

(b) At $x = x_P$ $\quad M_z = -P(L - x_P) = -M_Y = -\frac{PL}{1.2}$

$$x_P = L - \frac{L}{1.2} = \frac{L}{6} \quad \boxed{\text{ANS}}$$

(7.75) $$\frac{h_P}{c} = 1 - \sqrt{3 - \frac{(2)(1.3M_Y)}{M_Y}} = 0.3675\text{ in} \Rightarrow h_P = 0.3675(2) = 0.735\text{ in}$$

(a) $$\sigma = \begin{cases} \frac{-60}{2-0.735}y = -47.43y\ \text{kpsi} & -1.265 < y < 1.265\text{ in} \\ -60\ \text{kpsi} & y > 1.265\text{ in} \\ 60\ \text{kpsi} & y < -1.265\text{ in} \end{cases} \quad \boxed{\text{ANS}}$$

(continued)

(b) $\epsilon_{max} = \dfrac{S_y c}{E(c-h_p)} = \dfrac{60(10^3)(2)}{30(10^6)(2-0.735)} = \underline{3162\mu}$

(c) $\underline{-1.265 < y < 1.265}$

$$\sigma_{res} = \left(\frac{1.3M_Y}{I} - \frac{S_Y}{c-h_p}\right)y = \left\{\frac{1.3\left[\frac{2}{3}(2)(2)^2(60)\right]}{\frac{2}{3}(2)(2)^3} - \frac{60}{1.265}\right\}y$$

$= -8.43y$ kpsi **ANS**

$\underline{1.265 < y < 2}$

$$\sigma_{res} = \frac{1.3\left[\frac{2}{3}(2)(2)^2(60)\right]}{\frac{2}{3}(2)(2)^3}y - 60 = 39y - 60 \text{ kpsi}$$ **ANS**

$\underline{-1.265 > y > -2}$ $\quad \sigma_{res} = 60 - 39y$ kpsi **ANS**

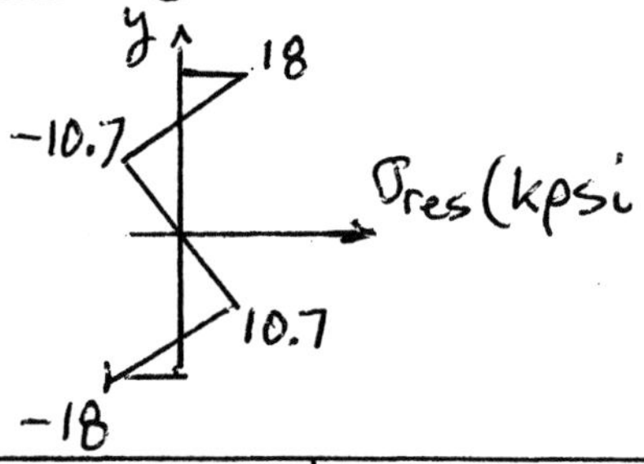

7.76

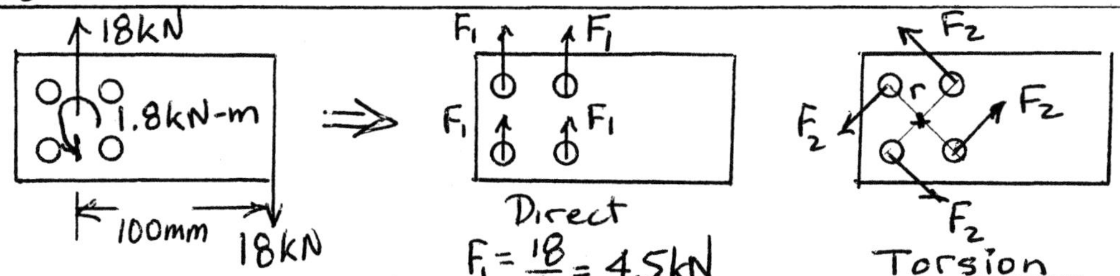

$F_1 = \dfrac{18}{4} = 4.5$ kN

$r = \sqrt{2(25^2)}$
$= 35.35$ mm

$4F_2(35.35)(10^{-3}) = 1.8$

$F_2 = 12.71$ kN

Inner rivets are loaded the greatest. Consider, for example the lower right rivet.

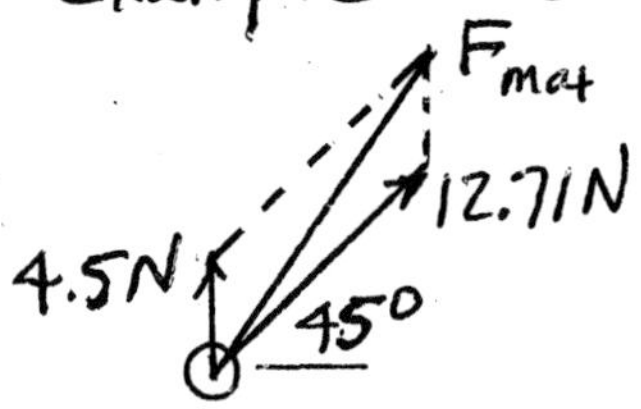

$F_{max} = \sqrt{(12.71\cos 45)^2 + (12.71\sin 45 + 4.5)^2}$

$= 16.21$ kN

$\tau_{max} = \dfrac{16.21(10^3)}{250} = \underline{64.8 \text{ MPa}}$

7.77

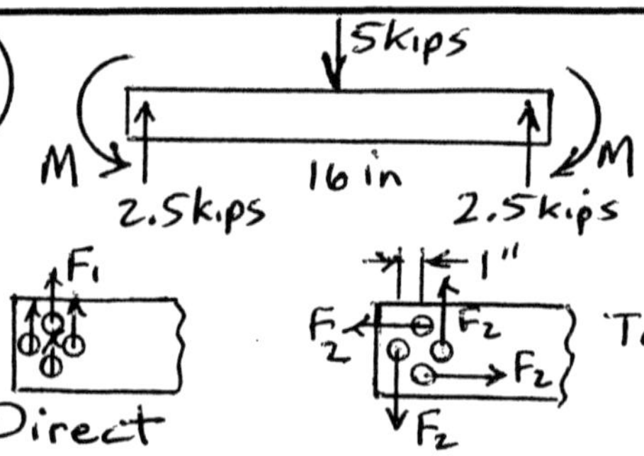

From Appendix C

$$M = \frac{5(8)^3}{(16)^2} = 10 \text{ kip-in}$$

Torsion: $4F_2(1) = 10$

$$F_2 = 2.5 \text{ kip}$$

$$F_1 = \frac{2500}{4} = 625 \text{ lb}$$

Worst case is right rivet:

$$F_{max} = 625 + 2500 = 3125 \text{ lb}$$

$$\tau_{max} = \frac{3125}{0.4} = 7810 \text{ psi} = \underline{7.81 \text{ kpsi}}$$

7.78

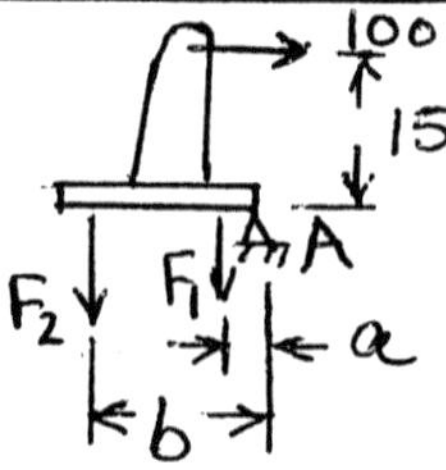

$$\frac{F_1}{a} = \frac{F_2}{b} \Rightarrow F_2 = \frac{b}{a}F_1 = \frac{9}{1}F_1 = 9F_1$$

$$\Sigma M_A = 0 \Rightarrow 2(F_2 b + F_1 a) = 1000(15)$$

$$2[9F_1(9) + F_1(1)] = 1000(15)$$

$F_1 = 395$ lb, $F_2 = 9F_1 = 3560$ lb, $F_{max} = 3560 + 5000 = 8560$ lb

$$\sigma_{max} = \frac{8560}{0.2} = 42.8(10^3) \text{ psi} = \underline{42.8 \text{ kpsi}}$$

7.79 (a)

$$T = \frac{63000 \, hp}{N} = \frac{63000(200)}{800} = 15.75(10^3) \text{ lb-in}$$

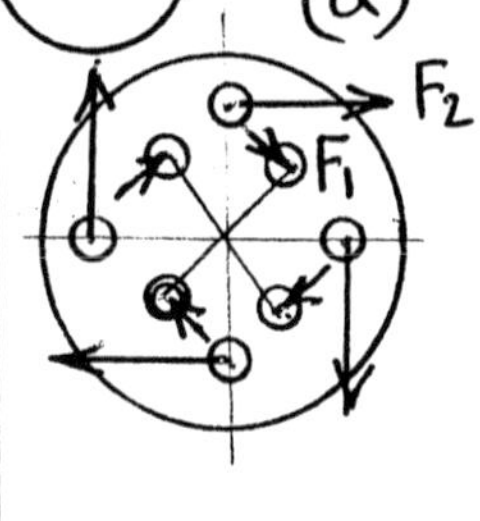

$$\theta \propto \frac{F/A}{r} \Rightarrow \frac{F_1/A}{r_1} = \frac{F_2/A}{r_2} \Rightarrow F_1 = \frac{r_1}{r_2}F_2 = \frac{5}{10}F_2 = \frac{F_2}{2}$$

$$\Sigma T = 4F_1(5) + 4F_2(10) = 15.75(10^3)$$

$$(20)\frac{F_2}{2} + 40F_2 = 15.75(10^3) \Rightarrow F_2 = 315 \text{ lb}$$

Since $F_2 > F_1$, $\tau_{max} = \dfrac{315}{\frac{\pi}{4}d^2} = 20(10^3) \Rightarrow \underline{d = 0.1416 \text{ in}}$

Use UNC 8-24 bolts with $d = 0.164$ in

(b) $F_2 = \dfrac{15.75(10^3)}{10n}$ $\quad n = \#$ of bolts

$$20(10^3) = \frac{1575/n}{\frac{\pi}{4}(0.164)^2} \Rightarrow n = 3.73 \quad \text{use } \underline{4 \text{ bolts}}$$

7.80 (a)

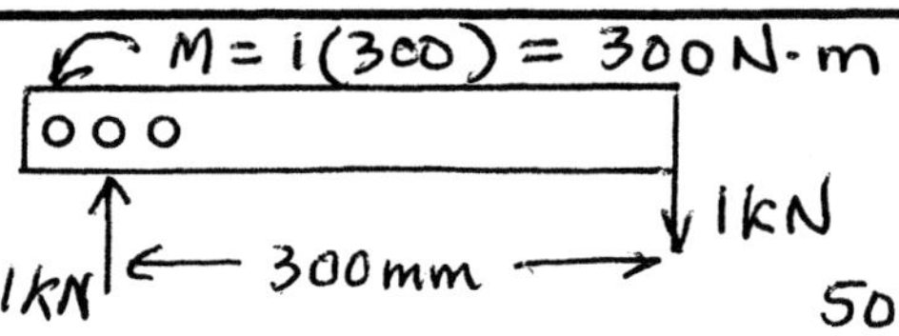

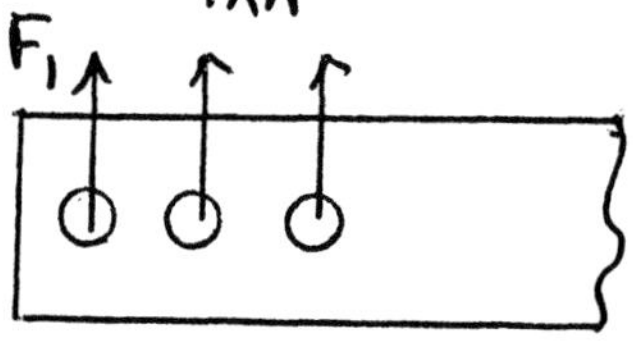

Direct

$$F_1 = \frac{1000}{3} = 333\text{ N}$$

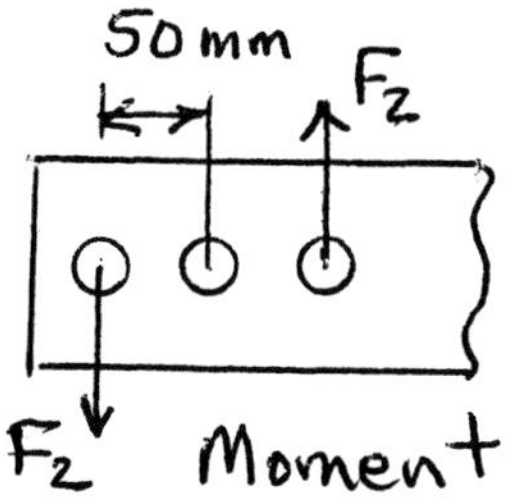

Moment

$$2F_2(0.050) = 300$$
$$F_2 = 3000\text{ N}$$

F_{max} on right bolt

$$F_{max} = 3000 + 333 = 3333\text{ N}$$

$$120 = \frac{3333}{\frac{\pi}{4}d^2} \Rightarrow \underline{d = 5.95\text{ mm}} \quad \text{use } d = 6\text{ mm}$$

(b) With 2 bolts $F_{max} = 3000 + 500 = 3500\text{ N}$

$$120 = \frac{3500}{\frac{\pi}{4}d^2} \Rightarrow \underline{d = 6.09\text{ mm}}$$

7.81 From the solution of Prob. 3.46 the max force is $F_D = 12\text{ kN}$. Thus,

(a) $$\frac{S_Y}{n} = \frac{420}{2} = \frac{12(10^3)}{\frac{\pi}{4}d^2} \Rightarrow \underline{d = 8.53\text{ mm}}$$

(b) limit analysis assumes both cables yield. Since the areas are equal the yield force is the same.

$$\Sigma M_A = F(500) + F(1500) - 20(1000) = 0 \Rightarrow F = 10\text{ kN}$$

$$\frac{420}{2} = \frac{10(10^3)}{\frac{\pi}{4}d^2} \Rightarrow \underline{d = 7.79\text{ mm}}$$

7.82

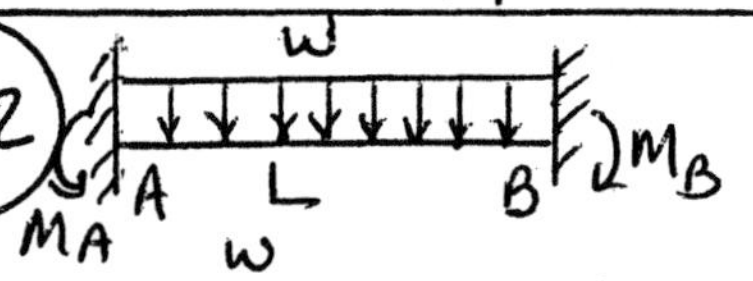

Due to symmetry $M_A = M_B$. Collapse does not occur until $M_A = M_B = M_p$ and $M_{x=\frac{L}{2}} = M_p$

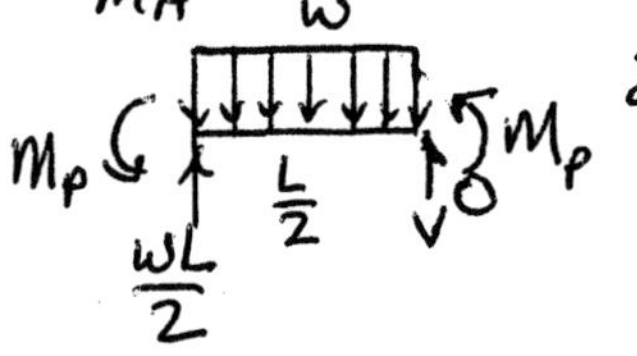

$$\Sigma M_o = 0 \Rightarrow \frac{wL}{2}\left(\frac{L}{2}\right) - 2M_p - \frac{wL}{2}\left(\frac{L}{4}\right) = 0$$

$$w_p = \frac{16M_p}{L^2} \quad \boxed{\text{ANS}}$$

(7.83) 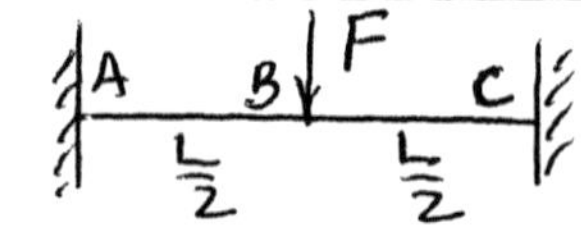

Since due to symmetry $M_A = M_C$ and collapse does not occur until $M_A = M_B = M_C = M_p$

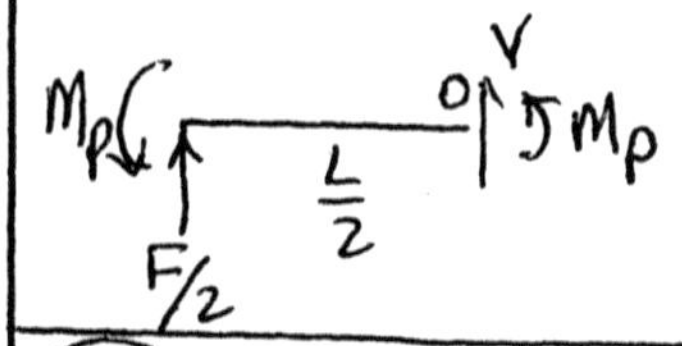

$\Sigma M_O = 0 \Rightarrow \frac{F}{2}\left(\frac{L}{2}\right) - 2M_p = 0 \Rightarrow F_p = \frac{8 M_p}{L}$ **ANS**

(7.84) From Appendix C, $M_A = M_C = \frac{F\left(\frac{L}{2}\right)\left(\frac{L}{2}\right)^2}{L^2} = \frac{FL}{8}$

At B, $M_B = \frac{F\left(\frac{L}{2}\right)^2}{L^3}\left[\frac{L}{2}\left(3\frac{L}{2} + \frac{L}{2}\right) - L\left(\frac{L}{2}\right)\right] = \frac{FL}{8}$

Thus, $M_{max} = \frac{FL}{8} = \frac{2}{3}M_p \Rightarrow F_Y = \frac{16}{3}\frac{M_p}{L} = 5.33\frac{M_p}{L}$ **ANS**

F_Y is 33% less than F_p determined in Prob 7.83

(7.85) 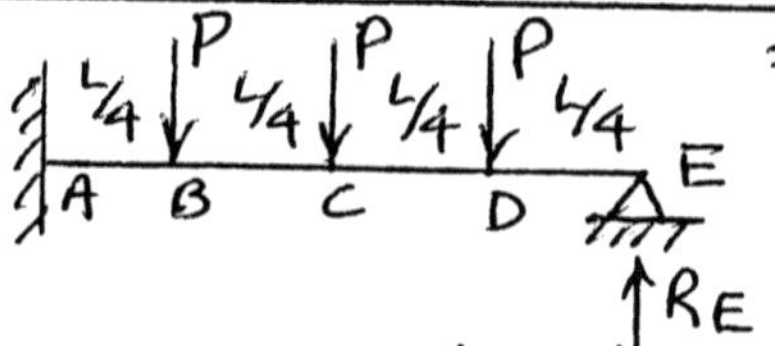

Possible modes $= \frac{4(3)}{2} = 6$

<u>Mode 1</u>: M_p at A and B

$\Sigma M_A: R_E L - \frac{3}{4}PL - \frac{1}{2}PL - \frac{1}{4}PL + M_p = 0 \Rightarrow R_E = 1.5P - \frac{M_p}{L}$ ①

$\Sigma M_B: \frac{3}{4}R_E L - \frac{1}{2}PL - \frac{1}{4}PL - M_p = 0$ ② sub in ①

$\frac{3}{4}L\left(1.5P - \frac{M_p}{L}\right) - \frac{3}{4}PL - M_p = 0 \Rightarrow P = \frac{14}{3}\frac{M_p}{L}$

<u>Mode 2</u>: M_p at A and C. Eq ① still applies.

$\Sigma M_C: \frac{1}{2}R_E L - \frac{1}{4}PL - M_p = 0$ ③ sub in ①

$\frac{L}{2}\left(1.5P - \frac{M_p}{L}\right) - \frac{1}{4}PL - M_p = 0 \Rightarrow P = \frac{3M_p}{L}$

<u>Mode 3</u>: M_p at A and D. Eq ① still applies.

$\Sigma M_D: \frac{1}{4}R_E L - M_p = 0$ ④ Sub in ①

$\frac{L}{4}\left(1.5P - \frac{M_p}{L}\right) - M_p = 0 \Rightarrow P = \frac{10}{3}\frac{M_p}{L}$

<u>Mode 4</u>: M_p at B and C

ΣM_B : Eq ② $R_E = P + \frac{4}{3}M_p/L$ ⑤

ΣM_C: Eq ③ sub in ⑤ $\frac{L}{2}\left(P + \frac{4}{3}M_p/L\right) - \frac{1}{4}PL - M_p = 0 \Rightarrow P = \frac{4}{9}\frac{M_p}{L}$

(continued)

Mode 5: M_p at B and D. Eq ⑤ still applies

ΣM_D: Eq ④ sub in ⑤ $\frac{L}{4}\left(P + \frac{4}{3} M_p/L\right) - M_p = 0 \Rightarrow P = \frac{8}{3}\frac{M_p}{L}$

Mode 6: M_p at C and D

ΣM_C: Eq ③ $R_E = \frac{P}{2} + 2M_p/L$ ⑥

ΣM_D: Eq ④ sub in ⑥ $\frac{L}{4}\left(\frac{P}{2} + 2M_p/L\right) - M_p = 0 \Rightarrow P = 4\frac{M_p}{L}$

Mode 4 has smallest value of P. Thus this is the collapse mode and $P_p = \frac{4}{9} M_p/L$ **ANS**

(7.86) a) Elastic analysis:

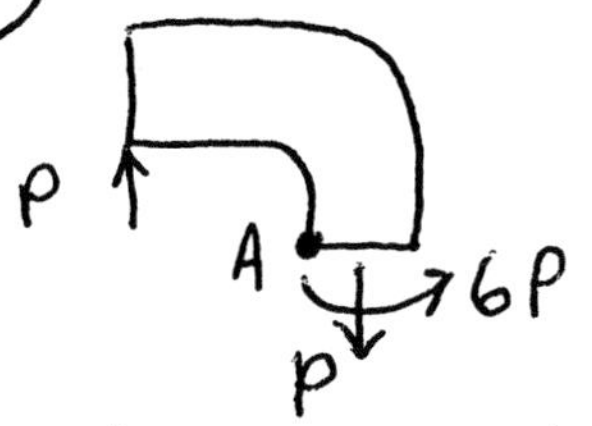

$$\sigma_A = \frac{P}{A} + \frac{6M}{bh^2} = \frac{P}{2(1)} + \frac{6P}{\frac{1}{12}1(2)^3} = 40 \text{ kpsi}$$

$P = 4.21$ kips

(b) Limit analysis. Neutral axis shifts

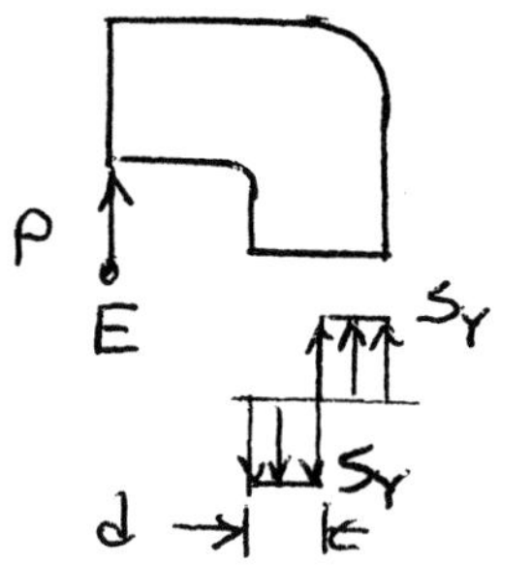

$\Sigma F = 0 \Rightarrow P = (1)[S_Y d - S_Y(2-d)]$ $\quad S_Y = 40$ kpsi

$P = 80(d-1)$ ①

$\Sigma M_E = 0 \Rightarrow (1)\left[\left(5+\frac{d}{2}\right)S_Y d - \left(\frac{5+d+7}{2}\right)S_Y(2-d)\right] = 0$

reduces to

$$d^2 + 10d - 12 = 0$$

Acceptable root: $d = 1.0828$ in sub into ①

$P = 6.62$ kips

(7.87) (a) Elastic analysis: $F_1 = F_3 \Rightarrow 2F_1 + F_2 = 100$ ①

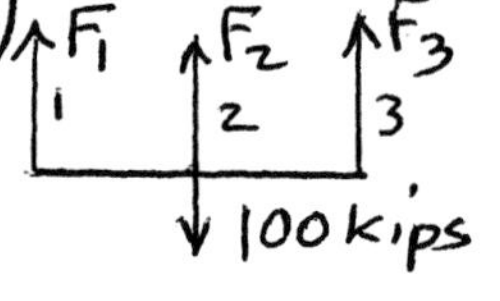

$\delta_2 = \delta_1 \Rightarrow \frac{F_2(50)}{AE} = \frac{F_1(60)}{AE} \Rightarrow F_2 = 1.2F_1$ sub into ①

$2F_1 + 1.2F_1 = 100 \Rightarrow F_1 = F_3 = 31.25$ kips

$F_2 = 1.2F_1 = 37.5$ kips

$S_Y = \frac{F_2}{\pi/4\, d^2}$ $\quad 40 = \frac{37.5}{\pi/4\, d^2} \Rightarrow d = 1.093$ in

(b) Limit analysis: $F_1 = F_2 = F_3 = \frac{100}{3} = 33.3$ kips

$40 = \frac{33.3}{\pi/4\, d^2} \Rightarrow d = 1.030$ in

8.1 $L_r = 0.1$, $F_r = 0.2$, $E_r = \frac{1}{0.9} = 1.11$

$\sigma_r = \frac{F_r}{L_r} = 20 \quad \therefore \sigma_p = 20(1.2) = \underline{24\text{ kpsi}}$

$\delta_r = \frac{F_r}{E_r L_r} = \frac{0.2}{(1.11)(0.1)} = 1.8 \quad \therefore \delta_p = 1.8(0.025) = \underline{0.045\text{ in}}$

8.2 $L_r = \frac{3}{0.5} = 6$

a) Model width $= \frac{75}{6} = \underline{12.5\text{ mm}}$, depth $= \frac{150}{6} = \underline{25\text{ mm}}$

(b) $w_r = \frac{w_p}{w_m} = \frac{6000}{240} = 25 = \frac{F_r}{L_r} \quad \therefore F_r = 25(6) = \underline{150}$

$\sigma_r = P_r = \frac{150}{(6)^2} = \underline{4.17}$, $E_r = \frac{20(10^{10})}{2(10^{10})} = \underline{10}$, $M = F_r L_r = 150(6) = \underline{900}$

$\varepsilon_r = \frac{\sigma_r}{E_r} = \frac{4.17}{10} = \underline{0.417}$, $\delta_r = \varepsilon_r L_r = 0.417(6) = \underline{2.5}$

(c) $\sigma_p = 4.167(1.4) = \underline{5.83\text{ MPa}}$, $\delta_p = 2.5(0.050) = \underline{0.125\text{ mm}}$

8.3 (a) $\sigma_m = E_m \epsilon_m = 71(10^3)(300)(10^{-6}) = 21.3\text{ MPa}$

$L_r = 4$, $\frac{F_r}{L_r} = \frac{800}{80} = 10$, $\therefore F_r = 10(4) = 40$

$\sigma_r = \frac{F_r}{L_r^2} = \frac{40}{(4)^2} = 2.5$, $\sigma_p = 2.5(21.3) = \underline{53.3\text{ MPa}}$

(b) $E_r = \frac{200}{71} = 2.82$, $\epsilon_r = \frac{F_r}{L_r^2 E_r} = \frac{2.5}{2.82} = 0.8875$

$\epsilon_p = 0.8875(300) = 266\mu$

$\sigma_p = \frac{E_p}{1-\nu_p^2}\epsilon_p = \frac{200(10^3)}{1-0.29^2}(266)(10^{-6}) = \underline{58.1\text{ MPa}}$

8.4

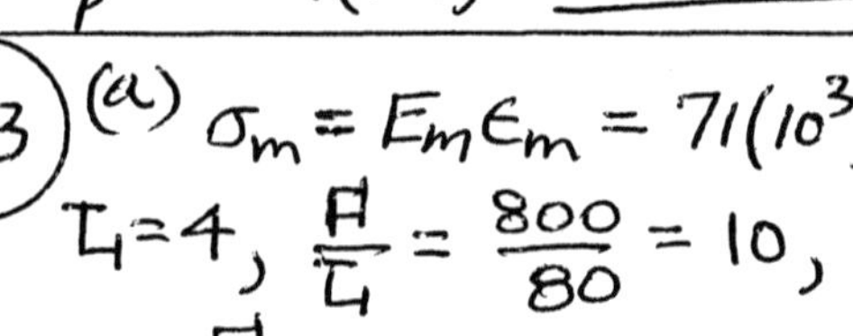

1, 3, 4, 8, 11: $\leftarrow\square\rightarrow \sigma$ one unknown

2: $\leftarrow\square\rightarrow \sigma$ one unknown; $100/0.25 = 400\text{ psi}$

5, 10: no stress

7: σ_1, σ_2 two unknowns

6: σ, $30°$ one unknown

9: $30°$, σ one unknown

8.5 Shear stress is zero due to symmetry.

8.6 Shear stress is zero due to symmetry.

8.7 From Example 8.4-1, $\epsilon_x = \epsilon_A$, $\epsilon_y = \epsilon_c$, $\gamma_{xy} = 2\epsilon_B - \epsilon_A - \epsilon_c$

$$\epsilon_{p1,p2} = \frac{\epsilon_x+\epsilon_y}{2} \pm \sqrt{\left(\frac{\epsilon_x-\epsilon_y}{2}\right)^2 + \left(\frac{\gamma_{xy}}{2}\right)^2} = \frac{\epsilon_A+\epsilon_c}{2} \pm \sqrt{\left(\frac{\epsilon_A-\epsilon_c}{2}\right)^2 + \left(\frac{2\epsilon_B-\epsilon_A-\epsilon_c}{2}\right)^2}$$

$$= \frac{1}{2}\left(\epsilon_A+\epsilon_c \pm \sqrt{(\epsilon_A-\epsilon_c)^2 + (2\epsilon_B-\epsilon_A-\epsilon_c)^2}\right) \quad \boxed{\text{ANS}}$$

$$\sigma_{p1} = \frac{E}{1-\nu^2}(\epsilon_{p1}+\nu\epsilon_{p2}), \quad \sigma_{p2} = \frac{E}{1-\nu^2}(\epsilon_{p2}+\nu\epsilon_{p1})$$

$$\sigma_{p1,p2} = \frac{E}{2(1-\nu^2)}\left[(1+\nu)(\epsilon_A+\epsilon_c) \pm (1-\nu)\sqrt{(\epsilon_A-\epsilon_c)^2 + (2\epsilon_B-\epsilon_A-\epsilon_c)^2}\right]$$

$$= \frac{E}{2}\left[\frac{\epsilon_A+\epsilon_c}{1-\nu} \pm \frac{1}{1+\nu}\sqrt{(\epsilon_A+\epsilon_c)^2 + (2\epsilon_B-\epsilon_A-\epsilon_c)^2}\right] \quad \boxed{\text{ANS}}$$

8.8 From Eq. (3.9-7) $\theta_p = \frac{1}{2}\tan^{-1}\left(\frac{2\tau_{xy}}{\sigma_x-\sigma_y}\right)$ replace σ with ϵ and τ with $\gamma/2$. Thus,

$$\theta_p = \frac{1}{2}\tan^{-1}\left(\frac{\gamma_{xy}}{\epsilon_x-\epsilon_y}\right).$$ From Example 8.4-1, $\epsilon_x = \epsilon_A$, $\epsilon_y = \epsilon_c$, and $\gamma_{xy} = 2\epsilon_B - \epsilon_A - \epsilon_c$

Thus, $$\theta_p = \frac{1}{2}\tan^{-1}\left(\frac{2\epsilon_B-\epsilon_A-\epsilon_c}{\epsilon_A-\epsilon_c}\right) \quad \boxed{\text{ANS}}$$

8.9 $\epsilon_x = \epsilon_A$. For B and C use Eq. (3.9-11). For B, $\theta = -60°$

$$\epsilon_B = \epsilon_A\cos^2(-60) + \epsilon_y\sin^2(-60) + \gamma_{xy}\sin(-60)\cos(-60)$$

$$= 0.25\epsilon_A + 0.75\epsilon_y - \frac{\sqrt{3}}{4}\gamma_{xy} \quad ①$$

For C, $\theta = 60°$. This yields

$$\epsilon_C = 0.25 + 0.75\epsilon_y + \frac{\sqrt{3}}{4}\gamma_{xy} \quad ②$$

Add ①+② yields $\epsilon_B+\epsilon_c = \frac{1}{2}\epsilon_A + \frac{3}{2}\epsilon_y \Rightarrow \epsilon_y = \frac{1}{3}(2\epsilon_B+2\epsilon_c-\epsilon_A)$ ③

②−① yields $\epsilon_c-\epsilon_B = \frac{\sqrt{3}}{2}\gamma_{xy} \Rightarrow \gamma_{xy} = \frac{2}{\sqrt{3}}(\epsilon_c-\epsilon_B)$ ④

$$\epsilon_{p1,p2} = \frac{\epsilon_x+\epsilon_y}{2} \pm \sqrt{\left(\frac{\epsilon_x-\epsilon_y}{2}\right)^2 + \left(\frac{\gamma_{xy}}{2}\right)^2}$$ sub in ③ & ④

(Continued)

$$\epsilon_{p1,p2} = \frac{\epsilon_A + \frac{1}{3}(2\epsilon_B + 2\epsilon_C - \epsilon_A)}{2} \pm \sqrt{\left[\frac{\epsilon_A - \frac{1}{3}(2\epsilon_B + 2\epsilon_C - \epsilon_A)}{2}\right]^2 + \frac{(\epsilon_C - \epsilon_B)^2}{3}}$$

$$= \frac{\epsilon_A + \epsilon_B + \epsilon_C}{3} \pm \sqrt{\left(\frac{2\epsilon_A - \epsilon_B - \epsilon_C}{3}\right)^2 + \frac{\epsilon_C^2 - 2\epsilon_B\epsilon_C + \epsilon_B^2}{3}}$$

$$= \frac{\epsilon_A + \epsilon_B + \epsilon_C}{3} \pm \frac{1}{3}\sqrt{4\epsilon_A^2 - 4\epsilon_A\epsilon_B - 4\epsilon_A\epsilon_C + \epsilon_B^2 + 2\epsilon_B\epsilon_C + \epsilon_C^2 + 3\epsilon_C^2 - 6\epsilon_B\epsilon_C + 3\epsilon_B^2}$$

$$= \frac{\epsilon_A + \epsilon_B + \epsilon_C}{3} \pm \frac{1}{3}\sqrt{4\epsilon_A^2 + 4\epsilon_B^2 + 4\epsilon_C^2 - 4\epsilon_A\epsilon_B - 4\epsilon_B\epsilon_C - 4\epsilon_C\epsilon_A}$$

$$= \frac{\epsilon_A + \epsilon_B + \epsilon_C}{3} \pm \frac{\sqrt{2}}{3}\sqrt{(\epsilon_A^2 - 2\epsilon_A\epsilon_B + \epsilon_B^2) + (\epsilon_B^2 - 2\epsilon_B\epsilon_C + \epsilon_C^2) + (\epsilon_C^2 - 2\epsilon_A\epsilon_C + \epsilon_A^2)}$$

$$= \frac{\epsilon_A + \epsilon_B + \epsilon_C}{3} \pm \frac{\sqrt{2}}{3}\sqrt{(\epsilon_A - \epsilon_B)^2 + (\epsilon_B - \epsilon_C)^2 + (\epsilon_C - \epsilon_A)^2}$$ **ANS**

$$\sigma_{p1} = \frac{E}{1-\nu^2}(\epsilon_{p1} + \nu\epsilon_{p2}), \quad \sigma_{p2} = \frac{E}{1-\nu^2}(\epsilon_{p2} + \nu\epsilon_{p1})$$

$$\sigma_{p1,p2} = \frac{E}{1-\nu^2}\left[(1+\nu)\left(\frac{\epsilon_A + \epsilon_B + \epsilon_C}{3}\right) \pm \frac{\sqrt{2}}{3}(1-\nu)\sqrt{(\epsilon_A - \epsilon_B)^2 + (\epsilon_B - \epsilon_C)^2 + (\epsilon_C - \epsilon_A)^2}\right]$$

$$= \frac{E}{3}\left[\frac{\epsilon_A + \epsilon_B + \epsilon_C}{1-\nu} \pm \frac{\sqrt{2}}{1+\nu}\sqrt{(\epsilon_A - \epsilon_B)^2 + (\epsilon_B - \epsilon_C)^2 + (\epsilon_C - \epsilon_A)^2}\right]$$ **ANS**

8.10 Eq. (3.9-7) for strain is $\theta_p = \frac{1}{2}\tan^{-1}\left(\frac{\gamma_{xy}}{\epsilon_x - \epsilon_y}\right)$

From the solution of Prob 8.11,

$\epsilon_x = \epsilon_A$, $\epsilon_y = \frac{1}{3}(2\epsilon_B + 2\epsilon_C - \epsilon_A)$, $\gamma_{xy} = \frac{2}{\sqrt{3}}(\epsilon_C - \epsilon_B)$. Thus,

$$\theta_p = \frac{1}{2}\tan^{-1}\frac{2}{\sqrt{3}}\frac{(\epsilon_C - \epsilon_B)}{\left[\epsilon_A - \frac{1}{3}(2\epsilon_B + 2\epsilon_C - \epsilon_A)\right]} = \frac{1}{2}\tan^{-1}\frac{2\sqrt{3}(\epsilon_C - \epsilon_B)}{3\epsilon_A - 2\epsilon_B - 2\epsilon_C + \epsilon_A}$$

$$= \frac{1}{2}\tan^{-1}\left[\frac{\sqrt{3}(\epsilon_C - \epsilon_B)}{2\epsilon_A - \epsilon_B - \epsilon_C}\right]$$ **ANS**

8.11 $\epsilon_{p1,p2} = \frac{1}{2}\left(0 + 0 \pm \sqrt{(0-0)^2 + [2(500) - 0 - 0]^2}\right) = \underline{500\mu, -500\mu}$

$\sigma_{p1,p2} = \frac{71(10^3)}{2}\left[0 \pm \frac{1}{1.34}\sqrt{[2(500)]^2}\right](10^{-6}) = \underline{26.5, -26.5\ MPa}$

$\theta_{p1} = \frac{1}{2}\tan^{-1}\left[\frac{2(500)}{0}\right] = 45^\circ$

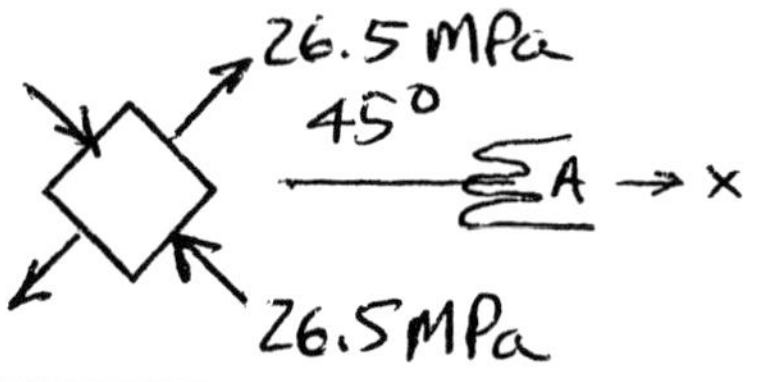

(Continued)

(b) $\epsilon_x = 0,\ \epsilon_y = 0,\ \gamma_{xy} = 2(500) = 1000\mu \Rightarrow \frac{\gamma_{xy}}{2} = 500\mu$

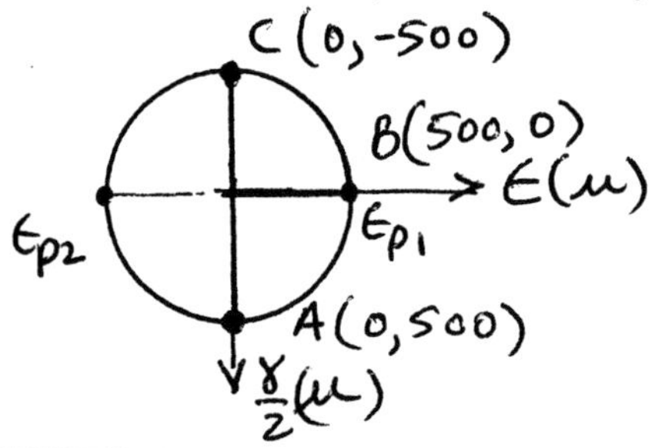

8.12 (a) $\epsilon_{p1,p2} = \frac{1}{2}\left(220+780 \pm \sqrt{(220-780)^2 + 2[2(500)-220-780]^2}\right) = \underline{780\mu,\ 220\mu}$

$$\sigma_{p1,p2} = \frac{29(10^3)}{2}\left[\frac{220+780}{1-0.285} \pm \frac{1}{1.285}\sqrt{(220-780)^2 + 2[2(500)-220-780]^2}\right](10^{-6})$$

$= \underline{25.6,\ 13.96 \text{ kpsi}}$

$$\theta_{p1} = \frac{1}{2}\tan^{-1}\left[\frac{2(500)-220-780}{220-780}\right] = \frac{1}{2}\tan^{-1}\left(\frac{0}{-560}\right) = \underline{90^\circ}$$

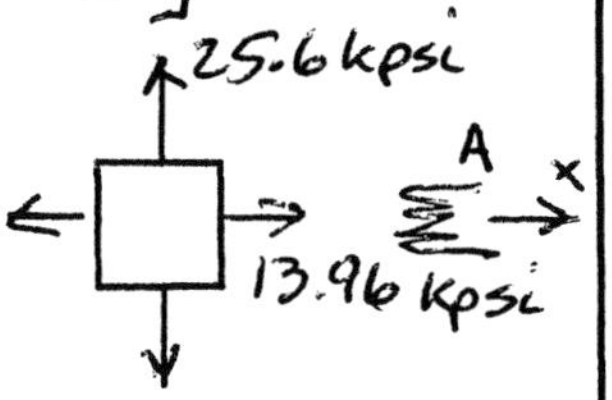

(b) $\epsilon_x = 220\mu,\ \epsilon_y = 780\mu,\ \gamma_{xy} = 2(500) - 220 - 780 = 0$

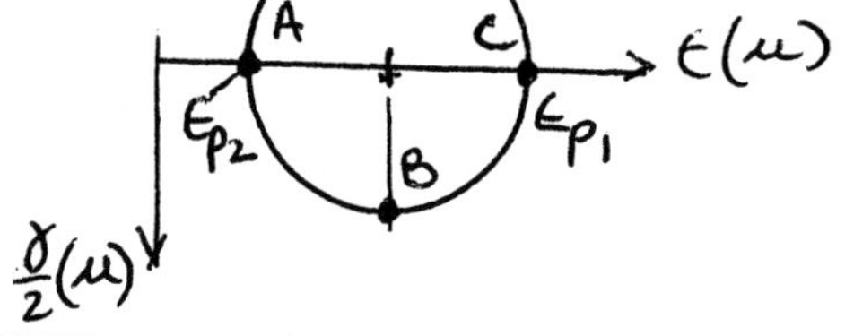

8.13 (a) $\epsilon_{p1,p2} = \frac{1}{2}\left(-400+200 \pm \sqrt{(-400-200)^2 + [2(-400)+400-200]^2}\right) = \underline{324\mu,\ -524\mu}$

$$\sigma_{p1,p2} = \frac{105(10^3)}{2}\left[\frac{-400+200}{1-0.324} \pm \frac{1}{1.324}\sqrt{(-400-200)^2 + [2(-400)+400-200]^2}\right](10^{-6})$$

$= \underline{18.1,\ -49.2 \text{ MPa}}$

$$\theta_{p1} = \frac{1}{2}\tan^{-1}\left[\frac{2(-400)+400-200}{-400-200}\right] = \frac{1}{2}\tan^{-1}\underbrace{\left(\frac{-600}{-600}\right)}_{\text{3rd quad}} = \underline{112.5^\circ}$$

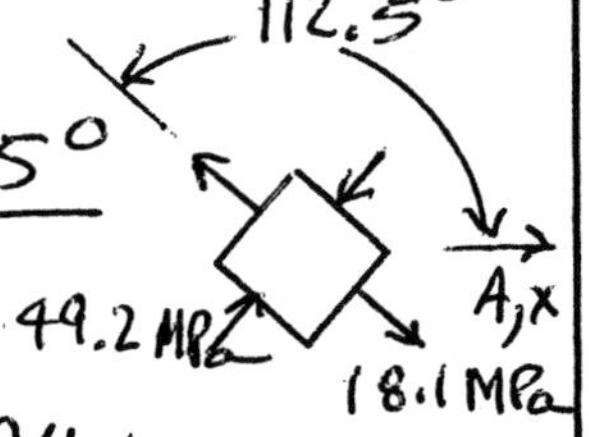

(b) $\epsilon_x = -400\mu,\ \epsilon_y = 200\mu,\ \frac{\gamma_{xy}}{2} = \frac{2(-400)+400-200}{2} = -300\mu$.

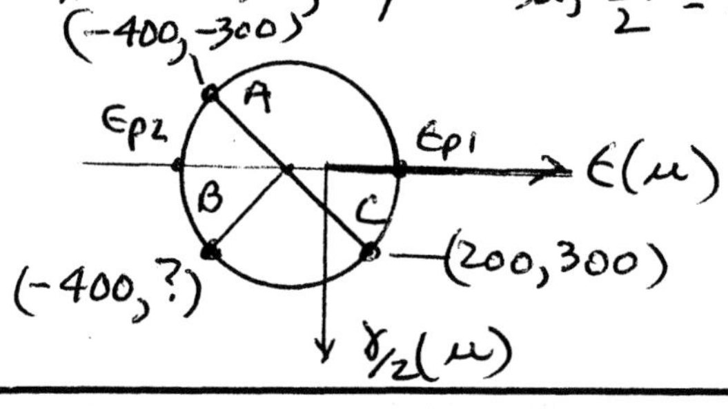

Can determine $\frac{\gamma}{2}$ for gage B but it is unnecessary here.

8.14 (a)

$$\epsilon_{p1,p2} = \frac{1}{2}\left(-250+750 \pm \sqrt{(-250-750)^2+[2(-616)+250-750]^2}\right)$$

$$= \underline{1250\mu, -750\mu}$$

$$\sigma_{p1,p2} = \frac{10.3(10^3)}{2}\left[\frac{-250+750}{1-0.334} \pm \frac{1}{1.334}\sqrt{(-250-750)^2+[2(-616)+250-750]^2}\right](10^{-6})$$

$$= \underline{11.6, -3.86 \text{ kpsi}}$$

$$\theta_{p1} = \frac{1}{2}\tan^{-1}\left[\frac{2(-616)+250-750}{-250-750}\right] = \frac{1}{2}\tan^{-1}\left(\frac{-1732}{-1000}\right) = \underline{120^\circ}$$

(3rd quadrant)

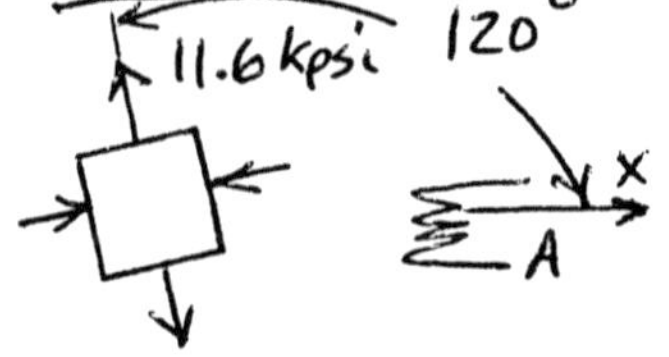

(b) $\epsilon_x = -250\mu$, $\epsilon_y = 750\mu$

$$\frac{\gamma_{xy}}{2} = \frac{2(-616)+250-750}{2} = -866\mu$$

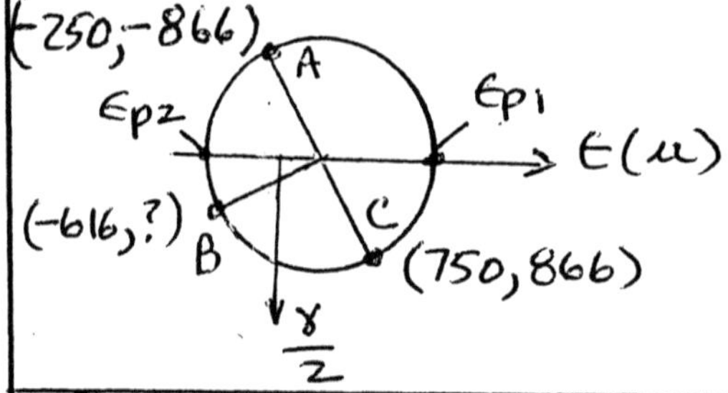

can determine $\frac{\gamma}{2}$ for gage B but it is unnecessary here.

8.15 (a)

$$\epsilon_{p1,p2} = \frac{1}{2}\left(131+269 \pm \sqrt{(131-269)^2+[2(594)-131-269]^2}\right) = \underline{600\mu, -200\mu}$$

$$\sigma_{p1,p2} = \frac{100(10^3)}{2}\left[\frac{131+269}{1-0.211} \pm \frac{1}{1.211}\sqrt{(131-269)^2+[2(594)-131-269]^2}\right](10^{-6})$$

$$= \underline{58.4, -7.68 \text{ MPa}}$$

$$\theta_{p1} = \frac{1}{2}\tan^{-1}\left[\frac{2(594)-131-269}{131-269}\right] = \frac{1}{2}\tan^{-1}\left(\frac{788}{-138}\right) = \underline{50^\circ}$$

(2nd quadrant)

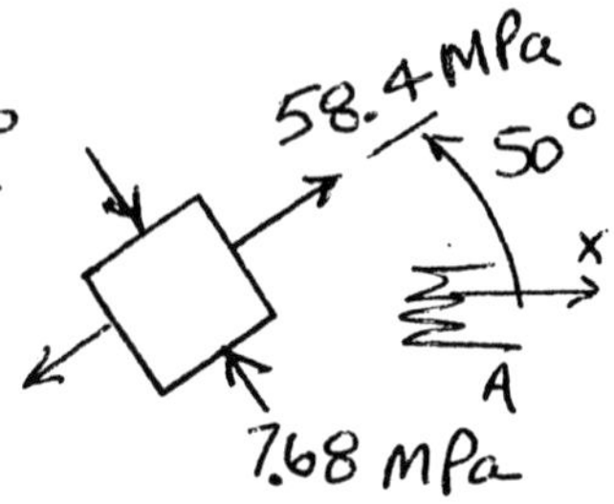

(b) $\epsilon_x = 131\mu$, $\epsilon_y = 269\mu$

$$\frac{\gamma_{xy}}{2} = \frac{2(594)-131-269}{2} = 394\mu$$

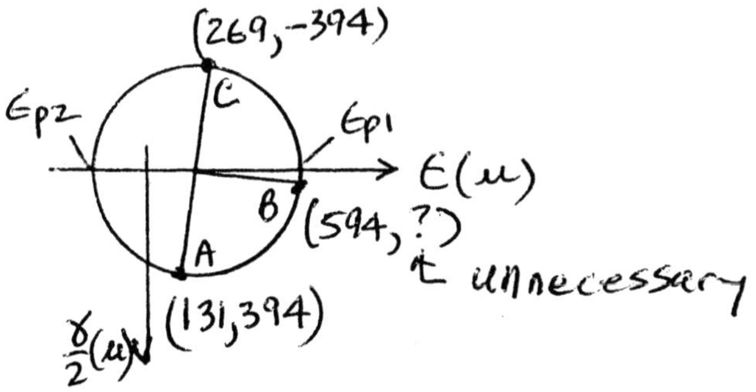

8.16 (a)

$$\epsilon_{p1,p2} = \frac{-150-150+300}{3} \pm \frac{\sqrt{2}}{3}\sqrt{(-150+150)^2+(-150-300)^2+(300+150)^2}$$

$$= \underline{300\mu, -300\mu}$$

$$\sigma_{p1,p2} = \frac{200(10^3)}{3}\left[\frac{-150-150+300}{1-0.285} \pm \frac{\sqrt{2}}{1.285}\sqrt{(-150+150)^2+(-150-300)^2+(300+150)^2}\right]\cdot 10^{-6}$$

$$= \underline{46.7, -46.7 \text{ MPa}}$$

2nd quadrant

$$\theta_{p1} = \frac{1}{2}\tan^{-1}\left[\frac{\sqrt{3}(300+150)}{2(-150)+150-300}\right] = \frac{1}{2}\tan^{-1}\left(\frac{779.4}{-450}\right) = \underline{60^\circ}$$

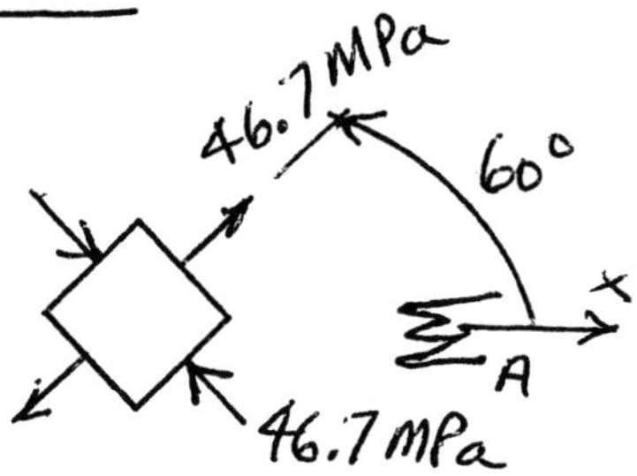

(b) $\epsilon_x = -150\mu$

$$\frac{\gamma_{xy}}{2} = \frac{1}{\sqrt{3}}(300+150) = 259.8\mu$$

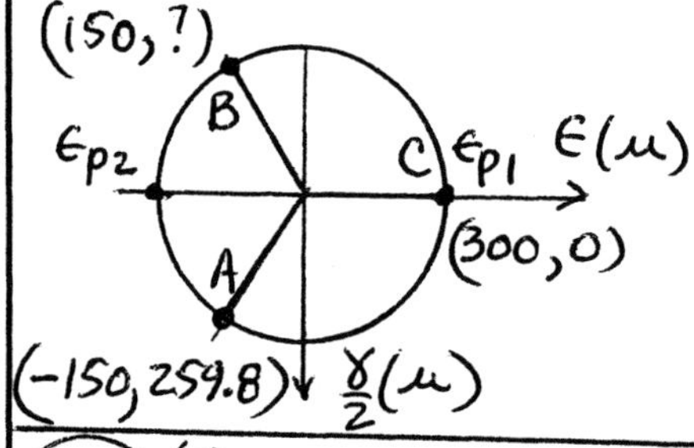

8.17 (a)

$$\epsilon_{p1,p2} = \frac{220+100+340}{3} \pm \frac{\sqrt{2}}{3}\sqrt{(220-100)^2+(100-340)^2+(340-220)^2}$$

$$= \underline{358.6\mu, 81.4\mu}$$

$$\sigma_{p1,p2} = \frac{10.3(10^3)}{3}\left[\frac{220+100+340}{1-0.334} \pm \frac{\sqrt{2}}{1.334}\sqrt{(220-100)^2+(100-340)^2+(340-220)^2}\right]10^{-6}$$

$$= \underline{4.47, 2.33 \text{ kpsi}}$$

$$\theta_{p1} = \frac{1}{2}\tan^{-1}\left[\frac{\sqrt{3}(340-100)}{2(220)-100-340}\right] = \frac{1}{2}\tan^{-1}\left(\frac{415.7}{0}\right) = \underline{45^\circ}$$

(b) $\epsilon_x = 220\mu$, $\frac{\gamma_{xy}}{2} = \frac{340-100}{\sqrt{3}} = 138.6\mu$

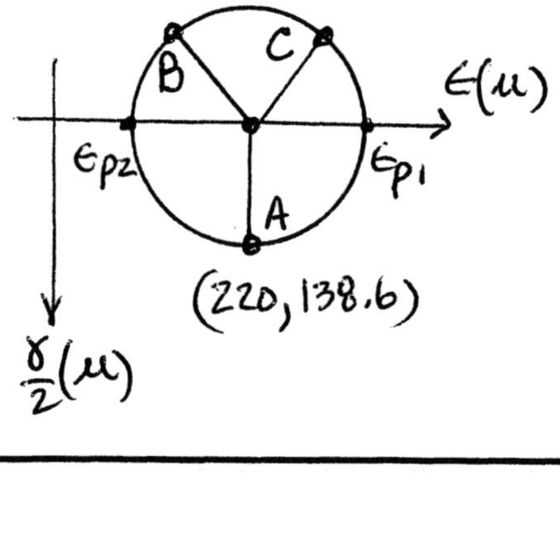

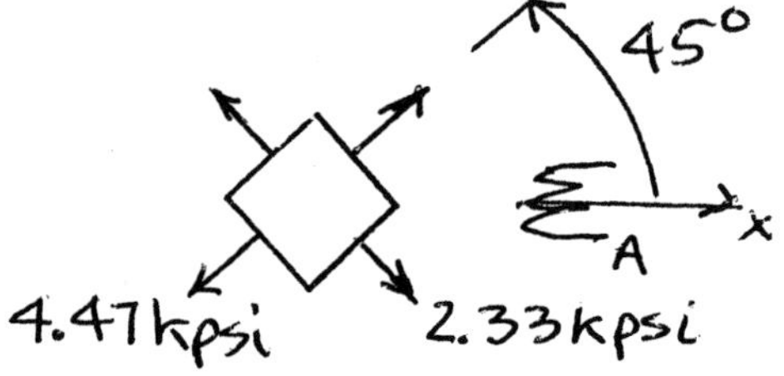

8.18 (a)

$$\epsilon_{p1,p2} = \frac{-215+442+440}{3} \pm \frac{\sqrt{2}}{3}\sqrt{(-215-442)^2+(442-440)^2+(440+215)^2}$$

$$= \underline{660\mu, -215\mu}$$

$$\sigma_{p1,p2} = \frac{71(10^3)}{3}\left[\frac{-215+442-440}{1-0.34} \pm \frac{\sqrt{2}}{1.34}\sqrt{(-215-442)^2+(442-440)^2+(440+215)^2}\right](10^{-6})$$

$$= \underline{47.1,\ 0.75\text{ MPa}}$$

$$\theta_{p1} = \frac{1}{2}\tan^{-1}\left[\frac{\sqrt{3}(440-442)}{2(-215)-442-440}\right] = \frac{1}{2}\tan^{-1}\left(\frac{-3.46}{-1312}\right) = \underline{90.1^\circ}$$

(3rd quadrant)

(b) $\epsilon_x = -215\mu$, $\frac{\gamma_{xy}}{2} = \frac{440-442}{\sqrt{3}} = -1.2\mu$

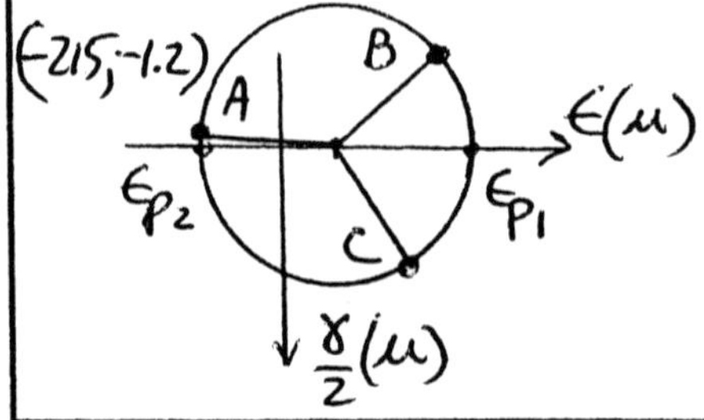

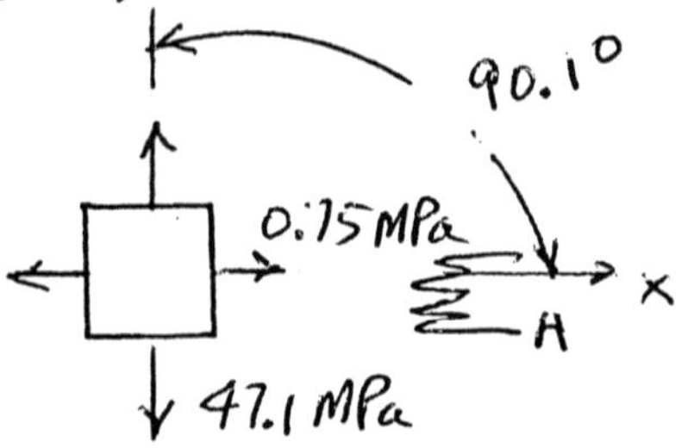

8.19 (a)

$$\epsilon_{p1,p2} = \frac{56-125-89}{3} \pm \frac{\sqrt{2}}{3}\sqrt{(56+125)^2+(-125+89)^2+(-89-56)^2}$$

$$= \underline{58.0\mu, -163.3\mu}$$

$$\sigma_{p1,p2} = \frac{29(10^3)}{3}\left[\frac{56-125-89}{1-0.285} \pm \frac{\sqrt{2}}{1.285}\sqrt{(56+125)^2+(-125+89)^2+(-89-56)^2}\right](10^{-6})$$

$$= \underline{0.361,\ -4.63\text{ kpsi}}$$

$$\theta_{p1} = \frac{1}{2}\tan^{-1}\left[\frac{\sqrt{3}(-89+125)}{2(56)+125+89}\right] = \frac{1}{2}\tan^{-1}\left(\frac{62.4}{326}\right) = 5.4^\circ$$

(b) $\epsilon_x = 56\mu$, $\frac{\gamma_{xy}}{2} = \frac{-89+125}{\sqrt{3}} = 20.8\mu$

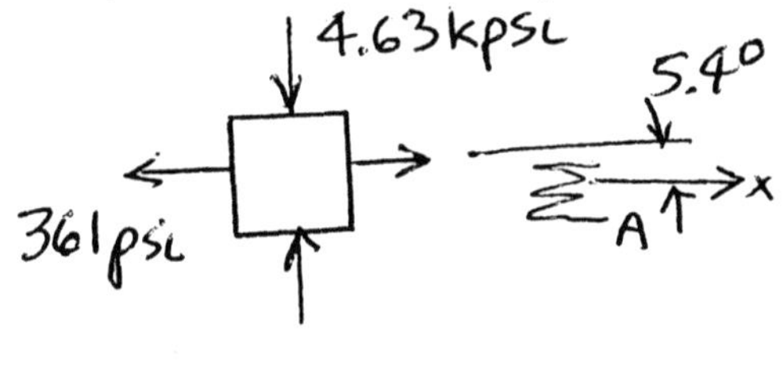

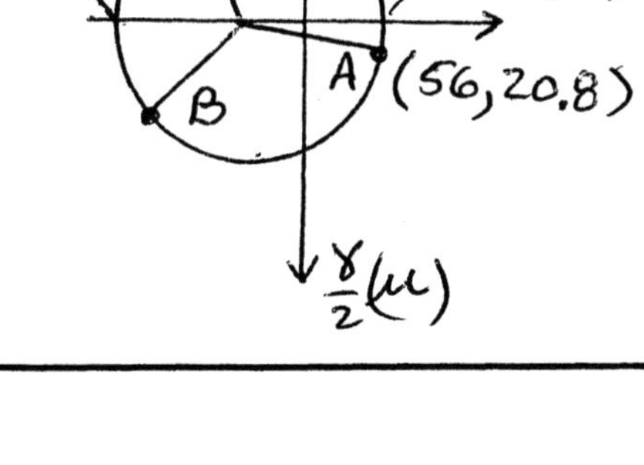

8.20) Considering Eq.(8.5-2) apply ΔR_i to Eq(8.5-1)

$$\Delta E = \frac{(R_1+\Delta R_1)(R_3+\Delta R_3)-(R_2+\Delta R_2)(R_4+\Delta R_4)}{(R_1+\Delta R+R_2+\Delta R_2)(R_3+\Delta R_3+R_4+\Delta R_4)} V \quad (1)$$

The denominator can be written as

$$(R_1+R_2)(R_3+R_4)\left(1+\frac{\Delta R_1+\Delta R_2}{R_1+R_2}\right)\left(1+\frac{\Delta R_3+\Delta R_4}{R_3+R_4}\right)$$

Place the last two terms ( )( ) into the numerator of ΔE as

$$\left(1+\frac{\Delta R_1+\Delta R_2}{R_1+R_2}\right)^{-1}\left(1+\frac{\Delta R_3+\Delta R_4}{R_3+R_4}\right)^{-1}$$

Expand these terms binomially and retain only the linear terms as

$$\left(1-\frac{\Delta R_1+\Delta R_2}{R_1+R_2}\right)\left(1-\frac{\Delta R_3+R_4}{R_3+R_4}\right) \quad (2)$$

The terms in the numerator can be written as

$$\left[R_1R_3\left(1+\frac{\Delta R_1}{R_1}\right)\left(1+\frac{\Delta R_3}{R_3}\right)-R_2R_4\left(1+\frac{\Delta R_2}{R_2}\right)\left(1+\frac{\Delta R_4}{R_4}\right)\right]V \quad (3)$$

But from Eq.(8.5-2) $R_2R_4 = R_1R_3$ and (3) becomes

$$R_1R_3\left[\left(1+\frac{\Delta R_1}{R_1}\right)\left(1+\frac{\Delta R_3}{R_3}\right)-\left(1+\frac{\Delta R_2}{R_2}\right)\left(1+\frac{\Delta R_4}{R_4}\right)\right]V \quad (4)$$

Expand (4) retaining only linear terms in ΔR

$$R_1R_3\left(\frac{\Delta R_1}{R_1}-\frac{\Delta R_2}{R_2}+\frac{\Delta R_3}{R_3}-\frac{\Delta R_4}{R_4}\right)V$$

Multiply this by (2), divide by $(R_1+R_2)(R_3+R_4)$ and retain liner terms in ΔR.

$$\Delta E = \frac{R_1R_3}{(R_1+R_2)(R_3+R_4)}\left(\frac{\Delta R_1}{R_1}-\frac{\Delta R_2}{R_2}+\frac{\Delta R_3}{R_3}-\frac{\Delta R_4}{R_4}\right)V \quad (5)$$

The term $R_3+R_4 = R_3\left(1+\frac{R_4}{R_3}\right) = R_3\left(1+\frac{R_1}{R_2}\right)$ from Eq(8.5-2)

$= R_2R_3(R_1+R_2)$ sub into (5) results in

$$\Delta E = Vr\left(\frac{\Delta R_1}{R_1}-\frac{\Delta R_2}{R_2}+\frac{\Delta R_3}{R_3}-\frac{\Delta R_4}{R_4}\right)$$

ANS

where $r = R_1R_2/(R_1+R_2)^2$

Using Eq (8.5-2) it can be shown that $\frac{R_1R_2}{(R_1+R_2)^2} = \frac{R_3R_4}{(R_3+R_4)^2}$

8.21 $\Delta E = \dfrac{(R_1+\Delta R_1)R_3 - R_2R_4}{(R_1+\Delta R_1+R_2)(R_3+R_4)}V = \dfrac{(R_1R_3-R_2R_4)+R_3\Delta R_1}{(R_1+R_2)(R_3+R_4)\left(1+\dfrac{\Delta R_1}{R_1+R_2}\right)}V$

From Eq. (8.5-2) $R_1R_3 - R_2R_4 = 0$. From problem statement $R_2 = R_1$

Thus, $\Delta E = \dfrac{R_1R_3}{(R_1+R_2)(R_3+R_4)} \dfrac{\Delta R_1/R_1}{\left(1+\dfrac{\Delta R_1}{2R_1}\right)}V = rV\dfrac{S_A\epsilon}{1+\dfrac{S_A\epsilon}{2}}$ but $S_A = 2$

$\therefore \dfrac{\Delta E}{Vr} = \dfrac{2\epsilon}{1+\epsilon}$ exact **ANS** linear: $\dfrac{\Delta E}{Vr} = 2\epsilon$

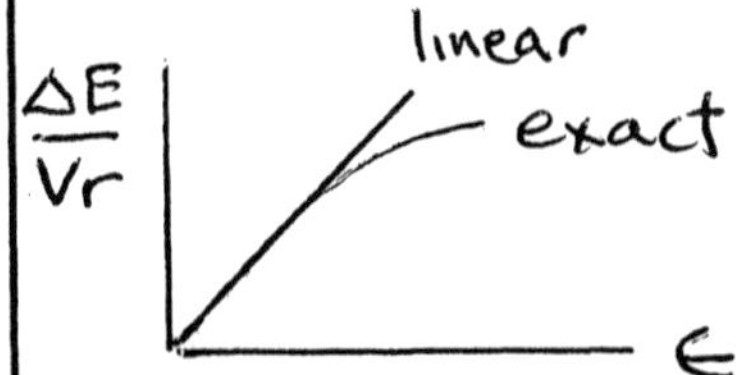

$\%\text{error} = \dfrac{2\epsilon - \dfrac{2\epsilon}{1+\epsilon}}{2\epsilon/1+\epsilon}(100) = 5$

$\dfrac{2\epsilon + 2\epsilon^2 - 2\epsilon}{2\epsilon} = 0.05 \Rightarrow \epsilon = 0.05$

8.22

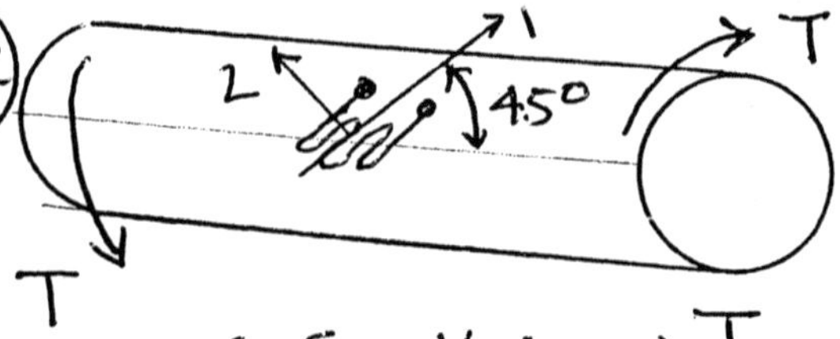

$\tau = \dfrac{16T}{\pi d^3} = \sigma_1 = -\sigma_2$

$\epsilon_1 = \dfrac{1}{E}(\sigma_1 - \nu\sigma_2) = \dfrac{1+\nu}{E}\sigma_1 = \dfrac{(1+\nu)16T}{E\pi d^3}$

$\therefore \epsilon = \dfrac{16}{\pi}(1+\nu)\dfrac{T}{Ed^3}$ **ANS**

8.23 Place a second gage in the 2 direction shown in the solution to Problem 8.22. From that solution.

$\epsilon_1 = \dfrac{16}{\pi}(1+\nu)\dfrac{T}{Ed^3}$. Similarly $\epsilon_2 = -\dfrac{16}{\pi}(1+\nu)\dfrac{T}{Ed^3}$

Placing gages 1 and 2 as R_1 and R_2 gages in the Wheatstone bridge, respectively, yields a reading of $\epsilon_1 - \epsilon_2 = \dfrac{32}{\pi}(1+\nu)\dfrac{T}{Ed^3}$ **ANS** With temperature compensation

8.24 Label the lower and upper gages as A and B, respect.

$\epsilon_A = \epsilon_x\cos^2 45 + \epsilon_y \sin^2 45 + \gamma_{xy}\sin 45\cos 45 = \frac{1}{2}(\epsilon_x + \epsilon_y + \gamma_{xy})$ ①

Likewise, $\epsilon_B = \frac{1}{2}(\epsilon_x + \epsilon_y - \gamma_{xy})$ ②. Adding ① & ② would give a measure of $\epsilon_x + \epsilon_y$. Subtracting ① & ② would give a measure of γ_{xy}. Thus connect terminals 1, 2, 3 to c, b, a respectively. **ANS** (Other combinations exist)

(b) Output $= \gamma_{xy}$ **ANS**

8.25 Note: use $\nu = 0.285$ in correction equations only

$$\epsilon_A = \epsilon_C = 0, \quad \epsilon_B = \frac{[1-0.285(0.035)]\,500}{1-0.035} = 513\mu$$

$$\epsilon_{p1,p2} = \pm\frac{1}{2}\sqrt{[2(513)]^2} = \underline{513\mu, -513\mu}$$

$$\sigma_{p1,p2} = \frac{71(10^3)}{2}\left[\pm\frac{1}{1.34}\sqrt{[2(513)]^2}\right](10^{-6}) = \underline{27.2, -27.2 \text{ MPa}}$$

$$\theta_{p1} = \frac{1}{2}\tan^{-1}\left[\frac{2(513)}{0-0}\right] = \underline{45^\circ}$$

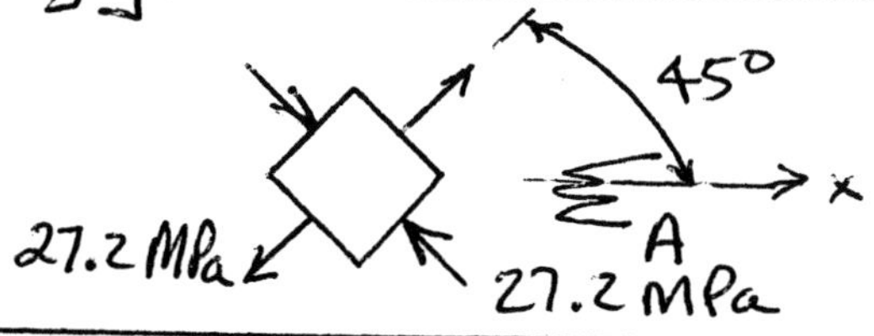

8.26

$$\epsilon_A = \frac{[1-0.285(0.07)](220)-(0.07)[1-0.285(0.07)]780}{1-(0.07)^2} = 162.9\mu$$

$$\epsilon_B = \frac{[1-0.285(0.05)](500)}{1-0.05} - \frac{0.05\{[1-0.285(0.07)](0.93)(220+780)\}}{(1-0.05)(1-0.07^2)} = 470.6\mu$$

$$\epsilon_C = \frac{[1-0.285(0.07)]780 - 0.07[1-0.285(0.07)]220}{1-(0.07)^2} = 753.0\mu$$

$$\epsilon_{p1,p2} = \frac{1}{2}\left(162.9+753 \pm \sqrt{(162.9-753)^2+[2(470.6)-162.9-753]^2}\right)$$

$$= \underline{753.3\mu, 162.6\mu}$$

$$\sigma_{p1,p2} = \frac{29(10^3)}{2}\left[\frac{162.9+753}{1-0.285} \pm \frac{1}{1.285}\sqrt{(162.9-753)^2+[2(470.6)-162.9-753]^2}\right](10^{-6})$$

$$= \underline{25.2, 11.9 \text{ kpsi}}$$

$$\theta_{p1} = \frac{1}{2}\tan^{-1}\left[\frac{2(470.6)-162.9-753}{162.9-753}\right] = \frac{1}{2}\underbrace{\tan^{-1}\left(\frac{25.3}{-590}\right)}_{\text{2nd quad.}} = \underline{88.8^\circ}$$

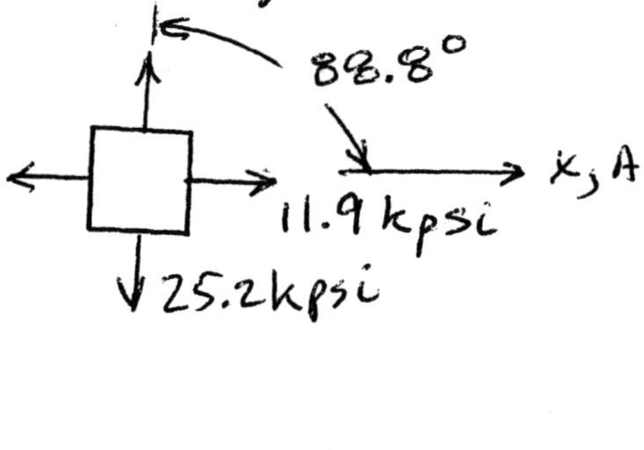

8.27 $\epsilon_A = \dfrac{[1-0.285(0.05)]}{1-0.05^2}\left[-400-(0.05)(200)\right] = -405.2\mu$

$\epsilon_B = \dfrac{[1-0.285(0.05)]}{1-0.05^2}\left[1.05(-400)-(0.05)(-400+200)\right] = -405.2\mu$

$\epsilon_C = \dfrac{[1-0.285(0.05)]}{1-0.05^2}\left[200-0.05(-400)\right] = 217.4\mu$

$\epsilon_{p_1,p_2} = \frac{1}{2}\left(-405.2+217.4 \pm \sqrt{(-405.2-217.4)^2+\left[2(-405.2)+405.2-217.4\right]^2}\right)$

$= \underline{346.3\mu, -534.1\mu}$

$\sigma_{p_1,p_2} = \dfrac{105(3)}{2}\left[\dfrac{-405.2+217.4}{1-0.324} \pm \dfrac{1}{1.324}\sqrt{(-405.2-217.4)^2+\left(2(-405.2)+405.2-217.4\right)^2}\right](10^{-6})$

$= \underline{20.3, -49.5 \text{ MPa}}$

$\theta_{p_1} = \frac{1}{2}\tan^{-1}\left[\dfrac{2(-405.2)+405.2-217.4}{-405.2-217.4}\right] = \frac{1}{2}\underbrace{\tan^{-1}\left(\dfrac{-622.6}{-622.6}\right)}_{\text{3rd quadrant}} = \underline{112.5^\circ}$

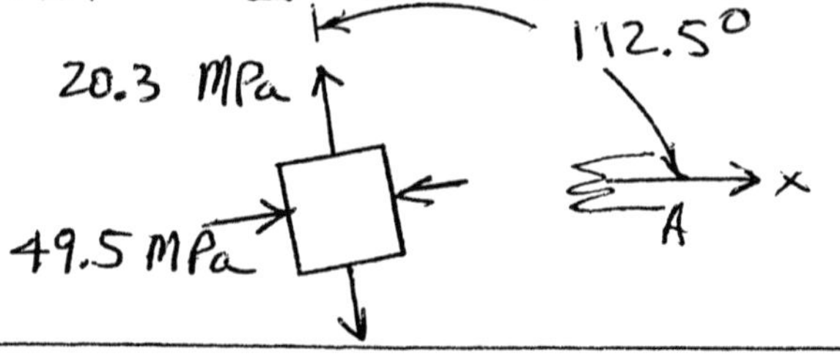

8.28 $\epsilon_A = \dfrac{[1-0.285(0.08)]}{1-0.08^2}\left[-250-(0.08)(750)\right] = -304.9\mu$

$\epsilon_B = \dfrac{[1-0.285(0.08)]}{1-0.08^2}\left[1.08(-616)-0.08(-250+750)\right] = -693.6\mu$

$\epsilon_C = \dfrac{[1-0.285(0.08)]}{1-0.08^2}\left[750-0.08(-250)\right] = 757.3\mu$

$\epsilon_{p_1,p_2} = \frac{1}{2}\left(-304.9+757.3 \pm \sqrt{(-304.9-757.3)^2+\left[2(-693.6)+304.9-757.3\right]^2}\right)$

$= \underline{1288\mu, -836\mu}$

$\sigma_{p_1,p_2} = \dfrac{10.3(10^3)}{2}\left[\dfrac{-304.9+757.3}{1-0.334} \pm \dfrac{1}{1.334}\sqrt{(-304.9-757.3)^2+\left[2(-693.6)+304.9-757.3\right]^2}\right](10^{-6})$

$= \underline{11.7, -4.7 \text{ kpsi}}$

(continued)

$$\theta_{p1} = \frac{1}{2}\tan^{-1}\left[\frac{2(-693.6)+304.9-757.3}{-304.9-757.3}\right]$$

$$= \frac{1}{2}\tan^{-1}\left(\frac{-1839.6}{-1062.2}\right) = \underline{120^\circ}$$

3rd quad.

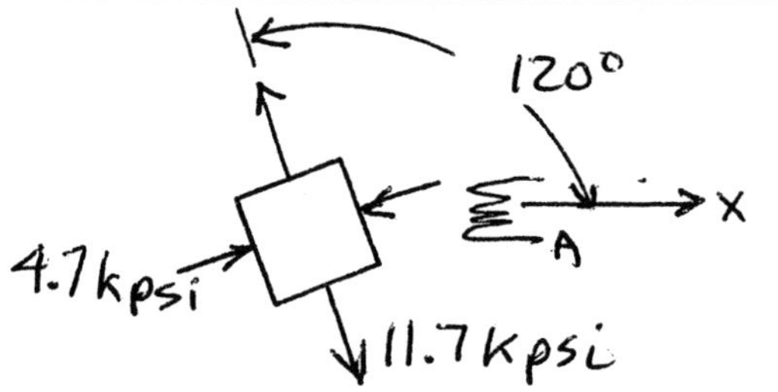

(8.29) $\epsilon_A = \dfrac{[1-0.285(0.08)]131-0.08[1-0.285(0.08)](269)}{1-0.08^2} = 107.7\mu$

$$\epsilon_B = \frac{[1-0.285(0.05)](594)}{1-0.05} - \frac{0.05\{[1-0.285(0.08)](0.92)(131+269)\}}{(1-0.05)(1-0.08^2)} = 597.3\mu$$

$$\epsilon_C = \frac{[1-0.285(0.08)](269)-0.08[1-0.285(0.8)](131)}{1-0.08^2} = 254.3\mu$$

$$\epsilon_{p1,p2} = \frac{1}{2}\left(107.7+254.3 \pm \sqrt{(107.7-254.3)^2+[2(597.3)-107.7-254.3]^2}\right)$$

$$\sigma_{p1,p2} = \frac{100(10^3)}{2}\left[\frac{107.7+254.3}{1-0.211} \pm \frac{1}{1.211}\sqrt{(107.7-254.3)^2+[2(597.3)-107.7-254.3]^2}\right](10^{-6})$$

$$= \underline{57.8,\ -12.0\ MPa}$$

$$\theta_p = \frac{1}{2}\tan^{-1}\left[\frac{2(597.3)-107.7-254.3}{107.7-254.3}\right]$$

$$= \frac{1}{2}\tan^{-1}\left(\frac{832.7}{-146.6}\right) = 50^\circ$$

2nd quad

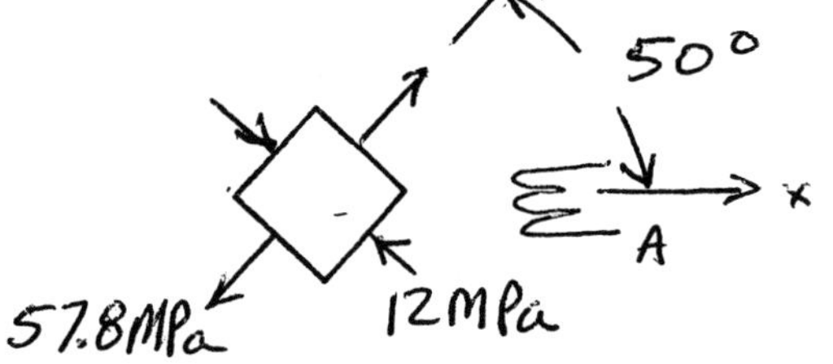

(8.30) Eqs. (H.3-1) were programmed in Excel for calculations
Corrected strains: $\epsilon_A = -155.7\mu$, $\epsilon_B = -153.9\mu$, $\epsilon_C = 311.2\mu$

$$\epsilon_{p1,p2} = \frac{-155.7-153.9+311.2}{3} \pm \frac{\sqrt{2}}{3}\sqrt{(-155.7+153.9)^2+(-153.9-311.2)^2+(311.2+155.7)^2}$$

$$= \underline{311.2\mu,\ -310.1\mu}$$

$$\sigma_{p1,p2} = \frac{200(10^3)}{3}\left[\frac{-155.7-153.9+311.2}{1-0.285} \pm \frac{\sqrt{2}}{1.285}\sqrt{(-155.7+153.9)^2+(-153.9-311.2)^2+(311.2+155.7)^2}\right](10^{-6})$$

$$= \underline{48.5,\ -48.2\ MPa}$$

$$\theta_{p1} = \frac{1}{2}\tan^{-1}\left(\frac{\sqrt{3}(311.2+153.9)}{2(-155.7)+153.9-311.2}\right)$$

$$= \frac{1}{2}\tan^{-1}\left(\frac{805.6}{-468.7}\right) = 60.1^\circ$$

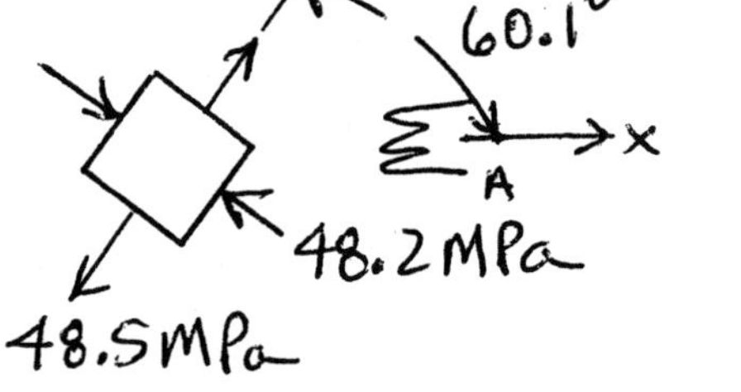

8.31 Eqs. (H.3-1) were programmed in Excel with $\nu = 0.285$

Corrected strains: $\epsilon_A = 201.2\mu$, $\epsilon_B = 82.3\mu$, $\epsilon_C = 327.6\mu$

$$\epsilon_{p1,p2} = \frac{201.2+82.3+327.6}{3} \pm \frac{\sqrt{2}}{3}\sqrt{(201.2-82.3)^2+(82.3-327.6)^2+(327.6-201.2)^2}$$

$$= \underline{345.3\mu,\ 62.1\mu}$$

$$\sigma_{p1,p2} = \frac{10.3(10^3)}{3}\left[\frac{201.2+82.3+327.6}{1-0.334} \pm \frac{\sqrt{2}}{1.334}\sqrt{(201.2-82.3)^2+(82.3-327.6)^2+(327.6-201.2)^2}\right](10^{-6})$$

$$= \underline{4.24,\ 2.06 \text{ kpsi}}$$

$$\theta_{p1} = \frac{1}{2}\tan^{-1}\left[\frac{\sqrt{3}(327.6-82.3)}{2(201.2)-82.3-327.6}\right]$$

$$= \frac{1}{2}\tan^{-1}\left(\frac{424.9}{-7.5}\right) = 45.5^\circ$$

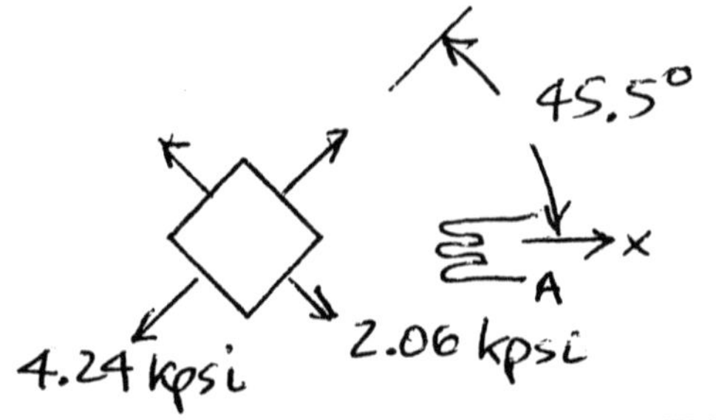

8.32 $$\epsilon_A = \frac{[1-(0.285)(0.05)]}{3(1-0.05^2)}\left[3.05(-215)-2(0.05)(442+440)\right] = -245.1\mu$$

$$\epsilon_B = \frac{[1-0.285(0.05)]}{3(1-0.05^2)}\left[3.05(442)-2(0.05)(-215+440)\right] = 436.7\mu$$

$$\epsilon_C = \frac{[1-0.285(0.05)]}{3(1-0.05^2)}\left[3.05(440)-2(0.05)(-215+442)\right] = 434.6\mu$$

$$\epsilon_{p1,p2} = \frac{-245.1+436.7+434.6}{3} \pm \frac{\sqrt{2}}{3}\sqrt{(-245.1-436.7)^2+(436.7-434.6)^2+(436.4+245.1)^2}$$

$$= \underline{663.2\mu,\ -245.8\mu}$$

$$\sigma_{p1,p2} = \frac{71(10^3)}{3}\left[\frac{-245.1+436.7+434.6}{1-0.34} \pm \frac{\sqrt{2}}{1.34}\sqrt{(-245.1-436.7)^2+(436.7-434.6)^2+(436.4+245.1)^2}\right]\cdot(10^{-6})$$

$$= \underline{46.5,\ -1.63 \text{ MPa}}$$

$$\theta_{p1} = \frac{1}{2}\tan^{-1}\left[\frac{\sqrt{3}(434.6-436.7)}{2(-245.1)-436.7-434.6}\right]$$

$$= \frac{1}{2}\tan^{-1}\left(\frac{-3.64}{-1361.5}\right) = 90.1^\circ$$

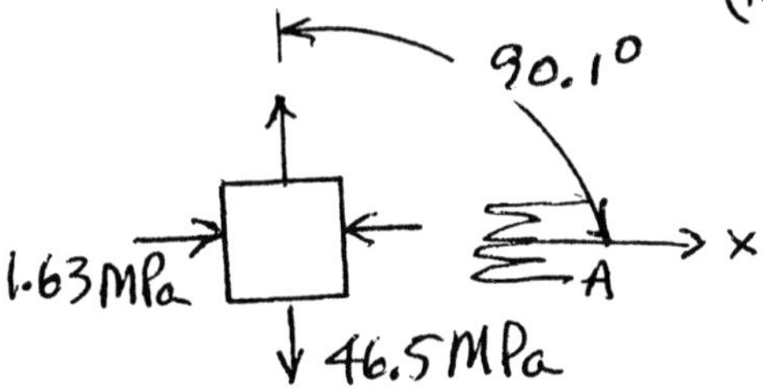

8.33 $$\epsilon_A = \frac{[1-0.285(0.08)]}{3(1-0.08^2)}\left[3.08(56)-2(0.08)(-125-89)\right] = 67.8\mu$$

$$\epsilon_B = \frac{[1-0.285(0.08)]}{3(1-0.08^2)}\left[3.08(-125)-2(0.08)(56-89)\right] = -124.5\mu$$

(Continued)

$$\epsilon_c = \frac{[1-0.285(0.08)]}{3(1-0.08^2)}\left[3.08(-89)-2(0.08)(56-125)\right] = -86.2\mu$$

$$\epsilon_{p1,p2} = \frac{67.8-124.5-86.2}{3} \pm \frac{\sqrt{2}}{3}\sqrt{(67.8+124.5)^2+(-124.5+86.2)^2+(-86.2-67.8)^2}$$

$$= 69.9\mu, -165.2\mu$$

$$\sigma_{p1,p2} = \frac{29(10^3)}{3}\left[\frac{67.8-124.5-86.2}{1-0.285} \pm \frac{\sqrt{2}}{1.285}\sqrt{(67.8+124.5)^2+(-124.5+86.2)^2+(-86.2-67.8)^2}\right](10^{-6})$$

$$= 0.72, -4.58 \text{ kpsi}$$

$$\theta_{p1} = \frac{1}{2}\tan^{-1}\left[\frac{\sqrt{3}(-86.2+124.5)}{2(67.8)+124.5+86.2}\right]$$

$$= \frac{1}{2}\tan^{-1}\left(\frac{66.3}{346.3}\right) = 5.4^\circ$$

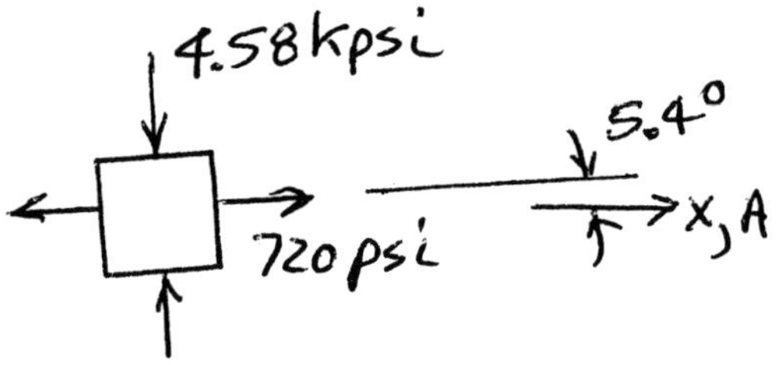

(8.34) Select axes in the direction of and ⊥ to each gage

$$S_{gA}\hat{\epsilon}_A = S_{aA}(\epsilon_A + K_{tA}\epsilon_y) \quad (1)$$

$$S_{gB}\hat{\epsilon}_B = S_{aB}(\epsilon_B + K_{tB}\epsilon_{y'}) \quad (2) \quad y' \perp \text{ to } B$$

$$S_{gC}\hat{\epsilon}_C = S_{aC}(\epsilon_C + K_{tC}\epsilon_{y''}) \quad (3) \quad y'' \perp \text{ to } C$$

For B, $\theta = -60^\circ$

$$\epsilon_B = \epsilon_x\cos^2(-60)+\epsilon_y\sin^2(-60)+\gamma_{xy}\sin(-60)\cos(-60) = \frac{\epsilon_A}{4}+\frac{3}{4}\epsilon_y-\frac{\sqrt{3}}{4}\gamma_{xy} \quad (4)$$

$$\epsilon_{y'} = \epsilon_x\cos^2(30)+\epsilon_y\sin^2(30)+\gamma_{xy}\sin 30\cos 30 = \frac{3}{4}\epsilon_A+\frac{1}{4}\epsilon_y+\frac{\sqrt{3}}{4}\gamma_{xy} \quad (5)$$

For C, $\theta = 60^\circ$, results in

$$\epsilon_C = \frac{\epsilon_A}{4}+\frac{3}{4}\epsilon_y+\frac{\sqrt{3}}{4}\gamma_{xy} \quad (6), \quad \epsilon_{y''} = \frac{3}{4}\epsilon_A+\frac{1}{4}\epsilon_y-\frac{\sqrt{3}}{4}\gamma_{xy} \quad (7)$$

(4)+(6) yields $\epsilon_B+\epsilon_C = \frac{\epsilon_A}{2}+\frac{3}{2}\epsilon_y \Rightarrow \epsilon_y = \frac{1}{3}(2\epsilon_B+2\epsilon_C-\epsilon_A)$ (8)

(6)−(4) yields $\epsilon_C-\epsilon_B = \frac{\sqrt{3}}{2}\gamma_{xy} \Rightarrow \gamma_{xy} = \frac{2}{\sqrt{3}}(\epsilon_C-\epsilon_B)$ (9)

Sub eqs (4)−(9) into (1)−(3)

$$\frac{S_{gA}}{S_{aA}}\hat{\epsilon}_A = \epsilon_A + \frac{K_{tA}}{3}(2\epsilon_B+2\epsilon_C-\epsilon_A)$$

$$\frac{S_{gB}}{S_{aB}}\hat{\epsilon}_B = \epsilon_B + \frac{K_{tB}}{4}\left[3\epsilon_A+\frac{1}{3}(2\epsilon_B+2\epsilon_C-\epsilon_A)+2(\epsilon_C-\epsilon_B)\right]$$

$$\frac{S_{gC}}{S_{aC}}\hat{\epsilon}_C = \epsilon_C + \frac{K_{tC}}{4}\left[3\epsilon_A+\frac{1}{3}(2\epsilon_B+2\epsilon_C-\epsilon_A)-2(\epsilon_C-\epsilon_B)\right]$$

(continued)

Let $\frac{S_{gA}}{S_{aA}}\hat{\epsilon}_A = A$, $\frac{S_{gB}}{S_{aB}}\hat{\epsilon}_B = B$, $\frac{S_{gc}}{S_{ac}} = C$.

$$\left(1-\frac{K_{tA}}{3}\right)\epsilon_A + \frac{2}{3}K_{tA}\epsilon_B + \frac{2}{3}K_{tA}\epsilon_C = A$$

$$\frac{K_{tB}}{4}\left(\frac{8}{3}\right)\epsilon_A + \left(1-\frac{K_{tB}}{3}\right)\epsilon_B + \frac{K_{tB}}{4}\left(\frac{8}{3}\right)\epsilon_C = B$$

$$\frac{K_{tC}}{4}\left(\frac{8}{3}\right)\epsilon_A + \frac{K_{tC}}{4}\left(\frac{8}{3}\right)\epsilon_B + \left(1-\frac{K_{tC}}{3}\right)\epsilon_C = C$$

$$(3-K_{tA})\epsilon_A + 2K_{tA}\epsilon_B + 2K_{tA}\epsilon_C = 3A \qquad (10)$$

$$2K_{tB}\epsilon_A + (3-K_{tB})\epsilon_B + 2K_{tB}\epsilon_C = 3B \qquad (11)$$

$$2K_{tC}\epsilon_A + 2K_{tC}\epsilon_B + (3-K_{tC})\epsilon_C = 3C \qquad (12)$$

Because of symmetry we will only solve for ϵ_A. The remaining strains will follow from the same procedure.

$$\epsilon_A = \frac{\begin{vmatrix} 3A & 2K_{tA} & 2K_{tA} \\ 3B & (3-K_{tB}) & 2K_{tB} \\ 3C & 2K_{tC} & (3-K_{tC}) \end{vmatrix}}{\begin{vmatrix} (3-K_{tA}) & 2K_{tA} & 2K_{tA} \\ 2K_{tB} & (3-K_{tB}) & 2K_{tB} \\ 2K_{tC} & 2K_{tC} & (3-K_{tC}) \end{vmatrix}} = \frac{N}{D} \qquad (13)$$

$$N = 3A\left[(3-K_{tB})(3-K_{tC}) - 4K_{tB}K_{tC}\right] - 3B\left[2K_{tA}(3-K_{tC}) - 4K_{tA}K_{tC}\right]$$
$$+3C\left[4K_{tA}K_{tB} - 2K_{tA}(3-K_{tB})\right]$$
$$= 3A\left[9-3K_{tB}-3K_{tC}-3K_{tB}K_{tC}\right] - 3B\left[6K_{tA}-6K_{tA}K_{tC}\right] - 3C\left[6K_{tA}-6K_{tA}K_{tB}\right]$$

But $A = \frac{S_{gA}}{S_{aA}}\hat{\epsilon}_A = (1-\nu_0 K_{tA})\hat{\epsilon}_A$. Likewise, $B = (1-\nu_0 K_{tB})\hat{\epsilon}_B$, $C = (1-\nu_0 K_{tC})\hat{\epsilon}_C$

Thus,

$$N = 9\Big\{(1-\nu_0 K_{tA})(3-K_{tB}-K_{tC}-K_{tB}K_{tC})\hat{\epsilon}_A - 2K_{tA}\big[(1-\nu_0 K_{tB})(1-K_{tC})\hat{\epsilon}_B$$
$$+ (1-\nu_0 K_{tC})(1-K_{tC})\hat{\epsilon}_C\big]\Big\} \qquad (14)$$

D in Eq. (13) can be shown to be

$$D = 9\left(3K_{tA}K_{tB}K_{tC} - K_{tA}K_{tB} - K_{tB}K_{tC} - K_{tC}K_{tA} - K_{tA} - K_{tB} - K_{tC} + 3\right) = 9K^{-1} \qquad (15)$$

Sub (14) & (15) into (13) which results in the first equation of Eq. (H.3-1)

8.35 For the gage, $\epsilon_t = -\frac{1}{2}\epsilon_a$. From Eq. (8.6-8)

$$e = \frac{0.08}{1-0.285(0.08)}\left(0.285\epsilon_a - \frac{1}{0.29}\epsilon_a\right) = -0.259\,\epsilon_a$$

Error is -25.9% **ANS**

8.36 (a) $\epsilon_{indicated} = \underline{200\mu}$

(b) $$\epsilon_{ind} = \frac{1}{0.5}\int_{-0.25}^{0.25} 200(1-2x)\,dx = \frac{200}{0.5}\left(x - \frac{2x^2}{2}\right)\Bigg|_{-0.25}^{0.25} = \underline{200\mu}$$

(c) $$\epsilon_{ind} = \frac{200}{0.5}\left(x - \frac{2x^2}{2} - \frac{3}{3}x^3\right)\Bigg|_{-0.25}^{0.25} = \underline{187.5\mu}$$

8.37 Eq. (4.2-27b) at $r=a$, $\sigma_\theta = \frac{\sigma}{2}(1+1-4\cos 2\theta) = \sigma(1-2\cos 2\theta)$

At $\theta = \pm 90°$, $\sigma_\theta = 3\sigma = \underline{30\text{ MPa}}$

For a 10 mm gage, $\theta_g = \frac{10}{40\pi}(2\pi) = 0.5\text{ rad}$. The indicated stress from the gage is

$$\sigma_{ind} = \frac{1}{0.5}\int_{\frac{\pi}{2}-0.25}^{\pi/2+0.25} \sigma(1-2\cos\theta)\,d\theta = \frac{\sigma}{0.5}\left(0.5 - \sin 2\theta\Big|_{\pi/2-0.25}^{\frac{\pi}{2}+0.25}\right) = 2.918\sigma$$

$$= \underline{29.18\text{ MPa}}$$

For a 2 mm gage, $\theta_g = 0.1\text{ rad}$

$$\sigma_{ind} = \frac{\sigma}{0.1}\left(0.1 - \sin 2\theta\Big|_{\pi/2-0.05}^{\pi/2+0.05}\right) = 2.997\sigma = \underline{29.97\text{ MPa}}$$

8.38 $\epsilon_{max} = l_p\epsilon_p/l_g = ct_p\epsilon_p/l_g$ ① $l/c = 4.8(10^{-6})$. Thus,

$c = 1.0/4.8(10^{-6}) = 2.083(10^5)$ ips. From ①

$$\epsilon_{max} = 500(10^{-6}) = \frac{2.083(10^5)(2)(10^{-6})}{1.0}\epsilon_p \Rightarrow \epsilon_p = 1200(10^{-6}) = \underline{1200\mu}$$

8.39 $l_p = ct_p$ $\quad$ $\underline{0 < x < ct_p/2}$: $\epsilon = 2\epsilon_p x/ct_p$

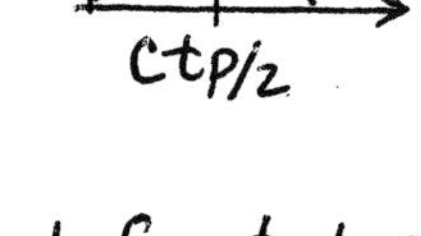

Assuming $ct_p < l_g$ as pulse enters gage area

$$\epsilon_{ind} = \frac{1}{l_g}\int \frac{2\epsilon_p x}{ct_p}dx = \frac{\epsilon_p x^2}{ct_p l_g}, \text{ at } t = t_p/2$$

$$\epsilon_{ind}\Big|_{t_p/2} = \frac{\epsilon_p(ct_p/2)^2}{ct_p l_g} = \frac{ct_p\epsilon_p}{4l_g}.$$ Due to symmetry,

ϵ_{ind} for $t_p/2 < t < t_p$ is mirrored

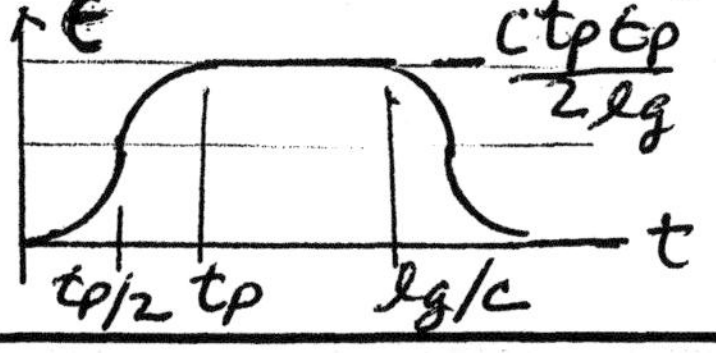

8.40 From curve, $(S_g)_T = \left[1 + \frac{(-0.45)}{100}\right](2.1) = 2.09$

$$\varepsilon_{apparent} = -48.85 + 3.86(-30) - 7.85(10^{-2})(-30)^2 + 4.05(10^{-4})(-30)^3 - 5.28(10^{-7})(-30)^4 = -246.7\,\mu$$

From Eq. (8.6-17) $\varepsilon_{actual} = \frac{2.0}{2.09}(800 + 246.7) = \underline{1001\,\mu}$

8.41 For fig 8.7-17 $N \approx 3$ at the center.

From Eqs. (4.2-38) at $x = 0$, $\sigma_x = \frac{P}{\pi R t}$, $\sigma_y = -\frac{3}{\pi}\frac{P}{Rt}$

Thus, $\sigma_1 - \sigma_2 = \frac{4P}{\pi R t} = \frac{CN}{t} \Rightarrow \frac{4(200)}{\pi(1.5)(0.25)} = \frac{C(3)}{0.25}$

$\therefore$ $\underline{C \approx 57 \text{ lb/in-fringe}}$

8.42 $M = Pe$, $\sigma_1 = \frac{6M}{th^2}$, $\sigma_2 = 0$ $\therefore$ $\sigma_1 - \sigma_2 = \frac{6Pe}{th^2} = \frac{CN}{t}$

Thus, $C = \frac{6Pe}{Nh^2}$. From the figure, $N = 2$

$$C = \frac{6(80)(0.020)}{2(0.025)^2} = 7.68(10^3)\text{ N/m-fringe} = \underline{7.68\text{ kN/m-fringe}}$$

8.43 Point A: $\sigma_x = \tau_{xy} = 0$. Since σ_y is tension, $\sigma_y = \sigma_1$

$\sigma_y = \sigma_1 = \frac{NC}{t} \approx \frac{(2.4)(60)}{1/4} = \underline{580\text{ psi}}$

Point B: $\sigma_x = \tau_{xy} = 0$. Since σ_y is compression, $\sigma_y = \sigma_2$

$\sigma_y = \sigma_2 \approx -\frac{NC}{t} = -\frac{7(60)}{1/4} = \underline{-1680\text{ psi}}$

Point C: $\sigma_y = \tau_{xy} = 0$. Since σ_x is tension, $\sigma_x = \sigma_1$

$\sigma_x = \sigma_1 \approx \frac{7(60)}{1/4} = \underline{1680\text{ psi}}$

8.44 $N_0 = 2$, $N_\theta = 1.64$, $\theta = 30°$. Eqs. (8.8-11)

$$\sigma_1 = \frac{60\cos 30}{0.25\sin^2 30}(1.64 - 2\cos 30) = \underline{-77\text{ psi} = \sigma_x}$$

$$\sigma_2 = \frac{60}{0.25\sin^2 30}(1.64\cos 30 - 2) = \underline{-557\text{ psi} = \sigma_y}$$

Due to symmetry, $\underline{\tau_{xy} = 0}$

8.45 (a) Away from the fillet where the stress is well-behaved $N = 6$ and $\sigma_{nom} = \frac{6M}{th^2} = \frac{6(100)}{0.25(2)^2} = \frac{CN}{t} = \frac{C(6)}{0.25}$

$\therefore$ $C = \underline{25 \text{ lb/in-fringe}}$

(b) At fillet, $N \approx 8$, $\sigma_{max} \approx \frac{CN}{t} = \frac{25(8)}{0.25} = \underline{800 \text{ psi}}$

(c) $K_t = \frac{\sigma_{max}}{\sigma_{nom}} = \frac{N_{max}}{N_{nom}} = \frac{8}{6} = \underline{1.33}$

(d)

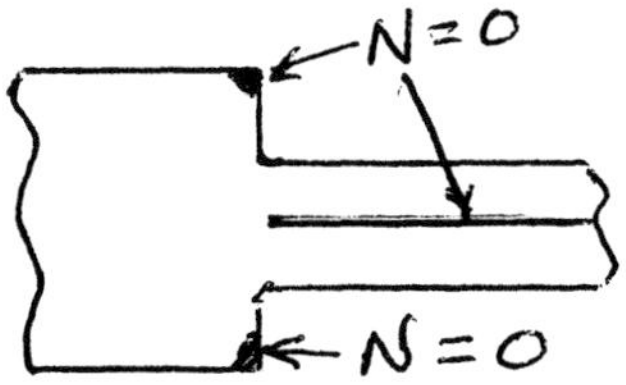

8.46 Eqs (8.8-26)

$\epsilon_1 = 1015[1.5(1.02) - 1.07] = 467\mu$

$\epsilon_2 = 1015[1.5(1.02) - 2(1.07)] = -619\mu$

$\sigma_1 = \frac{E}{1-\nu^2}(\epsilon_1 + \nu\epsilon_2) = \frac{200(10^3)}{1-0.29^2}[467 + 0.29(-619)](10^{-6}) = \underline{63 \text{ MPa}}$

$\sigma_2 = \frac{200(10^3)}{1-0.29^2}[-619 + 0.29(467)](10^{-6}) = \underline{-106 \text{ MPa}}$

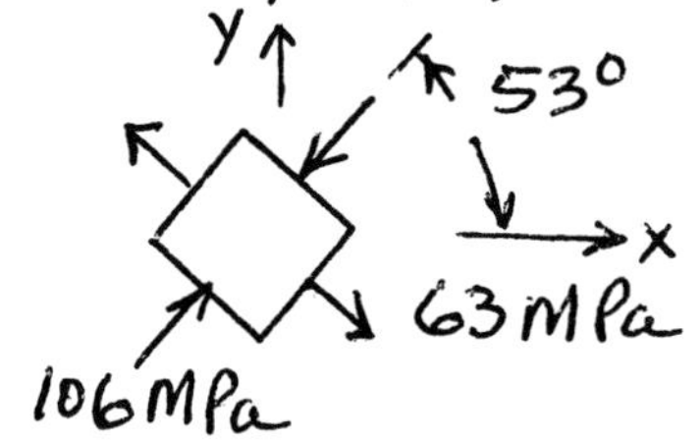

$\sigma_x = -106\cos^2(-53) + 63\sin^2(-53) = \underline{1.8 \text{ MPa}}$

$\sigma_y = -106\sin^2(-53) + 63\cos^2(-53) = \underline{-44.8 \text{ MPa}}$

$\tau_{xy} = -(-106 - 63)\sin(-53)\cos(-53) = \underline{-81.2 \text{ MPa}}$

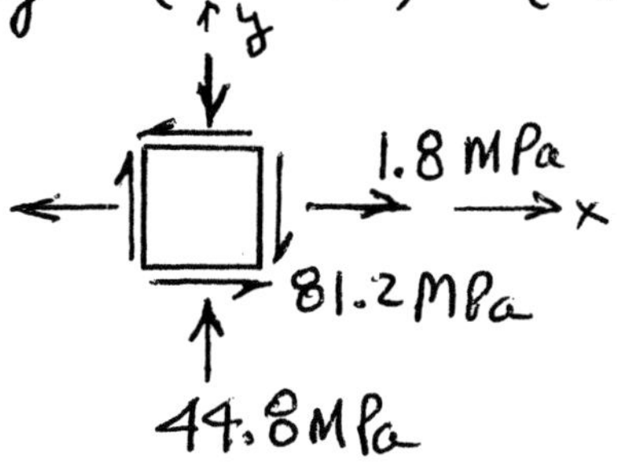

(8.47) $N = 2 + \frac{30}{180} = \underline{2.17}$

(8.48)

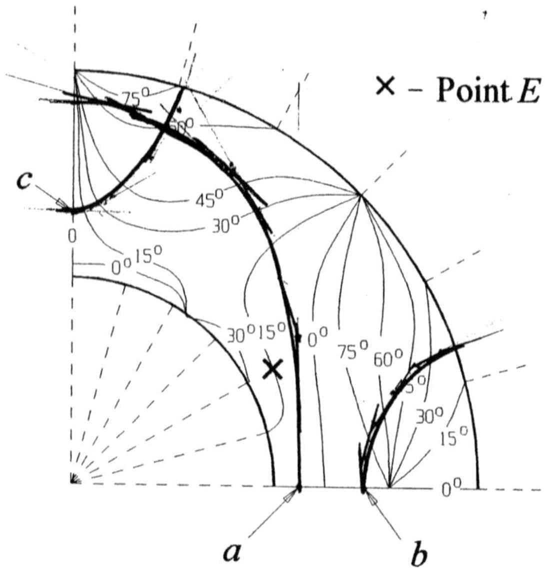

(8.49) From Prob. 8.43 figure $N_E \approx 3$,

$$\sigma_1 - \sigma_2 = \frac{CN}{t} = \frac{60(3)}{0.25} = 720 \text{ psi}$$

$$\tau_{xy} = \frac{\sigma_1 - \sigma_2}{2} \sin 2\theta = \frac{720}{2} \sin(2 \cdot 15) = \underline{180 \text{ psi}}$$

(8.50) (a) $\frac{\partial \tau_{xy}}{\partial x} + \frac{\partial \sigma_y}{\partial y} = 0 \Rightarrow \left.\frac{\Delta \tau_{xy}}{\Delta x}\right|_{y=const} + \left.\frac{\Delta \sigma}{\Delta y}\right|_{x=const} = 0$

$$\Delta \sigma_y = -\left.\frac{\Delta \tau_{xy}}{\Delta x}\right|_{y=const} \Delta y \quad \boxed{ANS}$$

(b)

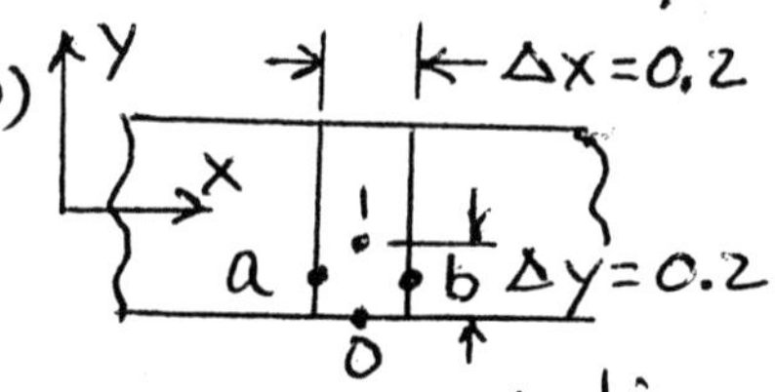

$(\sigma_y)_0 = 0$ (free surface)

Sample calculation to obtain $(\sigma_y)_1$.

$$(\tau_{xy})_a = \frac{5.63(60)}{2(0.25)} \sin 2(1) = 23.58 \text{ psi}$$

$$(\tau_{xy})_b = \frac{6.09(60)}{2(0.25)} \sin 2(1) = 25.50 \text{ psi}$$

(continued)

$$\Delta\sigma_y = -\frac{(25.50-23.58)(0.2)}{0.2} = -1.92\text{ psi}$$

$$(\sigma_y)_1 = (\sigma_y)_0 + \Delta\sigma_y = 0 - 1.92 = -1.92\text{ psi}$$

y (in)	τ_{xy} (psi) $x = 5.0$ in	τ_{xy} (psi) $x = 5.2$ in	$\Delta\tau_{xy}$ (psi)	$\Delta\sigma_y$ (psi)
−0.9	23.58	25.50	1.92	−1.92
−0.7	73.65	79.66	6.01	−6.01
−0.5	107.2	115.4	8.2	−8.2
−0.3	136.7	138.6	1.9	−1.9
−0.1	149.1	153.2	4.1	−4.1
0.1	148.6	152.6	4.0	−4.0
0.3	138.0	138.0	0	0
0.5	112.5	120.7	8.2	−8.2
0.7	76.49	82.33	5.84	−5.84
0.9	24.29	26.17	1.88	−1.88

$x = 5.1$ in

y (in)	σ_y (psi) finite diff.	σ_y (psi) Exact*
−1	0	0
−0.8	−1.92	−1.12
−0.6	−7.93	−4.16
−0.4	−16.1	−8.64
−0.2	−18.0	−14.1
0	−22.1	−20.0
0.2	−26.1	−25.9
0.4	−26.1	−31.4
0.6	−34.3	−35.8
0.8	−40.2	−38.9
1.0	−42.1	−40.0

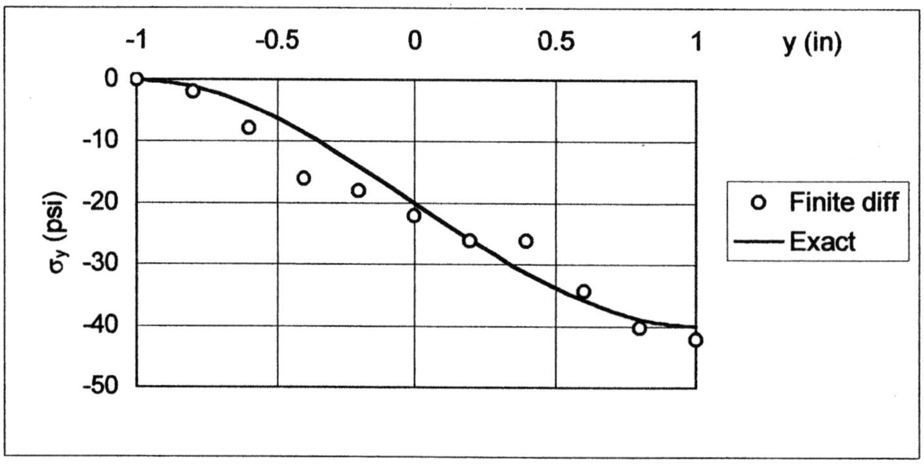

* See Eq. (f) of Example 2.4-1, $\sigma_y = -\frac{w}{6I_z}(2c^3 + 3c^2y - y^3)$

$I_z = \frac{1}{12}(0.25)(2)^3 = 1/6\text{ in}^4$, $w = 10\text{ lb/in}$, $c = 1 \Rightarrow \sigma_y = -10(2 + 3y - y^3)$

(9.1) Using 2 elements,

$$AB:\ k_1 = \frac{AE}{L} = \frac{200(10^{-3})^2(200)(10^9)}{0.300} = 133.3(10^6)\ \text{N/m}$$

$$BC:\ k_2 = \frac{400(10^{-3})^2(200)(10^9)}{0.500} = 160(10^6)\ \text{N/m}$$

$$[K_{sys}] = \begin{bmatrix} 1.333 & -1.333 & 0 \\ -1.333 & 2.933 & -1.6 \\ 0 & -1.6 & 1.6 \end{bmatrix}(10^8)$$

(a)

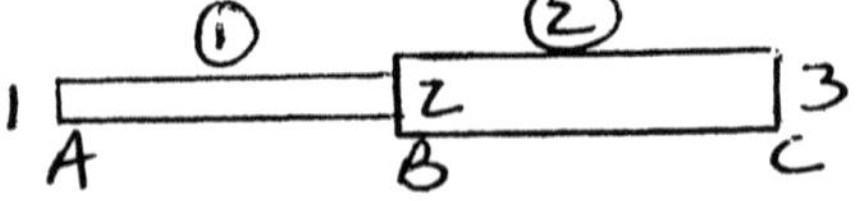

$$\begin{Bmatrix} F_1 \\ F \\ F_3 + 10(10^3) \end{Bmatrix} = [K_{sys}]\begin{Bmatrix} u_1 \\ u_2 \\ u_3 \end{Bmatrix} \qquad u_1 = 0,\ u_3 = 0.25(10^{-3})\ \text{m},\ F_3 = 0$$

Partition 1st & 3rd eqs.

$$F_1 = -1.333(10^8)u_2 \quad (1)$$

$$10(10^3) = 1.6(10^8)\left[-u_2 + 0.25(10^{-3})\right] \quad (2)$$

Remaining eq.

$$F = (2.933u_2 - 1.6u_3)(10^8) \quad (3)$$

Solve (2) for $u_2 = 1.875(10^{-4})$ m [ANS] Sub into (3)

$$F = \left[2.933(1.875)(10^{-4}) - 1.6(2.5)(10^{-4})\right](10^8) = 15(10^3)\text{N} = \underline{15\text{ kN}}$$

(b) $F = 35$ kN. Sub into (3) and solve for $u_2 = \underline{2.557(10^{-4})\text{m}}$

$$F_1 = F_A = -1.333(10^8)(2.557)(10^{-4}) = -34.1(10^3)\text{N} = \underline{-34.1\text{ kN}}$$

Including F_3 in Eq (2) $F_3 = 1.6(10^8)\left[-u_2 + 0.25(10^{-3})\right] - 10(10^3)$

Sub in u_2 and $F_3 = F_C = -10.9(10^3)\text{N} = \underline{-10.9\text{ kN}}$

(c) Part (a) $\sigma_{AB} = \frac{200(10^9)}{0.3}\{-1\ \ 1\}\begin{Bmatrix} 0 \\ 1.875(10^{-4}) \end{Bmatrix} = 125(10^6)\text{N/m}^2 = \underline{125\text{ MPa}}$

$$\sigma_{BC} = \frac{200(10^9)}{0.5}\{-1\ \ 1\}\begin{Bmatrix} 1.875(10^{-4}) \\ 2.5(10^{-4}) \end{Bmatrix} = 25(10^6)\text{N/m}^2 = \underline{25\text{ MPa}}$$

Part (b) $\sigma_{AB} = \frac{200(10^9)}{0.3}\{-1\ \ 1\}\begin{Bmatrix} 0 \\ 2.557(10^{-4}) \end{Bmatrix} = 170(10^6)\text{N/m}^2 = \underline{170\text{ MPa}}$

$$\sigma_{BC} = \frac{200(10^9)}{0.5}\{-1\ \ 1\}\begin{Bmatrix} 2.557(10^{-4}) \\ 2.5(10^{-4}) \end{Bmatrix} = -2.28(10^6)\text{N/m}^2 = \underline{-2.28\text{ MPa}}$$

9.2 First check to see if point C reaches wall.
From the solution of Prob. 9.1 $k_1 = k_{AB} = 1.333(10^8)$ N/m
$k_2 = k_{BC} = 1.6(10^8)$ N/m. For AB, $q_0 L_e = 65(0.3) = 19.5$ kN

Thus

$$\begin{Bmatrix} F_1 + \frac{19.5}{2}(10^3) \\ \frac{19.5}{2}(10^3) \\ 10(10^3) \end{Bmatrix} = \begin{bmatrix} 1.333 & -1.333 & 0 \\ -1.333 & 2.933 & -1.6 \\ 0 & -1.6 & 1.6 \end{bmatrix}(10^8)\begin{Bmatrix} u_1 \\ u_2 \\ u_3 \end{Bmatrix}$$

$u_1 = 0$, partition 1st eq, $F_1 = -1.333(10^8)u_2 - 9.75(10^3)$ ①

Remaining eqs. with $u_1 = 0$

$$\begin{Bmatrix} 9.75(10^3) \\ 10(10^3) \end{Bmatrix} = \begin{bmatrix} 2.933 & -1.6 \\ -1.6 & 1.6 \end{bmatrix}(10^8)\begin{Bmatrix} u_2 \\ u_3 \end{Bmatrix}$$ Take the inverse

$$\begin{Bmatrix} u_2 \\ u_3 \end{Bmatrix} = \begin{bmatrix} 0.75 & 0.75 \\ 0.75 & 1.375 \end{bmatrix}\begin{Bmatrix} 9.75 \\ 10 \end{Bmatrix}(10^{-5}) = \begin{Bmatrix} 1.481 \\ 2.106 \end{Bmatrix}(10^{-4}) \text{ m}$$ **ANS**

Since $u_3 = 2.106(10^{-4})$ m < 0.25 mm point C does not hit wall.

From ① $F_1 = -29(10^3)$ N $= \underline{-29.5 \text{ kN}}$

__Stresses:__ Element 1, AB, use Eq. (9.2-46)

$$\sigma_{AB} = \frac{200(10^9)}{0.3}\{-1 \quad 1\}\begin{Bmatrix} 0 \\ 1.481(10^{-4}) \end{Bmatrix} + \frac{65(10^3)}{2(200)(10^{-3})^2}(0.3 - 2x)$$

$$= (147.5 - 325x)(10^6) \text{ Pa} = \underline{147.5 - 325x \text{ MPa}} \quad (x \text{ in m})$$

Element 2, BC, $\sigma_{BC} = \frac{200(10^9)}{0.5}\{-1 \quad 1\}\begin{Bmatrix} 1.481 \\ 2.106 \end{Bmatrix}(10^{-6}) = 25(10^6)$ Pa

$\underline{\sigma_{BC} = 25 \text{ MPa}}$

__Forces:__ AB: $$\begin{Bmatrix} f_{11} \\ f_{12} \end{Bmatrix} = 1.333(10^8)\begin{bmatrix} 1 & -1 \\ -1 & 1 \end{bmatrix}\begin{Bmatrix} 0 \\ 1.481 \end{Bmatrix}(10^{-4}) - \begin{Bmatrix} 9.75 \\ 9.75 \end{Bmatrix}(10^3)$$

$$= \begin{Bmatrix} -29.5 \\ 10 \end{Bmatrix}(10^3) \text{ N} = \begin{Bmatrix} -29.5 \\ 10 \end{Bmatrix} \text{ kN}$$ **ANS**

BC: $$\begin{Bmatrix} f_{22} \\ f_{23} \end{Bmatrix} = 1.6(10^8)\begin{bmatrix} 1 & -1 \\ -1 & 1 \end{bmatrix}\begin{Bmatrix} 1.481 \\ 2.106 \end{Bmatrix}(10^{-4}) = \begin{Bmatrix} -10 \\ 10 \end{Bmatrix}(10^3) \text{ N} = \begin{Bmatrix} -10 \\ 10 \end{Bmatrix} \text{ kN}$$ **ANS**

9.3 From the solution of Prob 9.1, $k_1 = k_{AB} = 1.333(10^8)$ N/m
$k_2 = k_{BC} = 1.6(10^8)$ N/m.

Element 1: $EA\alpha\Delta T = 200(10^9)(200)(10^{-6})(11.7)(10^{-6})(120) = 56.16(10^3)$ N

Element 2: $EA\alpha\Delta T = 200(10^9)(400)(10^{-6})(11.7)(10^{-6})(120) = 112.32(10^3)$ N

(continued)

$$\begin{Bmatrix} F_1 - 56.16(10^3) \\ 20(10^3) + 56.16(10^3) - 112.32(10^3) \\ F_3 + 10(10^3) + 112.32(10^3) \end{Bmatrix} = \begin{bmatrix} 1.333 & -1.333 & 0 \\ -1.333 & 2.933 & -1.6 \\ 0 & -1.6 & .6 \end{bmatrix} (10^8) \begin{Bmatrix} u_1 \\ u_2 \\ u_3 \end{Bmatrix}$$

Point C will hit wall. Thus, $u_1 = 0$ and $u_3 = 2.5(10^{-4})$ m

Partition 1st and 3rd equations,

$F_1 = 56.16(10^3) - 1.333(10^8)u_2$ ①, $F_3 = 1.6(10^8)\left[2.5(10^{-4}) - u_2\right] - 122.32(10^3)$ ②

Remaining Eq. $-36.16(10^3) = \left[2.933u_2 - 1.6(2.5)(10^{-4})\right](10^8)$

$\therefore$ $u_2 = \underline{1.309(10^{-5})\text{m} = u_B}$. Sub into ① & ②

$F_1 = F_A = 54.4(10^3)\text{N} = \underline{54.4\text{ kN}}$, $F_3 = F_C = -84.4(10^3)\text{N} = \underline{-84.4\text{ kN}}$

Stresses: Element 1, AB,

$$\sigma_{AB} = \frac{200(10^9)}{0.300}\{-1 \;\; 1\}\begin{Bmatrix} 0 \\ 1.309(10^{-5}) \end{Bmatrix} - 200(10^9)(11.7)(10^{-6})(120)$$

$$= -272(10^6)\text{ Pa} = \underline{-272\text{ MPa}}$$

Element 2, BC,

$$\sigma_{BC} = \frac{200(10^9)}{0.500}\{-1 \;\; 1\}\begin{Bmatrix} 1.309(10^{-5}) \\ 2.5(10^{-4}) \end{Bmatrix} - 200(10^9)(11.7)(10^{-6})(120)$$

$$= -186(10^6)\text{ Pa} = \underline{-186\text{ MPa}}$$

9.4 (a) $A = 0.125\left(0.6 + \frac{0.4}{6}x\right) = (75 + 8.333x)(10^{-3})$ ①

$\sigma_x = \frac{2000}{A}$, $\epsilon_x = \frac{2000}{EA}$,

$u = \int \epsilon_x dx = \frac{2000\left[\ln(75 + 8.333x)\right]}{30(10^6)(8.333)(10^{-3})} + C$. At $x = 0$, $u = 0$

Thus, $u = \frac{2000}{30(10^6)(8.333)(10^{-3})}\ln\left(\frac{75 + 8.333x}{75}\right) = \underline{8(10^{-3})\ln(1 + 0.111}$

(b) $k_1 = \frac{(0.125)(0.7)(30)(10^6)}{3} = 8.75(10^5)$, $k_2 = \frac{0.125(0.9)(30)(10^6)}{3} = 1.125(10^6)$

$$\begin{Bmatrix} F_1 \\ 0 \\ 2000 \end{Bmatrix} = \begin{bmatrix} 8.75 & -8.75 & 0 \\ -8.75 & 20 & -11.25 \\ 0 & -11.25 & 11.25 \end{bmatrix}(10^5)\begin{Bmatrix} u_1 \\ u_2 \\ u_3 \end{Bmatrix},$$

$u_1 = 0$ so partition 1st eq.

$F_1 = -8.75(10^5)u_2$ ② (Continued)

Remaining eqs. with $u_1 = 0$

$$\begin{Bmatrix} 0 \\ 2000 \end{Bmatrix} = \begin{bmatrix} 20 & -11.25 \\ -11.25 & 11.25 \end{bmatrix} (10^5) \begin{Bmatrix} u_2 \\ u_3 \end{Bmatrix}$$ Take inverse

$$\begin{Bmatrix} u_2 \\ u_3 \end{Bmatrix} = \begin{bmatrix} 1.143 & 1.143 \\ 1.143 & 2.032 \end{bmatrix} (10^{-6}) \begin{Bmatrix} 0 \\ 2000 \end{Bmatrix} = \begin{Bmatrix} 2.286 \\ 4.063 \end{Bmatrix} (10^{-3}) \text{ in} \quad \boxed{\text{ANS}}$$

Stress: Elements 1 & 2:

$$(\sigma_x)_1 = \frac{30(10^6)}{3} \{-1 \quad 1\} \begin{Bmatrix} 0 \\ 2.286 \end{Bmatrix} (10^{-3}) = 22.86(10^3)\text{psi} = \underline{22.86 \text{ kpsi}}$$

$$(\sigma_x)_2 = \frac{30(10^6)}{3} \{-1 \quad 1\} \begin{Bmatrix} 2.286 \\ 4.063 \end{Bmatrix} (10^{-3}) = 17.77(10^3)\text{psi} = \underline{17.77 \text{ kpsi}}$$

(c) $k_1 = \dfrac{0.125(0.667)(30)(10^6)}{2} = 1.25(10^6)$

$k_2 = \dfrac{0.125(0.80)(30)(10^6)}{2} = 1.5(10^6)$

$k_3 = \dfrac{0.125(0.933)(30)(10^6)}{2} = 1.75(10^6)$

$$\begin{Bmatrix} F_1 \\ 0 \\ 0 \\ 2000 \end{Bmatrix} = \begin{bmatrix} 1.25 & -1.25 & 0 & 0 \\ -1.25 & 2.75 & -1.5 & 0 \\ 0 & -1.5 & 3.25 & -1.75 \\ 0 & 0 & -1.75 & 1.75 \end{bmatrix} (10^6) \begin{Bmatrix} u_1 \\ u_2 \\ u_3 \\ u_4 \end{Bmatrix}$$

Again $u_1 = 0$ so partition the 1st eq which leaves,

$$\begin{Bmatrix} 0 \\ 0 \\ 2000 \end{Bmatrix} = \begin{bmatrix} 2.75 & -1.5 & 0 \\ -1.5 & 3.25 & -1.75 \\ 0 & -1.75 & 1.75 \end{bmatrix} (10^6) \begin{Bmatrix} u_2 \\ u_3 \\ u_4 \end{Bmatrix}$$ Take inverse

$$\begin{Bmatrix} u_2 \\ u_3 \\ u_4 \end{Bmatrix} = \begin{bmatrix} 8 & 8 & 8 \\ 8 & 14.667 & 14.667 \\ 8 & 14.667 & 20.381 \end{bmatrix} (10^{-7}) \begin{Bmatrix} 0 \\ 0 \\ 2000 \end{Bmatrix} = \begin{Bmatrix} 1.6 \\ 2.933 \\ 4.076 \end{Bmatrix} (10^{-3}) \text{ in} \quad \boxed{\text{ANS}}$$

Stresses:

$$(\sigma_x)_1 = \frac{30(10^6)}{2} \{-1 \quad 1\} \begin{Bmatrix} 0 \\ 1.6 \end{Bmatrix} (10^{-3}) = 24(10^3)\text{psi} = \underline{24 \text{ kpsi}}$$

$$(\sigma_x)_2 = \frac{30(10^6)}{2} \{-1 \quad 1\} \begin{Bmatrix} 1.6 \\ 2.933 \end{Bmatrix} (10^{-3}) = 20(10^3)\text{psi} = \underline{20 \text{ kpsi}}$$

$$(\sigma_x)_3 = \frac{30(10^6)}{2} \{-1 \quad 1\} \begin{Bmatrix} 2.933 \\ 4.076 \end{Bmatrix} (10^{-3}) = 17.14(10^3)\text{psi} = \underline{17.14 \text{ kpsi}}$$

(Continued)

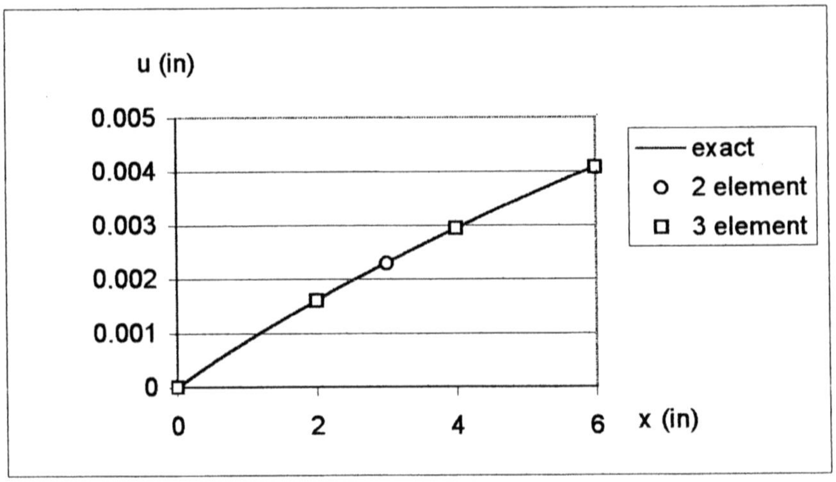

Note: At $x=0$ $u=0$ for all three cases. At $x=6$, the 2 element value does not show as it is directly under the 3 element value

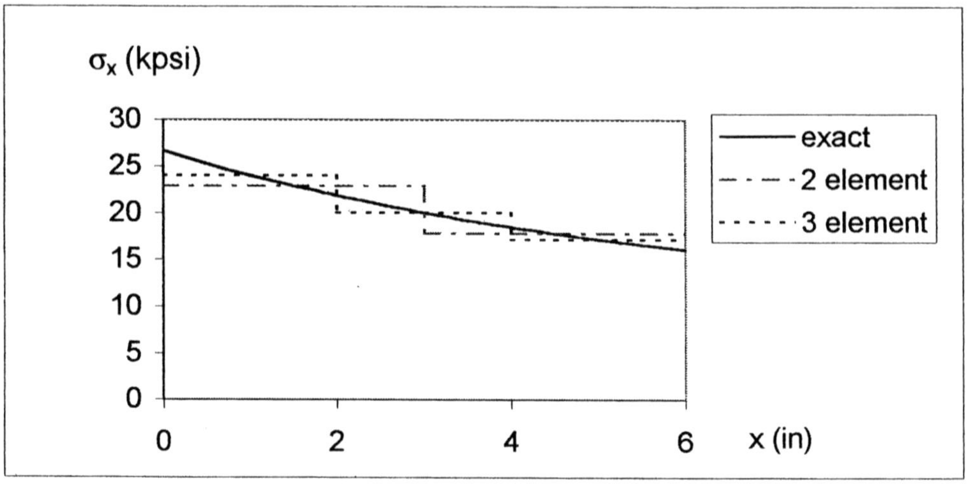

9.5 From Example 3.2-2, $du = \frac{\rho g}{E}(L-x)dx$. Thus $u = \frac{\rho g}{E}(Lx - \frac{x^2}{2}) + C$. At $x=0$, $u=0$ $\therefore$ $C=0$. Note that $\rho g = W/AL$ where W = total weight. Thus,

$$u = \frac{W}{2AE}x(2 - x/L) \quad \text{Exact displacement}$$

(a) One element of length L, $qL = W$

$$\begin{Bmatrix} F_1 + \frac{W}{2} \\ F_2 + \frac{W}{2} \end{Bmatrix} = \frac{AE}{L}\begin{bmatrix} 1 & -1 \\ -1 & 1 \end{bmatrix}\begin{Bmatrix} u_1 \\ u_2 \end{Bmatrix}$$

$u_1 = 0$, partition 1st eq. The remaining eq. with $u_1 = 0$ and $F_2 = 0$ is

$$\frac{W}{2} = \frac{AE}{L}u_2 \Rightarrow u_2 = \frac{WL}{2AE} \quad \boxed{ANS}$$

$$u = \frac{W}{2AE}x \quad \boxed{ANS}$$

(Continued)

(b) Two elements each of length, $L/2$. $k_1 = k_2 = \frac{AE}{L/2} = \frac{2AE}{L}$

$q_0 L_e = W/2$

$$\begin{Bmatrix} F_1 + \frac{W}{4} \\ \frac{W}{4} + \frac{W}{4} \\ W/4 \end{Bmatrix} = \frac{2AE}{L}\begin{bmatrix} 1 & -1 & 0 \\ -1 & 2 & -1 \\ 0 & -1 & 1 \end{bmatrix}\begin{Bmatrix} u_1 \\ u_2 \\ u_3 \end{Bmatrix}$$

$u_1 = 0$. Partition 1st eq. Remaining eqs with $u_1 = 0$ are

$$\begin{Bmatrix} W/2 \\ W/4 \end{Bmatrix} = \frac{2AE}{L}\begin{bmatrix} 2 & -1 \\ -1 & 1 \end{bmatrix}\begin{Bmatrix} u_2 \\ u_3 \end{Bmatrix}$$

Solve for u_2, u_3

$$\begin{Bmatrix} u_2 \\ u_3 \end{Bmatrix} = \frac{L}{2AE}\begin{bmatrix} 1 & 1 \\ 1 & 2 \end{bmatrix}\begin{Bmatrix} W/2 \\ W/4 \end{Bmatrix} = \begin{Bmatrix} 3 \\ 4 \end{Bmatrix}\frac{WL}{8AE}$$ **ANS**

$0 < x < L/2$: $u = \frac{3}{4}\frac{W}{AE}x$ $\frac{L}{2} < x < L$: $u = \frac{W}{4AE}x + \frac{WL}{4AE}$

(c) Three elements of length $L/3$. $k_1 = k_2 = k_3 = 3AE/L$

$$\begin{Bmatrix} F_1 + W/6 \\ W/6 + W/6 \\ W/6 + W/6 \\ W/6 \end{Bmatrix} = \frac{3AE}{L}\begin{bmatrix} 1 & -1 & 0 & 0 \\ -1 & 2 & -1 & 0 \\ 0 & -1 & 2 & -1 \\ 0 & 0 & -1 & 1 \end{bmatrix}\begin{Bmatrix} u_1 \\ u_2 \\ u_3 \\ u_4 \end{Bmatrix}$$

$u_1 = 0$, partition 1st eq. Remaining eqs with $u_1 = 0$ are

$$\begin{Bmatrix} W/3 \\ W/3 \\ W/6 \end{Bmatrix} = \frac{3AE}{L}\begin{bmatrix} 2 & -1 & 0 \\ -1 & 2 & -1 \\ 0 & -1 & 1 \end{bmatrix}\begin{Bmatrix} u_2 \\ u_3 \\ u_4 \end{Bmatrix}$$

Solve inverse eq.

$$\begin{Bmatrix} u_2 \\ u_3 \\ u_4 \end{Bmatrix} = \frac{L}{3AE}\begin{bmatrix} 1 & 1 & 1 \\ 1 & 2 & 2 \\ 1 & 2 & 3 \end{bmatrix}\begin{Bmatrix} W/3 \\ W/3 \\ W/6 \end{Bmatrix} = \begin{Bmatrix} 5 \\ 8 \\ 9 \end{Bmatrix}\frac{WL}{18AE}$$

$0 < x < \frac{L}{3}$: $u = \frac{5}{6}\frac{W}{AE}x$, $\frac{L}{3} < x < \frac{2L}{3}$: $u = \frac{1}{2}\frac{W}{AE}x + \frac{1}{9}\frac{WL}{AE}$

$\frac{2L}{3} < x < L$: $u = \frac{1}{6}\frac{Wx}{AE} + \frac{1}{3}\frac{WL}{AE}$

(d) Four elements of Length $L/4$

$$\begin{Bmatrix} F_1 + W/8 \\ W/8 + W/8 \\ W/8 + W/8 \\ W/8 + W/8 \\ W/8 \end{Bmatrix} = \frac{4AE}{L}\begin{bmatrix} 1 & -1 & 0 & 0 & 0 \\ -1 & 2 & -1 & 0 & 0 \\ 0 & -1 & 2 & -1 & 0 \\ 0 & 0 & -1 & 2 & -1 \\ 0 & 0 & 0 & -1 & 1 \end{bmatrix}\begin{Bmatrix} u_1 \\ u_2 \\ u_3 \\ u_4 \\ u_5 \end{Bmatrix}$$

$u_1 = 0$, partition

(Continued)

$$\begin{Bmatrix} W/4 \\ W/4 \\ W/4 \\ W/8 \end{Bmatrix} = \frac{4AE}{L}\begin{bmatrix} 2 & -1 & 0 & 0 \\ -1 & 2 & -1 & 0 \\ 0 & -1 & 2 & -1 \\ 0 & 0 & -1 & 1 \end{bmatrix}\begin{Bmatrix} u_2 \\ u_3 \\ u_4 \\ u_5 \end{Bmatrix}$$ Solve inverse

$$\begin{Bmatrix} u_2 \\ u_3 \\ u_4 \\ u_5 \end{Bmatrix} = \frac{L}{4AE}\begin{bmatrix} 1 & 1 & 1 & 1 \\ 1 & 2 & 2 & 2 \\ 1 & 2 & 3 & 3 \\ 1 & 2 & 3 & 4 \end{bmatrix}\begin{Bmatrix} W/4 \\ W/4 \\ W/4 \\ W/8 \end{Bmatrix} = \begin{Bmatrix} 7 \\ 12 \\ 15 \\ 16 \end{Bmatrix}\frac{WL}{32AE}$$

$0<x<\frac{L}{4}$: $u = \frac{7}{8}\frac{Wx}{AE}$, $\frac{L}{4}<x<\frac{L}{2}$: $u = \frac{5}{8}\frac{Wx}{AE} + \frac{WL}{16AE}$

$\frac{L}{2}<x<\frac{3L}{4}$: $u = \frac{3}{8}\frac{Wx}{AE} + \frac{3}{16}\frac{WL}{AE}$ $\frac{3}{4}L<x<L$: $u = \frac{Wx}{8AE} + \frac{3}{8}\frac{WL}{AE}$

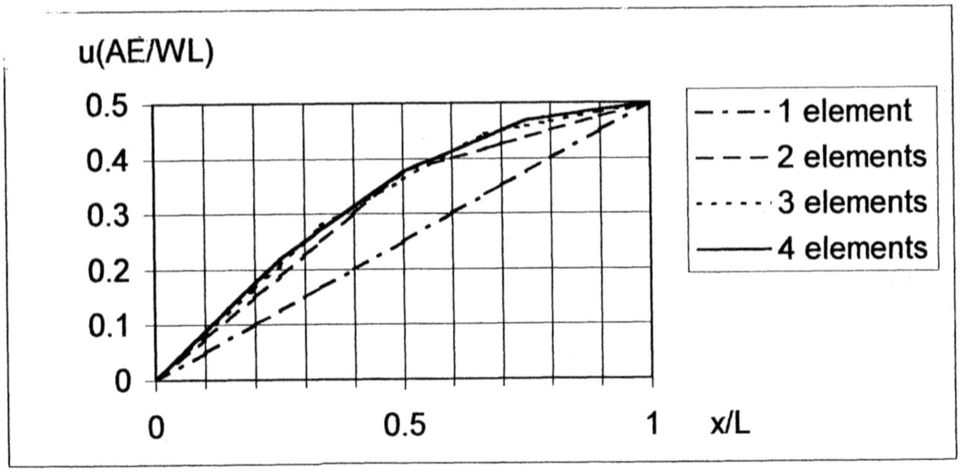

Note: The ends of each line segments intersect the exact values of u as the exact strain is linear and the finite element method is an averaging technique.

(9.6) 

The "dynamic" force on each element acts at the element center of mass and is $(\rho A L/3)\omega^2(x_{cm})i$

$F_{①} = \left(\rho A\frac{L}{3}\right)(\omega^2)\left(\frac{L}{6}\right) = \frac{\rho A L^2\omega^2}{18}$, $F_{②} = \frac{\rho A L^2\omega^2}{6}$, $F_{③} = \frac{5}{18}\rho A L^2\omega^2$

$k_1 = k_2 = k_3 = 3AE/L$. Let $\rho A L^2\omega^2 = D$

$$\begin{Bmatrix} F_1 + D/36 \\ D/36 + D/12 \\ D/12 + 5/36 D \\ 5/36 D \end{Bmatrix} = \frac{3AE}{L}\begin{bmatrix} 1 & -1 & 0 & 0 \\ -1 & 2 & -1 & 0 \\ 0 & -1 & 2 & -1 \\ 0 & 0 & -1 & 1 \end{bmatrix}\begin{Bmatrix} u_1 \\ u_2 \\ u_3 \\ u_4 \end{Bmatrix}$$

$u_1 = 0$, partition 1st eq. The remaining eqs with $u_1 = 0$ are written

(Continued)

$$\begin{Bmatrix} D/9 \\ 2D/9 \\ 5D/36 \end{Bmatrix} = \frac{3AE}{L}\begin{bmatrix} 2 & -1 & 0 \\ -1 & 2 & -1 \\ 0 & -1 & 1 \end{bmatrix}\begin{Bmatrix} u_2 \\ u_3 \\ u_4 \end{Bmatrix}$$ Solve inverse,

$$\begin{Bmatrix} u_2 \\ u_3 \\ u_4 \end{Bmatrix} = \frac{LD}{108AE}\begin{bmatrix} 1 & 1 & 1 \\ 1 & 2 & 2 \\ 1 & 2 & 3 \end{bmatrix}\begin{Bmatrix} 4 \\ 8 \\ 5 \end{Bmatrix} = \frac{LD}{108AE}\begin{Bmatrix} 17 \\ 30 \\ 35 \end{Bmatrix}$$

$L = 30$ in, $E = 29(10^6)$ psi, $\omega = 100$ rad/s, $\rho = \frac{\gamma}{g} = \frac{0.284}{386.4}$

$$\frac{LD}{108AE} = \frac{L(\rho A L^2 \omega^2)}{108AE} = \frac{\rho L^3 \omega^2}{108E} = \frac{0.284(30)^3(100)^2}{386.4(108)(29)(10^6)} = 6.336(10^{-5})$$

$$\begin{Bmatrix} u_2 \\ u_3 \\ u_4 \end{Bmatrix} = \begin{bmatrix} 1.077 \\ 1.901 \\ 2.218 \end{bmatrix}(10^{-3})\text{ in}$$ **ANS**

<u>Stresses:</u>

$$(\sigma_x)_1 = \frac{29(10^6)}{10}\{-1\ \ 1\}\begin{Bmatrix} 0 \\ 1.077 \end{Bmatrix}(10^{-3}) = 3124\text{ psi} = \underline{3.12\text{ kpsi}}$$

$$(\sigma_x)_2 = \frac{29(10^6)}{10}\{-1\ \ 1\}\begin{Bmatrix} 1.077 \\ 1.901 \end{Bmatrix}(10^{-3}) = 2389\text{ psi} = \underline{2.39\text{ kpsi}}$$

$$(\sigma_x)_3 = \frac{29(10^6)}{10}\{-1\ \ 1\}\begin{Bmatrix} 1.901 \\ 2.218 \end{Bmatrix}(10^{-3}) = \underline{919\text{ psi}}$$

From the solution of Prob. 3.3,

$$u = \frac{\rho\omega^2}{6E}x(3L^2 - x^2) = \frac{0.284(100^2)}{(386.4)(6)(29)(10^6)}x(2700 - x^2) = 4.224(10^{-8})x(2700 - x^2)$$

$$\sigma_x = \frac{\rho\omega^2}{2}(L^2 - x^2) = \frac{0.284(100^2)}{(386.4)(2)}(900 - x^2) = 3.675(900 - x^2)$$

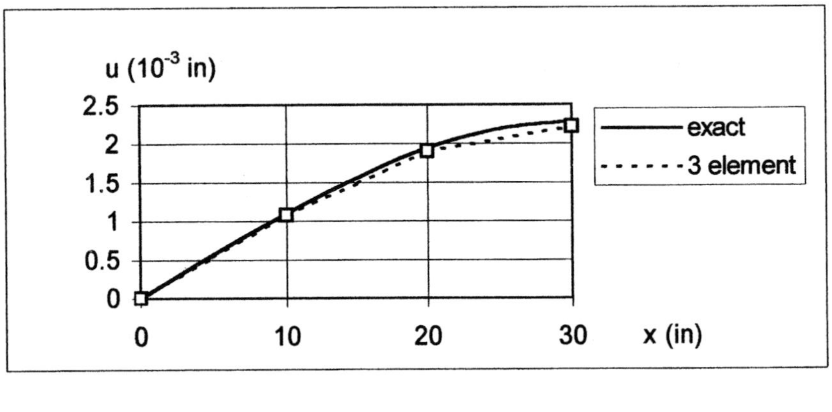

(Continued)

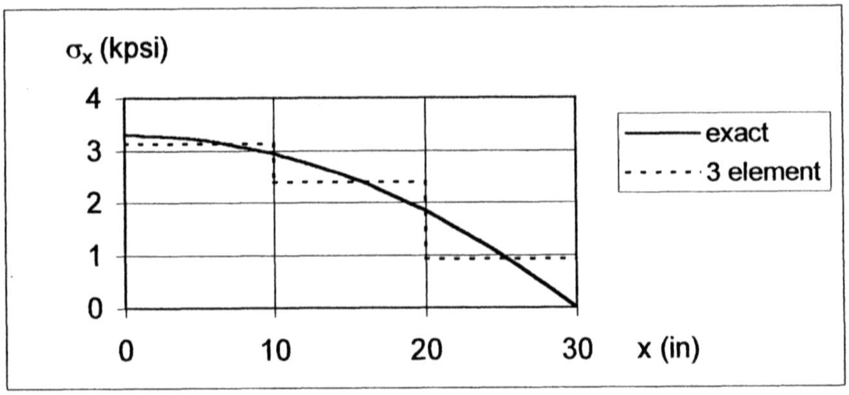

9.7

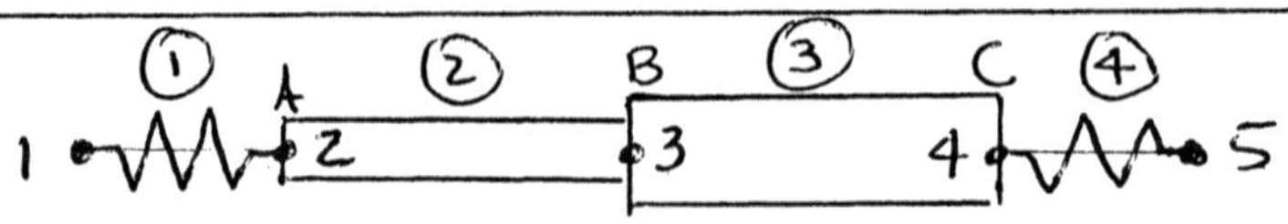

$$[k]_1 = [k]_4 = \begin{bmatrix} 1.5 & -1.5 \\ -1.5 & 1.5 \end{bmatrix}(10^9)\ \text{N/m}$$

$$k_2 = \left(\frac{AE}{L}\right)_{AB} = \frac{\frac{\pi}{4}(0.012)^2(207)(10^9)}{0.250} = 93.64(10^6)\ \text{N/m}$$

$$k_3 = \frac{\frac{\pi}{4}(0.025)^2(207)(10^9)}{0.400} = 254.03(10^6)\ \text{N/m}$$

Element ②: $EA\alpha\Delta T = 207(10^9)\frac{\pi}{4}(0.012)^2(11.7)(10^{-6})(5) = 1369.6$ N

③: $EA\alpha\Delta T = 207(10^9)\frac{\pi}{4}(0.025)^2(11.7)(10^{-6})(5) = 5944.2$ N

$$\begin{Bmatrix} F_1 \\ -1369.6 \\ 15(10^3)+1369.6-5944.2 \\ 5944.2 \\ F_5 \end{Bmatrix} = \begin{bmatrix} 1.5 & -1.5 & 0 & 0 & 0 \\ -1.5 & 1.50936 & -0.09364 & 0 & 0 \\ 0 & -0.09364 & 0.34767 & -0.254 & 0 \\ 0 & 0 & -0.254 & 1.754 & -1.5 \\ 0 & 0 & 0 & -1.5 & 1.5 \end{bmatrix}(10^9)\begin{Bmatrix} u_1 \\ u_2 \\ u_3 \\ u_4 \\ u_5 \end{Bmatrix}$$

$u_1 = u_5 = 0$ Partition 1st & 5th eqs. yields

$F_1 = -1.5(10^9)u_2$ ①, $F_5 = -1.5(10^9)u_4$ ② and,

$$\begin{Bmatrix} -1369.6 \\ 10.425(10^3) \\ 5944.2 \end{Bmatrix} = \begin{bmatrix} 1.50936 & -0.09364 & 0 \\ -0.09364 & 0.34767 & -0.254 \\ 0 & -0.254 & 1.754 \end{bmatrix}(10^9)\begin{Bmatrix} u_2 \\ u_3 \\ u_4 \end{Bmatrix}$$

Solve the inverse

(continued)

$$\begin{Bmatrix} u_2 \\ u_3 \\ u_4 \end{Bmatrix} = (10^{-9}) \begin{bmatrix} 0.6751 & 0.2034 & 0.02945 \\ 0.2034 & 3.278 & 0.4747 \\ 0.02945 & 0.4747 & 0.6389 \end{bmatrix} \begin{Bmatrix} -1369.6 \\ 10.425(10^3) \\ 5944.2 \end{Bmatrix} = \begin{Bmatrix} 1.297 \\ 36.69 \\ 8.703 \end{Bmatrix} (10^{-6}) \text{ m}$$

From Eqs ① & ②

$F_1 = -1.5(10^9)(1.297)(10^{-6}) = -1945\text{ N} = \underline{-1.95\text{ kN}}$

$F_5 = -1.5(10^9)(8.703)(10^{-6}) = -13050\text{ N} = \underline{-13.05\text{ kN}}$

(9.8) $u = u_i$ at $\bar{x} = 0 \Rightarrow a_1 = u_i$

$u = u_j$ at $\bar{x} = 1 \Rightarrow u_j = a_1 + a_2 = u_i + a_2 \Rightarrow a_2 = u_j - u_i$

Thus, $u = u_i + (u_j - u_i)\bar{x} = (1-\bar{x})u_i + \bar{x}u_j$. Therefore,

$N_i = 1 - \bar{x}, \quad N_j = \bar{x}$ **ANS**

<u>Stiffness matrix</u>: $\bar{x} = \frac{x}{L} \Rightarrow \frac{d\bar{x}}{dx} = \frac{1}{L}$

$$\epsilon_x = \frac{\partial u}{\partial x} = \frac{\partial u}{\partial \bar{x}}\frac{\partial \bar{x}}{\partial x} = \frac{1}{L}\frac{\partial u}{\partial \bar{x}} = \frac{1}{L}(-u_i + u_j) = \frac{1}{L}\{-1 \;\; 1\}\begin{Bmatrix} u_i \\ u_j \end{Bmatrix}$$

$= [B]\begin{Bmatrix} u_i \\ u_j \end{Bmatrix}$. where $[B] = \frac{1}{L}\{-1 \;\; 1\}$

$$[k] = E\int_0^L [B]^T[B]A\,dx = \frac{EA}{L^2}\begin{bmatrix} 1 & -1 \\ -1 & 1 \end{bmatrix}\int_0^L dx = \frac{EA}{L}\begin{bmatrix} 1 & -1 \\ -1 & 1 \end{bmatrix}$$ **ANS**

(9.9) $u = u_i$ at $\bar{x} = 0 \Rightarrow a_1 = u_i$

$u = u_j$ at $\bar{x} = 0.5 \Rightarrow u_j = u_i + 0.5a_2 + 0.25a_3$ ①

$u = u_k$ at $\bar{x} = 1 \Rightarrow u_k = u_i + a_2 + a_3$ ②

solve ① & ② for a_2 & a_3

$a_2 = -3u_i + 4u_j - u_k, \quad a_3 = 2(u_i - 2u_j + u_k)$

$u = u_i + (-3u_i + 4u_j - u_k)\bar{x} + 2(u_i - 2u_j + u_k)\bar{x}^2$

$\quad = (1 - 3\bar{x} + 2\bar{x}^2)u_i + 4\bar{x}(1-\bar{x})u_j - \bar{x}(1-2\bar{x})u_k$

$\therefore N_i = 1 - 3\bar{x} + 2\bar{x}^2, \quad N_j = 4\bar{x}(1-\bar{x}), \quad N_k = -\bar{x}(1-2\bar{x})$ **ANS**

<u>Stiffness matrix</u>: $\bar{x} = x/L$, $d\bar{x}/dx = 1/L$. Thus

$$\epsilon_x = \frac{\partial u}{\partial x} = \frac{\partial u}{\partial \bar{x}}\frac{\partial \bar{x}}{\partial x} = \frac{1}{L}\frac{\partial u}{\partial \bar{x}}$$

(continued)

$$\epsilon_x = \frac{1}{L}\left[(-3+4\bar{x})u_i + (4-8\bar{x})u_j + (-1+4\bar{x})u_k\right]$$

$$= \frac{1}{L}\left\{(-3+4\bar{x}) \quad (4-8\bar{x}) \quad (-1+4\bar{x})\right\}\begin{Bmatrix}u_i\\u_j\\u_k\end{Bmatrix} = [B]\begin{Bmatrix}u_i\\u_j\\u_k\end{Bmatrix}$$

where $[B] = \{B_1 \; B_2 \; B_3\} = \{(-3+4\bar{x}) \; (4-8\bar{x}) \; (-1+4\bar{x})\}$

$$[k] = E\int_0^L [B]^T[B]A\,dx = E\left(\frac{1}{L}\right)^2 AL\int_0^1 \begin{bmatrix} B_1^2 & B_1B_2 & B_1B_3 \\ B_1B_2 & B_2^2 & B_2B_3 \\ B_1B_3 & B_2B_3 & B_3^2 \end{bmatrix} d\bar{x}$$

$$\int_0^1 B_1^2 d\bar{x} = \int_0^1 (9-24\bar{x}+16\bar{x}^2)d\bar{x} = \frac{7}{3}, \quad \int_0^1 B_1B_2 d\bar{x} = \int_0^1(-12+40\bar{x}-32\bar{x}^2)d\bar{x} = -\frac{8}{3}$$

$$\int_0^1 B_1B_3 d\bar{x} = \int_0^1(3-16\bar{x}+16\bar{x}^2)d\bar{x} = \frac{1}{3}, \quad \int_0^1 B_2^2 d\bar{x} = \int_0^1(16-64\bar{x}+64\bar{x}^2)d\bar{x} = 16/3$$

$$\int_0^1 B_2B_3 d\bar{x} = \int_0^1(-4+24\bar{x}-32\bar{x}^2)d\bar{x} = -\frac{8}{3}, \quad \int_0^1 B_3^2 d\bar{x} = \int_0^1(1-8\bar{x}+16\bar{x}^2)d\bar{x} = 7/3$$

Thus,

$$[k] = \frac{AE}{3L}\begin{bmatrix} 7 & -8 & 1 \\ -8 & 16 & -8 \\ 1 & -8 & 7 \end{bmatrix} \quad \boxed{\text{ANS}}$$

9.10 $$W_p = -\{u_i \; u_j \; u_k\}\begin{Bmatrix}F_i\\F_j\\F_k\end{Bmatrix} = -q_0\int_{Le} u(x)dx \quad (1)$$

But from the solution of Prob 9.9, $dx = L_e d\bar{x}$ and $u = N_i u_i + N_j u_j + N_k u_k$ where $N_i = 1-3\bar{x}+2\bar{x}^2$, $N_j = 4\bar{x}(1-\bar{x})$, and $N_k = -\bar{x}(1-2\bar{x})$. Thus (1) is

$$-\{u_i \; u_j \; u_k\}\begin{Bmatrix}F_i\\F_j\\F_k\end{Bmatrix} = -\{u_i \; u_j \; u_k\}\, q_0 L_e \begin{Bmatrix}\int_0^1 N_i d\bar{x}\\ \int_0^1 N_j d\bar{x} \\ \int_0^1 N_k d\bar{x}\end{Bmatrix}$$

$$\int_0^1 N_i d\bar{x} = \int_0^1 (1-3\bar{x}+2\bar{x}^2)d\bar{x} = 1/6, \quad \int_0^1 N_j d\bar{x} = \int_0^1 4\bar{x}(1-\bar{x})d\bar{x} = 2/3$$

$$\int_0^1 -\bar{x}(1-2\bar{x})d\bar{x} = 1/6. \text{ Thus,}$$

$$\begin{Bmatrix}F_i\\F_j\\F_k\end{Bmatrix} = \begin{Bmatrix}1\\4\\1\end{Bmatrix}\frac{q_0 L_e}{6} \quad \boxed{\text{ANS}}$$

9.11

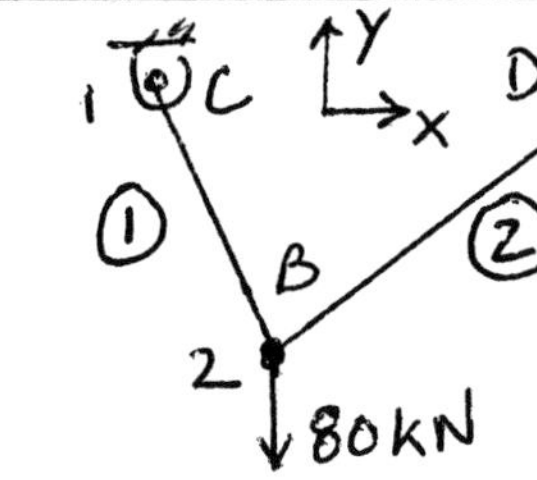

$L_{BD} = 734.85\text{mm}$

$$k_1 = \frac{AE}{L} = \frac{120(10^{-6})(207)(10^9)}{0.600} = 41.4(10^6)\ N/m$$

$$k_2 = \frac{120(10^{-6})(207)(10^9)}{0.73485} = 33.803(10^6)\ N/m$$

Element 1: $i, j = 1, 2$ $\quad \phi_x = -60°$, $\ell = 0.5$, $m = -\sqrt{3}/2$

$$[k]_1 = 41.4(10^6)\begin{bmatrix} 1/4 & -\sqrt{3}/4 & -1/4 & \sqrt{3}/4 \\ -\sqrt{3}/4 & 3/4 & \sqrt{3}/4 & -3/4 \\ -1/4 & \sqrt{3}/4 & 1/4 & -\sqrt{3}/4 \\ \sqrt{3}/4 & -3/4 & -\sqrt{3}/4 & 3/4 \end{bmatrix} \begin{matrix} u_1 \\ v_1 \\ u_2 \\ v_2 \end{matrix}$$

Element 2: $i, j = 2, 3$ $\quad \phi_x = 45°$, $\ell = m = \sqrt{2}/2$

$$[k]_2 = \frac{33.803(10^6)}{2}\begin{bmatrix} 1 & 1 & -1 & -1 \\ 1 & 1 & -1 & -1 \\ -1 & -1 & 1 & 1 \\ -1 & -1 & 1 & 1 \end{bmatrix} \begin{matrix} u_2 \\ v_2 \\ u_3 \\ v_3 \end{matrix}$$

Assemble adding common dof terms

$$[K_{sys}] = \begin{bmatrix} 1.035 & -1.793 & -1.035 & 1.793 & 0 & 0 \\ -1.793 & 3.105 & 1.793 & -3.105 & 0 & 0 \\ -1.035 & 1.793 & 2.725 & -0.1025 & -1.690 & -1.690 \\ 1.793 & -3.105 & -0.1025 & 4.795 & -1.690 & -1.690 \\ 0 & 0 & -1.690 & -1.690 & 1.690 & 1.690 \\ 0 & 0 & -1.690 & -1.690 & 1.690 & 1.690 \end{bmatrix}(10^7)$$

$u_1 = v_1 = u_3 = v_3 = 0$ $\therefore$ partition 1st, 2nd, 5th, and 6th eqs.

$F_{1x} = (-1.035u_2 + 1.793v_2)(10^7)$ ①, $F_{1y} = (1.793u_2 - 3.105v_2)(10^7)$ ②

$F_{3x} = (-1.69u_2 - 1.69v_2)(10^7)$ ③, $F_{3y} = (-1.69u_2 - 1.69v_2)(10^7)$ ④

The remaining eqs with $u_1 = v_1 = u_3 = v_3 = 0$ are

$$\begin{Bmatrix} 0 \\ -80(10^3) \end{Bmatrix} = (10^7)\begin{bmatrix} 2.725 & -0.1025 \\ -0.1025 & 4.795 \end{bmatrix}\begin{Bmatrix} u_2 \\ v_2 \end{Bmatrix}$$ Solve inverse,

$$\begin{Bmatrix} u_2 \\ v_2 \end{Bmatrix} = (10^{-7})\begin{bmatrix} 0.3672 & 0.007853 \\ 0.007853 & 0.2087 \end{bmatrix}\begin{Bmatrix} 0 \\ -80(10^3) \end{Bmatrix} = \begin{Bmatrix} -6.282(10^{-5}) \\ -1.670(10^{-3}) \end{Bmatrix} m$$ **ANS**

Sub u_2 & v_2 into ①–④ yielding

$F_{1x} = -29.3(10^3)\ N = \underline{-29.3\ kN}$, $F_{1y} = \underline{50.7\ kN}$

$F_{3x} = \underline{29.3\ kN = F_{3y}}$ (continued)

$F_1 = \sqrt{F_{1x}^2 + F_{1y}^2} = \underline{58.6\ \text{kN}}, \quad F_3 = \sqrt{F_{3x}^2 + F_{3y}^2} = \underline{41.4\ \text{kN}}$

Stresses. Element 1:

$$(\sigma)_1 = \frac{207(10^9)}{0.600}\{-1 \quad 1\}\begin{bmatrix} 0.5 & -\sqrt{3}/2 & 0 & 0 \\ 0 & 0 & 0.5 & -\sqrt{3}/2 \end{bmatrix}\begin{Bmatrix} 0 \\ 0 \\ -6.282(10^{-5}) \\ -1.670(10^{-3}) \end{Bmatrix} = 488(10^6)\ \text{Pa}$$

$$= \underline{488\ \text{MPa}}$$

Element 2:

$$(\sigma)_2 = \frac{207(10^9)}{0.73485}\{-1 \quad 1\}\frac{\sqrt{2}}{2}\begin{bmatrix} 1 & 1 & 0 & 0 \\ 0 & 0 & 1 & 1 \end{bmatrix}\begin{Bmatrix} -6.282(10^{-5}) \\ -1.670(10^{-3}) \\ 0 \\ 0 \end{Bmatrix} = 345(10^6)\ \text{Pa} = \underline{345\ \text{MPa}}$$

Nodal forces: Element 1:

$$\begin{Bmatrix} f_1 \\ f_2 \end{Bmatrix} = 41.4(10^6)\begin{bmatrix} 1 & -1 \\ -1 & 1 \end{bmatrix}\begin{bmatrix} 0.5 & -\sqrt{3}/2 & 0 & 0 \\ 0 & 0 & 0.5 & -\sqrt{3}/2 \end{bmatrix}\begin{Bmatrix} 0 \\ 0 \\ -6.282(10^{-5}) \\ -1.67(10^{-3}) \end{Bmatrix} = \begin{Bmatrix} -58.6 \\ 58.6 \end{Bmatrix}(10^3)\ \text{N}$$

$$= \begin{Bmatrix} -58.6 \\ 58.6 \end{Bmatrix}\ \text{kN} \quad \boxed{\text{ANS}}$$

Element 2:

$$\begin{Bmatrix} f_2 \\ f_3 \end{Bmatrix} = 33.803(10^6)\begin{bmatrix} 1 & -1 \\ -1 & 1 \end{bmatrix}\frac{\sqrt{2}}{2}\begin{bmatrix} 1 & 1 & 0 & 0 \\ 0 & 0 & 1 & 1 \end{bmatrix}\begin{Bmatrix} -6.282(10^{-5}) \\ -1.67(10^{-3}) \\ 0 \\ 0 \end{Bmatrix} = \begin{Bmatrix} -41.4 \\ 41.4 \end{Bmatrix}(10^3)\ \text{N} = \begin{Bmatrix} -41.4 \\ 41.4 \end{Bmatrix}\ \text{kN}$$

$\boxed{\text{ANS}}$

9.12

(a)

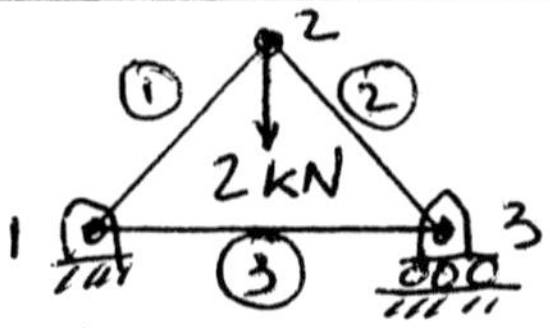

$$k_1 = k_2 = \frac{AE}{L} = \frac{950(10^{-6})(70)(10^9)}{0.500} = 133(10^6)\ \text{N/m}$$

$$k_3 = \frac{950(10^{-6})(70)(10^9)}{0.500\sqrt{2}} = 94.05(10^6)\ \text{N/m}$$

Element 1: $i, j = 1, 2 \quad \phi_x = 45° \quad l = m = \sqrt{2}/2$

$$[k]_1 = \frac{133(10^6)}{2}\begin{bmatrix} 1 & 1 & -1 & -1 \\ 1 & 1 & -1 & -1 \\ -1 & -1 & 1 & 1 \\ -1 & -1 & 1 & 1 \end{bmatrix} \begin{matrix} u_1 \\ v_1 \\ u_2 \\ v_2 \end{matrix}$$

Element 2: $i, j = 2, 3 \quad \phi_x = -45°, \ l = \sqrt{2}/2, \ m = -\sqrt{2}/2$

$$[k]_2 = \frac{133(10^6)}{2}\begin{bmatrix} 1 & -1 & -1 & 1 \\ -1 & 1 & 1 & -1 \\ -1 & 1 & 1 & -1 \\ 1 & -1 & -1 & 1 \end{bmatrix} \begin{matrix} u_2 \\ v_2 \\ u_3 \\ v_3 \end{matrix}$$

Element 3: $i, j = 1, 3 \quad \phi_x = 0, \ l = 1, m = 0$

$$[k]_3 = 94.05(10^6)\begin{bmatrix} 1 & 0 & -1 & 0 \\ 0 & 0 & 0 & 0 \\ -1 & 0 & 1 & 0 \\ 0 & 0 & 0 & 0 \end{bmatrix} \begin{matrix} u_1 \\ v_1 \\ u_3 \\ v_3 \end{matrix}$$

(Continued)

Assemble

$$[K_{sys}] = \begin{bmatrix} 16.055 & 6.65 & -6.65 & -6.65 & -9.405 & 0 \\ 6.65 & 6.65 & -6.65 & -6.65 & 0 & 0 \\ -6.65 & -6.65 & 13.3 & 0 & -6.65 & 6.65 \\ -6.65 & -6.65 & 0 & 13.3 & 6.65 & -6.65 \\ -9.405 & 0 & -6.65 & 6.65 & 16.055 & -6.65 \\ 0 & 0 & 6.65 & -6.65 & -6.65 & 6.65 \end{bmatrix} (10^7)$$

$u_1 = v_1 = v_3 = 0$. Partition 1st, 2nd, and 6th eqs

$$F_{1x} = (-6.65 u_2 - 6.65 v_2 - 9.405 u_3)(10^7) \quad (1)$$

$$F_{1y} = (-6.65 u_2 - 6.65 v_2)(10^7) \quad (2)$$

$$F_{3y} = (6.65 u_2 - 6.65 v_2 - 6.65 u_3)(10^7) \quad (3)$$

The remaining eqs with $u_1 = v_1 = v_3 = 0$ are

$$\begin{Bmatrix} 0 \\ -2000 \\ 0 \end{Bmatrix} = (10^7) \begin{bmatrix} 13.3 & 0 & -6.65 \\ 0 & 13.3 & 6.65 \\ -6.65 & 6.65 & 16.055 \end{bmatrix} \begin{Bmatrix} u_2 \\ v_2 \\ u_3 \end{Bmatrix} \quad \text{Solve inverse}$$

$$\begin{Bmatrix} u_2 \\ v_2 \\ u_3 \end{Bmatrix} = (10^{-7}) \begin{bmatrix} 0.1018 & -0.02658 & 0.05317 \\ -0.02658 & 0.1018 & -0.05317 \\ 0.05317 & -0.05317 & 0.1063 \end{bmatrix} \begin{Bmatrix} 0 \\ -2000 \\ 0 \end{Bmatrix} = \begin{Bmatrix} 5.317(10^{-6}) \\ -2.035(10^{-5}) \\ 1.0633(10^{-5}) \end{Bmatrix} \text{m} \quad \boxed{\text{ANS}}$$

$$F_{1x} = [-6.65(5.317)(10^{-6}) - 6.65(-2.035)(10^{-5}) - 9.405(1.0633)(10^{-5})](10^7) = \underline{\quad}$$

$$F_{1y} = [-6.65(5.317)(10^{-6}) - 6.65(-2.035)(10^{-5})](10^7) = \underline{1000 \text{ N}}$$

$$F_{3y} = [6.65(5.317)(10^{-6}) - 6.65(-2.035)(10^{-5}) - 6.65(1.0633)(10^{-5})](10^7)$$
$$= \underline{1000 \text{N}}$$

Stresses. Element 1:

$$(\sigma)_1 = \frac{70(10^9)}{0.5}\{-1 \ 1\} \frac{\sqrt{2}}{2} \begin{bmatrix} 1 & 1 & 0 & 0 \\ 0 & 0 & 1 & 1 \end{bmatrix} \begin{Bmatrix} 0 \\ 0 \\ 5.317(10^{-6}) \\ -2.035(10^{-5}) \end{Bmatrix} = -1.49(10^6) \text{Pa} = \underline{-1.49 \text{ MPa}}$$

Element 2:

$$(\sigma)_2 = \frac{70(10^9)}{0.5}\{-1 \ 1\} \frac{\sqrt{2}}{2} \begin{bmatrix} 1 & -1 & 0 & 0 \\ 0 & 0 & 1 & -1 \end{bmatrix} \begin{Bmatrix} 5.317(10^{-6}) \\ -2.035(10^{-5}) \\ 1.0633(10^{-5}) \\ 0 \end{Bmatrix} = -1.49(10^6) \text{Pa} = \underline{-1.49 \text{ MPa}}$$

Element 3:

$$(\sigma)_3 = \frac{70(10^9)}{0.5\sqrt{2}}\{-1 \ 1\} \begin{bmatrix} 1 & 0 & 0 & 0 \\ 0 & 0 & 1 & 0 \end{bmatrix} \begin{Bmatrix} 0 \\ 0 \\ 1.0633(10^{-5}) \\ 0 \end{Bmatrix} = 1.05(10^6) \text{Pa} = \underline{1.05 \text{ MPa}}$$

(continued)

Nodal forces. Element 1:

$$\begin{Bmatrix} f_1 \\ f_2 \end{Bmatrix}_1 = 133(10^6)\begin{bmatrix} 1 & -1 \\ -1 & 1 \end{bmatrix}\frac{\sqrt{2}}{2}\begin{bmatrix} 1 & 1 & 0 & 0 \\ 0 & 0 & 1 & 1 \end{bmatrix}\begin{Bmatrix} 0 \\ 0 \\ 5.317(10^{-6}) \\ -2.035(10^{-5}) \end{Bmatrix} = \begin{Bmatrix} 1414 \\ -1414 \end{Bmatrix} \text{ N} \quad \boxed{\text{ANS}}$$

Element 2:

$$\begin{Bmatrix} f_2 \\ f_3 \end{Bmatrix}_2 = 133(10^6)\begin{bmatrix} 1 & -1 \\ -1 & 1 \end{bmatrix}\frac{\sqrt{2}}{2}\begin{bmatrix} 1 & -1 & 0 & 0 \\ 0 & 0 & 1 & -1 \end{bmatrix}\begin{Bmatrix} 5.317(10^{-6}) \\ -2.035(10^{-5}) \\ 1.0633(10^{-5}) \\ 0 \end{Bmatrix} = \begin{Bmatrix} 1414 \\ -1414 \end{Bmatrix} \text{ N} \quad \boxed{\text{ANS}}$$

Element 3:

$$\begin{Bmatrix} f_1 \\ f_3 \end{Bmatrix}_3 = 94.05(10^6)\begin{bmatrix} 1 & -1 \\ -1 & 1 \end{bmatrix}\begin{bmatrix} 1 & 0 & 0 & 0 \\ 0 & 0 & 1 & 0 \end{bmatrix}\begin{Bmatrix} 0 \\ 0 \\ 1.0633(10^{-5}) \\ 0 \end{Bmatrix} = \begin{Bmatrix} -1000 \\ 1000 \end{Bmatrix} \text{ N} \quad \boxed{\text{ANS}}$$

(b) See the solution to Prob. 6.9. Element forces:

$F_{CB} = F_{CD} = -\frac{P}{\sqrt{2}} = -\frac{2000}{\sqrt{2}} = \underline{-1414 \text{ N}}$ checks

$F_{BD} = \frac{P}{2} = \underline{1000 \text{ N}}$ checks

Reactions: $\Sigma F_x = 0 \Rightarrow F_{Bx} = \underline{0}$, $\Sigma M = 0 \Rightarrow F_{BY} = F_{DY} = \underline{1000 \text{ N}}$

Stresses: $\sigma_{CB} = \sigma_{CD} = \frac{-1414}{950} = \underline{-1.49 \text{ MPa}}$ checks

$\sigma_{BD} = \frac{1000}{950} = \underline{1.05 \text{ MPa}}$ checks

Deflections: $(\delta_C)_V = \left(1 + \frac{\sqrt{2}}{4}\right)\frac{PL}{AE} = \left(1 + \frac{\sqrt{2}}{4}\right)\frac{2000(0.5)}{950(10^{-6})(70)(10^9)} = \underline{2.035(10^{-5}) \text{ m}}$

$(\delta_D)_H = \frac{\sqrt{2}}{2}\frac{PL}{AE} = \frac{\sqrt{2}}{2}\frac{(2000)(0.5)}{950(10^{-6})(70)(10^9)} = \underline{1.063(10^{-5}) \text{ m}}$

From symmetry, $(\delta_C)_H = (\delta_D)_H/2 = \underline{5.317(10^{-5}) \text{ m}}$

All answers check

9.13 $k_1 = k_2 = k_3 = k_4 = \frac{AE}{L} = \frac{500(10^{-6})(210)(10^9)}{0.400} = 262.5(10^6) \text{ N/m}$

$k_5 = \frac{500(10^{-6})(210)(10^9)}{0.4\sqrt{2}} = 185.6(10^6) \text{ N/m}$

Element 1: $i, j = 1, 2 \quad \phi_x = 0, \ l = 1, \ m = 0$

$$[k]_1 = 262.5(10^6)\begin{bmatrix} 1 & 0 & -1 & 0 \\ 0 & 0 & 0 & 0 \\ -1 & 0 & 1 & 0 \\ 0 & 0 & 0 & 0 \end{bmatrix} \begin{matrix} u_1 \\ v_1 \\ u_2 \\ v_2 \end{matrix}$$

Element 2: $i, j = 2, 4 \quad \phi_x = 90^\circ, \ l = 0, \ m = 1$

(Continued)

$$[k]_2 = 262.5(10^6)\begin{bmatrix} 0 & 0 & 0 & 0 \\ 0 & 1 & 0 & -1 \\ 0 & 0 & 0 & 0 \\ 0 & -1 & 0 & 1 \end{bmatrix}\begin{matrix} u_2 \\ v_2 \\ u_4 \\ v_4 \end{matrix}$$

Element 3: $i, j = 3, 4 \quad \phi_x = 0, \ l = 1, \ m = 0$

$$[k]_3 = 262.5(10^6)\begin{bmatrix} 1 & 0 & -1 & 0 \\ 0 & 0 & 0 & 0 \\ -1 & 0 & 1 & 0 \\ 0 & 0 & 0 & 0 \end{bmatrix}\begin{matrix} u_3 \\ v_3 \\ u_4 \\ v_4 \end{matrix}$$

Element 4: $i, j = 1, 3 \quad \phi_x = 90°, \ l = 0, \ m = 1$

$$[k]_4 = 262.5(10^6)\begin{bmatrix} 0 & 0 & 0 & 0 \\ 0 & 1 & 0 & -1 \\ 0 & 0 & 0 & 0 \\ 0 & -1 & 0 & 1 \end{bmatrix}\begin{matrix} u_1 \\ v_1 \\ u_3 \\ v_3 \end{matrix}$$

Element 5: $i, j = 2, 3 \quad \phi_x = 135°, \ l = -\sqrt{2}/2, \ m = \sqrt{2}/2$

$$[k]_5 = \frac{185.6(10^6)}{2}\begin{bmatrix} 1 & -1 & -1 & 1 \\ -1 & 1 & 1 & -1 \\ -1 & 1 & 1 & -1 \\ 1 & -1 & -1 & 1 \end{bmatrix}\begin{matrix} u_2 \\ v_2 \\ u_3 \\ v_3 \end{matrix}$$

Assemble

$$[K_{sys}] = \begin{bmatrix} 26.25 & 0 & -26.25 & 0 & 0 & 0 & 0 & 0 \\ 0 & 26.25 & 0 & 0 & 0 & -26.25 & 0 & 0 \\ -26.25 & 0 & 35.53 & -9.28 & -9.28 & 9.28 & 0 & 0 \\ 0 & 0 & -9.28 & 35.53 & 9.28 & -9.28 & 0 & -26.25 \\ 0 & 0 & -9.28 & 9.28 & 35.53 & -9.28 & -26.25 & 0 \\ 0 & -26.25 & 9.28 & -9.28 & -9.28 & 35.53 & 0 & 0 \\ 0 & 0 & 0 & 0 & -26.25 & 0 & 26.25 & 0 \\ 0 & 0 & 0 & -26.25 & 0 & 0 & 0 & 26.25 \end{bmatrix}(10^7) \text{ N/m}$$

$u_1 = u_2 = v_3 = u_4 = 0$. Partition 1st, 2nd, 6th, and 7th eqs.

$F_{1x} = -26.25(10^7)u_2$ ①, $F_{1y} = 0$ ②, $F_{3y} = (9.28u_2 - 9.28v_2 - 9.28u_3)(10^7)$ ③

$F_{4x} = -26.25(10^7)u_3$ ④ The remaining eqs with $u_1 = u_2 = v_3 = u_4 = 0$

$$\begin{Bmatrix} 1000 \\ -2000 \\ 0 \\ 0 \end{Bmatrix} = (10^7)\begin{bmatrix} 35.53 & -9.28 & -9.28 & 0 \\ -9.28 & 35.53 & 9.28 & -26.25 \\ -9.28 & 9.28 & 35.53 & 0 \\ 0 & -26.25 & 0 & 26.25 \end{bmatrix}\begin{Bmatrix} u_2 \\ v_2 \\ u_3 \\ v_4 \end{Bmatrix} \quad \text{Solve inverse}$$

$$\begin{Bmatrix} u_2 \\ v_2 \\ u_3 \\ v_4 \end{Bmatrix} = (10^{-7})\begin{bmatrix} 0.0381 & 0.0381 & 0 & 0.0381 \\ 0.0381 & 0.1839 & -0.0381 & 0.1839 \\ 0 & -0.0381 & 0.0381 & -0.0381 \\ 0.0381 & 0.1839 & -0.0381 & 0.222 \end{bmatrix}\begin{Bmatrix} 1000 \\ -2000 \\ 0 \\ 0 \end{Bmatrix} = \begin{Bmatrix} -3.81(10^{-6}) \\ -3.298(10^{-5}) \\ 7.619(10^{-6}) \\ -3.298(10^{-5}) \end{Bmatrix} \text{m} \quad \boxed{\text{ANS}}$$

(continued)

Sub. displacements into eqs ① – ④, resulting in

$F_{1x} = 1000\,N$, $F_{1y} = 0$, $F_{3y} = 2000\,N$, $F_{4x} = -2000\,N$

Forces in element 5:

$$\begin{Bmatrix} f_2 \\ f_3 \end{Bmatrix}_5 = 185.6(10^6)\begin{bmatrix} 1 & -1 \\ -1 & 1 \end{bmatrix}\frac{\sqrt{2}}{2}\begin{bmatrix} -1 & 1 & 0 & 0 \\ 0 & 0 & -1 & 1 \end{bmatrix}\begin{Bmatrix} -3.81(10^{-6}) \\ -3.298(10^{-5}) \\ 7.619(10^{-6}) \\ 0 \end{Bmatrix} = \begin{Bmatrix} -2828 \\ 2828 \end{Bmatrix} N \quad \boxed{\text{ANS}}$$

Stress in element 5:

$$(\sigma)_5 = \frac{210(10^3)}{0.4\sqrt{2}}\{-1 \quad 1\}\frac{\sqrt{2}}{2}\begin{bmatrix} -1 & 1 & 0 & 0 \\ 0 & 0 & -1 & 1 \end{bmatrix}\begin{Bmatrix} -3.81(10^{-6}) \\ -3.298(10^{-5}) \\ 7.619(10^{-6}) \\ 0 \end{Bmatrix} = \underline{5.66\ MPa}$$

(9.14) See the solution of Prob. 9.13 for stiffness matrices up to $[K_{sys}]$. B.C. $u_1 = v_1 = u_4 = 0$. Partition 1st, 2nd, and 7th eqs.

$$\begin{Bmatrix} 1000 \\ -2000 \\ F_{3x} \\ F_{3y} \\ 0 \end{Bmatrix} = (10^7)\begin{bmatrix} 35.53 & -9.28 & -9.28 & 9.28 & 0 \\ -9.28 & 35.53 & 9.28 & -9.28 & -26.25 \\ -9.28 & 9.28 & 35.53 & -9.28 & 0 \\ 9.28 & -9.28 & -9.28 & 35.53 & 0 \\ 0 & -26.25 & 0 & 0 & 26.25 \end{bmatrix}\begin{Bmatrix} u_2 \\ v_2 \\ u_3 \\ v_3 \\ v_4 \end{Bmatrix} = [K_{red}]\{u\}$$

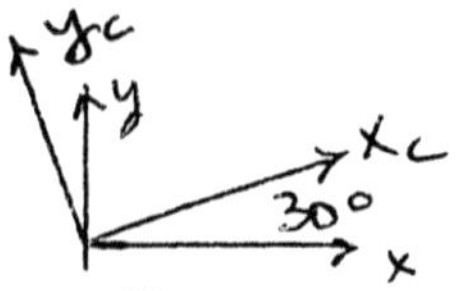

At node 3,

$$\begin{Bmatrix} u_3 \\ v_3 \end{Bmatrix}_{Global} = \begin{bmatrix} \cos 30 & -\sin 30 \\ \sin 30 & \cos 30 \end{bmatrix}\begin{Bmatrix} u_3 \\ v_3 \end{Bmatrix}_C = \frac{1}{2}\begin{bmatrix} \sqrt{3} & -1 \\ 1 & \sqrt{3} \end{bmatrix}\begin{Bmatrix} u_3 \\ v_3 \end{Bmatrix}_c \quad ①$$

The transformation matrix in terms of the reduced dof is

$$[T]_c = \begin{bmatrix} 1 & 0 & 0 & 0 & 0 \\ 0 & 1 & 0 & 0 & 0 \\ 0 & 0 & \sqrt{3}/2 & -1/2 & 0 \\ 0 & 0 & 1/2 & \sqrt{3}/2 & 0 \\ 0 & 0 & 0 & 0 & 1 \end{bmatrix}$$

$$[T]_c^T[K_{red}][T_c] = 10^7\begin{bmatrix} 35.53 & -9.281 & -3.397 & 12.678 & 0 \\ -9.281 & 35.53 & 3.397 & -12.678 & -26.25 \\ -3.397 & 3.397 & 27.493 & -4.64 & 0 \\ 12.678 & -12.678 & -4.64 & 43.568 & 0 \\ 0 & -26.25 & 0 & 0 & 26.25 \end{bmatrix}$$

$$\{F\}_c = \begin{Bmatrix} 1000 \\ -2000 \\ (F_{3x})_c \\ (F_{3y})_c \\ 0 \end{Bmatrix}$$

But $(v_3)_c = 0$. Thus partition 4th eq.

$$(F_{3y})_c = [12.678 u_2 - 12.678 v_2 - 4.64 (u_3)_c](10^7) \quad ②$$

(continued)

Note $(F_{3x})_c = 0$ and eq ② in the solution of Prob 9.13 changes since $v_3 \neq 0$. This is $F_{1y} = -26.25(10^7)\,v_3$ ③. The remaining eqs are,

$$\begin{Bmatrix} 1000 \\ -2000 \\ 0 \\ 0 \end{Bmatrix} = (10^7)\begin{bmatrix} 35.53 & -9.281 & -3.397 & 0 \\ -9.281 & 35.53 & 3.397 & -26.25 \\ -3.397 & 3.397 & 27.493 & 0 \\ 0 & -26.25 & 0 & 26.25 \end{bmatrix}\begin{Bmatrix} u_2 \\ v_2 \\ (u_3)_c \\ v_4 \end{Bmatrix} \quad \text{invert}$$

$$\begin{Bmatrix} u_2 \\ v_2 \\ (u_3)_c \\ v_4 \end{Bmatrix} = (10^{-7})\begin{bmatrix} 0.0381 & 0.0381 & 0 & 0.0381 \\ 0.0381 & 0.151 & -0.0139 & 0.151 \\ 0 & -0.0381 & 0.0381 & -0.0139 \\ 0.0381 & 0.151 & -0.0139 & 0.189 \end{bmatrix}\begin{Bmatrix} 1000 \\ -2000 \\ 0 \\ 0 \end{Bmatrix} = \begin{Bmatrix} -3.81(10^{-6}) \\ -2.638(10^{-5}) \\ 2.789(10^{-6}) \\ -2.638(10^{-5}) \end{Bmatrix} \text{m}$$

ANS

For the displacement of node 3 in xy coord.
Sub values into eq ①

$$\begin{Bmatrix} u_3 \\ v_3 \end{Bmatrix} = \frac{1}{2}\begin{bmatrix} \sqrt{3} & -1 \\ 1 & \sqrt{3} \end{bmatrix}\begin{Bmatrix} 2.789(10^{-6}) \\ 0 \end{Bmatrix} = \begin{Bmatrix} 2.415(10^{-6}) \\ 1.395(10^{-6}) \end{Bmatrix} \text{m} \quad \text{ANS}$$

Eqs. ① & ④ from Prob. 9.13 and ③ above

$$F_{1x} = -26.25(10^7)(-3.81)(10^{-6}) = \underline{1000\text{ N}}$$

$$F_{1y} = -26.25(10^7)(1.395)(10^{-6}) = \underline{-366\text{ N}}$$

$$F_{4x} = -26.25(10^7)(2.415)(10^{-6}) = \underline{-634\text{ N}}$$

From ② (previous page)

$$(F_{3y})_c = \left[12.68(-3.81+26.38)(10^{-6}) - 4.64(2.789)(10^{-6})\right](10^7) = \underline{2732\text{ N}}$$

The force components in the xy coord. sys are

$$\begin{Bmatrix} F_{3x} \\ F_{3y} \end{Bmatrix} = \frac{1}{2}\begin{bmatrix} \sqrt{3} & -1 \\ 1 & \sqrt{3} \end{bmatrix}\begin{Bmatrix} 0 \\ 2732 \end{Bmatrix} = \begin{Bmatrix} -1366 \\ 2366 \end{Bmatrix} \text{N} \quad \text{ANS}$$

<u>Force in element 5:</u>

$$\begin{Bmatrix} f_2 \\ f_3 \end{Bmatrix}_5 = 185.6(10^6)\begin{bmatrix} 1 & -1 \\ -1 & 1 \end{bmatrix}\frac{\sqrt{2}}{2}\begin{bmatrix} -1 & 1 & 0 & 0 \\ 0 & 0 & -1 & 1 \end{bmatrix}\begin{Bmatrix} -3.81(10^{-6}) \\ -2.638(10^{-5}) \\ 2.415(10^{-6}) \\ 1.395(10^{-6}) \end{Bmatrix} = \begin{Bmatrix} -2828 \\ 2828 \end{Bmatrix} \text{N} \quad \text{ANS}$$

<u>Stress in element 5:</u>

$$(\sigma)_5 = \frac{210(10^3)}{0.4\sqrt{2}}\{-1 \;\; 1\}\frac{\sqrt{2}}{2}\begin{bmatrix} -1 & 1 & 0 & 0 \\ 0 & 0 & -1 & 1 \end{bmatrix}\begin{Bmatrix} -3.81(10^{-6}) \\ -2.638(10^{-5}) \\ 2.415(10^{-6}) \\ 1.395(10^{-6}) \end{Bmatrix} = \underline{5.66\text{ MPa}} \quad \text{ANS}$$

9.15 $L_e = \left[(2-1)^2 + (1+1)^2 + (-2-0)^2\right]^{0.5} = 3 \text{ in}$

$\ell = \dfrac{2-1}{3} = \dfrac{1}{3}, \quad m = \dfrac{1+1}{3} = \dfrac{2}{3}, \quad n = -\dfrac{2}{3}$

$$[T] = \frac{1}{3}\begin{bmatrix} 1 & 2 & -2 & 0 & 0 & 0 \\ 0 & 0 & 0 & 1 & 2 & -2 \end{bmatrix}$$

$$\begin{Bmatrix} f_i \\ f_j \end{Bmatrix} = \frac{AE}{L}\begin{bmatrix} 1 & -1 \\ -1 & 1 \end{bmatrix}[T]\begin{Bmatrix} u_i \\ v_i \\ w_i \\ u_j \\ v_j \\ w_j \end{Bmatrix} = \frac{0.2(10)(10^6)}{3}\begin{bmatrix} 1 & -1 \\ -1 & 1 \end{bmatrix}\frac{1}{3}\begin{bmatrix} 1 & 2 & -2 & 0 & 0 & 0 \\ 0 & 0 & 0 & 1 & 2 & -2 \end{bmatrix}\begin{Bmatrix} -2.5 \\ 1.8 \\ 2.1 \\ -1.5 \\ 1.2 \\ 2.8 \end{Bmatrix}(10^{-2})$$

$$= \begin{Bmatrix} 3560 \\ -3560 \end{Bmatrix} \text{lb} \quad \boxed{\text{ANS}}$$

$$\sigma = \frac{10(10^6)}{3}\{-1 \quad 1\}\frac{1}{3}\begin{bmatrix} 1 & 2 & -2 & 0 & 0 & 0 \\ 0 & 0 & 0 & 1 & 2 & -2 \end{bmatrix}\begin{Bmatrix} -2.5 \\ 1.8 \\ 2.1 \\ -1.5 \\ 1.2 \\ 2.8 \end{Bmatrix}(10^{-2}) = 17.8(10^3)\text{psi} = \underline{17.8 \text{ kpsi}}$$

9.16 $L_2 = L_3 = \sqrt{(0-200)^2 + (100-0)^2 + (200-0)^2} = 300 \text{ mm}$

$$k_1 = \frac{AE}{L_1} = \frac{600(10^{-6})(70)(10^9)}{0.200} = 210(10^6) \text{ N/m}$$

$$k_2 = k_3 = \frac{600(10^{-3})(70)(10^9)}{0.300} = 140(10^6) \text{ N/m}$$

__Element 1:__ $i, j = 1, 2 \quad \phi_x = 0, \ell = 1, m = 0, n = 0$

$$[k]_1 = 210(10^6)\begin{bmatrix} 1 & 0 & 0 & -1 & 0 & 0 \\ 0 & 0 & 0 & 0 & 0 & 0 \\ 0 & 0 & 0 & 0 & 0 & 0 \\ -1 & 0 & 0 & 1 & 0 & 0 \\ 0 & 0 & 0 & 0 & 0 & 0 \\ 0 & 0 & 0 & 0 & 0 & 0 \end{bmatrix} \begin{matrix} u_1 \\ v_1 \\ w_1 \\ u_2 \\ v_2 \\ w_2 \end{matrix}$$

__Element 2:__ $i, j = 2, 3 \quad \ell = \dfrac{0-200}{300} = -\dfrac{2}{3}, \; m = \dfrac{1}{3}, \; n = \dfrac{2}{3}$

$$[k]_2 = \frac{140(10^6)}{9}\begin{bmatrix} 4 & -2 & -4 & -4 & 2 & 4 \\ -2 & 1 & 2 & 2 & -1 & -2 \\ -4 & 2 & 4 & 4 & -2 & -4 \\ -4 & 2 & 4 & 4 & -2 & -4 \\ 2 & -1 & -2 & -2 & 1 & 2 \\ 4 & -2 & -4 & -4 & 2 & 4 \end{bmatrix} \begin{matrix} u_2 \\ v_2 \\ w_2 \\ u_3 \\ v_3 \\ w_3 \end{matrix}$$

__Element 3:__ $i, j = 2, 4 \quad \ell = -\dfrac{2}{3}, \; m = \dfrac{1}{3}, \; n = -\dfrac{2}{3}$

(continued)

$$[k]_3 = \frac{140(10^6)}{9}\begin{bmatrix} 4 & -2 & 4 & -4 & 2 & -4 \\ -2 & 1 & -2 & 2 & -1 & 2 \\ 4 & -2 & 4 & -4 & 2 & -4 \\ -4 & 2 & -4 & 4 & -2 & 4 \\ 2 & -1 & 2 & -2 & 1 & -2 \\ -4 & 2 & -4 & 4 & -2 & 4 \end{bmatrix} \begin{matrix} u_2 \\ v_2 \\ w_2 \\ u_4 \\ v_4 \\ w_4 \end{matrix}$$

$$[K_{sys}] = \begin{bmatrix} 21 & 0 & 0 & -21 & 0 & 0 & 0 & 0 & 0 & 0 & 0 & 0 \\ 0 & 0 & 0 & 0 & 0 & 0 & 0 & 0 & 0 & 0 & 0 & 0 \\ 0 & 0 & 0 & 0 & 0 & 0 & 0 & 0 & 0 & 0 & 0 & 0 \\ -21 & 0 & 0 & 33.444 & -6.222 & 0 & -6.222 & 3.111 & 6.222 & -6.222 & 3.111 & -6.222 \\ 0 & 0 & 0 & -6.222 & 3.111 & 0 & 3.111 & -1.556 & -3.111 & 3.111 & -1.556 & 3.111 \\ 0 & 0 & 0 & 0 & 0 & 12.444 & 6.222 & -3.111 & -6.222 & -6.222 & 3.111 & -6.222 \\ 0 & 0 & 0 & -6.222 & 3.111 & 6.222 & 6.222 & -3.111 & -6.222 & 0 & 0 & 0 \\ 0 & 0 & 0 & 3.111 & -1.556 & -3.111 & -3.111 & 1.556 & 3.111 & 0 & 0 & 0 \\ 0 & 0 & 0 & 6.222 & -3.111 & -6.222 & -6.222 & 3.111 & 6.222 & 0 & 0 & 0 \\ 0 & 0 & 0 & -6.222 & 3.111 & -6.222 & 0 & 0 & 0 & 6.222 & -3.111 & 6.222 \\ 0 & 0 & 0 & 3.111 & -1.556 & 3.111 & 0 & 0 & 0 & -3.111 & 1.556 & -3.111 \\ 0 & 0 & 0 & -6.222 & 3.111 & -6.222 & 0 & 0 & 0 & 6.222 & -3.111 & 6.222 \end{bmatrix}(10^7)$$

$u_1 = v_1 = w_1 = u_3 = v_3 = w_3 = u_4 = v_4 = w_4 = 0$. Partition all but 4th, 5th, and 6th eqs which are (with $u_1 = v_1 = w_1 = u_3 = v_3 = w_3 = u_4 = v_4 = w_4 = 0$)

$$\begin{Bmatrix} 0 \\ -2000 \\ 0 \end{Bmatrix} = (10^7)\begin{bmatrix} 33.444 & -6.222 & 0 \\ -6.222 & 3.111 & 0 \\ 0 & 0 & 12.444 \end{bmatrix}\begin{Bmatrix} u_2 \\ v_2 \\ w_2 \end{Bmatrix} \qquad \text{Invert}$$

$$\begin{Bmatrix} u_2 \\ v_2 \\ w_2 \end{Bmatrix} = (10^{-7})\begin{bmatrix} 0.04762 & -0.09524 & 0 \\ -0.09524 & 0.5119 & 0 \\ 0 & 0 & 0.08036 \end{bmatrix}\begin{Bmatrix} 0 \\ -2000 \\ 0 \end{Bmatrix} = \begin{Bmatrix} -1.905(10^{-5}) \\ -1.024(10^{-4}) \\ 0 \end{Bmatrix} \text{m} \quad \boxed{\text{ANS}}$$

Element 2 nodal forces:

$$\begin{Bmatrix} f_2 \\ f_3 \end{Bmatrix}_2 = \frac{140(10^6)}{3}\begin{bmatrix} 1 & -1 \\ -1 & 1 \end{bmatrix}\begin{bmatrix} -2 & 1 & 2 & 0 & 0 & 0 \\ 0 & 0 & 0 & -2 & 1 & 2 \end{bmatrix}\begin{Bmatrix} -1.905(10^{-5}) \\ -1.024(10^{-4}) \\ 0 \\ 0 \\ 0 \\ 0 \end{Bmatrix} = \begin{Bmatrix} -3000 \\ 3000 \end{Bmatrix} \text{N} \quad \boxed{\text{ANS}}$$

Element 2 stress:

$$(\sigma)_2 = \frac{70(10^3)}{(0.3)}\{-1 \;\; 1\}\frac{1}{3}\begin{bmatrix} -2 & 1 & 2 & 0 & 0 & 0 \\ 0 & 0 & 0 & -2 & 1 & 2 \end{bmatrix}\begin{Bmatrix} -1.905(10^{-5}) \\ -1.024(10^{-4}) \\ 0 \\ 0 \\ 0 \\ 0 \end{Bmatrix} = \underline{5 \text{ MPa}} \quad \boxed{\text{ANS}}$$

9.17 Beam 1,

$$\begin{Bmatrix} V_1 \\ 2M_1/L \\ V_2 \\ 2M_2/L \end{Bmatrix} = \frac{E(2I)}{(L/2)^3}\begin{bmatrix} 12 & 6 & -12 & 6 \\ 6 & 4 & -6 & 2 \\ -12 & -6 & 12 & -6 \\ 6 & 2 & -6 & 4 \end{bmatrix}\begin{Bmatrix} v_1 \\ \theta_1 L/2 \\ v_2 \\ \theta_2 L/2 \end{Bmatrix}$$

Beam 2,

$$\begin{Bmatrix} V_2 \\ 2M_2/L \\ V_3 \\ 2M_3/L \end{Bmatrix} = \frac{EI}{(L/2)^3}\begin{bmatrix} 12 & 6 & -12 & 6 \\ 6 & 4 & -6 & 2 \\ -12 & -6 & 12 & -6 \\ 6 & 2 & -6 & 4 \end{bmatrix}\begin{Bmatrix} v_2 \\ \theta_2 L/2 \\ v_3 \\ \theta_3 L/2 \end{Bmatrix}$$

Assemble

$$\begin{Bmatrix} R_1 \\ 2M_1/L \\ 0 \\ 0 \\ -P \\ 0 \end{Bmatrix} = \frac{16EI}{L^3}\begin{bmatrix} 12 & 6 & -12 & 6 & 0 & 0 \\ 6 & 4 & -6 & 2 & 0 & 0 \\ -12 & -6 & 18 & -3 & -6 & 3 \\ 6 & 2 & -3 & 6 & -3 & 1 \\ 0 & 0 & -6 & -3 & 6 & -3 \\ 0 & 0 & 3 & 1 & -3 & 2 \end{bmatrix}\begin{Bmatrix} v_1 \\ \theta_1 L/2 \\ v_2 \\ \theta_2 L/2 \\ v_3 \\ \theta_3 L/2 \end{Bmatrix}$$

$v_1 = \theta_1 = 0$. Partition 1st and 2nd eqs. Remaining eqs are

$$\begin{Bmatrix} 0 \\ 0 \\ -P \\ 0 \end{Bmatrix} = \frac{16EI}{L^3}\begin{bmatrix} 18 & -3 & -6 & 3 \\ -3 & 6 & -3 & 1 \\ -6 & -3 & 6 & -3 \\ 3 & 1 & -3 & 2 \end{bmatrix}\begin{Bmatrix} v_2 \\ \theta_2 L/2 \\ v_3 \\ \theta_3 L/2 \end{Bmatrix} \quad \text{Invert}$$

$$\begin{Bmatrix} v_2 \\ \theta_2 L/2 \\ v_3 \\ \theta_3 L/2 \end{Bmatrix} = \frac{L^3}{96EI}\begin{bmatrix} 2 & 3 & 5 & 3 \\ 3 & 6 & 9 & 6 \\ 5 & 9 & 18 & 15 \\ 3 & 6 & 15 & 18 \end{bmatrix}\begin{Bmatrix} 0 \\ 0 \\ -P \\ 0 \end{Bmatrix} = -\begin{Bmatrix} 5 \\ 9 \\ 18 \\ 15 \end{Bmatrix}\frac{PL^3}{96EI}$$ **ANS**

9.18 See the solution of Prob 9.17 for element equations. The same partitioning and inversion apply as in Prob. 9.17. Thus, with $2M_3/L = -2M/L$

$$\begin{Bmatrix} v_2 \\ \theta_2 L/2 \\ v_3 \\ \theta_3 L/2 \end{Bmatrix} = \frac{L^3}{96EI}\begin{bmatrix} 2 & 3 & 5 & 3 \\ 3 & 6 & 9 & 6 \\ 5 & 9 & 18 & 15 \\ 3 & 6 & 15 & 18 \end{bmatrix}\begin{Bmatrix} 0 \\ 0 \\ 0 \\ -2M/L \end{Bmatrix} = -\begin{Bmatrix} 1 \\ 2 \\ 5 \\ 6 \end{Bmatrix}\frac{ML^2}{16EI}$$ **ANS**

9.19

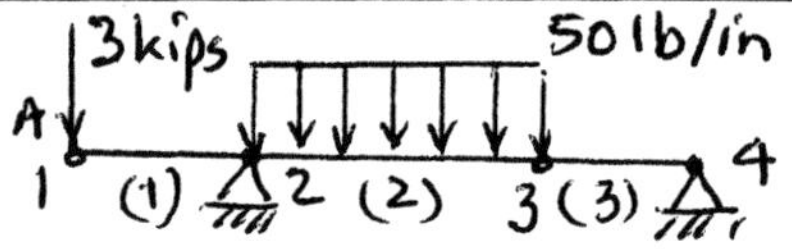

From the solution of Prob 3.39
$I = 136 \text{ in}^4$

Elements 1 & 3:

$$[k]_1 = [k]_3 = \frac{30(10^6)(136)}{(60)^3}\begin{bmatrix} 12 & 360 & -12 & 360 \\ 360 & 14400 & -360 & 7200 \\ -12 & -360 & 12 & -360 \\ 360 & 7200 & -360 & 14400 \end{bmatrix}$$

Element 2:

$$[k]_2 = \frac{30(10^6)(136)}{(120)^3}\begin{bmatrix} 12 & 720 & -12 & 720 \\ 720 & 57600 & -720 & 28800 \\ -12 & -720 & 12 & -720 \\ 720 & 28800 & -720 & 57600 \end{bmatrix}$$

Assemble. Use as common factor $\frac{30(10^6)(136)}{(120)^3} = 2361.1$

$$[K_{sys}] = 2361.1\begin{bmatrix} 96 & 2880 & -96 & 2880 & 0 & 0 & 0 & 0 \\ 2880 & 115200 & -2880 & 57600 & 0 & 0 & 0 & 0 \\ -96 & -2880 & 108 & -2160 & -12 & 720 & 0 & 0 \\ 2880 & 57600 & -2160 & 172800 & -720 & 28800 & 0 & 0 \\ 0 & 0 & -12 & -720 & 108 & 2160 & -96 & 2880 \\ 0 & 0 & 720 & 28800 & 2160 & 172800 & -2880 & 57600 \\ 0 & 0 & 0 & 0 & -96 & -2880 & 96 & -2880 \\ 0 & 0 & 0 & 0 & 2880 & 57600 & -2880 & 115200 \end{bmatrix}$$

$v_2 = v_4 = 0$. Partition 3rd & 7th eqs. Remaining eqs with $q_o L_e/2 = -50(120)/2 = -3000 \text{ lb}$, $q_o L_e^2/12 = -50(120^2)/12 = -60{,}000 \text{ lb-in}$ for element 2 are

$$\begin{Bmatrix} -3000 \\ 0 \\ -60{,}000 \\ -3000 \\ 60{,}000 \\ 0 \end{Bmatrix} = 2361.1\begin{bmatrix} 96 & 2880 & 2880 & 0 & 0 & 0 \\ 2880 & 115200 & 57600 & 0 & 0 & 0 \\ 2880 & 57600 & 172800 & -720 & 28800 & 0 \\ 0 & 0 & -720 & 108 & 2160 & 2880 \\ 0 & 0 & 28800 & 2160 & 172800 & 57600 \\ 0 & 0 & 0 & 2880 & 57600 & 115200 \end{bmatrix}\begin{Bmatrix} v_1 \\ \theta_1 \\ \theta_2 \\ v_2 \\ \theta_3 \\ \theta_4 \end{Bmatrix}$$

$$\begin{Bmatrix} v_1 \\ \theta_1 \\ v_2 \\ \theta_2 \\ v_3 \\ \theta_3 \end{Bmatrix} = (2361.1)^{-1}\begin{bmatrix} 0.167 & -3.13E-03 & -2.08E-03 & -5.56E-02 & 6.94E-04 & 1.04E-03 \\ -3.13E-03 & 6.94E-05 & 3.47E-05 & 9.26E-04 & -1.16E-05 & -1.74E-05 \\ -2.08E-03 & 3.47E-05 & 3.47E-05 & 9.26E-04 & -1.16E-05 & -1.74E-05 \\ -5.56E-02 & 9.26E-04 & 9.26E-04 & 5.56E-02 & -4.63E-04 & -1.16E-03 \\ 6.94E-04 & -1.16E-05 & -1.16E-05 & -4.63E-04 & 1.16E-05 & 5.79E-06 \\ 1.04E-03 & -1.74E-05 & -1.74E-05 & -1.16E-03 & 5.79E-06 & 3.47E-05 \end{bmatrix}\begin{Bmatrix} -3000 \\ 0 \\ -60{,}000 \\ -3000 \\ 60{,}000 \\ 0 \end{Bmatrix}$$

(continued)

$$\begin{Bmatrix} v_1 \\ \theta_1 \\ v_2 \\ \theta_2 \\ v_3 \\ \theta_3 \end{Bmatrix} = \begin{Bmatrix} -7.06E\text{-}02 \text{ in} \\ 1.62E\text{-}03 \text{ rad} \\ 2.94E\text{-}04 \text{ rad} \\ -3.53E\text{-}02 \text{ in} \\ 2.94E\text{-}04 \text{ rad} \\ 7.35E\text{-}04 \text{ rad} \end{Bmatrix} = \begin{Bmatrix} -7.06E\text{-}02 \text{ in} \\ 9.27E\text{-}02 \text{ deg} \\ 1.69E\text{-}02 \text{ deg} \\ -3.53E\text{-}02 \text{ in} \\ 1.69E\text{-}02 \text{ deg} \\ 4.21E\text{-}02 \text{ deg} \end{Bmatrix}$$

Thus,

$v_A = -0.0706 \text{ in}$

$\theta_A = 0.0927^\circ$ **ANS**

(9.20)

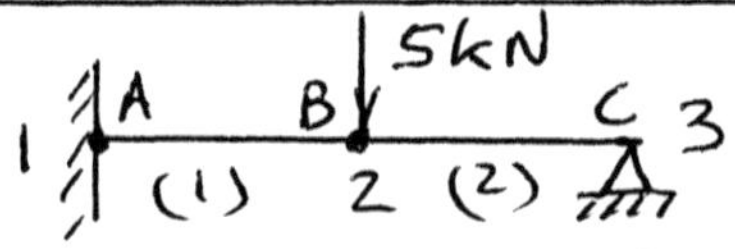

$L_1 = L_2 = 1\text{m}$

(a) For each beam

$$[k] = \frac{200(10^3)}{1^3}\begin{bmatrix} 12 & 6 & -12 & 6 \\ 6 & 4 & -6 & 2 \\ -12 & -6 & 12 & -6 \\ 6 & 2 & -6 & 4 \end{bmatrix}$$

<u>Assemble</u>

$$[K_{sys}] = 400(10^3)\begin{bmatrix} 6 & 3 & -6 & 3 & 0 & 0 \\ 3 & 2 & -3 & 1 & 0 & 0 \\ -6 & -3 & 12 & 0 & -6 & 3 \\ 3 & 1 & 0 & 4 & -3 & 1 \\ 0 & 0 & -6 & -3 & 6 & -3 \\ 0 & 0 & 3 & 1 & -3 & 2 \end{bmatrix}$$

$v_1 = \theta_1 = v_3 = 0$, partition 1st, 2nd, and 5th eqs. Remaining eqs.

$$\begin{Bmatrix} -5000 \\ 0 \\ 0 \end{Bmatrix} = 400(10^3)\begin{bmatrix} 12 & 0 & 3 \\ 0 & 4 & 1 \\ 3 & 1 & 2 \end{bmatrix}\begin{Bmatrix} v_2 \\ \theta_2 \\ \theta_3 \end{Bmatrix}$$

Invert eq.

$$\begin{Bmatrix} v_2 \\ \theta_2 \\ \theta_3 \end{Bmatrix} = 2.5(10^{-6})\begin{bmatrix} 0.1458 & 0.0625 & -0.25 \\ 0.0625 & 0.3125 & -0.25 \\ -0.25 & -0.25 & 1 \end{bmatrix}\begin{Bmatrix} 5000 \\ 0 \\ 0 \end{Bmatrix} = \begin{Bmatrix} -1.823(10^{-3})\text{ m} \\ -7.81(10^{-4})\text{ rad} \\ 3.13(10^{-3})\text{ rad} \end{Bmatrix}$$

Thus, $(v_2)_B = \underline{-1.823 \text{ mm}}$

For the stress at A and B determine the bending moments in element 1:

$$\begin{Bmatrix} V_1 \\ M_1 \\ V_2 \\ M_2 \end{Bmatrix}_1 = 200(10^3)\begin{bmatrix} 12 & 6 & -12 & 6 \\ 6 & 4 & -6 & 2 \\ -12 & -6 & 12 & -6 \\ 6 & 2 & -6 & 4 \end{bmatrix}\begin{Bmatrix} 0 \\ 0 \\ -1.823(10^{-3}) \\ -7.81(10^{-4}) \end{Bmatrix} = \begin{Bmatrix} 3438 \text{ N} \\ 1875 \text{ N}\cdot\text{m} \\ -3438 \text{ N} \\ 1563 \text{ N}\cdot\text{m} \end{Bmatrix}$$

$$\sigma_A = \frac{1875(10^3)}{35(10^3)} = \underline{53.6 \text{ MPa}}$$

$$\sigma_B = \frac{1563(10^3)}{35(10^3)} = \underline{44.6 \text{ MPa}}$$

(continued)

(b)

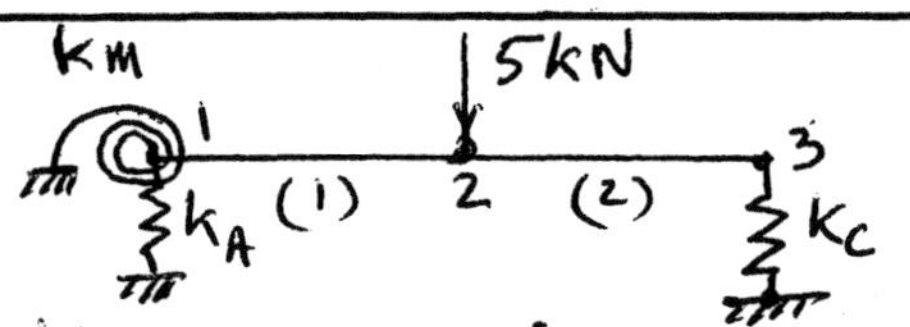

$k_A = 10(10^6)$ N/m

$k_C = 5(10^6)$ N/m

$k_M = 8(10^6)$ N-m

Considering the $[K_{sys}]$ matrix of part a add k_A to the k_{11} term, k_M to the k_{22} term, and k_C to the k_{55} term. Note: $10(10^6)/400(10^3) = 25$, $8(10^6)/400(10^3) = 20$, and $5(10^6)/400(10^3) = 12.5$ to add within $[K_{sys}]$ on previous page.

Thus

$$\begin{Bmatrix} 0 \\ 0 \\ -5000 \\ 0 \\ 0 \\ 0 \end{Bmatrix} = 400(10^3)\begin{bmatrix} 31 & 3 & -6 & 3 & 0 & 0 \\ 3 & 22 & -3 & 1 & 0 & 0 \\ -6 & -3 & 12 & 0 & -6 & 3 \\ 3 & 1 & 0 & 4 & -3 & 1 \\ 0 & 0 & -6 & -3 & 18.5 & -3 \\ 0 & 0 & 3 & 1 & -3 & 2 \end{bmatrix}\begin{Bmatrix} v_1 \\ \theta_1 \\ v_2 \\ \theta_2 \\ v_3 \\ \theta_3 \end{Bmatrix}$$

Invert.

$$\begin{Bmatrix} v_1 \\ \theta_1 \\ v_2 \\ \theta_2 \\ v_3 \\ \theta_3 \end{Bmatrix} = [400(10^3)]^{-1}\begin{bmatrix} 3.97E-02 & -7.08E-04 & 2.72E-02 & -2.19E-02 & 5.66E-04 & -2.90E-02 \\ -7.08E-04 & 4.82E-02 & 1.80E-02 & -4.83E-03 & 1.42E-03 & -2.25E-02 \\ 2.72E-02 & 1.80E-02 & 1.79E-01 & 5.93E-02 & 2.56E-02 & -2.60E-01 \\ -2.19E-02 & -4.83E-03 & 5.93E-02 & 3.50E-01 & 4.39E-02 & -1.98E-01 \\ 5.66E-04 & 1.42E-03 & 2.56E-02 & 4.39E-02 & 7.89E-02 & 5.80E-02 \\ -2.90E-02 & -2.25E-02 & -2.60E-01 & -1.98E-01 & 5.80E-02 & 1.08E+00 \end{bmatrix}\begin{Bmatrix} 0 \\ 0 \\ -5000 \\ 0 \\ 0 \\ 0 \end{Bmatrix}$$

$$\begin{Bmatrix} v_1 \\ \theta_1 \\ v_2 \\ \theta_2 \\ v_3 \\ \theta_3 \end{Bmatrix} = \begin{Bmatrix} -3.402E-04 \text{ m} \\ -2.255E-04 \text{ rad} \\ -2.241E-03 \text{ m} \\ -7.415E-04 \text{ rad} \\ -3.196E-04 \text{ m} \\ 3.253E-03 \text{ rad} \end{Bmatrix}$$

ANS

$(v_c)_B = v_2 = -2.241$ mm

Element 1:

$$\begin{Bmatrix} V_1 \\ M_1 \\ V_2 \\ M_2 \end{Bmatrix} = 200(10^3)\begin{bmatrix} 12 & 6 & -12 & 6 \\ 6 & 4 & -6 & 2 \\ -12 & -6 & 12 & -6 \\ 6 & 2 & -6 & 4 \end{bmatrix}\begin{Bmatrix} -3.402(10^{-4}) \\ -2.255(10^{-4}) \\ -2.241(10^{-3}) \\ -7.415(10^{-4}) \end{Bmatrix} = \begin{Bmatrix} 3402 \text{ N} \\ 1804 \text{ N}\cdot\text{m} \\ -3402 \text{ N} \\ 1598 \text{ N}\cdot\text{m} \end{Bmatrix}$$

$$\sigma_A = \frac{1804(10^3)}{35(10^3)} = 51.5 \text{ MPa}$$

$$\sigma_B = \frac{1598(10^3)}{35(10^3)} = 45.7 \text{ MPa}$$

9.21 (a) $L_e = 0.6\text{ m}$, $q_o L_e/2 = -2000(0.6)/2 = -600\text{ N}$

For each element,

$$[k] = \frac{320(10^3)}{(0.6)^3}\begin{bmatrix} 12 & 3.6 & -12 & 3.6 \\ 3.6 & 1.44 & -3.6 & 0.72 \\ -12 & -3.6 & 12 & -3.6 \\ 3.6 & 0.72 & -3.6 & 1.44 \end{bmatrix}$$

Assemble

$$[K_{sys}] = 1.482(10^6)\begin{bmatrix} 12 & 3.6 & -12 & 3.6 & 0 & 0 \\ 3.6 & 1.44 & -3.6 & 0.72 & 0 & 0 \\ -12 & -3.6 & 24 & 0 & -12 & 3.6 \\ 3.6 & 0.72 & 0 & 2.88 & -3.6 & 0.72 \\ 0 & 0 & -12 & -3.6 & 12 & -3.6 \\ 0 & 0 & 3.6 & 0.72 & -3.6 & 1.44 \end{bmatrix}$$

$u_1 = v_1 = 0$. Partition first 2 eqs. Remaining eqs with $u_1 = v_1 = 0$ are

$$\begin{Bmatrix} -1200 \\ 0 \\ -600 \\ 0 \end{Bmatrix} = 1.482(10^6)\begin{bmatrix} 24 & 0 & -12 & 3.6 \\ 0 & 2.88 & -3.6 & 0.72 \\ -12 & -3.6 & 12 & -3.6 \\ 3.6 & 0.72 & -3.6 & 1.44 \end{bmatrix}\begin{Bmatrix} v_2 \\ \theta_2 \\ v_3 \\ \theta_3 \end{Bmatrix} \quad \text{Invert.}$$

$$\begin{Bmatrix} v_2 \\ \theta_2 \\ v_3 \\ \theta_3 \end{Bmatrix} = \left[1.482(10^6)\right]^{-1}\begin{bmatrix} 0.333 & 0.833 & 0.833 & 0.833 \\ 0.833 & 2.778 & 2.5 & 2.778 \\ 0.833 & 2.5 & 2.667 & 3.333 \\ 0.833 & 2.778 & 3.333 & 5.556 \end{bmatrix}\begin{Bmatrix} -1200 \\ 0 \\ -600 \\ 0 \end{Bmatrix} = (10^{-3})\begin{Bmatrix} -0.607\text{ m} \\ -1.687\text{ rad} \\ -1.755\text{ m} \\ -2.025\text{ rad} \end{Bmatrix}$$

ANS

(b) $L_e = 0.4\text{ m}$, $q_o L_e/2 = -2000(0.4)/2 = -400\text{ N}$

For each element

$$[k] = \frac{320(10^3)}{(0.4)^3}\begin{bmatrix} 12 & 2.4 & -12 & 2.4 \\ 2.4 & 0.64 & -2.4 & 0.32 \\ -12 & -2.4 & 12 & -2.4 \\ 2.4 & 0.32 & -2.4 & 0.64 \end{bmatrix}$$

Assemble

$$[K_{sys}] = 5(10^6)\begin{bmatrix} 12 & 2.4 & -12 & 2.4 & 0 & 0 & 0 & 0 \\ 2.4 & 0.64 & -2.4 & 0.32 & 0 & 0 & 0 & 0 \\ -12 & -2.4 & 24 & 0 & -12 & 2.4 & 0 & 0 \\ 2.4 & 0.32 & 0 & 1.28 & -2.4 & 0.32 & 0 & 0 \\ 0 & 0 & -12 & -2.4 & 24 & 0 & -12 & 2.4 \\ 0 & 0 & 2.4 & 0.32 & 0 & 1.28 & -2.4 & 0.32 \\ 0 & 0 & 0 & 0 & -12 & -2.4 & 12 & -2.4 \\ 0 & 0 & 0 & 0 & 2.4 & 0.32 & -2.4 & 0.64 \end{bmatrix}$$

Again, partition the first two eqs with $v_1 = \theta_1 = 0$

(Continued)

$$\begin{Bmatrix} -800 \\ 0 \\ -800 \\ 0 \\ -400 \\ 0 \end{Bmatrix} = 5(10^6)\begin{bmatrix} 24 & 0 & -12 & 2.4 & 0 & 0 \\ 0 & 1.28 & -2.4 & 0.32 & 0 & 0 \\ -12 & -2.4 & 24 & 0 & -12 & 2.4 \\ 2.4 & 0.32 & 0 & 1.28 & -2.4 & 0.32 \\ 0 & 0 & -12 & -2.4 & 12 & -2.4 \\ 0 & 0 & 2.4 & 0.32 & -2.4 & 0.64 \end{bmatrix}\begin{Bmatrix} v_2 \\ \theta_2 \\ v_3 \\ \theta_3 \\ v_4 \\ \theta_4 \end{Bmatrix} \quad \text{Invert}$$

$$\begin{Bmatrix} v_2 \\ \theta_2 \\ v_3 \\ \theta_3 \\ v_4 \\ \theta_4 \end{Bmatrix} = \frac{1}{5(10^6)}\begin{bmatrix} 0.333 & 1.25 & 0.833 & 1.25 & 1.333 & 1.25 \\ 1.25 & 6.25 & 3.75 & 6.25 & 6.25 & 6.25 \\ 0.833 & 3.75 & 2.667 & 5 & 4.667 & 5 \\ 1.25 & 6.25 & 5 & 12.5 & 10 & 12.5 \\ 1.333 & 6.25 & 4.667 & 10 & 9 & 11.25 \\ 1.25 & 6.25 & 5 & 12.5 & 11.25 & 18.75 \end{bmatrix}\begin{Bmatrix} -800 \\ 0 \\ -800 \\ 0 \\ -400 \\ 0 \end{Bmatrix} = (10^{-3})\begin{Bmatrix} -0.293\text{ m} \\ -1.300\text{ rad} \\ -0.933\text{ m} \\ -1.800\text{ rad} \\ -1.680\text{ m} \\ -1.900\text{ rad} \end{Bmatrix}$$

ANS

From Appendix C, $v_c = \dfrac{wx^2}{24EI}(4Lx - x^2 - 6L^2)$, $\theta = \dfrac{wx}{6EI}(3Lx - x^2 - 3L^2)$

v_c (mm)

x (m)	Exact	2 element	3 element	% error
0.4	-0.2867		-0.2933	2.3
0.6	-0.5738	-0.6075		5.9
0.8	-0.9067		-0.9333	2.9
1.2	-1.62	-1.755		8.3
1.2	-1.62		-1.68	3.7

θ (10^{-3} rad)

x (m)	Exact	2 element	3 element	% error
0.4	-1.267		-1.3	2.6
0.6	-1.575	-1.687		7.1
0.8	-1.733		-1.8	3.9
1.2	-1.8	-2.025		12.5
1.2	-1.8		-1.9	5.6

9.22 (a) $L_1 = 0.4$, $q_0 L_1/2 = -2000(0.4)/2 = -400\text{N}$

$L_2 = 0.8$, $q_0 L_2/2 = -2000(0.8)/2 = -800\text{N}$

$$[k]_1 = \frac{320(10^3)}{(0.4)^3}\begin{bmatrix} 12 & 2.4 & -12 & 2.4 \\ 2.4 & 0.64 & -2.4 & 0.32 \\ -12 & -2.4 & 12 & -2.4 \\ 2.4 & 0.32 & -2.4 & 0.64 \end{bmatrix}$$

$$[k]_2 = \frac{320(10^3)}{(0.8)^3}\begin{bmatrix} 12 & 4.8 & -12 & 4.8 \\ 4.8 & 2.56 & -4.8 & 1.28 \\ -12 & -4.8 & 12 & -4.8 \\ 4.8 & 1.28 & -4.8 & 2.56 \end{bmatrix}$$

Assemble

(Continued)

$$[K_{sys}] = 5(10^6)\begin{bmatrix} 12 & 2.4 & -12 & 2.4 & 0 & 0 \\ 2.4 & 0.64 & -2.4 & 0.32 & 0 & 0 \\ -12 & -2.4 & 13.5 & -1.8 & -1.5 & 0.6 \\ 2.4 & 0.32 & -1.8 & 0.96 & -0.6 & 0.16 \\ 0 & 0 & -1.5 & -0.6 & 1.5 & -0.6 \\ 0 & 0 & 0.6 & 0.16 & -0.6 & 0.32 \end{bmatrix}$$

$v_1 = \theta_1 = 0$. Partition 1st & 2nd eqs.

$$\begin{Bmatrix} -400-800 \\ 0 \\ -800 \\ 0 \end{Bmatrix} = 5(10^6)\begin{bmatrix} 13.5 & -1.8 & -1.5 & 0.6 \\ -1.8 & 0.96 & -0.6 & 0.16 \\ -1.5 & -0.6 & 1.5 & -0.6 \\ 0.6 & 0.16 & -0.6 & 0.32 \end{bmatrix}\begin{Bmatrix} v_2 \\ \theta_2 \\ v_3 \\ \theta_3 \end{Bmatrix} \quad \text{Invert.}$$

$$\begin{Bmatrix} v_2 \\ \theta_2 \\ v_3 \\ \theta_3 \end{Bmatrix} = \frac{1}{5(10^6)}\begin{bmatrix} 0.333 & 1.25 & 1.333 & 1.25 \\ 1.25 & 6.25 & 6.25 & 6.25 \\ 1.333 & 6.25 & 9 & 11.25 \\ 1.25 & 6.25 & 11.25 & 18.75 \end{bmatrix}\begin{Bmatrix} -1200 \\ 0 \\ -800 \\ 0 \end{Bmatrix} = -\begin{Bmatrix} 0.293\text{ m} \\ 1.30\text{ rad} \\ 1.76\text{ m} \\ 2.10 \end{Bmatrix}(10^{-3})$$ **ANS**

(b) $L_1 = 0.2667$ m, $q_0L_1/2 = -266.7$ N
$L_2 = 0.4$ m, $q_0L_2/2 = -400$ N
$L_3 = 0.5333$ m, $q_0L_3/2 = -533.3$ N

$$[k]_1 = \frac{320(10^3)}{(0.2667)^3}\begin{bmatrix} 12 & 1.6 & -12 & 1.6 \\ 1.6 & 0.2844 & -1.6 & 0.1422 \\ -12 & -1.6 & 12 & -1.6 \\ 1.6 & 0.1422 & -1.6 & 0.2844 \end{bmatrix}$$

$[k]_2$ is the same as $[k]_1$ in part (a)

$$[k_3] = \frac{320(10^3)}{(0.5333)^3}\begin{bmatrix} 12 & 3.2 & -12 & 3.2 \\ 3.2 & 1.1378 & -3.2 & 0.5689 \\ -12 & -3.2 & 12 & -3.2 \\ 3.2 & 0.5689 & -3.2 & 1.1378 \end{bmatrix}$$

Assemble

(Continued)

$$[K_{sys}] = 16.875(10^6)\begin{bmatrix} 12 & 1.6 & -12 & 1.6 & 0 & 0 & 0 & 0 \\ 1.6 & 0.2844 & -1.6 & 0.1422 & 0 & 0 & 0 & 0 \\ -12 & -1.6 & 15.556 & -0.8889 & -3.556 & 0.7111 & 0 & 0 \\ 1.6 & 0.1422 & -0.8889 & 0.4741 & -0.7111 & 0.0948 & 0 & 0 \\ 0 & 0 & -3.556 & -0.7111 & 5.056 & -0.3111 & -1.5 & 0.4 \\ 0 & 0 & 0.7111 & 0.0948 & -0.3111 & 0.3319 & -0.4 & 0.0711 \\ 0 & 0 & 0 & 0 & -1.5 & -0.4 & 1.5 & -0.4 \\ 0 & 0 & 0 & 0 & 0.4 & 0.0711 & -0.4 & 0.1422 \end{bmatrix}$$

Partition 1st & 2nd eqs.

$$\begin{Bmatrix} -266.7-400 \\ 0 \\ -400-533.3 \\ 0 \\ -533.3 \\ 0 \end{Bmatrix} = 16.875(10^6)\begin{bmatrix} 15.556 & -0.8889 & -3.556 & 0.7111 & 0 & 0 \\ -0.8889 & 0.4741 & -0.7111 & 0.0948 & 0 & 0 \\ -3.556 & -0.7111 & 5.056 & -0.3111 & -1.5 & 0.4 \\ 0.7111 & 0.0948 & -0.3111 & 0.3319 & -0.4 & 0.0711 \\ 0 & 0 & -1.5 & -0.4 & 1.5 & -0.4 \\ 0 & 0 & 0.4 & 0.0711 & -0.4 & 0.1422 \end{bmatrix}\begin{Bmatrix} v_2 \\ \theta_2 \\ v_3 \\ \theta_3 \\ v_4 \\ \theta_4 \end{Bmatrix}$$

Invert

$$\begin{Bmatrix} v_2 \\ \theta_2 \\ v_3 \\ \theta_3 \\ v_4 \\ \theta_4 \end{Bmatrix} = \left[16.875(10^6)\right]^{-1}\begin{bmatrix} 0.333 & 1.873 & 1.082 & 1.872 & 2.08 & 1.872 \\ 1.873 & 14.05 & 7.491 & 14.04 & 14.98 & 14.04 \\ 1.082 & 7.491 & 5.201 & 11.70 & 11.44 & 11.70 \\ 1.872 & 14.04 & 11.70 & 35.097 & 30.42 & 35.097 \\ 2.080 & 14.98 & 11.44 & 30.42 & 30.33 & 37.92 \\ 1.872 & 14.04 & 11.70 & 35.097 & 37.92 & 63.22 \end{bmatrix}\begin{Bmatrix} -666.7 \\ 0 \\ -933.3 \\ 0 \\ -533.3 \\ 0 \end{Bmatrix}$$

$$= -\begin{Bmatrix} 0.139 \text{ m} \\ 0.962 \text{ rad} \\ 0.692 \text{ m} \\ 1.682 \text{ rad} \\ 1.673 \text{ m} \\ 1.919 \text{ rad} \end{Bmatrix}(10^{-3}) \qquad \boxed{\text{ANS}}$$

v_c (mm)

x	exact	2 elements	3 elements	% error
0.4	-0.2867		-0.1388	-51.6
0.6	-0.5738	-0.2933		-48.9
0.8	-0.9067		-0.6920	-23.7
1.2	-1.6200	-1.7600		8.6
1.2	-1.6200		-1.6735	3.3

θ (10^{-3}) rad

x	exact	2 elements	3 elements	% error
0.4	-1.2667		-0.9617	-24.1
0.6	-1.5750	-1.3000		-17.5
0.8	-1.7333		-1.6823	-2.9
1.2	-1.8000	-2.1000		16.7
1.2	-1.8000		-1.9193	6.6

9.23 (a) $L_e = L$, $q_0L/2 = -wL/2$, $q_0L^2/12 = -wL^2/12$

Considering the hint given for Prob 9.17,

$$[K_{sys}] = \frac{2EI}{L^3}\begin{bmatrix} 6 & 3 & -6 & 3 \\ 3 & 2 & -3 & 1 \\ -6 & -3 & 6 & -3 \\ 3 & 1 & -3 & 2 \end{bmatrix}$$

$v_1 = v_2 = 0$. Partition 1st & 3rd eqs. The remaining eqs are

$$\begin{Bmatrix} -wL/12 \\ wL/12 \end{Bmatrix} = \frac{2EI}{L^3}\begin{bmatrix} 2 & 1 \\ 1 & 2 \end{bmatrix}\begin{Bmatrix} \theta_1 L \\ \theta_2 L \end{Bmatrix} \qquad \text{Invert.}$$

$$\begin{Bmatrix} \theta_1 \\ \theta_2 \end{Bmatrix} = \frac{L^2}{2EI}\left(\frac{1}{3}\right)\begin{bmatrix} 2 & -1 \\ -1 & 2 \end{bmatrix}\frac{wL}{12}\begin{Bmatrix} -1 \\ 1 \end{Bmatrix} = \begin{Bmatrix} -1 \\ 1 \end{Bmatrix}\frac{wL^3}{24EI}$$

Substitute $v_1 = v_2 = 0$ and $\theta_1 = -\theta_2 = -wL^3/24EI$ into Eq. (9.3-4)

$$v(x) = \left(x - 2\frac{x^2}{L} + \frac{x^3}{L^2}\right)\left(-\frac{wL^3}{24EI}\right) + \left(-\frac{x^2}{L} + \frac{x^3}{L^2}\right)\left(\frac{wL^3}{24EI}\right)$$

$$= -\frac{wL^2}{24EI}x(L-x) \quad \boxed{\text{ANS}}$$

$$\theta(x) = \frac{dv}{dx} = -\frac{wL^2}{24EI}(L-2x) \quad \boxed{\text{ANS}}$$

<u>Element eq:</u>

$$\begin{Bmatrix} V_1 \\ M_1/L \\ V_2 \\ M_2/L \end{Bmatrix} = \frac{2EI}{L^3}\begin{bmatrix} 6 & 3 & -6 & 3 \\ 3 & 2 & -3 & 1 \\ -6 & -3 & 6 & -3 \\ 3 & 1 & -3 & 2 \end{bmatrix}\begin{Bmatrix} 0 \\ -wL^4/24EI \\ 0 \\ wL^4/24EI \end{Bmatrix} - \begin{Bmatrix} -wL/2 \\ -wL/12 \\ -wL/2 \\ wL/12 \end{Bmatrix} = \begin{Bmatrix} wL/2 \\ 0 \\ wL/2 \\ 0 \end{Bmatrix}$$

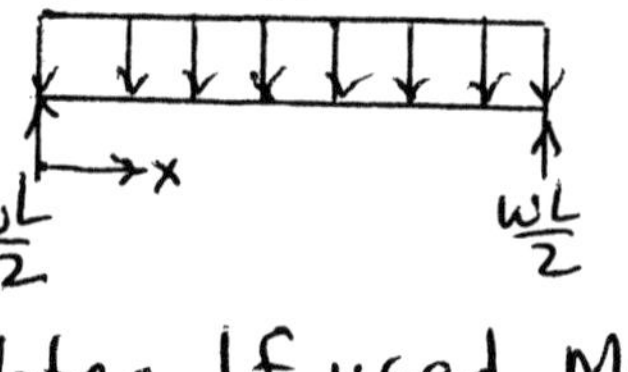

By standard str. of mat'ls methods

$$M(x) = \frac{wL}{2}x - \frac{wx^2}{2} \quad \boxed{\text{ANS}}$$

Note: If used $M(x) = EI\frac{d^2v}{dx^2} = \frac{wL^3}{24}\left(\frac{2}{L}\right) = \frac{wL^2}{12}$. This is only the average of the moment throughout the beam.

(b) $L_e = L/2$, $q_0L_e/2 = -wL/4$, $q_0L_e^2/12 = -wL^2/48$

(continued)

For each element,

$$[k]_e = \frac{16EI}{L^3}\begin{bmatrix} 6 & 3 & -6 & 3 \\ 3 & 2 & -3 & 1 \\ -6 & -3 & 6 & -3 \\ 3 & 1 & -3 & 2 \end{bmatrix}$$

Assemble two elements

$$[K_{sys}] = \frac{16EI}{L^3}\begin{bmatrix} 6 & 3 & -6 & 3 & 0 & 0 \\ 3 & 2 & -3 & 1 & 0 & 0 \\ -6 & -3 & 12 & 0 & -6 & 3 \\ 3 & 1 & 0 & 4 & -3 & 1 \\ 0 & 0 & -6 & -3 & 6 & -3 \\ 0 & 0 & 3 & 1 & -3 & 2 \end{bmatrix}$$

$v_1 = v_3 = 0$. Partition 1st & 5th eqs. The remaining eqs are

$$\begin{Bmatrix} -wL/24 \\ -wL/2 \\ wL/24 - wL/24 \\ wL/24 \end{Bmatrix} = \frac{16EI}{L^3}\begin{bmatrix} 2 & -3 & 1 & 0 \\ -3 & 12 & 0 & 3 \\ 1 & 0 & 4 & 1 \\ 0 & 3 & 1 & 2 \end{bmatrix}\begin{Bmatrix} \theta_1 L/2 \\ v_2 \\ \theta_2 L/2 \\ \theta_3 L/2 \end{Bmatrix}$$

Invert eq.

$$\begin{Bmatrix} \theta_1 L/2 \\ v_2 \\ \theta_2 L/2 \\ \theta_3 L/2 \end{Bmatrix} = \left(\frac{L^3}{16EI}\right)\left(\frac{wL}{24}\right)\left(\frac{1}{6}\right)\begin{bmatrix} 8 & 3 & -1 & -4 \\ 3 & 2 & 0 & -3 \\ -1 & 0 & 2 & -1 \\ -4 & -3 & -1 & 8 \end{bmatrix}\begin{Bmatrix} -1 \\ -12 \\ 0 \\ 1 \end{Bmatrix} = -\begin{Bmatrix} 8 \\ 5 \\ 0 \\ -8 \end{Bmatrix}\frac{wL^4}{384} \quad \boxed{\text{ANS}}$$

From Eq. (9.3-4) for $0 < x < L/2$

$$v(x) = \left(x - 4\frac{x^2}{L} + 4\frac{x^3}{L^2}\right)\left(-\frac{wL^3}{24EI}\right) + \left(12\frac{x^2}{L^2} - 16\frac{x^3}{L^3}\right)\left(-\frac{5}{384}\frac{wL^4}{EI}\right)$$

$$= -\frac{wLx}{96EI}\left(4L^2 - Lx - 4x^2\right) \quad \boxed{\text{ANS}}$$

$$\theta = \frac{dv}{dx} = -\frac{wL}{48EI}\left(2L^2 - Lx - 6x^2\right) \quad \boxed{\text{ANS}}$$

compare with Appendix C

Using $M(x) = EI\frac{d^2v}{dx^2} = \frac{wL}{48}(L + 12x)$ gives the linear average of the moment in the element. E.g.

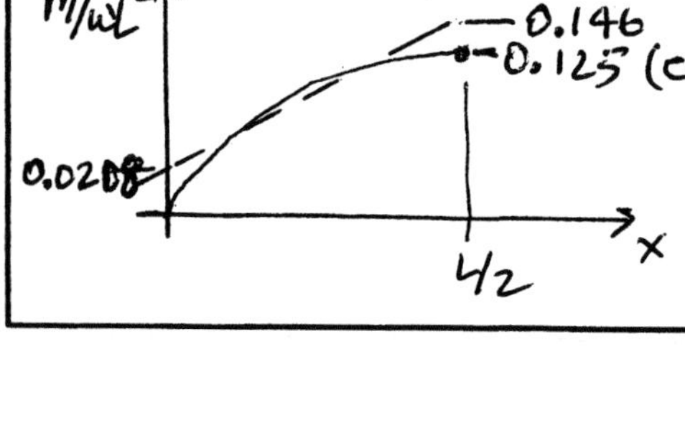

The exact $M(x)$ can only be arrived at using nodal values with basic strength of materials methods

(continued)

Element 1:

$$\begin{Bmatrix} V_1\\ M_1\\ V_2\\ M_2\end{Bmatrix}_1 = \frac{8EI}{L^3}\begin{bmatrix}12 & 3L & -12 & 3L\\ 3L & L^2 & -3L & L^2/2 \\ -12 & -3L & 12 & -3L \\ 3L & L^2/2 & -3L & L^2\end{bmatrix}\begin{Bmatrix}0\\ -\frac{wL^3}{24EI} \\ -\frac{5wL^4}{384EI} \\ 0\end{Bmatrix} - \begin{Bmatrix} -\frac{wL}{4} \\ -\frac{wL^2}{48} \\ -\frac{wL}{4} \\ \frac{wL^2}{48}\end{Bmatrix} = \begin{Bmatrix}\frac{wL}{2}\\ 0\\ 0\\ \frac{wL^2}{8}\end{Bmatrix}$$

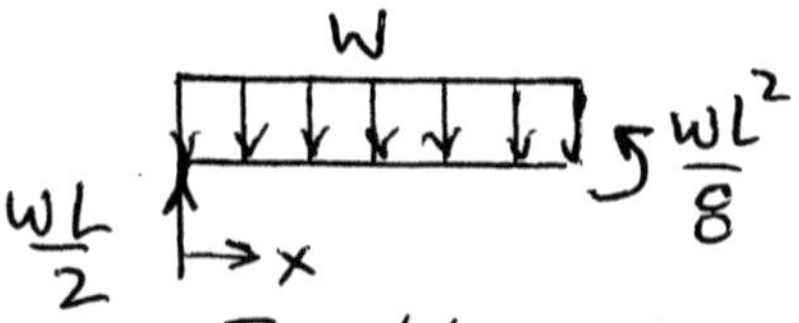

$$M(x) = \frac{wL}{2}x - \frac{wx^2}{2} \quad \boxed{\text{ANS}}$$

For $L/2 < x < L$ the results are the mirror of $0 \le x \le L/2$ due to symmetry. The form of eq. (9.3-4) uses a <u>local</u> x coordinate system

---

(9.24)

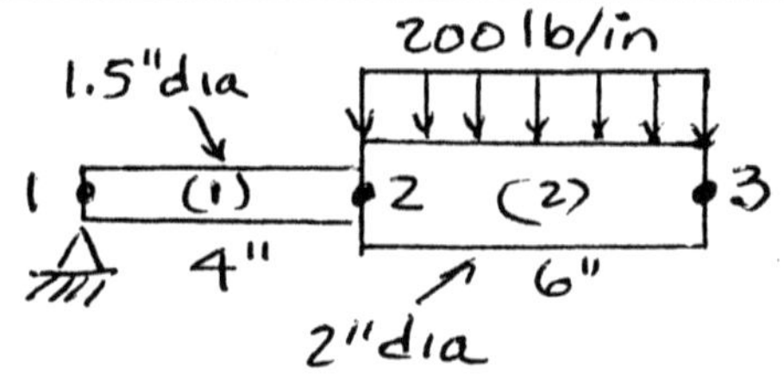

Element (1):

$$\frac{EI}{L^3} = \frac{30(10^6)\frac{\pi}{64}(1.5^4)}{(4)^3} = 116.5(10^3)$$

Element (2):

$$\frac{EI}{L^3} = \frac{30(10^6)\frac{\pi}{64}(2)^4}{(6)^3} = 109.1(10^3)$$

Element 1:

$$[k]_1 = 116.5(10^3)\begin{bmatrix}12 & 24 & -12 & 24\\ 24 & 64 & -24 & 32\\ -12 & -24 & 12 & -24\\ 24 & 32 & -24 & 64\end{bmatrix}$$

Element 2:

$$[k]_2 = 109.1(10^3)\begin{bmatrix}12 & 36 & -12 & 36\\ 36 & 144 & -36 & 72\\ -12 & -36 & 12 & -36\\ 36 & 72 & -36 & 144\end{bmatrix}$$

$$q_0L/2 = -200(6)/2 = -600 \text{ lb}$$

$$q_0L^2/12 = -200(6)^2/12 = -600 \text{ lb-in}$$

Assemble

$$[K_{sys}] = 116.5(10^3)\begin{bmatrix}12 & 24 & -12 & 24 & 0 & 0\\ 24 & 64 & -24 & 32 & 0 & 0\\ -12 & -24 & 23.237 & 9.712 & -11.237 & 33.712\\ 24 & 32 & 9.712 & 198.85 & -33.712 & 67.424\\ 0 & 0 & -11.237 & -33.712 & 11.237 & -33.712\\ 0 & 0 & 33.712 & 67.424 & -33.712 & 134.85\end{bmatrix}$$

$v_1 = \theta_3 = 0$. Partition 1st and last eqs. The remaining eqs are

(continued)

$$\begin{Bmatrix} 0 \\ -600 \\ -600 \\ -600 \end{Bmatrix} = 116.5(10^3)\begin{bmatrix} 64 & -24 & 32 & 0 \\ -24 & 23.237 & 9.712 & -11.237 \\ 32 & 9.712 & 198.85 & -33.712 \\ 0 & -11.237 & -33.712 & 11.237 \end{bmatrix}\begin{Bmatrix} \theta_1 \\ v_2 \\ \theta_2 \\ v_3 \end{Bmatrix} \quad \text{Invert}$$

$$\begin{Bmatrix} \theta_1 \\ v_2 \\ \theta_2 \\ v_3 \end{Bmatrix} = \left[116.5(10^3)\right]^{-1}\begin{bmatrix} 0.09216 & 0.2437 & 0.02966 & 0.3327 \\ 0.2437 & 0.8080 & 0.1187 & 1.1639 \\ 0.02966 & 0.1187 & 0.02966 & 0.2077 \\ 0.3327 & 1.1639 & 0.2077 & 1.8759 \end{bmatrix}\begin{Bmatrix} 0 \\ -600 \\ -600 \\ -600 \end{Bmatrix} = -\begin{Bmatrix} 3.12\text{ rad} \\ 10.77\text{ in} \\ 1.83\text{ rad} \\ 16.73\text{ in} \end{Bmatrix}(10^{-3})$$

$\therefore\ \delta_{midpoint} = \underline{-0.0167 \text{ in}}$

9.25 Element 1: $q_0L/2 = -2000(0.5)/2 = -500\text{ N}$

$q_0L^2/12 = -2000(0.5)^2/12 = -41.667\text{ N}\cdot\text{m}$

$EI/L^3 = 210(10^9)(1.5)(10^{-6})/(0.5)^3 = 2.52(10^6)\text{ N/m}$

$$[k]_1 = 2.52(10^6)\begin{bmatrix} 12 & 3 & -12 & 3 \\ 3 & 1 & -3 & 0.5 \\ -12 & -3 & 12 & -3 \\ 3 & 0.5 & -3 & 1 \end{bmatrix}$$

Element 2: $q_0L/2 = -1000\text{ N}$, $q_0L^2/12 = -166.667\text{ N}\cdot\text{m}$

$EI/L^3 = 210(10^9)(1.5)(10^{-6})/(1)^3 = 315(10^3)\text{ N/m}$

$$[k_2] = 315(10^3)\begin{bmatrix} 12 & 6 & -12 & 6 \\ 6 & 4 & -6 & 2 \\ -12 & -6 & 12 & -6 \\ 6 & 2 & -6 & 4 \end{bmatrix}$$

Assemble

$$[K_{sys}] = 2.52(10^6)\begin{bmatrix} 12 & 3 & -12 & 3 & 0 & 0 \\ 3 & 1 & -3 & 0.5 & 0 & 0 \\ -12 & -3 & 13.5 & -2.25 & -1.5 & 0.75 \\ 3 & 0.5 & -2.25 & 1.5 & -0.75 & 0.25 \\ 0 & 0 & -1.5 & -0.75 & 1.5 & -0.75 \\ 0 & 0 & 0.75 & 0.25 & -0.75 & 0.5 \end{bmatrix}$$

B.C.s. $v_2 = v_3 = \theta_3 = 0$. Partition 3rd, 5th, and 6th eqs. These are,

$V_2 = 2.52(10^6)(-12v_1 - 3\theta_1 - 2.25\theta_2) + 500 + 1000$ ①

$V_2 = 2.52(10^6)(-0.75\theta_2) + 1000$ ②

$M_3 = 2.52(10^6)(0.25\theta_2) - 166.67$ ③

(Continued)

The remaining eqs are,

$$\begin{Bmatrix} -500 \\ -41.667 \\ 41.667-166.667 \end{Bmatrix} = 2.52(10^6)\begin{bmatrix} 12 & 3 & 3 \\ 3 & 1 & 0.5 \\ 3 & 0.5 & 1.5 \end{bmatrix}\begin{Bmatrix} V_1 \\ \theta_1 \\ \theta_2 \end{Bmatrix} \quad \text{Invert}$$

$$\begin{Bmatrix} V_1 \\ \theta_1 \\ \theta_2 \end{Bmatrix} = \left[2.52(10^6)\right]^{-1}\begin{bmatrix} 0.833 & -2 & -1 \\ -2 & 6 & 2 \\ -1 & 2 & 2 \end{bmatrix}\begin{Bmatrix} -500 \\ -41.667 \\ -125 \end{Bmatrix} = \begin{Bmatrix} -8.267 \text{ m} \\ 19.84 \text{ rad} \\ 6.614 \text{ rad} \end{Bmatrix}(10^{-5}) \quad \boxed{\text{ANS}}$$

Sub into eqs ① - ③ yields

$V_2 = \underline{2125 \text{ N}}$, $V_3 = \underline{875 \text{ N}}$, $M_3 = \underline{-125 \text{ N-m}}$

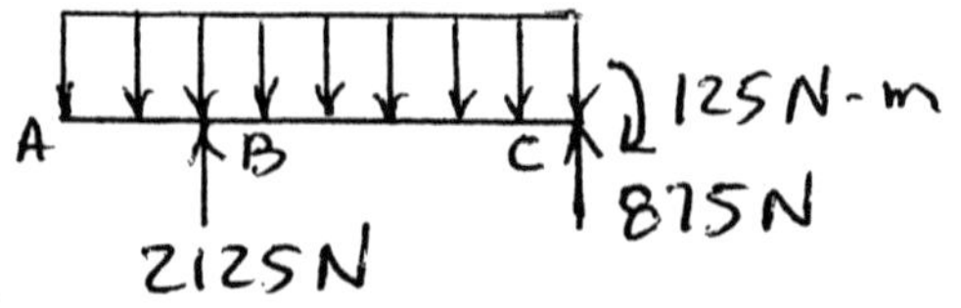

Element 1:

$$\begin{Bmatrix} V_1 \\ M_1 \\ V_2 \\ M_2 \end{Bmatrix}_1 = 2.52(10^6)\begin{bmatrix} 12 & 3 & -12 & 3 \\ 3 & 1 & -3 & 0.5 \\ -12 & -3 & 12 & -3 \\ 3 & 0.5 & -3 & 1 \end{bmatrix}\begin{Bmatrix} -8.267(10^{-5}) \\ 1.984(10^{-4}) \\ 0 \\ 6.614(10^{-5}) \end{Bmatrix} - \begin{Bmatrix} -500 \\ -41.667 \\ -500 \\ 41.667 \end{Bmatrix} = \begin{Bmatrix} 0 \\ 0 \\ 1000 \text{ N} \\ -250 \text{ N-m} \end{Bmatrix} \quad \boxed{\text{ANS}}$$

Element 2:

$$\begin{Bmatrix} V_2 \\ M_2 \\ V_3 \\ M_3 \end{Bmatrix}_2 = 315(10^3)\begin{bmatrix} 12 & 6 & -12 & 6 \\ 6 & 4 & -6 & 2 \\ -12 & -6 & 12 & -6 \\ 6 & 2 & -6 & 4 \end{bmatrix}\begin{Bmatrix} 0 \\ 6.614(10^{-5}) \\ 0 \\ 0 \end{Bmatrix} - \begin{Bmatrix} -1000 \\ -166.67 \\ -1000 \\ 166.67 \end{Bmatrix} = \begin{Bmatrix} 1125 \text{ N} \\ 250 \text{ N}\cdot\text{m} \\ 875 \text{ N} \\ -125 \text{ N}\cdot\text{m} \end{Bmatrix}$$

Max bending moment occurs at B

$$\sigma_B = \frac{250}{3.5(10^{-5})} = 7.14(10^6)\text{ Pa} = \underline{7.14 \text{ MPa}}$$

(9.26)

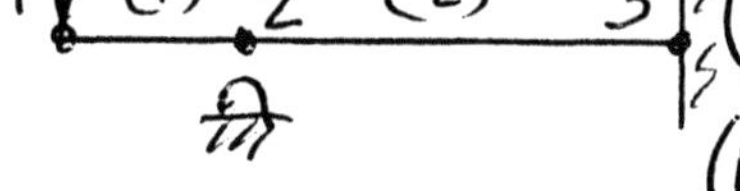

$$\left(\frac{EI}{L^3}\right)_1 = \frac{2.8(10^5)}{(2)^3} = 35(10^3)\text{ lb/in}$$

$$\left(\frac{EI}{L^3}\right)_2 = \frac{2.8(10^5)}{(8)^3} = 546.9 \text{ lb/in}$$

(Continued)

Element 1:

$$[k]_1 = 35(10^3)\begin{bmatrix} 12 & 12 & -12 & -12 \\ 12 & 16 & -12 & 8 \\ -12 & -12 & 12 & -12 \\ 12 & 8 & -12 & 16 \end{bmatrix}$$

Element 2:

$$[k]_2 = 546.9\begin{bmatrix} 12 & 48 & -12 & 48 \\ 48 & 256 & -48 & 128 \\ -12 & -48 & 12 & -48 \\ 48 & 128 & -48 & 256 \end{bmatrix}$$

Assemble

$$[K_{sys}] = 35(10^3)\begin{bmatrix} 12 & 12 & -12 & 12 & 0 & 0 \\ 12 & 16 & -12 & 8 & 0 & 0 \\ -12 & -12 & 12.188 & -11.25 & -0.1875 & 0.75 \\ 12 & 8 & -11.25 & 20 & -0.75 & 2 \\ 0 & 0 & -0.1875 & -0.75 & 0.1875 & -0.75 \\ 0 & 0 & 0.75 & 2 & -0.75 & 4 \end{bmatrix}$$

$v_3 = \theta_3 = 0$. Partition last two eqs. The remaining eqs are

$$\begin{Bmatrix} -P \\ 0 \\ 0 \\ 0 \end{Bmatrix} = 35(10^3)\begin{bmatrix} 12 & 12 & -12 & 12 \\ 12 & 16 & -12 & 8 \\ -12 & -12 & 12.188 & -11.25 \\ 12 & 8 & -11.25 & 20 \end{bmatrix}\begin{Bmatrix} v_1 \\ \theta_1 \\ -\epsilon \\ \theta_2 \end{Bmatrix} \qquad \text{Invert}$$

$$\begin{Bmatrix} v_1 \\ \theta_1 \\ -\epsilon \\ \theta_2 \end{Bmatrix} = \left[35(10^3)\right]^{-1}\begin{bmatrix} 41.667 & -6.25 & 29.333 & -6 \\ -6.25 & 1.25 & -4 & 1 \\ 29.333 & -4 & 21.333 & -4 \\ -6 & 1 & -4 & 1 \end{bmatrix}\begin{Bmatrix} -P \\ 0 \\ 0 \\ 0 \end{Bmatrix}$$

From the third eq.

$$-\epsilon = -0.125 = \frac{29.333P}{35(10^3)} \Rightarrow \underline{P = 149 \text{ lb}}$$

9.27

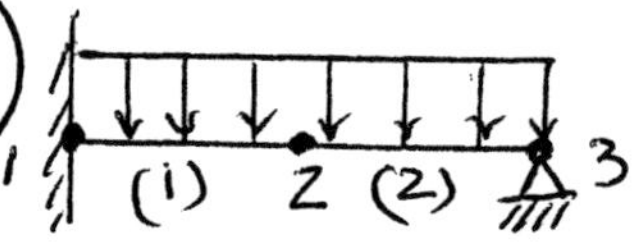

For each element

$\frac{EI}{L^3} = 160(10^3)$ N/m, $\frac{q_0 L}{2} = -1000$ N

$q_0 L^2/12 = -166.67$ N-m

$$[k]_e = 160(10^3)\begin{bmatrix} 12 & 6 & -12 & 6 \\ 6 & 4 & -6 & 2 \\ -12 & -6 & 12 & -6 \\ 6 & 2 & -6 & 4 \end{bmatrix}$$

(continued)

Placing a hinge on node 2 of element 1

$$\begin{Bmatrix} V_1 - 1000 \\ M_1 - 166.67 \\ V_2 - 1000 \\ M_2 + 166.67 \end{Bmatrix}_1 = 160(10^3)\begin{bmatrix} 12 & 6 & -12 & 6 \\ 6 & 4 & -6 & 2 \\ -12 & -6 & 12 & -6 \\ 6 & 2 & -6 & 4 \end{bmatrix}\begin{Bmatrix} v_1 \\ \theta_1 \\ v_2 \\ (\theta_2)_1 \end{Bmatrix} \qquad (1)$$

$M_2 = 0$ at hinge. From the last eq. of (1)

$$(\theta_2)_1 = -\frac{6}{4}v_1 - \frac{2}{4}\theta_1 + \frac{6}{4}v_2 + \frac{166.67}{4(160)(10^3)} = -1.5v_1 - 0.5\theta_1 + 1.5v_2 + \frac{41.667}{160(10^3)} \qquad (2)$$

Sub (2) back into (1)

$$\begin{Bmatrix} V_1 - 1000 - 250 \\ M_1 - 166.67 - 83.3 \\ V_2 - 1000 + 250 \end{Bmatrix}_1 = 160(10^3)\begin{bmatrix} 12-6(1.5) & 6-6(0.5) & -12+6(1.5) \\ 6-2(1.5) & 4-2(0.5) & -6+2(1.5) \\ -12+6(1.5) & -6+6(0.5) & 12-6(1.5) \end{bmatrix}\begin{Bmatrix} v_1 \\ \theta_1 \\ v_2 \end{Bmatrix}$$

Reduces to

$$\begin{Bmatrix} V_1 - 1250 \\ M_1 - 250 \\ V_2 - 750 \end{Bmatrix}_1 = 160(10^3)\begin{bmatrix} 3 & 3 & -3 \\ 3 & 3 & -3 \\ -3 & -3 & 3 \end{bmatrix}\begin{Bmatrix} v_1 \\ \theta_1 \\ v_2 \end{Bmatrix}$$

Element 2:

$$\begin{Bmatrix} V_2 - 1000 \\ M_2 - 166.67 \\ V_3 - 1000 \\ M_3 + 166.67 \end{Bmatrix}_2 = 160(10^3)\begin{bmatrix} 12 & 6 & -12 & 6 \\ 6 & 4 & -6 & 2 \\ -12 & -6 & 12 & -6 \\ 6 & 2 & -6 & 4 \end{bmatrix}\begin{Bmatrix} v_2 \\ (\theta_2)_2 \\ v_3 \\ \theta_3 \end{Bmatrix}$$

Assemble and partition 1st, 2nd and 5th eqs with $v_1 = \theta_1 = v_3 = 0$

$$V_1 = 1250 + 160(10^3)(-3v_2) \qquad (3)$$

$$M_1 = 250 + 160(10^3)(-3v_2) \qquad (4)$$

$$V_3 = 1000 + 160(10^3)\left[-12v_2 - 6(\theta_2)_2 - 6\theta_3\right] \qquad (5)$$

The remaining eqs are

$$\begin{Bmatrix} V_2 - 1750 \\ M_2 - 166.67 \\ M_3 + 166.67 \end{Bmatrix} = 160(10^3)\begin{bmatrix} 15 & 6 & 6 \\ 6 & 4 & 2 \\ 6 & 2 & 4 \end{bmatrix}\begin{Bmatrix} v_2 \\ (\theta_2)_2 \\ \theta_3 \end{Bmatrix} \qquad \text{Invert}$$

(Continued)

$$\begin{Bmatrix} v_2 \\ (\theta_2)_2 \\ \theta_3 \end{Bmatrix} = \left[160(10^3)\right]^{-1} \begin{bmatrix} 0.333 & -0.333 & -0.333 \\ -0.333 & 0.667 & 0.167 \\ -0.333 & 0.167 & 0.667 \end{bmatrix} \begin{Bmatrix} -1750 \\ -166.67 \\ 166.67 \end{Bmatrix} = \begin{Bmatrix} -3.646 \text{ m} \\ 3.125 \text{ rad} \\ 4.167 \text{ rad} \end{Bmatrix} (10^{-3}) \quad \boxed{\text{ANS}}$$

From eqs (2) – (5)

$(\theta_2)_1 = -5.208(10^{-3})$ rad, $V_1 = 3000$ N, $M_1 = 2000$ N-m, $V_3 = 1000$ N $\boxed{\text{ANS}}$

Element 1:

$$\begin{Bmatrix} V_1 \\ M_1 \\ V_2 \end{Bmatrix}_1 = 160(10^3) \begin{bmatrix} 3 & 3 & -3 \\ 3 & 3 & -3 \\ -3 & -3 & 3 \end{bmatrix} \begin{Bmatrix} 0 \\ 0 \\ -3.646(10^{-3}) \end{Bmatrix} + \begin{Bmatrix} 1250 \\ 250 \\ 750 \end{Bmatrix} = \begin{Bmatrix} 3000 \text{ N} \\ 2000 \text{ N·m} \\ -1000 \text{ N} \end{Bmatrix} \quad \boxed{\text{ANS}}$$

Element 2:

$$\begin{Bmatrix} V_2 \\ M_2 \\ V_3 \\ M_3 \end{Bmatrix}_2 = 160(10^3) \begin{bmatrix} 12 & 6 & -12 & 6 \\ 6 & 4 & -6 & 2 \\ -12 & -6 & 12 & -6 \\ 6 & 2 & -6 & 4 \end{bmatrix} \begin{Bmatrix} -3.646 \\ 3.125 \\ 0 \\ 4.167 \end{Bmatrix} (10^{-3}) + \begin{Bmatrix} 1000 \\ 166.67 \\ 1000 \\ -166.67 \end{Bmatrix} = \begin{Bmatrix} 1000 \text{ N} \\ 0 \\ 1000 \text{ N} \\ 0 \end{Bmatrix} \quad \boxed{\text{ANS}}$$

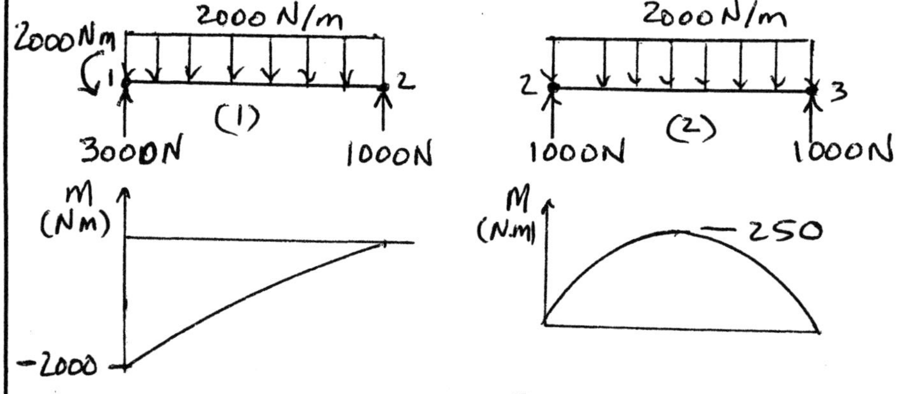

Element 1: $\sigma_{max} = \dfrac{2000(10^3)}{20(10^3)} = \underline{100 \text{ MPa}}$

Element 2: $\sigma_{max} = \dfrac{250(10^3)}{20(10^3)} = \underline{12.5 \text{ MPa}}$

9.28 Refer to the solution of Prob. 9.27 for matrix development. This problem is just another load case. Without the uniform load, Element 1 is given by

$$\begin{Bmatrix} V_1 \\ M_1 \\ V_2 \end{Bmatrix}_1 = 160(10^3) \begin{bmatrix} 3 & 3 & -3 \\ 3 & 3 & -3 \\ -3 & -3 & 3 \end{bmatrix} \begin{Bmatrix} v_1 \\ \theta_1 \\ v_2 \end{Bmatrix}$$

(continued)

and, $(\theta_2)_1 = -1.5v_1 - 0.5\theta_1 + 1.5v_2$ ①

$V_1 = 160(10^3)(-3v_2)$ ②, $M_1 = 160(10^3)(-3v_2)$ ③,

$V_3 = 1000 + 160(10^3)[-12v_2 - 6(\theta_2)_2 - 6\theta_3]$ ④

$$\begin{Bmatrix} v_2 \\ (\theta_2)_2 \\ \theta_3 \end{Bmatrix} = [160(10^3)]^{-1} \begin{bmatrix} 0.333 & -0.333 & -0.333 \\ -0.333 & 0.667 & 0.167 \\ -0.333 & 0.167 & 0.667 \end{bmatrix} \begin{Bmatrix} -1000 \\ -166.67 \\ 166.67 \end{Bmatrix} = \begin{Bmatrix} -2.803\text{m} \\ 1.563\text{ rad} \\ 2.604\text{ rad} \end{Bmatrix}(10^{-3})$$ ANS

Sub results into eqs ①-④ yielding

$(\theta_2)_1 = -3.125(10^{-3})$ rad, $V_1 = 1000$ N, $M_1 = 1000$ N·m, $V_3 = 1000$ N ANS

As in Prob 9.27 $\begin{Bmatrix} V_1 \\ M_1 \\ V_2 \end{Bmatrix}_1 = \begin{Bmatrix} 1000\text{N} \\ 1000\text{ N·m} \\ -1000\text{N} \end{Bmatrix}$, $\begin{Bmatrix} V_2 \\ M_2 \\ V_3 \\ M_3 \end{Bmatrix}_2 = \begin{Bmatrix} 1000\text{N} \\ 0 \\ 1000\text{N} \\ 0 \end{Bmatrix}$

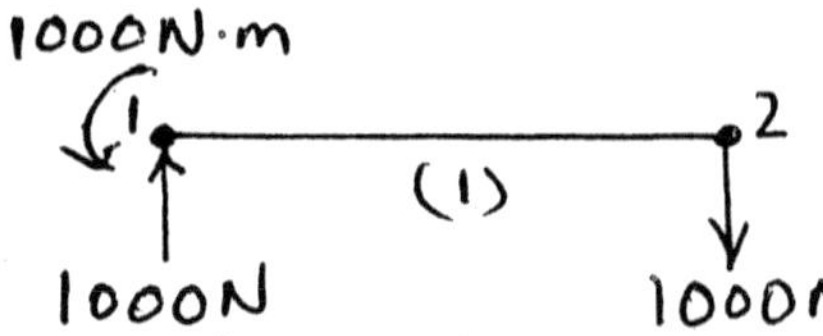

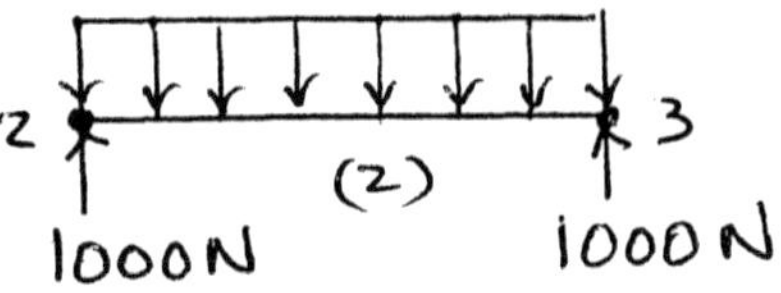

$(M_{max})_1 = 1000$ N·m $\quad (M_{max})_2 = 250$ N·m at center

Element 1: $(\sigma_{max})_1 = \dfrac{1000(10^3)}{20(10^3)} = \underline{50\text{ MPa}}$

Element 2: $(\sigma_{max})_2 = \dfrac{250(10^3)}{20(10^3)} = \underline{12.5\text{ MPa}}$

9.29 $q(x) = qx/L_e$

$$W_p = -\frac{q}{L_e}\int_0^{L_e}\left[\left(x - \frac{3x^3}{L_e^2} + 2\frac{x^4}{L_e^3}\right)v_i + \left(x^2 - 2\frac{x^3}{L_e} + \frac{x^4}{L_e^2}\right)\theta_i + \left(3\frac{x^3}{L_e^2} - 2\frac{x^4}{L_e^3}\right)v_j + \left(\frac{-x^3}{L_e} + \frac{x^4}{L_e^2}\right)\theta_j\right]dx$$

$$= -\frac{q}{L_e}\left[\left(\frac{L_e^2}{2} - \frac{3}{4}L_e^2 + \frac{2}{5}L_e^2\right)v_i + \left(\frac{L_e^3}{3} - \frac{L_e^3}{2} + \frac{L_e^3}{5}\right)\theta_i + \left(\frac{3}{4}L_e^2 - \frac{2}{5}L_e^2\right)v_j + \left(-\frac{L_e^3}{4} + \frac{L_e^3}{5}\right)\theta_j\right]$$

$$= -\frac{q}{L_e}\left(\frac{3}{20}L_e^2 v_i + \frac{1}{30}L_e^3\theta_i + \frac{7}{20}L_e^2 v_j - \frac{1}{20}L_e^3\theta_j\right)$$

(Continued)

$$W_p = -\{v_i\ \theta_i\ v_j\ \theta_j\}\left\{\tfrac{3}{20}qL_e\ \ \tfrac{1}{30}qL_e^2\ \ \tfrac{7}{20}qL_e\ \ -\tfrac{1}{20}qL_e^2\right\}^T$$

$$= -\{v_i\ \theta_i\ v_j\ \theta_j\}\{V_i\ M_i\ V_j\ M_j\}^T$$

$$\therefore\ \{V_i\ M_i\ V_j\ M_j\}^T = \left\{\tfrac{3}{20}qL_e\ \ \tfrac{1}{30}qL_e^2\ \ \tfrac{7}{20}qL_e\ \ -\tfrac{1}{20}qL_e^2\right\}^T \quad \boxed{\text{ANS}}$$

For the figure in text $\Sigma F_y = \frac{1}{2}qL_e$, $\circlearrowleft \Sigma M_{x=0} = \left(\frac{2}{3}L_e\right)\left(\frac{1}{2}qL_e\right) = \frac{1}{3}qL_e$

From the solution $\Sigma F_y = \frac{3}{20}qL_e + \frac{7}{20}qL_e = \frac{1}{2}qL_e$ ✓

$\Sigma M_{x=0} = \frac{1}{30}qL_e^2 - \frac{1}{20}qL_e^2 + (L_e)\left(\frac{7}{20}qL_e\right) = \frac{1}{3}qL_e$ ✓ Thus statically equivalent.

9.30 Viewing down the y axis positive nodal loads and displacements are

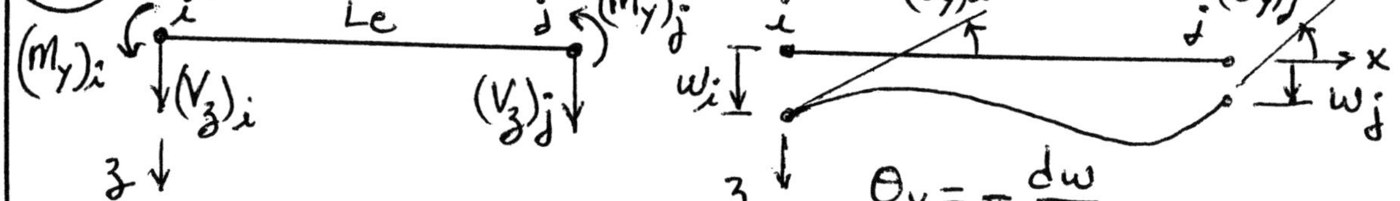

$$\theta_y = -\frac{dw}{dx}$$

$$w(x) = a_1 + a_2x + a_3x^2 + a_4x^3$$

$$\frac{dw}{dx} = a_2 + 2a_3x + 3a_4x^2$$

At $x = 0$, $w = w_i$, $\frac{dw}{dx} = -(\theta_y)_i$

$x = L_e$, $w = w_j$, $\frac{dw}{dx} = -(\theta_y)_j$

From this point on, the derivation follows the same procedure as in the text for the xy plane. The final form of w(x) is,

$$w(x) = \left[1 - 3\left(\frac{x}{L_e}\right)^2 + 2\left(\frac{x}{L_e}\right)^3\right]w_i + \left[-x + 2\frac{x^2}{L_e} - \frac{x^3}{L_e^2}\right](\theta_y)_i + \left[3\left(\frac{x}{L_e}\right)^2 - 2\left(\frac{x}{L_e}\right)^3\right]w_j$$
$$+ \left[\frac{x^2}{L_e} - \frac{x^3}{L_e^2}\right](\theta_y)_j$$

Again, the remainder of the derivation follows the text except the B_2 and B_4 terms are opposite in sign. This changes the signs of (row, column): (1,2), (2,1), (1,4), (4,1), (2,3), (3,2), (3,4), and (4,3) of Eq. (9.3-16). Thus

$$[k]_e = \left(\frac{EI_y}{L^3}\right)_e \begin{bmatrix} L & -6L & -12 & -6L \\ -6L & 4L^2 & 6L & 2L^2 \\ -12 & 6L & 12 & 6L \\ -6L & 2L^2 & 6L & 4L^2 \end{bmatrix} \quad \boxed{\text{ANS}}$$

(9.31) From Eq. (9.3-35) the torsional stiffness is $k_e = \left(\frac{EJ}{2(1+\nu)L}\right)_e$

$$k_1 = \frac{30(10^6)\frac{\pi}{32}(1)^4}{2(1.3)(12)} = 94.4(10^3)\text{ lb-in}$$

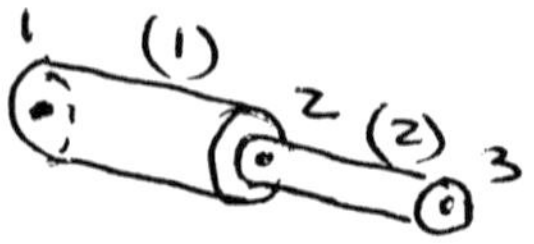

$$k_2 = \frac{30(10^6)\frac{\pi}{32}(0.75)^4}{2(1.3)(10)} = 35.84(10^3)\text{ lb-in}$$

$$[K_{sys}] = (10^4)\begin{bmatrix} 9.44 & -9.44 & 0 \\ -9.44 & 13.024 & -3.584 \\ 0 & -3.584 & 3.584 \end{bmatrix}$$

$\theta_1 = 0$. Partition 1st eq. The remaining eqs are

$$\begin{Bmatrix} 0 \\ 5000 \end{Bmatrix} = (10^4)\begin{bmatrix} 13.024 & -3.584 \\ -3.584 & 3.584 \end{bmatrix}\begin{Bmatrix} \theta_2 \\ \theta_3 \end{Bmatrix} \quad \text{Invert.}$$

$$\begin{Bmatrix} \theta_2 \\ \theta_3 \end{Bmatrix} = (10^{-4})\begin{bmatrix} 0.1059 & 0.1059 \\ 0.1059 & 0.3850 \end{bmatrix}\begin{Bmatrix} 0 \\ 5000 \end{Bmatrix} = \begin{Bmatrix} 0.0530 \\ 0.1925 \end{Bmatrix}\text{rad} = \begin{Bmatrix} 3.03^\circ \\ 11.03^\circ \end{Bmatrix} \quad \boxed{\text{ANS}}$$

(9.32) See the solution of Prob 9.31 for stiffnesses. Partition 1st and 3rd eqs of $[K_{sys}]$ which are (with $\theta_1 = \theta_3 = 0$)

$T_1 = -9.44(10^4)\theta_2$ ①, $T_2 = -3.584(10^4)\theta_2$ ②

The remaining eq is

$5000 = 13.024(10^4)\theta_2 \Rightarrow \theta_2 = 0.0384\text{ rad} = 2.2^\circ$

Sub θ_2 into ① & ②

$T_1 = -9.44(10^4)(0.0384) = -3624\text{ lb-in}$

$T_2 = -3.584(10^4)(0.0384) = -1376\text{ lb-in}$

Using Eq. (9.3-35)

Element 1:

$$\begin{Bmatrix} T_1 \\ T_2 \end{Bmatrix}_1 = 94.4(10^3)\begin{bmatrix} 1 & -1 \\ -1 & 1 \end{bmatrix}\begin{Bmatrix} 0 \\ 0.03839 \end{Bmatrix} = \begin{Bmatrix} -3624 \\ 3624 \end{Bmatrix}\text{lb-in} \quad \boxed{\text{ANS}}$$

Element 2:

$$\begin{Bmatrix} T_2 \\ T_3 \end{Bmatrix}_2 = 35.84(10^3)\begin{bmatrix} 1 & -1 \\ -1 & 1 \end{bmatrix}\begin{Bmatrix} 0.03839 \\ 0 \end{Bmatrix} = \begin{Bmatrix} 1376 \\ -1376 \end{Bmatrix}\text{lb-in} \quad \boxed{\text{ANS}}$$

Stresses ($\tau = 16T/\pi d^3$)

Element 1: $\tau_1 = \frac{16(3.624)}{\pi(1)^3} = 18.5\text{ kpsi}$, $\tau_2 = \frac{16(1.376)}{\pi(0.75)^3} = 16.6\text{ kpsi}$

9.33 (a)

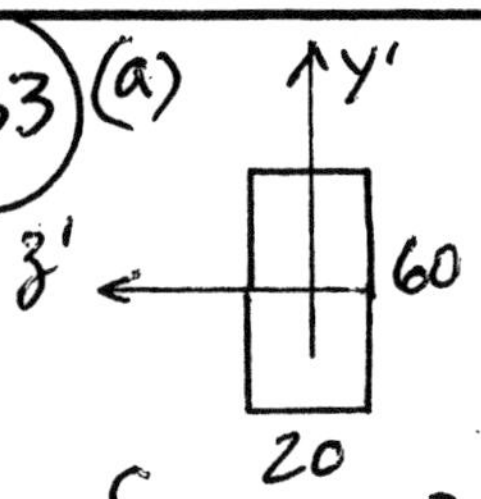

$$I_{y'} = \frac{1}{12}60(20)^3 = 40(10^3)\,\text{mm}^4$$

$$I_{z'} = \frac{1}{12}20(60)^3 = 360(10^3)\,\text{mm}^4$$

$$A = 20(60) = 1.2(10^3)\,\text{mm}^2$$

$$J = \left\{1 - 0.63\left(\frac{20}{60}\right)\left[1 - \frac{1}{12}\left(\frac{20}{60}\right)^4\right]\right\}\left[\frac{60(20)^3}{3}\right] = 126.4(10^3)\,\text{mm}^4$$

$$L_e = \left[(800-500)^2 + (500-200)^2 + (-300-0)^2\right]^{1/2} = 519.6\,\text{mm}$$

$$E/L = \frac{70(10^9)}{0.5196} = 134.72(10^9)\,\text{N/m}^3.$$ Expressing terms in N, m

$6I_{z'}/L = 4.157(10^{-6})$, $12I_{z'}/L = 8.314(10^{-6})$, $12I_{z'}/L^2 = 16(10^{-6})$,

$6I_{y'}/L = 461.88(10^{-9})$, $12I_{y'}/L = 923.76(10^{-9})$, $12I_{y'}/L^2 = 1.778(10^{-6})$

$2I_{y'} = 80(10^{-9})$, $4I_{y'} = 160(10^{-9})$, $2I_{z'} = 720(10^{-9})$, $4I_{z'} = 1.44(10^{-6})$,

$J/[2(1+\nu)] = 47.53(10^{-9})$, $A = 1.2(10^{-3})$

$$[k]_{x'y'z'} = 134.7\begin{bmatrix}
1.2E+6 & 0 & 0 & 0 & 0 & 0 & -1.2E+6 & 0 & 0 & 0 & 0 & 0 \\
0 & 1.6E+4 & 0 & 0 & 0 & 4157 & 0 & -1.6E+4 & 0 & 0 & 0 & 4157 \\
0 & 0 & 1778 & 0 & -461.9 & 0 & 0 & 0 & -1778 & 0 & -461.9 & 0 \\
0 & 0 & 0 & 47.5 & 0 & 0 & 0 & 0 & 0 & -47.5 & 0 & 0 \\
0 & 0 & -461.9 & 0 & 160 & 0 & 0 & 0 & 461.9 & 0 & 80 & 0 \\
0 & 4157 & 0 & 0 & 0 & 1440 & 0 & -4157 & 0 & 0 & 0 & 720 \\
-1.2E+6 & 0 & 0 & 0 & 0 & 0 & 1.2E+6 & 0 & 0 & 0 & 0 & 0 \\
0 & -1.6E+4 & 0 & 0 & 0 & -4157 & 0 & 1.6E+4 & 0 & 0 & 0 & -4157 \\
0 & 0 & -1778 & 0 & 461.9 & 0 & 0 & 0 & 1778 & 0 & 461.9 & 0 \\
0 & 0 & 0 & -47.5 & 0 & 0 & 0 & 0 & 0 & 47.5 & 0 & 0 \\
0 & 0 & -461.9 & 0 & 80 & 0 & 0 & 0 & 461.9 & 0 & 160 & 0 \\
0 & 4157 & 0 & 0 & 0 & 720 & 0 & -4157 & 0 & 0 & 0 & 1440
\end{bmatrix}$$

(b) $l_{x'} = \frac{300}{519.6} = 0.5774$, $m_{x'} = 0.5774$, $n_{x'} = -0.5774$

$$\bar{K} = (500-500)\bar{i} + (1000-200)\bar{j} + (100-0)\bar{k} = 800\bar{j} + 100\bar{k}$$

$$\bar{L} = 300\bar{i} + 300\bar{j} - 300\bar{k}$$

$$\bar{z}' = \bar{L}\times\bar{K} = (27\bar{i} - 3\bar{j} + 24\bar{k})(10^4),\quad \sqrt{27^2 + (-3)^2 + 24^2} = 36.249$$

$$l_{z'} = \frac{27}{36.249} = 0.7448,\quad m_{z'} = \frac{-3}{36.249} = -0.08276,\quad n_{z'} = \frac{24}{36.249} = 0.6621$$

$$\bar{y}' = \bar{z}'\times\bar{L} = (900)(9\bar{i} - \bar{j} + 8\bar{k})\times(\bar{i} + \bar{j} - \bar{k}) = (900)(-7\bar{i} + 17\bar{j} + 10\bar{k})$$

$$\sqrt{(-7)^2 + 17^2 + 10^2} = 20.928$$

$l_{y'} = \frac{-7}{20.928} = -0.3345$, $m_{y'} = \frac{17}{20.928} = 0.8123$, $n_{y'} = \frac{10}{20.928} = 0.4778$ (continued)

$$[T] = \begin{bmatrix} 0.5774 & 0.5774 & -0.5774 \\ -0.3345 & 0.8123 & 0.4778 \\ 0.7448 & -0.08276 & 0.6621 \end{bmatrix} \quad \boxed{\text{ANS}}$$

(c) Form $[T]$ according to Eq. (9.3-47) then perform

$$[k]_{global} = [T]^T [k]_{x'y'z'} [T]$$

$$= \begin{bmatrix}
5.43E+7 & 5.33E+7 & -5.41E+7 & -1.24E+5 & -2.21E+4 & -1.46E+5 \\
5.33E+7 & 5.53E+7 & -5.31E+7 & 3.37E+5 & -3.35E+4 & 3.04E+5 \\
-5.41E+7 & -5.31E+7 & 5.45E+7 & 2.13E+5 & -5.56E+4 & 1.57E+5 \\
-1.24E+5 & 3.37E+5 & 2.13E+5 & 1.12E+5 & -1.57E+4 & 9.01E+4 \\
-2.21E+4 & -3.35E+4 & -5.56E+4 & -1.57E+4 & 1.77E+4 & -4.40E+3 \\
-1.46E+5 & 3.04E+5 & 1.57E+5 & 9.01E+4 & -4.40E+3 & 9.21E+4 \\
-5.43E+7 & -5.33E+7 & 5.41E+7 & 1.24E+5 & 2.21E+4 & 1.46E+5 \\
-5.33E+7 & -5.53E+7 & 5.31E+7 & -3.37E+5 & 3.35E+4 & -3.04E+5 \\
5.41E+7 & 5.31E+7 & -5.45E+7 & -2.13E+5 & 5.56E+4 & -1.57E+5 \\
-1.24E+5 & 3.37E+5 & 2.13E+5 & 5.29E+4 & -1.10E+4 & 4.82E+4 \\
-2.21E+4 & -3.35E+4 & -5.56E+4 & -1.10E+4 & 5.64E+3 & 1.00E+3 \\
-1.46E+5 & 3.04E+5 & 1.57E+5 & 4.82E+4 & 1.00E+3 & 4.28E+4
\end{matrix}$$

$$\begin{matrix}
-5.43E+7 & -5.33E+7 & 5.41E+7 & -1.24E+5 & -2.21E+4 & -1.46E+5 \\
-5.33E+7 & -5.53E+7 & 5.31E+7 & 3.37E+5 & -3.35E+4 & 3.04E+5 \\
5.41E+7 & 5.31E+7 & -5.45E+7 & 2.13E+5 & -5.56E+4 & 1.57E+5 \\
1.24E+5 & -3.37E+5 & -2.13E+5 & 5.29E+4 & -1.10E+4 & 4.83E+4 \\
2.21E+4 & 3.35E+4 & 5.56E+4 & -1.10E+4 & 5.63E+3 & 1.01E+3 \\
1.46E+5 & -3.04E+5 & -1.57E+5 & 4.83E+4 & 1.01E+3 & 4.28E+4 \\
5.43E+7 & 5.33E+7 & -5.41E+7 & 1.24E+5 & 2.21E+4 & 1.46E+5 \\
5.33E+7 & 5.53E+7 & -5.31E+7 & -3.37E+5 & 3.35E+4 & -3.04E+5 \\
-5.41E+7 & -5.31E+7 & 5.45E+7 & -2.13E+5 & 5.56E+4 & -1.57E+5 \\
1.24E+5 & -3.37E+5 & -2.13E+5 & 1.12E+5 & -1.57E+4 & 9.01E+4 \\
2.21E+4 & 3.35E+4 & 5.56E+4 & -1.57E+4 & 1.77E+4 & -4.40E+3 \\
1.46E+5 & -3.04E+5 & -1.57E+5 & 9.01E+4 & -4.40E+3 & 9.21E+4
\end{bmatrix} \quad \text{ANS}$$

9.34 (a) $EI/L^3 = 16(10^3)$, $N/30L = -9000/30 = -300$

$$[k] = 16(10^3)\begin{bmatrix} 12 & 6 & -12 & 6 \\ 6 & 4 & -6 & 2 \\ -12 & -6 & 12 & -6 \\ 6 & 2 & -6 & 4 \end{bmatrix} - 300 \begin{bmatrix} 36 & 3 & -36 & 3 \\ 3 & 4 & -3 & -1 \\ -36 & -3 & 36 & -3 \\ 3 & -1 & -3 & 4 \end{bmatrix}$$

$v_1 = \theta_1 = 0$. Partition 1st & 2nd eqs. The remaining eqs are

(Continued)

$$\begin{Bmatrix} -2000 \\ 0 \end{Bmatrix} = \left\{ 16(10^3)\begin{bmatrix} 12 & -6 \\ -6 & 4 \end{bmatrix} - 300\begin{bmatrix} 36 & -3 \\ -3 & 4 \end{bmatrix} \right\}\begin{Bmatrix} v_2 \\ \theta_2 \end{Bmatrix} = (10^4)\begin{bmatrix} 18.12 & -9.51 \\ -9.51 & 6.28 \end{bmatrix}\begin{Bmatrix} v_2 \\ \theta_2 \end{Bmatrix}$$

Solve inverse

$$\begin{Bmatrix} v_2 \\ \theta_2 \end{Bmatrix} = (10^{-4})\begin{bmatrix} 0.2689 & 0.4072 \\ 0.4072 & 0.7759 \end{bmatrix}\begin{Bmatrix} -2000 \\ 0 \end{Bmatrix} = -\begin{Bmatrix} 0.05378\text{ m} \\ -0.08144\text{ rad} \end{Bmatrix}$$

$\underline{\delta = -53.78\text{ mm}}$

(b) $EI/L_e^3 = 128(10^3)$, $N/30L_e = -600$

$$[k]_e = 128(10^3)\begin{bmatrix} 12 & 3 & -12 & 3 \\ 3 & 1 & -3 & 0.5 \\ -12 & -3 & 12 & -3 \\ 3 & 0.5 & -3 & 1 \end{bmatrix} - 600\begin{bmatrix} 36 & 1.5 & -36 & 1.5 \\ 1.5 & 1 & -1.5 & -0.25 \\ -36 & -1.5 & 36 & -1.5 \\ 1.5 & -0.25 & -1.5 & 1 \end{bmatrix}$$

$$= (10^4)\begin{bmatrix} 151.44 & 38.31 & -151.44 & 38.31 \\ 38.31 & 12.74 & -38.31 & 6.415 \\ -151.44 & -38.31 & 151.44 & -38.31 \\ 38.31 & 6.415 & -38.31 & 12.74 \end{bmatrix}$$

$$[K_{sys}] = (10^4)\begin{bmatrix} 151.44 & 38.31 & -151.44 & 38.31 & 0 & 0 \\ 38.31 & 12.74 & -38.31 & 6.415 & 0 & 0 \\ -151.44 & -38.31 & 302.88 & 0 & -151.44 & 38.31 \\ 38.31 & 6.415 & 0 & 25.48 & -38.31 & 6.415 \\ 0 & 0 & -151.44 & -38.31 & 151.44 & -38.31 \\ 0 & 0 & 38.31 & 6.415 & -38.31 & 12.74 \end{bmatrix}$$

Partition 1st & 2nd eqs. The remaining eqs are

$$\begin{Bmatrix} 0 \\ 0 \\ -2000 \\ 0 \end{Bmatrix} = (10^4)\begin{bmatrix} 302.88 & 0 & -151.44 & 38.31 \\ 0 & 25.48 & -38.31 & 6.415 \\ -151.44 & -38.31 & 151.44 & -38.31 \\ 38.31 & 6.415 & -38.31 & 12.74 \end{bmatrix}\begin{Bmatrix} v_2 \\ \theta_2 \\ v_3 \\ \theta_3 \end{Bmatrix} \quad \text{Invert}$$

$$\begin{Bmatrix} v_2 \\ \theta_2 \\ v_3 \\ \theta_3 \end{Bmatrix} = (10^{-4})\begin{bmatrix} 0.03144 & 0.09819 & 0.08298 & 0.1055 \\ 0.09819 & 0.38815 & 0.3019 & 0.4171 \\ 0.08298 & 0.3019 & 0.2690 & 0.4074 \\ 0.1055 & 0.4171 & 0.4074 & 0.7763 \end{bmatrix}\begin{Bmatrix} 0 \\ 0 \\ -2000 \\ 0 \end{Bmatrix} = -\begin{Bmatrix} 0.0166\text{ m} \\ 0.06038\text{ rad} \\ 0.0538\text{ m} \\ 0.08149\text{ rad} \end{Bmatrix}$$

$\underline{\delta = -53.8\text{ mm}}$

From Example 6.13-3

$$\delta = \frac{5(2000)(1)^3}{3\left[5(16)(10^3) - 2(9000)(1)^2\right]} = 0.05376\text{ m} = \underline{53.76\text{ mm}}$$

9.35

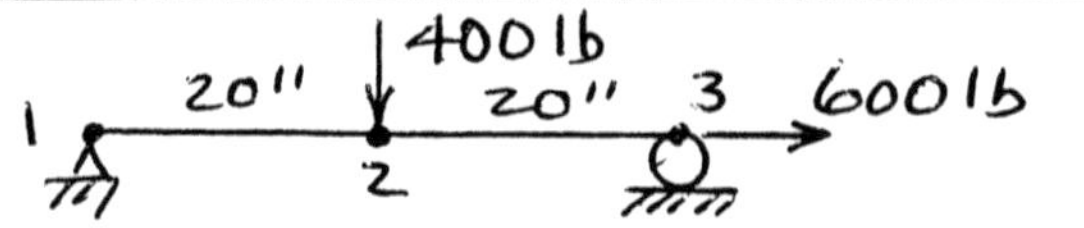

$$\frac{EI}{L_e^3} = \frac{1(10^6)}{(20)^3} = 125, \quad \frac{N}{30L_e} = \frac{600}{30(20)} = 1$$

$$[k]_e = (125)\begin{bmatrix} 12 & 120 & -12 & 120 \\ 120 & 1600 & -120 & 800 \\ -12 & -120 & 12 & -120 \\ 120 & 800 & -120 & 1600 \end{bmatrix} + (1)\begin{bmatrix} 36 & 60 & -36 & 60 \\ 60 & 1600 & -60 & -400 \\ -36 & -64 & 36 & -60 \\ 60 & -400 & -60 & 1600 \end{bmatrix}$$

$$= (10^3)\begin{bmatrix} 1.536 & 15.06 & -1.536 & 15.06 \\ 15.06 & 201.6 & -15.06 & 99.6 \\ -1.536 & -15.06 & 1.536 & -15.06 \\ 15.06 & 99.6 & -15.06 & 201.6 \end{bmatrix}$$

$$[K_{sys}] = (10^3)\begin{bmatrix} 1.536 & 15.06 & -1.536 & 15.06 & 0 & 0 \\ 15.06 & 201.6 & -15.06 & 99.6 & 0 & 0 \\ -1.536 & -15.06 & 3.072 & 0 & -1.536 & 15.06 \\ 15.06 & 99.6 & 0 & 403.2 & 15.06 & 99.6 \\ 0 & 0 & -1.536 & -15.06 & 1.536 & -15.06 \\ 0 & 0 & 15.06 & 99.6 & -15.06 & 201.6 \end{bmatrix}$$

$v_1 = v_3 = 0$. Partition 1st & 5th eqs.

$$\begin{Bmatrix} 0 \\ -400 \\ 0 \\ 0 \end{Bmatrix} = (10^3)\begin{bmatrix} 201.6 & -15.06 & 99.6 & 0 \\ -15.06 & 3.072 & 0 & 15.06 \\ 99.6 & 0 & 403.2 & 99.6 \\ 0 & 15.06 & 99.6 & 201.6 \end{bmatrix}\begin{Bmatrix} \theta_1 \\ v_2 \\ \theta_2 \\ \theta_3 \end{Bmatrix} \quad \text{Invert}$$

$$\begin{Bmatrix} \theta_1 \\ v_2 \\ \theta_3 \\ \theta_4 \end{Bmatrix} = (10^{-3})\begin{bmatrix} 0.01255 & 0.09088 & -0.00162 & -0.00599 \\ 0.09088 & 1.2166 & 0 & -0.09088 \\ -0.00162 & 0 & 0.003281 & -0.00162 \\ -0.00599 & -0.09088 & -0.00162 & 0.01255 \end{bmatrix}\begin{Bmatrix} 0 \\ -400 \\ 0 \\ 0 \end{Bmatrix}$$

$$= \begin{Bmatrix} -0.03635\text{ rad} \\ -0.4866\text{ in} \\ 0 \\ 0.03635\text{ rad} \end{Bmatrix} \qquad \therefore\ \underline{v_2 = -0.4866\text{ in}}$$

Element 1:

$$\begin{Bmatrix} V_1 \\ M_1 \\ V_2 \\ M_2 \end{Bmatrix}_1 = 10^3\begin{bmatrix} 1.536 & 15.06 & -1.536 & 15.06 \\ 15.06 & 201.6 & -15.06 & 99.6 \\ -1.536 & -15.06 & 1.536 & -15.06 \\ 15.06 & 99.6 & -15.06 & 201.6 \end{bmatrix}\begin{Bmatrix} 0 \\ -0.03635 \\ -0.4866 \\ 0 \end{Bmatrix} = \begin{Bmatrix} 200\text{ lb} \\ 0 \\ -200\text{ lb} \\ 3708\text{ lb-in} \end{Bmatrix}$$

(continued)

$$\sigma_{max} = \frac{M}{S} + \frac{F}{A} = \frac{3708}{1.85} + \frac{600}{0.85} = 2710\,psi = \underline{2.71\,kpsi}$$

(9.36)

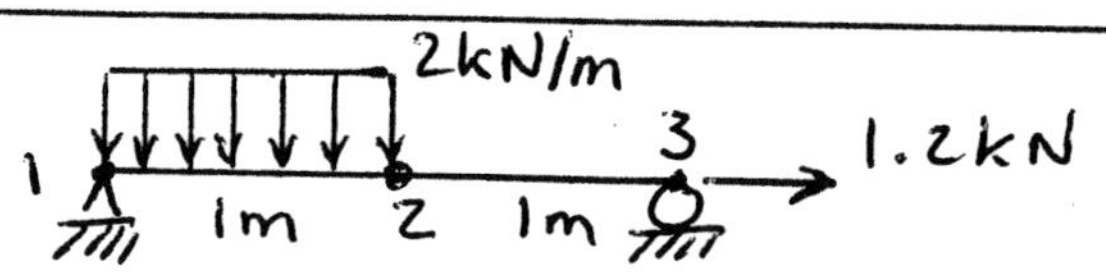

Element 1: $q_o L_e/2 = -1000\,N$

$q_o L_e^2/12 = -166.67\,N \cdot m$

$$\frac{EI}{L_e^3} = 12(10^3)\,N/m, \quad \frac{N}{30L_e} = \frac{1200}{30(1)} = 40\,N/m$$

$$[k]_e = 12(10^3)\begin{bmatrix} 12 & 6 & -12 & 6 \\ 6 & 4 & -6 & 2 \\ -12 & -6 & 12 & -6 \\ 6 & 2 & -6 & 4 \end{bmatrix} + (40)\begin{bmatrix} 36 & 3 & -36 & 3 \\ 3 & 4 & -3 & -1 \\ -36 & -3 & 36 & -3 \\ 3 & -1 & -3 & 4 \end{bmatrix}$$

$$= (10^4)\begin{bmatrix} 14.544 & 7.212 & -14.544 & 7.212 \\ 7.212 & 4.816 & -7.212 & 2.396 \\ -14.544 & -7.212 & 14.544 & -7.212 \\ 7.212 & 2.396 & -7.212 & 4.816 \end{bmatrix}$$

$$[K_{sys}] = (10^4)\begin{bmatrix} 14.544 & 7.212 & -14.544 & 7.212 & 0 & 0 \\ 7.212 & 4.816 & -7.212 & 2.396 & 0 & 0 \\ -14.544 & -7.212 & 29.088 & 0 & -14.544 & 7.212 \\ 7.212 & 2.396 & 0 & 9.632 & -7.212 & 2.396 \\ 0 & 0 & -14.544 & -7.212 & 14.544 & -7.212 \\ 0 & 0 & 7.212 & 2.396 & -7.212 & 4.816 \end{bmatrix}$$

$v_1 = v_3 = 0$. Partition 1st and 5th eqs. The remaining eqs are

$$\begin{Bmatrix} -166.67 \\ -1000 \\ 166.67 \\ 0 \end{Bmatrix} = (10^4)\begin{bmatrix} 4.816 & -7.212 & 2.396 & 0 \\ -7.212 & 29.088 & 0 & 7.212 \\ 2.396 & 0 & 9.632 & 2.396 \\ 0 & 7.212 & 2.396 & 4.816 \end{bmatrix}\begin{Bmatrix} \theta_1 \\ v_2 \\ \theta_2 \\ \theta_3 \end{Bmatrix} \quad \text{Invert.}$$

$$\begin{Bmatrix} \theta_1 \\ v_2 \\ \theta_2 \\ \theta_3 \end{Bmatrix} = 10^{-4}\begin{bmatrix} 0.5413 & 0.200 & -0.06864 & -0.2653 \\ 0.200 & 0.1335 & 0 & -0.200 \\ -0.06864 & 0 & 0.1380 & -0.06864 \\ -0.2653 & -0.200 & -0.06864 & 0.5413 \end{bmatrix}\begin{Bmatrix} -166.67 \\ -1000 \\ 166.67 \\ 0 \end{Bmatrix} = \begin{Bmatrix} -0.03016\,rad \\ -0.01669\,m \\ 0.00344\,rad \\ 0.02328\,rad \end{Bmatrix}$$

$$v_2 = \underline{-16.7\,mm}$$

Element 1:

$$\begin{Bmatrix} V_1 \\ M_1 \\ V_2 \\ M_2 \end{Bmatrix}_1 = 10^4\begin{bmatrix} 14.544 & 7.212 & -14.544 & 7.212 \\ 7.212 & 4.816 & -7.212 & 2.396 \\ -14.544 & -7.212 & 14.544 & -7.212 \\ 7.212 & 2.396 & -7.212 & 4.816 \end{bmatrix}\begin{Bmatrix} 0 \\ -0.03016 \\ -0.01669 \\ 0.00344 \end{Bmatrix} - \begin{Bmatrix} -1000 \\ -166.67 \\ -1000 \\ 166.67 \end{Bmatrix} = \begin{Bmatrix} 1500\,N \\ 0 \\ 500\,N \\ 480\,N\text{-}m \end{Bmatrix}$$

(Continued)

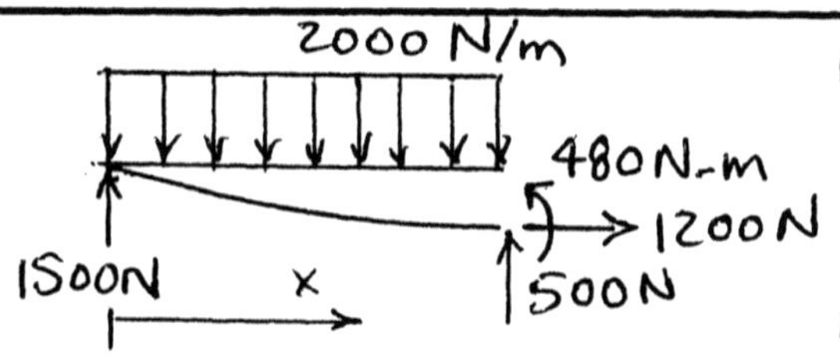

Maximum moment due to transverse loads occurs where $V = 0$

$x = \frac{1500}{2000} = 0.75\,m$

Assuming $v_{x=0.75} \approx 0.0167\,m$

$$M_{max} \approx 1500(0.75) - \frac{1}{2}(2000)(0.75)^2 - 1200(0.0167) = 540\,N\cdot m$$

$$\sigma_{max} = \frac{M}{S} + \frac{F}{A} \approx \frac{540(10^3)}{32(10^3)} + \frac{1200}{300} = \underline{21.0\,MPa}$$

9.37 In the element formulation per the hint in Prob. 9.17 divide the moment eqs by $L_e = L/2$ and factor L_e out with θ_i, θ_j.

$$[k]_e = \frac{EI}{(L/2)^3}\begin{bmatrix} 12 & 6 & -12 & 6 \\ 6 & 4 & -6 & 2 \\ -12 & -6 & 12 & -6 \\ 6 & 2 & -6 & 4 \end{bmatrix} - \frac{P_{cr}}{15L}\begin{bmatrix} 36 & 3 & -36 & 3 \\ 3 & 4 & -3 & -1 \\ -36 & -3 & 36 & -3 \\ 3 & -1 & -3 & 4 \end{bmatrix}$$

Assemble. Then with $v_1 = \theta_1 = v_3 = \theta_3 = 0$ partition 1st, 2nd, 5th and 6th eqs out. What remains is,

$$\frac{EI}{(L/2)^3}\begin{bmatrix} 24 & 0 \\ 0 & 8 \end{bmatrix} - \frac{P_{cr}}{15L}\begin{bmatrix} 72 & 0 \\ 0 & 8 \end{bmatrix} = [K_{red}]$$

For buckling $|K_{red}| = 0$ or,

$$\begin{vmatrix} (24-72C) & 0 \\ 0 & (8-8C) \end{vmatrix} = 0 \quad \text{where } C = \frac{P_{cr}(\frac{L}{2})^3}{(15L)EI} = \frac{P_{cr}L^2}{120EI} \quad (1)$$

$3C^2 - 4C + 1 = 0 \Rightarrow C = 1/3, 1$ lowest value yields P_{cr}

From (1) $\frac{P_{cr}L^2}{120EI} = \frac{1}{3} \Rightarrow P_{cr} = \frac{40EI}{L^2}$ **ANS**

From Table 3.10-1 $P_{cr} = \frac{\pi^2 EI}{(0.5L)^2} = 39.48\frac{EI}{L^2}$ F.E model is 1.32% high

---

9.38 See the solution of Prob. 9.37 for the element stiffness matrix. Assemble the two elements then with $v_1 = \theta_1 = v_3 = 0$ partition out 1st, 2nd, and 5 eqs. The reduced matrix is,

$$[K_{red}] = \frac{8EI}{L^3}\begin{bmatrix} 24 & 0 & 6 \\ 0 & 8 & 2 \\ 6 & 2 & 4 \end{bmatrix} - \frac{P_{cr}}{15L}\begin{bmatrix} 72 & 0 & 3 \\ 0 & 8 & -1 \\ 3 & -1 & 4 \end{bmatrix}$$

(Continued)

With $C = \frac{P_{cr}}{15L}\frac{L^3}{8EI} = \frac{P_{cr}L^2}{120EI}$ Set $|K_{red}| = 0$

$$\begin{vmatrix} (24-72C) & 0 & (6-3C) \\ 0 & (8-8C) & (2+C) \\ (6-3C) & (2+C) & (4-4C) \end{vmatrix} = 0$$

reduces to

$$45C^3 - 110C^2 + 64C - 8 = 0$$

The roots are: $C = 0.17257, 0.62585, 1.64603$

The lowest value yields P_{cr}

$$\frac{P_{cr}L^2}{120EI} = 0.17257 \Rightarrow P_{cr} = 20.71\frac{EI}{L^2} \quad \boxed{\text{ANS}}$$

Table 3.10-1 gives $P_{cr} = \frac{\pi^2 EI}{(0.699L)^2} = 20.20\frac{EI}{L^2}$ F.E. model is 2.5% high

---

(9.39) $\frac{EI}{L^3} = \frac{5(10^3)}{(0.25)^3} = 320(10^3)$ N/m

$$[k]_e = 320(10^3)\begin{bmatrix} 12 & 1.5 & -12 & 1.5 \\ 1.5 & 0.25 & -1.5 & 0.125 \\ -12 & -1.5 & 12 & -1.5 \\ 1.5 & 0.125 & -1.5 & 0.25 \end{bmatrix} - \frac{P_{cr}}{7.5}\begin{bmatrix} 36 & 0.75 & -36 & 0.75 \\ 0.75 & 0.25 & -0.75 & -0.125 \\ -36 & -0.75 & 36 & -0.75 \\ 0.75 & -0.125 & -0.75 & 0.25 \end{bmatrix}$$

Assemble

$$[K_{sys}] = 320(10^3)\begin{bmatrix} 12 & 1.5 & -12 & 1.5 & 0 & 0 \\ 1.5 & 0.25 & -1.5 & 0.125 & 0 & 0 \\ -12 & -1.5 & 24 & 0 & -12 & 1.5 \\ 1.5 & 0.125 & 0 & 0.5 & -1.5 & 0.125 \\ 0 & 0 & -12 & -1.5 & 12 & -1.5 \\ 0 & 0 & 1.5 & 0.125 & -1.5 & 0.25 \end{bmatrix}$$

$$- \frac{P_{cr}}{7.5}\begin{bmatrix} 36 & 0.75 & -36 & 0.75 & 0 & 0 \\ 0.75 & 0.25 & -0.75 & -0.125 & 0 & 0 \\ -36 & -0.75 & 72 & 0 & -36 & 0.75 \\ 0.75 & -0.125 & 0 & 0.5 & -0.75 & -0.125 \\ 0 & 0 & -36 & -0.75 & 36 & -0.75 \\ 0 & 0 & 0.75 & -0.125 & -0.75 & 0.25 \end{bmatrix}$$

$v_1 = \theta_1 = 0$. Partition 1st & 2nd eqs.

Let $x = \frac{P_{cr}}{7.5(320)(10^3)} = 4.1667(10^{-7})P_{cr}$ (continued)

$$|K_{red}| = \begin{vmatrix} (24-72x) & 0 & (-12+36x) & (1.5-0.75x) \\ 0 & (0.5-0.5x) & (-1.5+0.75x) & (0.125+0.125x) \\ (-12+36x) & (-1.5+0.75x) & (12-36x) & (-1.5+0.75x) \\ (1.5-0.75x) & (0.125+0.125x) & (-1.5+0.75x) & (0.25-0.25x) \end{vmatrix} = 0$$

reduces to

$$x^4 - 2.8835x^3 + 1.7754x^2 - 0.26016x + 4.8178(10^{-3}) = 0$$

The roots are: $x = 0.02159, 0.18208, 0.58512, 2.0947$

The lowest root yields P_{cr}

$$4.1667(10^{-7})P_{cr} = 0.02159 \Rightarrow P_{cr} = 51.8(10^3)\text{N} = \underline{51.8\text{kN}}$$

From Table 3.10-1, $P_{cr} = \dfrac{\pi^2(5)(10^3)}{[2(0.5)]^2} = 49.4(10^3)\text{N}$ F.E model is 5% high

9.40 (a) $\alpha_i = 3(2) - 3(-2) = 12$, $\beta_i = -2-2 = -4$, $\gamma_i = 3-3 = 0$

$\alpha_j = 3(0) - (0)(2) = 0$, $\beta_j = 2-0 = 2$, $\gamma_j = 0-3 = -3$

$\alpha_k = 0(-2) - 3(0) = 0$, $\beta_k = 0-(-2) = 2$, $\gamma_k = 3-0 = 3$

$2A = \Sigma\alpha = 12$

$$[B] = \frac{1}{12}\begin{bmatrix} -4 & 0 & 2 & 0 & 2 & 0 \\ 0 & 0 & 0 & -3 & 0 & 3 \\ 0 & -4 & -3 & 2 & 3 & 2 \end{bmatrix} \quad [B]^T = \frac{1}{12}\begin{bmatrix} -4 & 0 & 0 \\ 0 & 0 & -4 \\ 2 & 0 & -3 \\ 0 & -3 & 2 \\ 2 & 0 & 3 \\ 0 & 3 & 2 \end{bmatrix}$$

$$[D] = \frac{10(10^6)}{2(1-0.33^2)}\begin{bmatrix} 2 & 2(0.33) & 0 \\ 2(0.33) & 2 & 0 \\ 0 & 0 & (1-0.33) \end{bmatrix} = 5.611(10^6)\begin{bmatrix} 2 & 0.66 & 0 \\ 0.66 & 2 & 0 \\ 0 & 0 & 0.67 \end{bmatrix}$$

Eq(9.4-16) $[k]_e = 6(0.5)[B]^T[D][B]$

$$[k]_e = \begin{bmatrix} 3740.70 & 0.00 & -1870.35 & 925.82 & -1870.35 & -925.82 \\ 0.00 & 1253.13 & 939.85 & -626.57 & -939.85 & -626.57 \\ -1870.35 & 939.85 & 1640.06 & -932.84 & 230.29 & -7.01 \\ 925.82 & -626.57 & -932.84 & 2417.42 & 7.01 & -1790.86 \\ -1870.35 & -939.85 & 230.29 & 7.01 & 1640.06 & 932.84 \\ -925.82 & -626.57 & -7.01 & -1790.86 & 932.84 & 2417.42 \end{bmatrix}(10^3) \quad \boxed{\text{ANS}}$$

(continued)

(b) $\{f\} = [k]_e \{u\}$

$\{u\} = \{-5.45\ \ 8.24\ \ -3.15\ \ 6.10\ \ 1.21\ \ 5.56\}^T (10^{-3})$ in

Matrix multiplication yields

$\{f\}^T = \{-16.26\ \ -1.08\ \ 7.32\ \ -2.47\ \ 8.94\ \ 3.55\}(10^3)$ lb **ANS**

(c) $\{\sigma\} = [D][B]\{u\}$ Matrix multiplication yields

$$\{\sigma\} = \begin{Bmatrix} \sigma_x \\ \sigma_y \\ \tau_{xy} \end{Bmatrix} = \begin{Bmatrix} 16.26 \\ 4.02 \\ 1.08 \end{Bmatrix} \text{ kpsi}$$ **ANS**

9.41 $$[D] = \frac{30(10^6)}{2(1-0.3^2)}\begin{bmatrix} 2 & 0.6 & 0 \\ 0.6 & 2 & 0 \\ 0 & 0 & 0.7 \end{bmatrix} = 16.484(10^6)\begin{bmatrix} 2 & 0.6 & 0 \\ 0.6 & 2 & 0 \\ 0 & 0 & 0.7 \end{bmatrix}$$

<u>Element A:</u> $i, j, k = 1, 3, 5$ $(0,0)\ (6, 0.3)\ (0, 0.3)$

$\alpha_i = 6(0.3) - 0(0.3) = 1.8$, $\beta_i = 0.3 - 0.3 = 0$, $\gamma_i = 0 - 6 = -6$

$\alpha_j = 0(0) - 0(0.3) = 0$, $\beta_j = 0.3 - 0 = 0.3$, $\gamma_j = 0 - 0 = 0$

$\alpha_k = 0(0.3) - 6(0) = 0$, $\beta_k = 0 - 0.3 = -0.3$, $\gamma_k = 6 - 0 = 6$

$2A = \Sigma\alpha = 1.8$

$$[B]_A = \frac{1}{1.8}\begin{bmatrix} 0 & 0 & 0.3 & 0 & -0.3 & 0 \\ 0 & -6 & 0 & 0 & 0 & 6 \\ -6 & 0 & 0 & 0.3 & 6 & -0.3 \end{bmatrix}, \quad t = 0.125 \text{ in}$$

$[k]_A = At[B]_A^T[D][B]_A$ Performing the matrix multiplication and writing the results in terms of the complete system dof yields

$$[k]_A = \begin{bmatrix} 1442.31 & 0 & 0 & 0 & 0 & -72.12 & 0 & 0 & -1442.31 & 72.12 \\ 0 & 4120.88 & 0 & 0 & -61.81 & 0 & 0 & 0 & 61.81 & -4120.88 \\ 0 & 0 & 0 & 0 & 0 & 0 & 0 & 0 & 0 & 0 \\ 0 & 0 & 0 & 0 & 0 & 0 & 0 & 0 & 0 & 0 \\ 0 & -61.81 & 0 & 0 & 10.30 & 0 & 0 & 0 & -10.30 & 61.81 \\ -72.12 & 0 & 0 & 0 & 0 & 3.61 & 0 & 0 & 72.12 & -3.61 \\ 0 & 0 & 0 & 0 & 0 & 0 & 0 & 0 & 0 & 0 \\ 0 & 0 & 0 & 0 & 0 & 0 & 0 & 0 & 0 & 0 \\ -1442.31 & 61.81 & 0 & 0 & -10.30 & 72.12 & 0 & 0 & 1452.61 & -133.93 \\ 72.12 & -4120.88 & 0 & 0 & 61.81 & -3.61 & 0 & 0 & -133.93 & 4124.48 \end{bmatrix}[10^4]$$

(continued)

Element B $i, j, k = 1, 2, 3$ (0,0) (6,0) (6,0.3)

$\alpha_i = 6(0.3) - 6(0) = 1.8$, $\beta_i = 0 - 0.3 = -0.3$, $\gamma_i = 6 - 6 = 0$

$\alpha_j = 6(0) - 0(0.3) = 0$, $\beta_j = 0.3 - 0 = 0.3$, $\gamma_j = 0 - 6 = -6$

$\alpha_k = 0(0) - 6(0) = 0$, $\beta_k = 0 - 0 = 0$, $\gamma_k = 6 - 0 = 6$

$2A = \Sigma\alpha = 1.8$

$$[B]_B = \frac{1}{1.8}\begin{bmatrix} -0.3 & 0 & 0.3 & 0 & 0 & 0 \\ 0 & 0 & 0 & -6 & 0 & 6 \\ 0 & -0.3 & -6 & 0.3 & 6 & 0 \end{bmatrix}$$

$[k]_B = At[B]_B^T[D][B]_B$ The result, in system dof, is

$$[k]_B = \begin{bmatrix} 10.30 & 0 & -10.30 & 61.81 & 0 & -61.81 & 0 & 0 & 0 & 0 \\ 0 & 3.61 & 72.12 & -3.61 & -72.12 & 0 & 0 & 0 & 0 & 0 \\ -10.30 & 72.12 & 1452.61 & -133.93 & -1442.31 & 61.81 & 0 & 0 & 0 & 0 \\ 61.81 & -3.61 & -133.93 & 4124.48 & 72.12 & -4120.88 & 0 & 0 & 0 & 0 \\ 0 & -72.12 & -1442.31 & 72.12 & 1442.31 & 0 & 0 & 0 & 0 & 0 \\ -61.81 & 0 & 61.81 & -4120.88 & 0 & 4120.88 & 0 & 0 & 0 & 0 \\ 0 & 0 & 0 & 0 & 0 & 0 & 0 & 0 & 0 & 0 \\ 0 & 0 & 0 & 0 & 0 & 0 & 0 & 0 & 0 & 0 \\ 0 & 0 & 0 & 0 & 0 & 0 & 0 & 0 & 0 & 0 \\ 0 & 0 & 0 & 0 & 0 & 0 & 0 & 0 & 0 & 0 \end{bmatrix} [10^4]$$

Element C $i, j, k = 3, 4, 5$ (6,0.3) (6,0.5) (0,0.3)

$\alpha_i = 6(0.3) - 0(0.5) = 1.8$, $\beta_i = 0.5 - 0.3 = 0.2$, $\gamma_i = 0 - 6 = -6$

$\alpha_j = 0(0.3) - 6(0.3) = -1.8$, $\beta_j = 0.3 - 0.3 = 0$, $\gamma_j = 6 - 0 = 6$

$\alpha_k = 6(0.5) - 6(0.3) = 1.2$, $\beta_k = 0.3 - 0.5 = -0.2$, $\gamma_k = 6 - 6 = 0$

$2A = \Sigma\alpha = 1.2$

$$[B]_C = \frac{1}{1.2}\begin{bmatrix} 0.2 & 0 & 0 & 0 & -0.2 & 0 \\ 0 & -6 & 0 & 6 & 0 & 0 \\ -6 & 0.2 & 6 & 0 & 0 & -0.2 \end{bmatrix}$$

$[k]_C = At[B]_C^T[D][B]_C$

The result, in system dof, is

(Continued)

$$[k]_C = \begin{bmatrix} 0 & 0 & 0 & 0 & 0 & 0 & 0 & 0 & 0 & 0 \\ 0 & 0 & 0 & 0 & 0 & 0 & 0 & 0 & 0 & 0 \\ 0 & 0 & 0 & 0 & 0 & 0 & 0 & 0 & 0 & 0 \\ 0 & 0 & 0 & 0 & 0 & 0 & 0 & 0 & 0 & 0 \\ 0 & 0 & 0 & 0 & 3255.49 & -200.89 & -3245.19 & 92.72 & -10.30 & 108.17 \\ 0 & 0 & 0 & 0 & -200.89 & 9275.58 & 108.17 & -9271.98 & 92.72 & -3.61 \\ 0 & 0 & 0 & 0 & -3245.19 & 108.17 & 3245.19 & 0 & 0 & -108.17 \\ 0 & 0 & 0 & 0 & 92.72 & -9271.98 & 0 & 9271.98 & -92.72 & 0 \\ 0 & 0 & 0 & 0 & -10.30 & 92.72 & 0 & -92.72 & 10.30 & 0 \\ 0 & 0 & 0 & 0 & 108.17 & -3.61 & -108.17 & 0 & 0 & 3.61 \end{bmatrix} [10^4]$$

Assemble matrices by simply adding

B.C. $u_1 = v_1 = v_2 = u_5 = 0$. Partition 1st, 2nd, 4th, and 9th eqs.

For loads

Side 2,3: $F_{23} = (16{,}000)(\frac{1}{8})(0.3) = 600 \text{ lb}$

3,4: $F_{34} = (16{,}000)(\frac{1}{8})(0.2) = 400 \text{ lb}$

$F_{2x} = 300 \text{ lb}$, $F_{3x} = 300 + 200 = 500 \text{ lb}$, $F_{4x} = 200 \text{ lb}$

$$[K]_{red} = \begin{bmatrix} 1452.61 & -1442.31 & 61.81 & 0 & 0 & 0 \\ -1442.31 & 4708.10 & -200.89 & -3245.19 & 92.72 & 169.99 \\ 61.81 & -200.89 & 13400.07 & 108.17 & -9271.98 & -7.21 \\ 0 & -3245.19 & 108.17 & 3245.19 & 0 & -108.17 \\ 0 & 92.72 & -9271.98 & 0 & 9271.98 & 0 \\ 0 & 169.99 & -7.21 & -108.17 & 0 & 4128.09 \end{bmatrix} [10^4]$$

Solve for the inverse and multiply the force vector

$$\begin{Bmatrix} u_2 \\ u_3 \\ v_3 \\ u_4 \\ v_4 \\ v_5 \end{Bmatrix} = \begin{bmatrix} 3.59\text{E-}06 & 3.54\text{E-}06 & -5.38\text{E-}08 & 3.54\text{E-}06 & -8.92\text{E-}08 & -5.31\text{E-}08 \\ 3.54\text{E-}06 & 3.56\text{E-}06 & -5.31\text{E-}08 & 3.56\text{E-}06 & -8.87\text{E-}08 & -5.34\text{E-}08 \\ -5.38\text{E-}08 & -5.31\text{E-}08 & 2.51\text{E-}08 & -5.39\text{E-}08 & 2.56\text{E-}08 & 8.18\text{E-}10 \\ 3.54\text{E-}06 & 3.56\text{E-}06 & -5.39\text{E-}08 & 3.59\text{E-}06 & -8.95\text{E-}08 & -5.26\text{E-}08 \\ -8.92\text{E-}08 & -8.87\text{E-}08 & 2.56\text{E-}08 & -8.95\text{E-}08 & 3.73\text{E-}08 & 1.35\text{E-}09 \\ -5.31\text{E-}08 & -5.34\text{E-}08 & 8.18\text{E-}10 & -5.26\text{E-}08 & 1.35\text{E-}09 & 2.50\text{E-}08 \end{bmatrix} \begin{Bmatrix} 300 \\ 500 \\ 0 \\ 200 \\ 0 \\ 0 \end{Bmatrix} = \begin{Bmatrix} 3.55\text{E-}03 \\ 3.56\text{E-}03 \\ -5.35\text{E-}05 \\ 3.56\text{E-}03 \\ -8.90\text{E-}05 \\ -5.32\text{E-}05 \end{Bmatrix} \text{in}$$

ANS

For stresses in each element write the displacement vector for the element and perform the matrix multiplication

$\{\sigma\} = [D][B]\{u\}$ (continued)

For example

$$\{\sigma_A\} = \frac{16.484(10^6)}{1.8}\begin{bmatrix} 2 & 0.6 & 0 \\ 0.6 & 2 & 0 \\ 0 & 0 & 0.7 \end{bmatrix}\begin{bmatrix} 0 & 0 & 0.3 & 0 & -0.3 & 0 \\ 0 & -6 & 0 & 0 & 0 & 6 \\ -6 & 0 & 0 & 0.3 & 6 & -0.3 \end{bmatrix}\begin{Bmatrix} 0 \\ 0 \\ 3.56(10^{-3}) \\ -5.35(10^{-5}) \\ 0 \\ -5.32(10^{-5}) \end{Bmatrix}$$

$$= \begin{Bmatrix} 17.8(10^3) \\ 17.8 \\ -0.56 \end{Bmatrix} \text{psi} \quad \boxed{\text{ANS}}$$

Similarly for elements B & C

$$\{\sigma_B\} = \begin{Bmatrix} 17.8(10^3) \\ -17.8 \\ 88.2 \end{Bmatrix} \text{psi}, \quad \{\sigma_c\} = \begin{Bmatrix} 17.8(10^3) \\ 0 \\ 355.6 \end{Bmatrix} \text{psi} \quad \boxed{\text{ANS}}$$

9.42

$$[k]_A = \begin{bmatrix} k_{11} & k_{12} & \cdots & k_{16} \\ \vdots & & & \vdots \\ k_{61} & \cdots & & k_{66} \end{bmatrix}_A \quad \text{columns/rows: } u_1, v_1, u_4, v_4, u_3, v_3$$

$$[k]_B = \begin{bmatrix} k_{11} & \cdots & & k_{16} \\ k_{21} & & & \\ \vdots & & & \\ k_{61} & \cdots & & k_{66} \end{bmatrix}_B \quad \text{columns/rows: } u_3, v_3, u_4, v_4, u_5, v_5$$

$$[k]_C = \begin{bmatrix} k_{11} & k_{12} & \cdots & k_{16} \\ \vdots & & & \\ k_{61} & \cdots & & k_{66} \end{bmatrix}_C \quad \text{columns/rows: } u_2, v_2, u_3, v_3, u_5, v_5$$

$$[k]_D = \begin{bmatrix} k_{11} & \cdots & & k_{16} \\ k_{21} & & & \\ \vdots & & & \\ k_{61} & \cdots & & k_{66} \end{bmatrix}_D \quad \text{columns/rows: } u_1, v_1, u_3, v_3, u_2, v_2$$

(continued)

Resulting $[K_{sys}]$ is a 10×10 matrix.
The coefficients are shown below
where $k_{ji} = k_{ij}$.

$k_{1\,1} = (k_{11})_A + (k_{11})_D$
$k_{1\,2} = (k_{12})_A + (k_{12})_D$
$k_{1\,3} = (k_{15})_D$
$k_{1\,4} = (k_{16})_D$
$k_{1\,5} = (k_{15})_A + (k_{13})_D$
$k_{1\,6} = (k_{16})_A + (k_{14})_D$
$k_{1\,7} = (k_{13})_A$
$k_{1\,8} = (k_{14})_A$
$k_{1\,9} = 0$
$k_{1\,10} = 0$
$k_{2\,2} = (k_{22})_A + (k_{22})_D$
$k_{2\,3} = (k_{25})_D$
$k_{2\,4} = (k_{26})_D$
$k_{2\,5} = (k_{25})_A + (k_{23})_D$
$k_{2\,6} = (k_{26})_A + (k_{24})_D$
$k_{2\,7} = (k_{23})_A$
$k_{2\,8} = (k_{24})_A$
$k_{2\,9} = 0$
$k_{2\,10} = 0$
$k_{3\,3} = (k_{11})_C + (k_{55})_D$
$k_{3\,4} = (k_{12})_C + (k_{56})_D$
$k_{3\,5} = (k_{13})_C + (k_{35})_D$
$k_{3\,6} = (k_{14})_C + (k_{45})_D$
$k_{3\,7} = 0$
$k_{3\,8} = 0$
$k_{3\,9} = (k_{15})_C$
$k_{3\,10} = (k_{16})_C$
$k_{4\,4} = (k_{22})_C + (k_{66})_D$
$k_{4\,5} = (k_{23})_C + (k_{36})_D$
$k_{4\,6} = (k_{24})_C + (k_{46})_D$
$k_{4\,7} = 0$
$k_{4\,8} = 0$
$k_{4\,9} = (k_{25})_C$
$k_{4\,10} = (k_{26})_C$
$k_{5\,5} = (k_{55})_A + (k_{11})_B + (k_{33})_C + (k_{33})_D$
$k_{5\,6} = (k_{56})_A + (k_{12})_B + (k_{34})_C + (k_{34})_D$
$k_{5\,7} = (k_{35})_A + (k_{13})_B$
$k_{5\,8} = (k_{45})_A + (k_{14})_B$
$k_{5\,9} = (k_{15})_B + (k_{35})_C$
$k_{5\,10} = (k_{16})_B + (k_{36})_C$

$k_{6\,6} = (k_{66})_A + (k_{22})_B + (k_{44})_C + (k_{44})_D$
$k_{6\,7} = (k_{36})_A + (k_{23})_B$
$k_{6\,8} = (k_{46})_A + (k_{24})_B$
$k_{6\,9} = (k_{25})_B + (k_{45})_C$
$k_{6\,10} = (k_{26})_B + (k_{46})_C$
$k_{7\,7} = (k_{33})_A + (k_{33})_B$
$k_{7\,8} = (k_{34})_A + (k_{34})_B$
$k_{7\,9} = (k_{35})_B$
$k_{7\,10} = (k_{36})_B$
$k_{8\,8} = (k_{44})_A + (k_{44})_B$
$k_{8\,9} = (k_{45})_B$
$k_{8\,10} = (k_{46})_B$
$k_{9\,9} = (k_{55})_B + (k_{55})_C$
$k_{9\,10} = (k_{56})_B + (k_{56})_C$
$k_{10\,10} = (k_{66})_B + (k_{66})_C$

10.1

This is a crude mesh with 162 elements which can be made finer.

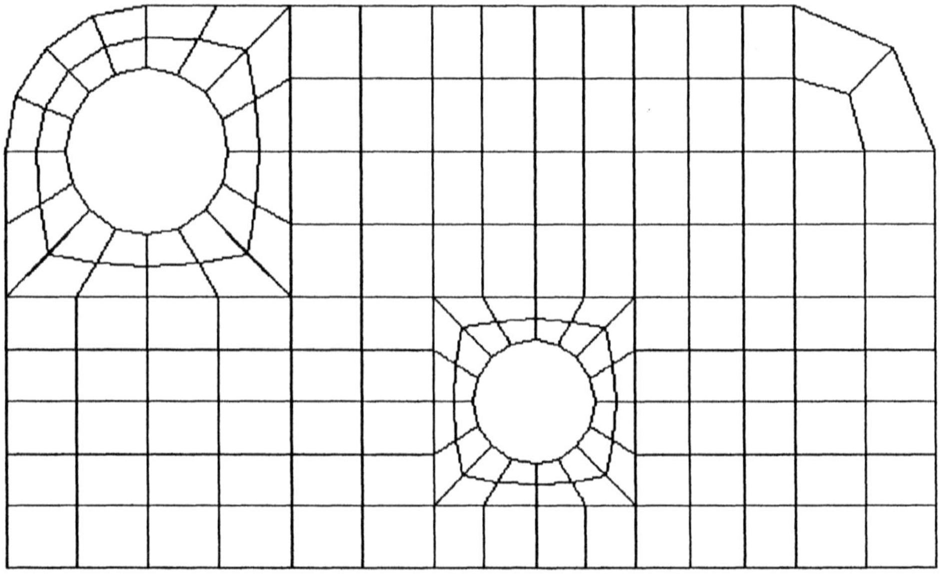

10.2 $(x_1, y_1) = (1,2)$, $(x_2, y_2) = (3,1)$, $(x_3, y_3) = (5,3)$, $(x_4, y_4) = (5.2, 9)$, $(x_5, y_5) = (x_6, y_6) = (x_7, y_7) = (1.5, 10)$, $(x_8, y_8) = (1.5, 6)$

Use same table as Table 10.2-2. Sample calculation for point $(\xi, \eta) = (0.5, 1/3)$. The values of N_i will be the same as in Example 10.2-1. That is, $N_1 = -0.15278$, $N_2 = 0.25$, $N_3 = -0.20833$, $N_4 = 2/3$, $N_5 = -0.08333$, $N_6 = 0.5$, $N_7 = -0.19444$, $N_8 = 0.22222$. The corresponding values of x and y come from Eqs. (10.2-2).

$$x = -0.15278(1) + (0.25)(3) + (-0.20833)(5) + (0.6667)(5.2) + (-0.08333)(1.5) + (0.5)(1.5) + (-0.19444)(1.5) + (0.2222)(1.5)$$
$$= 3.6889$$

$$y = (-0.15278)(2) + (0.25)(1) + (-0.20833)(3) + (0.6667)(9) + (-0.08333)(10) + (0.5)(10) + (-0.19444)(10) + (0.2222)(6)$$
$$= 8.875$$

(Continued)

The coordinates and meshed region are shown

x	y
1	2
2	1.125
3	1
4	1.625
5	3
1.388889	4.666667
2.433333	4.638889
3.477778	5.111111
4.522222	6.083333
5.566667	7.555556
1.555556	7.333333
2.266667	7.597222
2.977778	8.111111
3.688889	8.875
4.4	9.888889
1.5	10
1.5	10
1.5	10
1.5	10
1.5	10

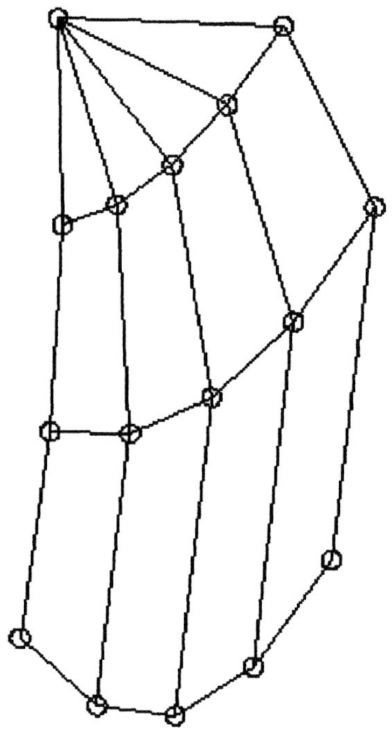

10.3

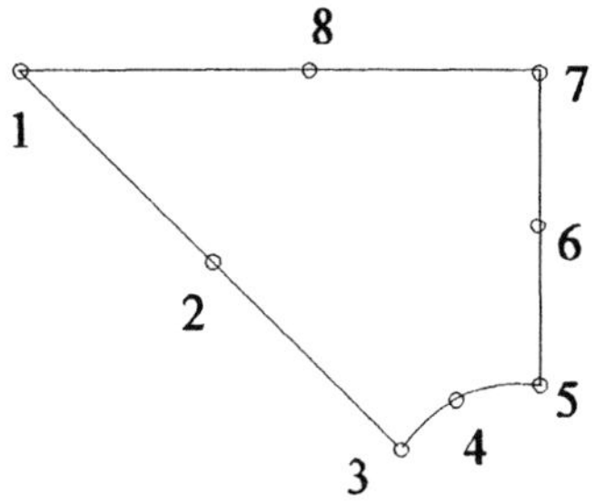

Ordering the region in this manner, Table 10.2-2 can be used.

	Bdry edge coordinates		Mesh coordinates	
No.	x	y	x	y
1	-2.5	2.5	-2.5	2.5
2	-1.6036	1.6036	-2.05178	2.051777
3	-0.707	0.707	-1.60355	1.603553
4	-0.3827	0.9238	-1.15533	1.15533
5	0	1	-0.70711	0.707107
6	0	1.75	-1.66667	2.5
7	0	2.5	-1.37432	2.091812
8	-1.25	2.5	-1.08198	1.683625
			-0.78964	1.275437
			-0.4973	0.86725
			-0.83333	2.5
			-0.6904	2.11622
			-0.54746	1.732441
			-0.40453	1.348661
			-0.2616	0.964881
			0	2.5
			0	2.125
			0	1.75
			0	1.375
			0	1

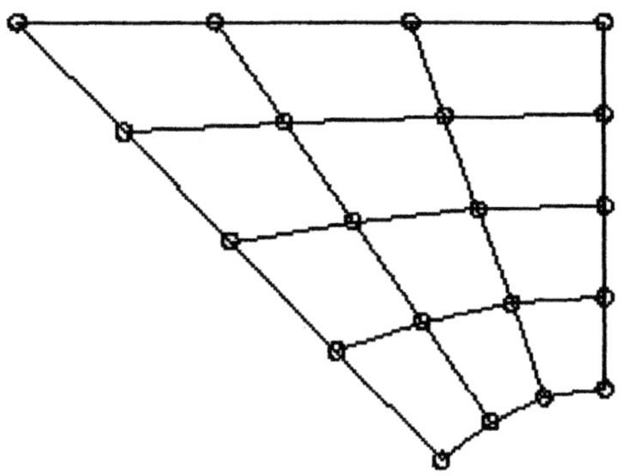

10.4 For the first mesh see Prob. 10.3 for the mesh coordinates of the left half of region. The right half of the region will be the mirror image.
For the second mesh, use the boundary edge points shown in the right figure.

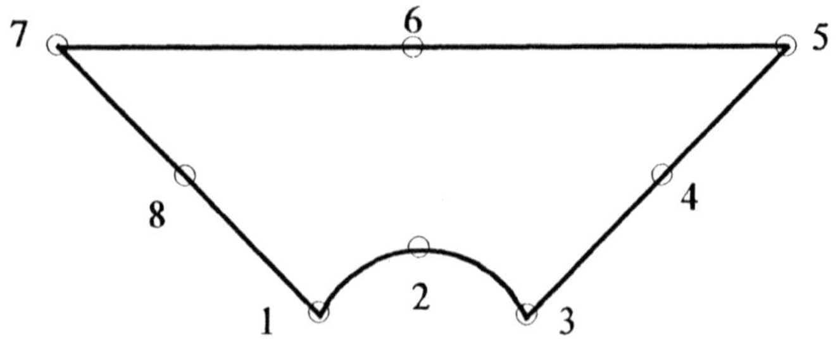

Boundary edge coordinates

No.	1	2	3	4	5	6	7	8
x	-0.707	0	0.707	1.6036	2.5	0	-2.5	-1.6036
y	0.707	1	0.707	1.6036	2.5	2.5	2.5	1.6036

For a 6×4 mesh let ξ: -1, -2/3, -1/3, 0, 1/3, 2/3, 1; η: -1, -1/2, 0, 1/2, 1.

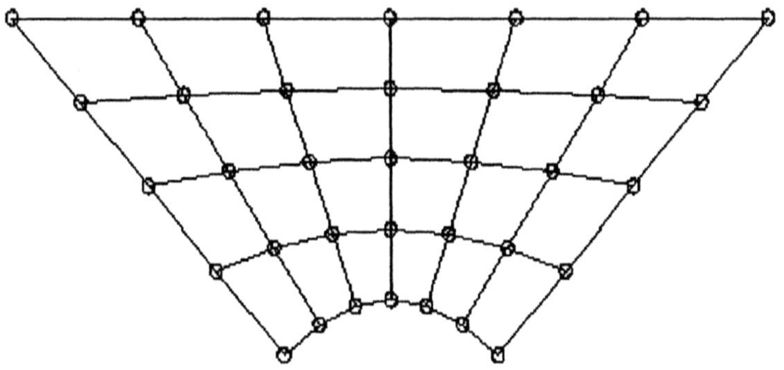

Mesh coordinates

x	y
-0.70711	0.707107
-0.4714	0.869825
-0.2357	0.967456
0	1
0.235702	0.967456
0.471405	0.869825
0.707107	0.707107
-1.15533	1.15533
-0.77022	1.277369
-0.38511	1.350592
0	1.375
0.38511	1.350592
0.77022	1.277369
1.15533	1.15533
-1.60355	1.603553
-1.06904	1.684913
-0.53452	1.733728
0	1.75
0.534518	1.733728
1.069036	1.684913
1.603553	1.603553
-2.05178	2.051777
-1.36785	2.092456
-0.68393	2.116864
0	2.125
0.683926	2.116864
1.367851	2.092456
2.051777	2.051777
-2.5	2.5
-1.66667	2.5
-0.83333	2.5
0	2.5
0.833333	2.5
1.666667	2.5
2.5	2.5

(continued)

The maximum deviation occurs along the arc.

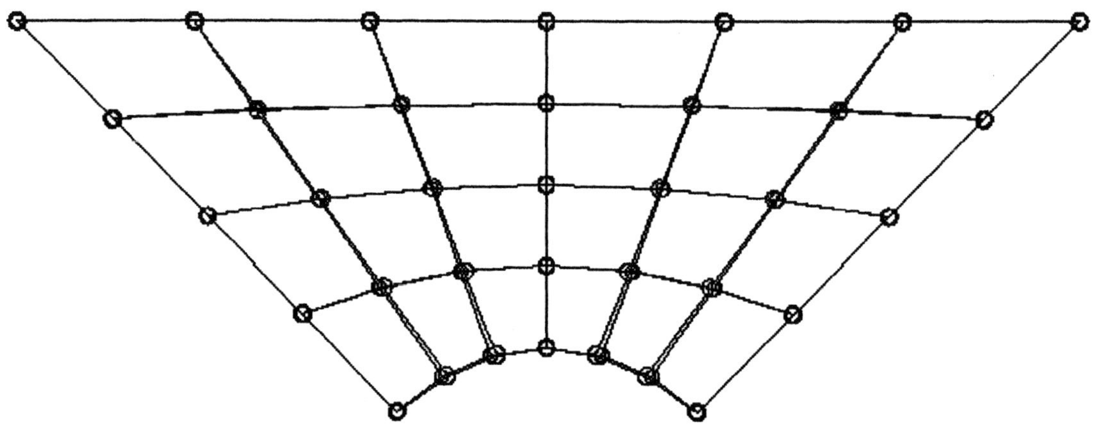

(10.5) If element 1 is rigid, $u_2 = u_1$.

$$\{u_G\} = [T]\{u_R\}$$

$$\begin{Bmatrix} u_1 \\ v_1 \\ u_2 \\ v_2 \\ u_3 \\ v_3 \end{Bmatrix} = \begin{bmatrix} 1 & 0 & 0 & 0 & 0 & 0 \\ 0 & 1 & 0 & 0 & 0 & 0 \\ 1 & 0 & 1 & 0 & 0 & 0 \\ 0 & 0 & 0 & 1 & 0 & 0 \\ 0 & 0 & 0 & 0 & 1 & 0 \\ 0 & 0 & 0 & 0 & 0 & 1 \end{bmatrix} \begin{Bmatrix} u_1 \\ v_1 \\ u_2 \\ v_2 \\ u_3 \\ v_3 \end{Bmatrix}$$

From Example 9.2-3 $\{F\} = [K]\{u\}$

$$\begin{Bmatrix} R_{1x} \\ R_{1y} \\ 2(10^3) \\ -3(10^3) \\ R_{3x} \\ 0 \end{Bmatrix} = \frac{1}{2} \begin{bmatrix} 2k_1 & 0 & -2k_1 & 0 & 0 & 0 \\ 0 & 2k_2 & 0 & 0 & 0 & -2k_2 \\ -2k_1 & 0 & (2k_1+k_3) & k_3 & -k_3 & -k_3 \\ 0 & 0 & k_3 & k_3 & -k_3 & -k_3 \\ 0 & 0 & -k_3 & -k_3 & k_3 & k_3 \\ 0 & -2k_2 & -k_3 & -k_3 & k_3 & (2k_2+k_3) \end{bmatrix} \begin{Bmatrix} u_1 \\ v_1 \\ u_2 \\ v_2 \\ u_3 \\ v_3 \end{Bmatrix}$$

$$[T]^T\{F\} = \begin{Bmatrix} R_{1x} + 2(10^3) \\ R_{1y} \\ -3(10^3) \\ R_{3x} \\ 0 \end{Bmatrix}$$

(continued)

$$[T]^T[K][T] = \frac{1}{2}\begin{bmatrix} k_3 & 0 & k_3 & -k_3 & -k_3 \\ 0 & 2k_2 & 0 & 0 & -k_2 \\ k_3 & 0 & k_3 & -k_3 & -k_3 \\ -k_3 & 0 & -k_3 & k_3 & k_3 \\ -k_3 & -2k_2 & -k_3 & k_3 & (2k_2+k_3) \end{bmatrix}$$

$u_1 = v_1 = u_3 = 0$. Partition 1st, 2nd, and 4th eqs. Remaining eqs are

$$\begin{Bmatrix} -3(10^3) \\ 0 \end{Bmatrix} = \frac{1}{2}\begin{bmatrix} k_3 & -k_3 \\ -k_3 & (2k_2+k_3) \end{bmatrix}\begin{Bmatrix} v_2 \\ v_3 \end{Bmatrix} = \frac{10^6}{2}\begin{bmatrix} 4\sqrt{2} & -4\sqrt{2} \\ -4\sqrt{2} & (16+4\sqrt{2}) \end{bmatrix}\begin{Bmatrix} v_2 \\ v_3 \end{Bmatrix}$$

$$= 2(10^6)\begin{bmatrix} \sqrt{2} & -\sqrt{2} \\ -\sqrt{2} & (4+\sqrt{2}) \end{bmatrix}\begin{Bmatrix} v_2 \\ v_3 \end{Bmatrix} \quad \text{Invert.}$$

$$\begin{Bmatrix} v_2 \\ v_3 \end{Bmatrix} = \frac{10^{-6}}{2}\begin{bmatrix} 0.9571 & 0.25 \\ 0.25 & 0.25 \end{bmatrix}\begin{Bmatrix} -3000 \\ 0 \end{Bmatrix} = \begin{Bmatrix} -1.436(10^{-3}) \\ -3.75\,(10^{-4}) \end{Bmatrix} \text{ m}$$

Thus,

$$\{u_1\ v_1\ u_2\ v_2\ u_3\ v_3\}^T = \{0\ \ 0\ \ 0\ -1.436\ \ 0\ -0.375\}^T \text{ mm} \quad \boxed{\text{ANS}}$$

(10.6) Eq. (a) of Example 9.2-4 with the constraints $u_1 = v_1 = u_3 = 0$, $\{F\} = [K_{red}]\{u\}$

$$\begin{Bmatrix} R_{2x} \\ R_{2y} \\ -2000 \end{Bmatrix} = 2(10^6)\begin{bmatrix} (4+\sqrt{2}) & \sqrt{2} & -\sqrt{2} \\ \sqrt{2} & \sqrt{2} & -\sqrt{2} \\ -\sqrt{2} & -\sqrt{2} & (4+\sqrt{2}) \end{bmatrix}\begin{Bmatrix} u_2 \\ v_2 \\ v_3 \end{Bmatrix}$$

Since $u_2 = 2v_2$ ① $\{u_G\} = [T]\{u_R\}$ or,

$$\begin{Bmatrix} u_2 \\ v_2 \\ v_3 \end{Bmatrix} = \begin{bmatrix} 2 & 0 \\ 1 & 0 \\ 0 & 1 \end{bmatrix}\begin{Bmatrix} v_2 \\ v_3 \end{Bmatrix} \quad \text{Thus, } [T] = \begin{bmatrix} 2 & 0 \\ 1 & 0 \\ 0 & 1 \end{bmatrix}, \ [T]^T = \begin{bmatrix} 2 & 1 & 0 \\ 0 & 0 & 1 \end{bmatrix}$$

$$[T]^T\{F\} = \begin{Bmatrix} 2R_{2x} + R_{2y} \\ -2(10^3) \end{Bmatrix} = \begin{Bmatrix} 0 \\ -2000 \end{Bmatrix} \ *$$

* Note: Since $u_2 = 2v_2$, $R_{2y} = -2R_{2x}$

(Continued)

$$[T]^T[K][T] = 2(10^6)\begin{bmatrix} (16+9\sqrt{2}) & -3\sqrt{2} \\ -3\sqrt{2} & (4+\sqrt{2}) \end{bmatrix}$$ Thus,

$$\begin{Bmatrix} 0 \\ -2000 \end{Bmatrix} = 2(10^6)\begin{bmatrix} (16+9\sqrt{2}) & -3\sqrt{2} \\ -3\sqrt{2} & (4+\sqrt{2}) \end{bmatrix}\begin{Bmatrix} v_2 \\ v_3 \end{Bmatrix}$$ Invert

$$\begin{Bmatrix} v_2 \\ v_3 \end{Bmatrix} = \frac{10^{-6}}{2}\begin{bmatrix} 0.039365 & 0.030847 \\ 0.030847 & 0.208871 \end{bmatrix}\begin{Bmatrix} 0 \\ -2000 \end{Bmatrix} = \begin{Bmatrix} -3.085(10^{-5}) \\ -2.089(10^{-4}) \end{Bmatrix} m$$

From ① $u_2 = 2v_2 = -6.169(10^{-5})\,m$. Thus,

$$\begin{Bmatrix} u_1 \\ v_1 \\ u_2 \\ v_2 \\ u_3 \\ v_3 \end{Bmatrix} = \begin{Bmatrix} 0 \\ 0 \\ -6.169 \\ -3.085 \\ 0 \\ -20.89 \end{Bmatrix}(10^{-5})\,m$$ **ANS**

This agrees with Example 9.2-4

The remaining problems in this chapter are open-ended design problems for which no solutions will be given here. Follow the recommendations given and compare your results with classical theory.